Basic Mathematics
for College Students

Books in the Tussy/Gustafson Series

Basic Mathematics for College Students
Prealgebra
Elementary Algebra
Intermediate Algebra

Basic Mathematics
for College Students

Alan S. Tussy
Citrus College

R. David Gustafson
Rock Valley College

Brooks/Cole Publishing Company

An International Thomson Publishing Company

Pacific Grove • Albany • Bonn • Boston • Cincinnati • Detroit • Johannesburg • London • Madrid • Melbourne
Mexico City • New York • Paris • San Francisco • Singapore • Tokyo • Toronto • Washington

A ROBERT W. PIRTLE BOOK

Sponsoring Editor: *Robert W. Pirtle*
Marketing Team: *Jennifer Huber, Christine Davis, Debra Johnston*
Editorial Assistant: *Erin Wickersham*
Production Editor: *Ellen Brownstein*
Production Service: *Hoyt Publishing Services*
Manuscript Editor: *David Hoyt*
Permissions Editor: *Carline Haga*

Interior & Cover Design: *Vernon T. Boes*
Interior Illustration: *Lori Heckelman*
Cover Image: *Pierre-Yves Goavec/Image Bank*; *Robert J. Western*
Art Coordinator: *David Hoyt*
Photo Editor: *Terry Powell*
Typesetting: *The Clarinda Company*
Cover Printing: *Phoenix Color Corp.*
Printing and Binding: *Banta Book Group*

For more information, contact:

BROOKS/COLE PUBLISHING COMPANY
511 Forest Lodge Road
Pacific Grove, CA 93950
USA

International Thomson Publishing Europe
Berkshire House 168-173
High Holborn
London WC1V 7AA
England

Thomas Nelson Australia
102 Dodds Street
South Melbourne, 3205
Victoria, Australia

Nelson Canada
1120 Birchmount Road
Scarborough, Ontario
Canada M1K 5G4

International Thomson Editores
Seneca 53
Col. Polanco
11560 México, D. F., México

International Thomson Publishing GmbH
Königswinterer Strasse 418
53227 Bonn
Germany

International Thomson Publishing Asia
60 Albert Street #15-01
Albert Complex
Singapore 189969

International Thomson Publishing Japan
Hirakawacho Kyowa Building, 3F
2-2-1 Hirakawacho
Chiyoda-ku, Tokyo 102
Japan

Printed in the United States of America

10 9 8 7 6 5 4 3 2 1

Library of Congress Cataloging-in-Publication Data

Tussy, Alan S., [date]
    Basic mathematics for college students / Alan S. Tussy, R. David
Gustafson.
        p.   cm.
    Includes index.
    ISBN 0-534-36493-4 (pbk. ; alk. paper)
    1. Mathematics.   I. Gustafson, R. David (Roy David), [date].
II. Title.
QA39.2.T878   1999
513′.14—dc21
                                        98-49874
                                        CIP

Photo credits: p. 1, Darrell Gulin/Corbis; **p. 59,** Darrel Gulin/Corbis; **p. 117,** Joel W. Rogers/Corbis; **p. 185,** Gary Braasch/Corbis; **p. 243,** Michael Fogden & Patricia Fogden/Corbis; **p. 287,** PNI; **p. 335,** Tom Bean/Corbis; **p. 359,** Phil Schermeister/Corbis; **p. 423,** Adam Woolfitt/Corbis.

PREFACE

For the Instructor

This book provides a comprehensive review of arithmetic and geometry, while introducing some fundamental algebraic concepts. It aims to improve the mathematical skills of students, giving them the confidence to apply those skills at home, in the marketplace, and on the job. Also, students planning to take an introductory algebra course in the future can use this text to build the mathematical foundation they will need. Our goal has been to write a book that is interesting and enjoyable to read—one that will attract and keep the attention of college students of all ages.

Basic Mathematics for College Students aims to teach students how to think. It develops basic mathematical skills in the context of solving meaningful and creative application problems. A variety of instructional approaches are used, reflecting the recommendations of NCTM and AMATYC. The text is suitable for use in various formats—lecture, laboratory, or self-study.

Features of the text

A Blend of Traditional and Reform Approaches
We have used a combination of instructional methods from the traditional and reform approaches, endeavoring to write a book that contains the best of both. You will find the vocabulary, practice, and well-defined pedagogy of a traditional basic mathematics book. The text also features problem solving, reasoning, communicating, and technology, as emphasized by the reform movement.

Thorough Coverage of Arithmetic
This book provides thorough coverage of the arithmetic of whole numbers, fractions, and decimals. Other topics traditionally taught in a basic mathematics course are also included, such as percent, ratio and proportion, and measurement.

Early Introduction of the Integers
In Chapter 2, integers are introduced, and a wide variety of applications are discussed. Then the procedures used to add, subtract, multiply, and divide integers are studied. The chapter concludes with evaluation of numerical expressions using the rule for the order of operations.

Constructing Charts, Tables, and Graphs; Statistics
Many problems require students to present their solutions in the form of a table, graph, or chart. Often, students must examine such data displays to obtain necessary information to solve a problem. Chapter 7, on descriptive statistics, covers the mean, median, and mode.

Introduction to Algebra
Chapter 8 introduces two fundamental algebraic concepts: variable and equation. Students learn how to simplify expressions, solve equations, and evaluate formulas.

Interactivity
Most worked examples in the text are accompanied by Self Checks. This feature allows students to practice skills discussed in the example by working a similar problem. Be-

cause the Self Check problems are adjacent to the worked examples, students can easily refer to the solution and author's notes of the example as they solve the Self Check. Author's notes are used to explain the steps in the solutions of examples. The notes are extensive so as to increase the students' ability to read and write mathematics.

Most examples ▶
have Self Checks.

Author's notes explain ▶
the steps in the solution
process.

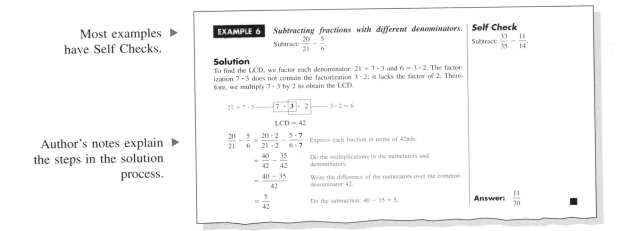

EXAMPLE 6   *Subtracting fractions with different denominators.*
Subtract: $\dfrac{20}{21} - \dfrac{5}{6}$.

Solution
To find the LCD, we factor each denominator: $21 = 7 \cdot 3$ and $6 = 3 \cdot 2$. The factorization $7 \cdot 3$ does not contain the factorization $3 \cdot 2$; it lacks the factor of 2. Therefore, we multiply $7 \cdot 3$ by 2 to obtain the LCD.

$$21 = 7 \cdot 3 \longrightarrow \boxed{7 \cdot 3 \cdot 2} \longleftarrow 3 \cdot 2 = 6$$
$$\text{LCD} = 42$$

$$\dfrac{20}{21} - \dfrac{5}{6} = \dfrac{20 \cdot 2}{21 \cdot 2} - \dfrac{5 \cdot 7}{6 \cdot 7} \quad \text{Express each fraction in terms of 42nds.}$$

$$= \dfrac{40}{42} - \dfrac{35}{42} \quad \text{Do the multiplications in the numerators and denominators.}$$

$$= \dfrac{40 - 35}{42} \quad \text{Write the difference of the numerators over the common denominator 42.}$$

$$= \dfrac{5}{42} \quad \text{Do the subtraction: } 40 - 35 = 5.$$

Self Check
Subtract: $\dfrac{33}{35} - \dfrac{11}{14}$.

Answer: $\dfrac{11}{70}$ ■

In-Depth Coverage of Geometry

The concepts of perimeter and area are introduced in Chapter 1 and revisited throughout the book. We have also included a wide variety of plane and solid figures in the Study Sets. Since many of the students taking a basic mathematics course did not take a geometry class in high school, Chapter 9 offers an overview of some of the most important geometric topics. The material is presented in a way that reinforces algebraic concepts such as formula, evaluation, and problem solving.

Geometric topics ▶
are presented in a
practical setting.

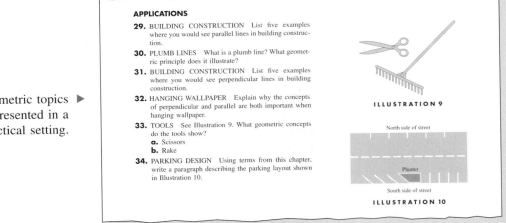

APPLICATIONS

29. BUILDING CONSTRUCTION List five examples where you would see parallel lines in building construction.

30. PLUMB LINES What is a plumb line? What geometric principle does it illustrate?

31. BUILDING CONSTRUCTION List five examples where you would see perpendicular lines in building construction.

32. HANGING WALLPAPER Explain why the concepts of perpendicular and parallel are both important when hanging wallpaper.

33. TOOLS See Illustration 9. What geometric concepts do the tools show?
  a. Scissors
  b. Rake

34. PARKING DESIGN Using terms from this chapter, write a paragraph describing the parking layout shown in Illustration 10.

ILLUSTRATION 9

North side of street

Planter

South side of street

ILLUSTRATION 10

Study Sets—More Than Just Exercises

The problems at the end of each section are called Study Sets. Each Study Set includes Vocabulary, Notation, and Writing problems designed to help students improve their ability to read, write, and communicate mathematical ideas. The problems in the Concepts section of the Study Sets encourage students to engage in independent thinking and reinforce major ideas through exploration. In the Practice section of the Study Sets, students get the drill necessary to master the material. In the Applications section, students deal with real-life situations that involve the topics being studied. Each Study Set concludes with a Review section consisting of problems similar to those in previous sections.

STUDY SET Section 2.6

VOCABULARY *In Exercises 1–4, fill in the blanks to make a true statement.*

1. When asked to evaluate expressions containing more than one operation, we should apply the rules for _____ of _____.

2. In situations where an exact answer is not needed, an approximation or _____ can be a quick way of obtaining an idea of what the size of the actual answer would be.

3. Absolute value symbols, parentheses, and brackets are types of _____ symbols.

4. If an expression involves two sets of grouping symbols, always begin working within the _____ symbols and then work to the _____.

CONCEPTS

5. Consider $5(-2)^2 - 1$. How man___ to be performed to evaluate this ___ in the order in which they shou___

7. Consider $\dfrac{5 + 5(7)}{2 + (4 - 8)}$. In the nu___ should be performed first? In the ___ eration should be performed firs___

9. Explain the difference between ___

NOTATION *In Exercises 11–14*

11. Evaluate: $-8 - 5(-2)^2$.
$-8 - 5(-2)^2 = -8 - 5(\underline{})$
$= -8 - \underline{}$
$= -8 + (\underline{})$
$= -28$

13 Evaluate: $[-4(2 + 7)] - 6$
$[-4(2 + 7)] - 6 = [-4(\underline{})]$
$= \underline{} - 6$
$= -42$

PRACTICE *In Exercises 15–34,*

15. $(-3)^2 - 4^2$ **16.**

19. $(2 - 5)(5 + 2)$ **20.**

23. $\dfrac{-6 - 8}{2}$ **24.**

APPLICATIONS

71. FLIGHT OF A BALL   A boy throws a ball from the top of a building, as shown in Illustration 1. At the instant he does this, his friend starts a stop watch and keeps track of the time as the ball rises to a peak and then falls to the ground. Use the vertical number line to complete the table by finding the position of the ball at the specified times.

Time	Position of duck
0 sec	
1 sec	
2 sec	
3 sec	
4 sec	

73. TECHNOLOGY   The readout from a testing device is shown in Illustration 3. It is important to know the value of each of the "peaks" and each of the "valleys" shown

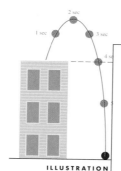

ILLUSTRATION

Time	Position of ball
1 sec	
2 sec	
3 sec	
4 sec	
5 sec	
6 sec	

72. SHOOTING GALLERY   At an ___ shooting gallery contains moving ___ one duck is shown in Illustration 2, ___ it takes the duck to reach certain p___ lery wall. Complete the table at the ___ ing the horizontal number line in t___

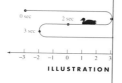

ILLUSTRATION

68   *Chapter 2   The Integers*

---

69. BASEBALL DIAMOND   Illustration 2 shows some dimensions of a major league baseball field. How far is it from home plate to second base?

ILLUSTRATION 2

70. SURVEYING   Use the imaginary triangles set up by a surveyor to find the length of each lake. (The measurements are in meters.)

a.  Length $\sqrt{118,656}$

b. Length $\sqrt{93,025}$

71. BIG-SCREEN TELEVISION   The picture screen on a television set is measured diagonally. What size screen is shown in Illustration 3?

$\sqrt{1,681}$ in.

ILLUSTRATION 3

72. LADDER   A painter's ladder is shown in Illustration 4. How long are the legs of the ladder?

$\sqrt{225}$ ft $\sqrt{169}$ ft

ILLUSTRATION 4

In Exercises 73–76, use a calculator to evaluate each radical expression. If an answer is not exact, round to the nearest ten thousandth.

73. $\sqrt{24,000,201}$ **74.** $-\sqrt{4.012009}$ **75.** $-\sqrt{0.00111}$ **76.** $\sqrt{\dfrac{27}{44}}$

WRITING   *Write a paragraph using your own words.*

77. When asked to find $\sqrt{16}$, one student's answer was 8. Explain his misunderstanding of the concept of square root.

78. Explain the difference between the square and the square root of a number.

79. What is a nonterminating decimal? Use an example in your explanation.

80. What do you think might be meant by the term *cube root*?

REVIEW

81. Multiply: $6.75 \cdot 12.2$.

82. Simplify: $\dfrac{\frac{-2}{3}}{8}$.

83. Evaluate: $5(-2)^2 - \dfrac{16}{4}$.

84. Divide: $5.7\overline{)18.525}$.

85. List the natural numbers.

86. Evaluate: $(3.4)^3$.

87. Simplify: $\left(\dfrac{2}{3}\right)^2 - \left(-\dfrac{3}{4}\right)^2$.

88. Insert the proper symbol, < or >, in the blank to make a true statement: -15 ___ -14.

232   *Chapter 4   Decimals*

A distinguishing feature of this book is its wealth of application problems. We have included numerous applications from disciplines such as science, economics, business, manufacturing, history, and entertainment, as well as mathematics.

57. BAROMETRIC PRESSURE Barometric pressure readings are recorded on the weather map in Illustration 4. In a low pressure area (L on the map), the weather is often stormy. The weather is usually fair in a high pressure area (H). What is the difference in readings between the areas of highest and lowest pressure? In what part of the country would you expect the weather to be fair?

ILLUSTRATION 4

58. QUALITY CONTROL An electronics company has strict specifications for silicon chips used in a computer. The company will install only chips that are within 0.05 centimeters of the specified thickness. Illustration 5 gives that specification for two types of chip. Fill in the blanks to complete the chart.

Chip type	Thickness specification	Acceptable range	
		Low	High
A	0.78 cm		
B	0.643 cm		

ILLUSTRATION 5

59. OFFSHORE DRILLING A company needs to construct a pipeline from an offshore oil well to a refinery located on the coast. Company engineers have come up with two plans for consideration, as shown in Illustration 6. Use the information in the illustration to complete the following table.

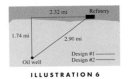

ILLUSTRATION 6

	Pipe underwater	Pipe underground	Total pipe
Design 1			
Design 2			

60. TV RATINGS The bar graph in Illustration 7 shows the prime-time ratings for the major television networks as of November 1995.

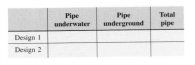

a. Which netw the lowest

b. What is the lowest ratin

c. How far wa

61. AMERICAN R Joyner holds Thompson hol freestyle swim Griffith-Joyner swam it?

62. FLIGHT PATH tance a plane storm.

14.57 mi

73. WEIGHTS AND MEASURES A consumer protection agency verifies the accuracy of butcher shop scales by placing a known three-quarter-pound weight on the scale and then comparing that to the scale's readout. According to Illustration 5, by how much is this scale off? Does it result in undercharging or overcharging customers on their meat purchases?

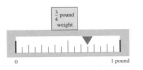

ILLUSTRATION 5

74. WRENCHES A mechanic likes to hang his wrenches above his tool bench in order of narrowest to widest. What is the proper order of the wrenches in Illustration 6?

$\frac{1}{4}$ in. $\frac{3}{8}$ in. $\frac{3}{16}$ in. $\frac{5}{32}$ in.

ILLUSTRATION 6

75. HIKING Illustration 7 shows the length of each part of a three-part hike. Rank the lengths from longest to shortest.

ILLUSTRATION 7

76. FIGURE DRAWING As an aid in drawing the human body, artists divide the body into three parts. Each part is then expressed as a fraction of the total body height. (See Illustration 8.) For example, the torso is $\frac{4}{15}$ of the body height. What fraction of body height is the head?

77. STUDY HABITS College students taking a full load were asked to give the average number of hours they

ILLUSTRATION 8

Head

Torso: $\frac{4}{15}$

Below the waist: $\frac{3}{5}$

studied each day. The results are shown in the pie chart in Illustration 9. What fraction of the students study 2 hours or more daily?

ILLUSTRATION 9

78. MUSICAL NOTES The notes used in music have fractional values. Their names and the symbols used to represent them are shown in Illustration 10(a). In common time, the values of the notes in each measure must add to 1. Is the measure in Illustration 10(b) complete?

(a)

(b)

ILLUSTRATION 10

79. GARAGE DOOR OPENER What is the difference in strength between a $\frac{1}{3}$-hp and a $\frac{1}{2}$-hp garage door opener?

80. DELIVERY TRUCK A truck can safely carry a one-ton load. Could it be used to deliver one-half ton of sand, one-third ton of gravel, and one-fifth ton of cement in one trip to a job site?

Every application ▶ problem has a title.

Estimation

Estimation is often used to check the reasonableness of answers. Special two-page Estimation features appear in the chapters dealing with whole numbers, decimals, and percent. In these features, estimation procedures are introduced and put to use in real-life situations that require only approximate answers.

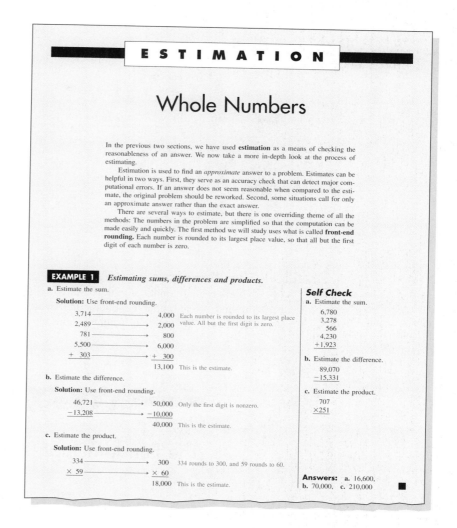

Building a Foundation for Graphing

Graphing on the number line is introduced in Chapter 2 when working with integers. The topic is revisited in Chapters 3 and 4 when the students work with signed fractions and decimals.

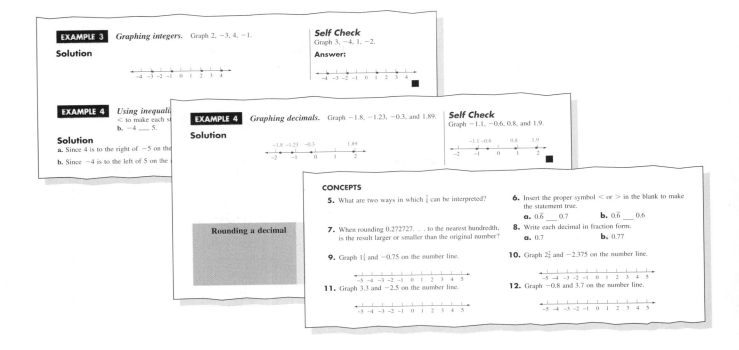

Key Concepts

Nine key mathematical concepts are highlighted in one-page features that appear near the end of each chapter. Each summarizes a concept and gives students the opportunity to review the role it plays in the big picture.

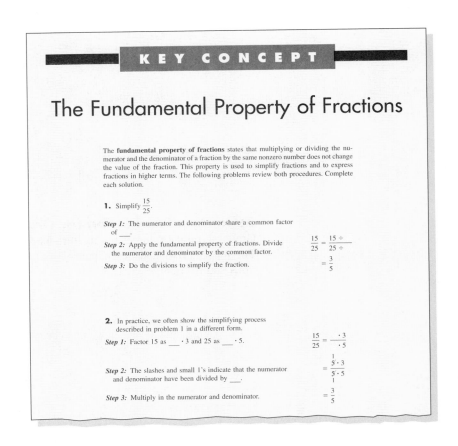

KEY CONCEPT

Proportions

A **proportion** is a statement that two ratios are equal.

Fill in the blanks as we set up a proportion to solve a problem.

1. TEACHER'S AIDES   For every 15 children on the playground, a child care center is required to have 2 teacher's aides supervising. How many teacher's aides will be needed to supervise 75 children?

Step 1: Let x = the number of _____.

The ratios of the number of children to the number of teacher's aides must be equal.

15 children are to ___ aides as ___ children are to ___ aides.

Expressing this as a proportion, we have $\dfrac{15}{} = \dfrac{75}{}$.

In the proportion $\frac{15}{2} = \frac{75}{x}$, 15 and x are the *extremes* and 2 and 75 are the *means*. After setting up the proportion, we solve it using the fact that the product of the extremes is equal to the product of the means.

Step 2: Solve the proportion: $\dfrac{15}{2} = \dfrac{75}{x}$.

$$\underline{} \cdot x = 2 \cdot \underline{} \qquad \text{The product of the extremes equals the product of the means.}$$

$$15x = \underline{} \qquad \text{Do the multiplication.}$$

$$\dfrac{15x}{} = \dfrac{150}{} \qquad \text{Divide both sides by 15.}$$

$$x = \underline{} \qquad \text{Simplify.}$$

Ten teacher's aides are needed to supervise 75 children.

Set up and solve each problem using a proportion. A calculator will be helpful.

2. PARKING   A city code requires that companies provide 10 parking spaces for every 12 employees. How many spaces will be needed if a company employs 450 people?

Calculators

For those instructors who wish to include calculators as part of the instruction in this course, there is an Accent on Technology feature that introduces keystrokes and shows how scientific calculators can be used to solve application problems. Selected Study Sets include problems that are to be solved using a calculator. These problems are marked with a calculator logo. For those instructors who do not wish to introduce calculators, that material can be skipped without interrupting the flow of ideas.

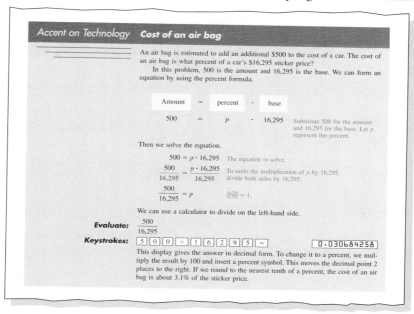

Systematic Review

Each Study Set ends with a Review section that contains problems similar to those in previous sections. Each chapter ends with a Chapter Review and a Chapter Test. The chapter reviews have been designed to be "user friendly." In a unique format, the reviews list the important concepts of each section of the chapter in one column, with appropriate review problems running parallel in a second column. In addition, Cumulative Review Exercises appear after Chapters 2, 4, 6, 8, and 9.

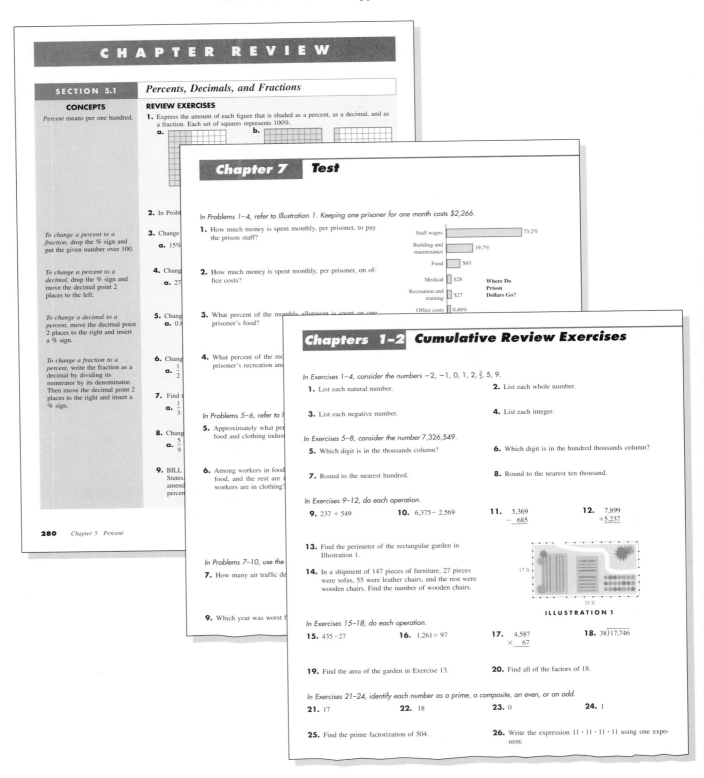

Student support

We have included many features that make *Basic Mathematics for College Students* very accessible to students.

Worked Examples

The text contains more than 300 worked examples, many with several parts. Explanatory notes make the examples easy to follow.

Self Checks

More than 275 Self Check problems allow students to practice the skills demonstrated in the worked examples.

Author's Notes

Author's notes, printed in red, are used to explain the steps in the solutions of examples. The notes are written using complete sentences so as to increase the students' ability to read and write mathematics.

Problems

The book includes more than 4,100 carefully graded exercises. In the Annotated Instructor's Edition, answers are printed in blue beside the problems. In the student edition, an appendix provides the answers to most of the odd-numbered exercises in the Study Sets as well as all the answers to the Key Concept, Chapter Review, Chapter Test, and Cumulative Review problems.

Functional Use of Color

For easy reference, definition boxes (in yellow), strategy boxes (in purple), and rule boxes (in blue) are color-coded. In addition, the book uses color to highlight terms and expressions that you would point to in a classroom discussion.

Study Skills and Math Anxiety

These two topics are discussed in detail in the section entitled "For the Student" at the end of this preface. In "Success in Mathematics," students are asked to design a personal strategy for studying and learning the material. "Taking a Math Test" helps students prepare for a test and then gives them suggestions for improving their performance.

Reading and Writing Mathematics

Also included (on pages xvii–xviii) are two features to help students improve their ability to read and write mathematics. "Reading Mathematics" helps students get the most out of the examples in this book by showing them how to read the solutions properly. "Writing Mathematics" highlights the characteristics of a well-written solution.

Videotapes

The videotape series that accompanies this book shows students the steps in solving many examples in the text. A video logo **oo** placed next to an example indicates that the example is taught on tape.

Ancillaries for the instructor

Annotated Instructor's Edition

In the Annotated Instructor's Edition, the answers to all exercises are printed in blue next to the exercises.

Thomson World Class Learning and Testing Tools
This integrated testing software package features algorithmic test generation, online testing, and class management capabilities.

Tutorial Video Series
These videotapes contain worked-out solutions to problems from the text and new problems, with added instruction.

Printed Test Items
This manual contains test exercises and answers, arranged according to the organization of the text.

Complete Solutions Manual
This instructor's manual includes complete solutions to both the even- and the odd-numbered problems, arranged according to the organization of the main text.

Ancillaries for the student

Student Video
This videotape contains worked-out examples of the concepts students have the most trouble understanding.

Student Solutions Manual
The manual provides the solutions for all the odd-numbered problems in the main text.

Web Site
The web site contains text-specific practice and study material, additional instruction, and online quizzes.

Acknowledgments

We are grateful to the following instructors, who have reviewed the text at various stages of its development. Their comments and suggestions have proven invaluable in making this a better book. We sincerely thank them for lending their time and talent to this project.

April Allen
Hartnell College

Laurette Blakey-Foster
Prarie View A & M

Julia Brown
Atlantic Community College

Mark Greenhalgh
Fullerton College

Mickey Levendusky
Pima Community College, Downtown

Donna E. Nordstrom
Pasadena City College

Angelo Segalla
Orange Coast College

Rita Sturgeon
San Bernardino Valley College

Terrie Teegarden
San Diego Mesa College

We want to express our gratitude to Peter Hurlimann, who read the entire manuscript and worked every problem. We would also like to thank Bob Pirtle, Jennifer Huber, Ellen Brownstein, David Hoyt, Lori Heckelman, Vernon Boes, Erin Wickersham, Melissa Henderson, and the Clarinda Company for their help in creating this book.

Alan S. Tussy
R. David Gustafson

For the Student

Congratulations. You now own a state-of-the-art textbook that has been written especially for you. To use this book properly, read it carefully, do the exercises, and check your progress with the Chapter Reviews, Chapter Tests, and Cumulative Review Exercises. To obtain additional help, a *Student Solutions Manual* is available at your college bookstore. It contains solutions to the odd-numbered exercises.

When you finish this course, consider keeping your book. It is a single reference source that will keep at your fingertips the information that you have learned. You may need this reference material in future courses in mathematics, science, or business.

The phrase "Practice makes perfect" is not quite true. It is *perfect* practice that makes perfect. For this reason, it is important that you learn how to study mathematics to get the most out of this course. Although we all learn differently, most students can benefit from certain hints on how to study. We have listed some hints along with other suggestions in four special features: *"Success in Mathematics," "Taking a Math Test," "Reading Mathematics,"* and *"Writing Mathematics."*

We wish you the best as you begin this course. We hope that this experience is a positive one and that you will continue to take more mathematics courses.

Success in mathematics

To be successful in mathematics, you need to know how to study it. The following checklist will help you develop your own personal strategy to study and learn the material. The suggestions listed below require some time and self-discipline on your part, but it will be worth the effort. This will help you get the most out of this course.

As you read each of the following statements, place a check mark in the box if you can truthfully answer Yes. If you can't answer Yes, think of what you might do to make the suggestion part of your personal study plan. You should go over this checklist several times during the semester to be sure you are following it.

Preparing for the Class

- [ ] I have made a commitment to myself to give this course my best effort.
- [ ] I have the proper materials: a pencil with an eraser, paper, a notebook, a ruler, and a calendar or day planner.
- [ ] I am willing to spend a minimum of two hours doing homework for every hour of class.
- [ ] I will try to work on this subject every day.
- [ ] I have a copy of the class syllabus. I understand the requirements of the course and how I will be graded.
- [ ] I have tried to schedule a free hour after the class to give me time to review my notes and begin the homework assignment.

Class Participation

- [ ] I will regularly attend the class sessions and be on time.
- [ ] When I am absent, I will find out what the class studied, get a copy of any notes or handouts, and make up the work that was assigned when I was gone.
- [ ] I will sit where I can hear the instructor and see the chalkboard.
- [ ] I will pay attention in class and take careful notes.
- [ ] I will ask the instructor questions when I don't understand the material.
- [ ] When tests, quizzes, or homework papers are passed back and discussed in class, I will write down the correct solutions for the problems I missed so that I can learn from my mistakes.

Study Sessions

- [ ] I will find a comfortable and quiet place to study.
- [ ] I realize that reading a math book is different than reading from a newspaper or a novel. Quite often, it will take more than one reading to understand the material.
- [ ] After studying an example in the textbook, I will work the accompanying Self Check.
- [ ] I will begin the homework assignment only after reading the assigned section.
- [ ] I will try to use the mathematical vocabulary mentioned in the book and used by my instructor when I am writing or talking about the topics studied in the course.
- [ ] I will look for opportunities to explain the material to others.
- [ ] I will check all of my answers to the problems with those provided in the back of the book (or with the *Student Solutions Manual*) and reconcile any differences.
- [ ] My homework will be organized and neat. My solutions will show all the necessary steps.
- [ ] I will try to work some review problems every day.
- [ ] After completing the homework assignment, I will read the next section to prepare for the coming class session.
- [ ] I will keep a notebook containing my class notes, homework papers, quizzes, tests, and any handouts—all in order by date.

Special Help

- [ ] I know my instructor's office hours and am willing to go in to ask for help.
- [ ] I have formed a study group with classmates that meets regularly to discuss the material and work on problems.
- [ ] When I need additional explanation of a topic, I view the video and check the web site.
- [ ] I take advantage of extra tutorial assistance that my school offers for mathematics courses.
- [ ] I have purchased the *Student Solutions Manual* that accompanies the text, and I use it.

To follow each of these suggestions will take time. It takes a lot of practice to learn mathematics, just as with any other skill.

No doubt, you will sometimes become frustrated along the way. This is natural. When it occurs, take a break and come back to the material after you have had time to clear your thoughts. Keep in mind that the skills and discipline you learn in this course will help make for a brighter future. Good luck!

Taking a math test

The best way to relieve anxiety about taking a mathematics test is to know that you are well-prepared for it and that you have a plan. Before any test, ask yourself three questions. When? What? How?

When Will I Study?

1. When is the test?

2. When will I begin to review for the test?

3. What are the dates and times that I will reserve for studying for the test?

What Will I Study?

1. What sections will the test cover?

2. Has the instructor indicated any types of problems that are guaranteed to be on the test?

How Will I Prepare for the Test?

Put a check mark by each method you will use to prepare for the test.

- ☐ Review the class notes.
- ☐ Outline the chapter(s) on a piece of poster board to see the big picture and to see how the topics relate to one another.
- ☐ Recite the important formulas, definitions, vocabulary, and rules into a tape recorder.
- ☐ Make flash cards for the important formulas, definitions, vocabulary, and rules.
- ☐ Rework problems from the homework assignments.
- ☐ Rework each of the Self Check problems in the text.
- ☐ Form a study group to discuss and practice the topics to be tested.
- ☐ Complete the appropriate Chapter Review(s) and the Chapter Test(s).
- ☐ Review the Warnings given in the text.
- ☐ Work on improving speed in answering questions.
- ☐ Review the methods that can be used to check answers.
- ☐ Write a sample test, trying to think of questions the instructor will ask.
- ☐ Complete the appropriate Cumulative Review Exercises.
- ☐ Get organized the night before the test. Have materials ready to go so that the trip to school will not be hurried.
- ☐ Take some time to relax immediately before the test. Don't study right up to the last minute.

Taking the Test

Here are some tips that can help improve your performance on a mathematics test.

- When you receive the test, scan it, looking for the types of problems you had expected to see. Do them first.
- Read the instructions carefully.
- Write down any formulas or rules as soon as you receive the test.
- Don't spend too much time on any one problem until you have attempted all the problems.
- If your instructor gives partial credit, at least try to begin a solution.
- Save the most difficult problems for last.
- Don't be afraid to skip a problem and come back to it later.
- If you finish early, go back over your work and look for mistakes.

Reading mathematics

To get the most out of this book, you need to learn how to read it correctly. A mathematics textbook must be read differently than a novel or a newspaper. For one thing, you need to read it slowly and carefully. At times, you will have to reread a section to understand its content. You should also have pencil and paper with you when reading a mathematics book, so that you can work along with the text to understand the concepts presented.

Perhaps the most informative parts of a mathematics book are its examples. Each example in this textbook consists of a problem and its corresponding solution. One form of solution that is used many times in this book is shown in the diagram on the next page. It is important that you follow the "flow" of its steps if you are to understand the mathematics involved. For this solution form, the basic idea is this:

- A property, rule, or procedure is applied to the original expression to obtain an equivalent expression. We show that the two expressions are equivalent by writing an equals sign between them. The property, rule, or procedure that was used is then listed next to the equivalent expression in the form of an author's note, printed in red.

- The process of writing equivalent expressions and explaining the reasons behind them continues, step by step, until the final result is obtained.

The solution in the following diagram consists of three steps, but solutions have varying lengths.

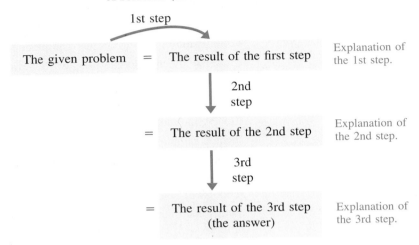

A solution (one of the basic forms)

1st step

The given problem $=$ The result of the first step — Explanation of the 1st step.

2nd step

$=$ The result of the 2nd step — Explanation of the 2nd step.

3rd step

$=$ The result of the 3rd step (the answer) — Explanation of the 3rd step.

Writing mathematics

One of the major objectives of this course is for you to learn how to write solutions to problems properly. A written solution to a problem should explain your thinking in a series of neat and organized mathematical steps. Think of a solution as a mathematical essay—one that your instructor and other students should be able to read and understand. Some solutions will be longer than others, but they must all be in the proper format and use the correct notation. To learn how to do this will take time and practice.

To give you an idea of what will be expected, let's look at two samples of student work. In the first, we have highlighted some important characteristics of a well-written solution. The second sample is poorly done and would not be acceptable.

Evaluate: $35 - 2^2 \cdot 3$.

A well-written solution:

The problem was copied ▶ from the textbook.

$$35 - 2^2 \cdot 3 = 35 - 4 \cdot 3$$ ◀ The first step of the solution is written here.
$$= 35 - 12$$ ◀ The steps are written under each other in
$$= 23$$ a neat, organized manner.

The equals signs are lined up vertically.

A poorly written solution:

The problem wasn't ▶ copied from the text.

$$2^2 = 4 = 35 - 4 \cdot 3$$ ◀ An equals sign is improperly used.
                                   12

SUB: 35
      -12
      $23 \rightarrow 23$ ◀ The work is disorganized and difficult to follow.

CONTENTS

Basic Mathematics
for College Students

Whole Numbers

1

THE NEED TO KNOW MATHEMATICS IS INCREASING AS TECHNOLOGY BECOMES A BIGGER PART OF OUR DAILY LIVES. IN THIS CHAPTER, WE BEGIN OUR MATHEMATICS STUDY BY EXAMINING THE PROCEDURES USED TO SOLVE PROBLEMS INVOLVING WHOLE NUMBERS.

1.1   *Whole Numbers, Place Value, and Rounding*

In this section, you will learn about

- **Sets of numbers**
- **Place value**
- **Expanded notation**
- **Ordering of the whole numbers**
- **Rounding whole numbers**

INTRODUCTION. In this section, we will discuss the natural numbers and the whole numbers. These numbers are used to answer questions such as "How many?" "How fast?" "How far?" "How heavy?"

- There are 151,732 persons living in Rockford, Illinois.
- The speed limit on interstate highways in Wisconsin is 65 mph.
- The distance between Chicago and Memphis is 530 miles.
- A statue weighs 132 pounds.

Sets of numbers

A **set** is a collection of objects (for example, a set of dishes or a set of golf clubs). Two important sets in mathematics are the natural numbers (the numbers that we count with) and the whole numbers.

Natural numbers	1, 2, 3, 4, 5, 6, 7, 8, 9, 10, 11, 12, . . .
Whole numbers	0, 1, 2, 3, 4, 5, 6, 7, 8, 9, 10, 11, 12, . . .

The three dots in the previous lists indicate that these sets continue on forever. There is no largest natural number or whole number.

Since every natural number is also a whole number, we say that the set of natural numbers is a **subset** of the set of whole numbers. However, not all whole numbers are natural numbers. Note that 0 is a whole number but not a natural number.

As an example of the use of whole numbers, we refer to Figure 1-1, which shows the growth of a chain of grocery stores from 1991 to 1996. Figure 1-1(a) shows the information in table form, Figure 1-1(b) shows the information in bar graph form, and Figure 1-1(c) shows the information in line graph form. For example, each part of the figure shows that the chain operated 12 stores in 1993.

Table

Year	Number of stores
1991	7
1992	10
1993	12
1994	16
1995	16
1996	20

(a)

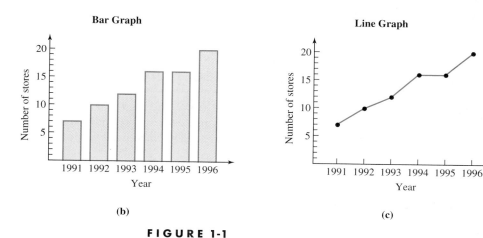

(b)

(c)

FIGURE 1-1

Place value

When we express numbers with *numerals* containing the *digits* 0, 1, 2, 3, 4, 5, 6, 7, 8, 9, we say that we have written the numbers in **standard notation**. The position of a digit in a numeral determines its value. In the numeral 325, 5 is in the *ones column*, 2 is in the *tens column*, and 3 is in the *hundreds column*.

To make a numeral easy to read, we use commas to separate its digits into groups of three, called **periods.** Each period has a name, such as *ones, thousands, millions,* and so on. The following table shows the place value of each digit in the numeral 345,576,475,897,431, which is read as

three hundred forty-five trillion, five hundred seventy-six billion, four hundred seventy-five million, eight hundred ninety-seven thousand, four hundred thirty-one

345 trillion			576 billion			475 million			897 thousand			4 hundred thirty-one		
3	4	5	5	7	6	4	7	5	8	9	7	4	3	1
Trillions			Billions			Millions			Thousands			Ones		
Hundreds	Tens	Ones	Hundreds	Tens	Ones	Hundreds	Tens	Ones	Hundreds	Tens	Ones	Hundreds	Tens	Ones

Note that as we move to the left in this table, the place value of each column is 10 times greater than the column to its right. This is why we call our number system a *base-10 number system.*

EXAMPLE 1 *CNN network news.* The cable network CNN is carried by approximately 11,698 cable systems. Which digit tells the number of hundreds?

Solution

Since the hundreds column is the third column from the right, 6 tells the number of hundreds.

Expanded notation

In the numeral 6,352, the digit 6 is in the thousands column, 3 is in the hundreds column, 5 is in the tens column, and 2 is in the ones (or units) column. The meaning of 6,352 becomes clear when we write it in **expanded notation.**

6 thousands + 3 hundreds + 5 tens + 2 ones

We read the numeral 6,352 as "six thousand, three hundred fifty-two."

EXAMPLE 2 *Writing numbers in expanded notation.*

a. Write 63,427 in expanded notation.

b. Write 1,254,709 in expanded notation.

Solution

a. 6 ten thousands + 3 thousands + 4 hundreds + 2 tens + 7 ones

We read this number as

sixty-three thousand, four hundred twenty-seven

b. 1 million + 2 hundred thousands + 5 ten thousands + 4 thousands + 7 hundreds + 0 tens + 9 ones

Since 0 tens is zero, the expanded notation can also be written as

1 million + 2 hundred thousands + 5 ten thousands + 4 thousands + 7 hundreds + 9 ones

We read this number as

one million, two hundred fifty-four thousand, seven hundred nine

EXAMPLE 3 *From words to standard notation.* Write twenty-three thousand forty in standard notation.

Solution

In expanded notation, the number is written as

2 ten thousands + 3 thousands + 4 tens

We write this as 23,040.

Ordering of the whole numbers

Whole numbers can be illustrated by drawing points on a **number line.** To construct a number line, we draw a line and mark off equal distances along it, as shown in Figure 1-2(a). The arrowhead at the right-hand side of the line indicates that the line goes on forever. Next, we label the left-hand point with 0 and the rest of the points as shown in Figure 1-2(b). Finally, we illustrate the whole numbers 0 through 7 by drawing a dot at each position on the number line that represents 0, 1, 2, 3, and so on, as shown in Figure 1-2(c).

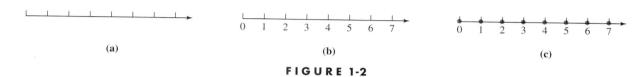

(a)   (b)   (c)

FIGURE 1-2

As we move to the right on the number line, the numbers get larger. Because 12 lies to the right of 7, we say that 12 is greater than 7, and we write

$12 > 7$ Read $>$ as "is greater than."

Since $12 > 7$, it is also true that $7 < 12$. (Read $<$ as "is less than.")

EXAMPLE 4   *Using the $<$ and $>$ symbols.*   Place an $<$ or an $>$ symbol between each pair of numerals to make a true statement:
a. 3 ___ 7   and   **b.** 18 ___ 16.

Solution

a. Since 3 is to the left of 7 on the number line, $3 < 7$.

b. Since 18 is to the right of 16 on the number line, $18 > 16$.

Self Check

Place an $<$ or an $>$ symbol between each pair of numerals to make a true statement:
a. 12 ___ 4   and   **b.** 7 ___ 10.

Answer: **a.** $>$,   **b.** $<$ ■

Rounding whole numbers

When we don't need exact results, we round numbers. When a teacher with 37 students says "I have 40 students," she has **rounded** the actual number to the *nearest ten,* because 37 is closer to 40 than it is to 30:

30, 31, 32, 33, 34, 35, 36, **37**, 38, 39, **40**

If a television set costs $1,285 and a customer says, "I can't afford $1,300," she has rounded the price to the nearest $100, because $1,285 is closer to $1,300 than it is to $1,200.

Rounding to the Nearest Ten

To round a whole number to the nearest 10, we find the digit in the tens column. If the *test digit* to the right of that column (the digit in the ones column) is 5 or greater, we **round up** by increasing the tens digit by 1 and placing a 0 in the ones column. If the test digit to the right of the tens column is less than 5, we **round down** by leaving the tens digit unchanged and placing a 0 in the ones column.

EXAMPLE 5 *Rounding to the nearest ten.*

a. Round 3,764 to the nearest ten.

b. Round 12,087 to the nearest ten.

Solution
a. We find the digit in the tens column, which is 6.

┌──── Test digit
3,764
└──── Rounding digit

We then look at the test digit to the right of 6, the 4 in the ones column. Since 4 < 5, we round down by leaving the 6 unchanged and replacing the test digit with 0. The rounded answer is 3,760.

b. We find the digit in the tens column, which is 8.

┌──── Test digit
12,087
└──── Rounding digit

We then look at the test digit to the right of 8, the 7 in the ones column. Because 7 > 5, we round up by adding 1 to 8 and replacing the test digit with 0. The rounded answer is 12,090.

Rounding to the Nearest Hundred

To round to the nearest 100, we find the digit in the hundreds column. If the test digit to the right of that column is 5 or more, we round up. If the test digit is less than 5, we round down.

EXAMPLE 6 *Rounding to the nearest hundred.*

a. Round 3,734 to the nearest hundred.

b. Round 7,960 to the nearest hundred.

Solution
a. We find the digit in the hundreds column, which is 7.

┌──── Test digit
3,734
└──── Rounding digit

We look at the 3 to the right of 7. Because 3 < 5, we round down by replacing the digits to the right of 7 with 0's. The rounded answer is 3,700.

b. Find the digit in the hundreds column, which is 9.

┌──── Test digit
7,960
└──── Rounding digit

We then look at the 6 to the right of 9. Because 6 > 5, we round up and increase 9 in the hundreds column by 1. Since the 9 in the hundreds column represents 900, increasing 9 by 1 represents increasing 900 to 1,000. Thus, we replace the 9 with a 0 and add 1 to the 7 in the thousands column. Finally, we replace the two rightmost digits with 0's. The rounded answer is 8,000.

Rounding to Other Places

A similar scheme is used to round numbers to the nearest thousand, the nearest ten thousand, and so on. To round a number, we follow the steps shown in the box.

Rounding a whole number	1. To round a number to a certain place, locate the digit in that place. We call this digit the *rounding digit*. 2. Look at the test digit to the right of the rounding digit. 3. If the test digit is 5 or greater, round up by adding 1 to the rounding digit and changing all of the digits to the right of the rounding digit to 0. If the test digit is less than 5, round down by keeping the rounding digit and changing all of the digits to the right of the rounding digit to 0.

EXAMPLE 7 *Rounding a whole number.* Round 257,125:

    a. To the nearest thousand.

    b. To the nearest ten thousand.

Solution **a.** We find the rounding digit in the thousands column of 257,125. It is 7. Because the test digit of 1 is less than 5, we round down to get 257,000.

    b. We find the rounding digit in the ten thousands column of 257,125. It is 5. Because the test digit of 7 is greater than 5, we round up to get 260,000. ■

STUDY SET   Section 1.1

VOCABULARY *Fill in the blanks to make a true statement.*

1. A _____ is a collection of objects.

2. The numbers 0, 1, 2, 3, 4, . . . form the set of _____.

3. The numbers 1, 2, 3, 4, 5, . . . form the set of _____.

4. When we separate the digits of a numeral into groups of three, the groups are called _____.

5. When 297 is written as 2 hundreds + 9 tens + 7 ones, it is written in _____.

6. To _____ 625 to the nearest ten, we write 630.

CONCEPTS *In Exercises 7–10, consider the numeral 57,634.*

7. What digit is in the tens column?

8. What digit is in the thousands column?

9. What digit is in the hundreds column?

10. What digit is in the ten thousands column?

11. What set of numbers is obtained when 0 is combined with the natural numbers?

12. Place the numbers 25, 17, 37, 15, 45 in order, from smallest to largest.

NOTATION *Fill in the blanks to make a true statement.*

13. The symbol > means _____.

14. The symbol < means _____.

PRACTICE *In Exercises 15–16, refer to the table shown in Illustration 1 and answer each question.*

15. How much of the average premium went for losses involving drunk drivers?

16. How much of the average premium went for losses involving theft and fire?

In Exercises 17–18, refer to the bar graph shown in Illustration 2 and answer each question.

17. How many rushing touchdowns did Emmitt Smith have through week 14 of the 1995 season?

18. Which team scored the most rushing touchdowns through week 14: the Seahawks or the Steelers?

Where the average automobile premium of $731 went in 1995	
Medical care, car repair, wage loss, legal expenses	$321
Uninsured motorists	$15
Drunk drivers	$61
Fraud	$64
Theft and fire	$54
Inflation	$49
Company operating costs, taxes, fees, commissions, dividends	$167
Total	**$731**

ILLUSTRATION 1

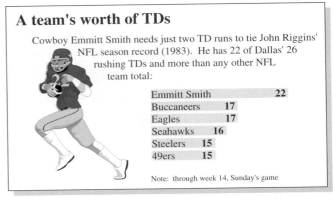

A team's worth of TDs

Cowboy Emmitt Smith needs just two TD runs to tie John Riggins' NFL season record (1983). He has 22 of Dallas' 26 rushing TDs and more than any other NFL team total:

Emmitt Smith	22
Buccaneers	17
Eagles	17
Seahawks	16
Steelers	15
49ers	15

Note: through week 14, Sunday's game

ILLUSTRATION 2

In Exercises 19–20, refer to the line graph shown in Illustration 3 and answer each question.

19. Find the closing value of the Dow Jones industrial average.

20. Did the market go up or down on Tuesday?

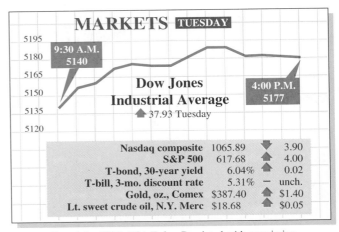

MARKETS TUESDAY

9:30 A.M. 5140

Dow Jones Industrial Average
▲ 37.93 Tuesday

4:00 P.M. 5177

Nasdaq composite	1065.89	▼	3.90
S&P 500	617.68	▲	4.00
T-bond, 30-year yield	6.04%	▲	0.02
T-bill, 3-mo. discount rate	5.31%	—	unch.
Gold, oz., Comex	$387.40	▲	$1.40
Lt. sweet crude oil, N.Y. Merc	$18.68	▲	$0.05

ILLUSTRATION 3

In Exercises 21–22, refer to the table shown in Illustration 4.

Natural gas reserves, 1997 (in trillion cubic feet)	
United States	162
Venezuela	129
Canada	95
Mexico	70
Argentina	27

ILLUSTRATION 4

21. Use the data to construct a bar graph.

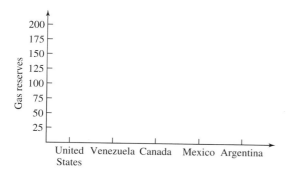

22. Use the data to construct a line graph.

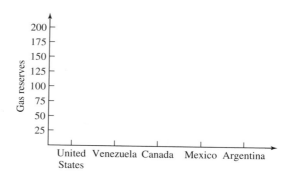

In Exercises 23–24, refer to the table shown in Illustration 5.

Crimes in Eagle River, WI (1995)	
Homicide	0
Rape	2
Robbery	12
Assault	15
Burglary	14
Auto theft	3
Larceny	8

ILLUSTRATION 5

23. Use the data to construct a line graph.

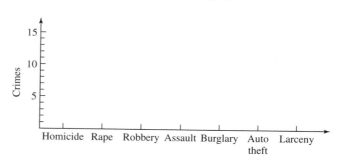

24. Use the data to construct a bar graph.

In Exercises 25–32, write each number in expanded notation and tell how the number is read.

25. 245

26. 508

27. 3,609

28. 3,960

29. 32,500

30. 73,009

31. 104,401

32. 570,003

In Exercises 33–46, write each number in standard notation.

33. 4 hundreds + 2 tens + 5 ones

34. 7 hundreds + 7 tens + 7 ones

35. 2 thousands + 7 hundreds + 3 tens + 6 ones

36. 7 ten thousands + 3 hundreds + 5 tens

37. Four hundred fifty-six

38. Three thousand seven hundred thirty-seven

39. Twenty-seven thousand five hundred ninety-eight

40. Seven million, four hundred fifty-two thousand, eight hundred sixty

41. There are six hundred sixty feet in one furlong.

42. One square foot is approximately nine hundred thirty square centimeters.

43. Hadrian, emperor of the Roman world, died in the year one hundred thirty-eight.

44. There are eighty-six thousand four hundred seconds in one day.

45. The United States shoreline of Lake Superior is eight hundred sixty-three miles long.

46. One thousand five hundred three passengers died when the Titanic hit an iceberg.

In Exercises 47–50, round 79,593 to the nearest . . .

47. ten **48.** hundred **49.** thousand **50.** ten thousand

In Exercises 51–54, round 5,925,830 to the nearest . . .

51. thousand **52.** ten thousand **53.** hundred thousand **54.** million

In Exercises 55–58, round $419,161 to the nearest . . .

55. $10 **56.** $100 **57.** $1,000 **58.** $10,000

APPLICATIONS *In Exercises 59–60, complete each check by writing the amount in words on the proper line.*

59.

No. 201	March 9 , 19 97
Payable to ___Davis Chevrolet___	$ 15,601 $\frac{00}{100}$
_____ DOLLARS	
	Don Smith
45-365-02	

60.

No. 202	Aug. 12 , 19 97
Payable to ___Dr. Anderson___	$ 3,433 $\frac{46}{100}$
_____ DOLLARS	
	Juan Decito
45-828-02	

61. FORMAL ANNOUNCEMENTS One style used when printing formal invitations and announcements is to write all numbers in words. Use this style to write each of the following phrases.

a. This diploma awarded this 27th day of June, 1996.

b. Charity fund raiser: The suggested contribution is $850 a plate, or an entire table may be purchased for $5,250.

62. EDITING Edit this excerpt from a history text by circling all numbers written in words and rewriting them using digits.

Abraham Lincoln was elected with a total of one million, eight hundred sixty-five thousand, five hundred ninety-three votes—four hundred eighty-two thousand, eight hundred eighty more than the runner-up, Stephen Douglas. He was assassinated after having served a total of one thousand five hundred three days in office. Lincoln's Gettysburg Address, a mere two hundred sixty-nine words long, was delivered at the battle site where forty-three thousand four hundred forty-nine casualties occurred.

63. SPEED OF LIGHT The speed of light in a vacuum is 299,792,458 meters per second. Round this number to the nearest hundred thousand meters per second.

64. SPEED OF LIGHT Round the speed of light to the nearest million meters per second. (See Exercise 63.)

65. Explain why the natural numbers are called the counting numbers.

66. Explain how you would round 687 to the nearest ten.

1.2 *Adding and Subtracting Whole Numbers*

In this section, you will learn about

- **Adding whole numbers less than 10**
- **Adding whole numbers greater than 10**
- **Finding the perimeter of a rectangle and a square**
- **Subtracting whole numbers**

INTRODUCTION.   The importance of whole numbers increases when we learn to add and subtract them. Mastering these operations will enable us to solve problems from geometry, business, and science. For example, to find the distance around a rectangle, we will need to add the lengths of its four sides. To prepare an annual budget, we will need to add separate line items. To find the difference between two temperatures, we will need to subtract them.

Adding whole numbers less than 10

Adding whole numbers corresponds to combining sets of objects. If a set of 4 objects is combined with a set of 5 objects, we get a total of 9 objects.

A set of four objects	A set of five objects	A set of nine objects
★★★★	★★★★★	★★★★★★★★★
We combine these two sets		to get this set.

This corresponds to the addition fact

$4 + 5 = 9$ Read as "4 plus 5 equals 9."

In this addition fact, 4 and 5 are called **addends,** and 9 is called the **sum.**
   If we combine the sets in the opposite order, we still get the same sum.

A set of five objects	A set of four objects	A set of nine objects
★★★★★	★★★★	★★★★★★★★★
We combine these two sets		to get this set.

This corresponds to the addition fact

$5 + 4 = 9$ Read as "5 plus 4 equals 9."

These examples illustrate that two whole numbers can be added in either order to get the same sum. This property is called the **commutative property of addition.**

| **Commutative property of addition** | The order in which whole numbers are added does not affect their sum. |

To find the sum of three whole numbers, we add two of them together and then add the third. For example, we can add $3 + 4 + 7$ in two ways. We will use the grouping symbols ( ), called **parentheses,** to show this. It is standard practice to do the operations within parentheses first.

Method 1: Group 3 and 4

$(3 + 4) + 7 = 7 + 7$ Because of the parentheses, add the 3 and 4 first to get 7.

$\qquad\qquad\quad = 14$ Then add 7 and 7 to get 14.

Method 2: Group 4 and 7

$3 + (4 + 7) = 3 + 11$ Because of the parentheses, add 4 and 7 to get 11.

$\qquad\qquad\quad = 14$ Then add 3 and 11 to get 14.

Either way, the sum is 14. It does not matter how we group or associate numbers in addition. This property is called the **associative property of addition.**

| **Associative property of addition** | The way in which whole numbers are grouped does not affect their sum. |

Whenever we add 0 to a number, the number remains the same. For example,

$$3 + 0 = 3, \qquad 5 + 0 = 5, \qquad \text{and} \qquad 9 + 0 = 9$$

These examples suggest the **addition property of 0.**

| **Addition property of 0** | The sum of any whole number and 0 is that whole number. |

EXAMPLE 1   *Adding whole numbers less than 10.*   Find each sum:

a. $8 + 9$ **b.** $9 + 8$ **c.** $5 + (7 + 8)$ **d.** $(5 + 7) + 8$ **e.** $(3 + 0) + 4$

Solution

a. $8 + 9 = 17$

b. $9 + 8 = 17$

c. $5 + (7 + 8) = 5 + 15$ Do the addition in parentheses first: $7 + 8 = 15$.

$\qquad\qquad\quad\;\; = 20$

d. $(5 + 7) + 8 = 12 + 8$ Do the addition in parentheses first: $5 + 7 = 12$.

$\qquad\qquad\quad\;\; = 20$

e. $(3 + 0) + 4 = 3 + 4$ Do the addition in parentheses first: $3 + 0 = 3$.

$\qquad\qquad\quad\;\; = 7$

Self Check

Find each sum:

a. $6 + 7$ **b.** $7 + 6$

c. $4 + (6 + 3)$ **d.** $(4 + 6) + 3$

e. $3 + (0 + 4)$

Answer: **a.** 13,   **b.** 13,
c. 13,   **d.** 13,   **e.** 7

Adding whole numbers greater than 10

We can add whole numbers greater than 10 by using a vertical format that adds the corresponding digits with the same place value. Because the additions within each column often exceed 9, it is sometimes necessary to *carry* the excess to the next column to the left. For example, to add 27 and 15, we write the numerals with the digits of the same place value aligned vertically.

$$\begin{array}{r} 2\,7 \\ +\ 1\,5 \\ \hline \end{array}$$

We begin by adding the two digits in the ones column: $7 + 5 = 12$. Because $12 = 1$ ten and 2 ones, we place a 2 in the ones column of the answer and carry 1 to the tens column.

$$\begin{array}{r} 1 \\ 2\,7 \\ +\ 1\,5 \\ \hline 2 \end{array}$$ Add the digits in the ones column: $7 + 5 = 12$. Carry 1 to the tens column.

Then we add the digits in the tens column.

$$\begin{array}{r} 1 \\ 2\,7 \\ +\ 1\,5 \\ \hline 4\,2 \end{array}$$ Add the digits in the tens column. Place the result, 4, in the tens column of the answer.

Thus, $27 + 15 = 42$.

EXAMPLE 2 *Adding whole numbers greater than 10.*
Add: $9,834 + 692$.

Solution

We write the numerals with their corresponding digits aligned. Then we add the numbers, one column at a time.

$$\begin{array}{r} 9{,}8\,3\,4 \\ +\ \ 6\,9\,2 \\ \hline \end{array}$$ Write the numerals in a column, with corresponding digits aligned vertically.

$$\begin{array}{r} 9{,}8\,3\,4 \\ +\ \ 6\,9\,2 \\ \hline 6 \end{array}$$ Add the digits in the ones column and place the result in the ones column of the answer.

$$\begin{array}{r} 1 \\ 9{,}8\,3\,4 \\ +\ \ 6\,9\,2 \\ \hline 2\,6 \end{array}$$ Add the digits in the tens column. The result, 12, exceeds 9. Place the 2 in the tens column of the answer and carry 1 to the hundreds column.

$$\begin{array}{r} 1\,1 \\ 9{,}8\,3\,4 \\ +\ \ 6\,9\,2 \\ \hline 5\,2\,6 \end{array}$$ Add the digits in the hundreds column. Since the result, 15, exceeds 9, place the 5 in the hundreds column of the answer and carry 1 to the thousands column.

$$\begin{array}{r} 1 \\ 9{,}8\,3\,4 \\ +\ \ 6\,9\,2 \\ \hline 1\,0{,}5\,2\,6 \end{array}$$ Since the sum of the digits in the thousands column is 10, write 0 in the thousands column and 1 in the ten thousands column of the answer.

Thus, $9,834 + 692 = 10,526$.

Self Check
Add: $675 + 1,497$.

Answer: 2,172 ■

To see that the answer in Example 2 is reasonable, we can round off and **estimate** the answer: 9,834 is a little less than 10,000, and 692 is a little less than 700. We estimate that the answer will be a little less than $10,000 + 700$, or 10,700. An answer of 10,526 is reasonable. We will study estimation in more detail later in this chapter.

Accent on Technology   Area of Lake Superior

The area of Lake Superior that is in the United States is 20,587 square miles, and the area that is in Canada is 11,093 square miles. To find the total area of the lake, we must add these two areas. We can do this addition using a calculator by pressing these keys.

Evaluate: 20,587 + 11,093.

Keystrokes: 2 0 5 8 7 + 1 1 0 9 3 =   31680

The total area is 31,680 square miles.

Words such as *increase, gain, credit, up, forward, rises, in the future,* and *to the right* are used to indicate addition.

EXAMPLE 3   *Calculating temperatures.*   At noon on Monday, the temperature was 31°. By 1:00 P.M., the temperature had increased 5°, and by 2:00 P.M., it had risen another 7°. Find the temperature at 2:00 P.M.

Solution   To the temperature at noon, we add the two increases.

$$31 + 5 + 7 = 43$$

The temperature at 2:00 P.M. was 43°.

EXAMPLE 4   *Finding the population of four colonies.*   In 1630, the populations of four American colonies were as shown in the figure. Find the total population.

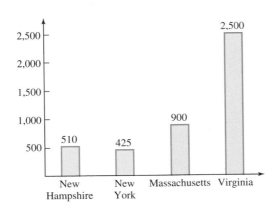

Self Check

A man went on a business trip. His airplane ticket cost $415, his rental car cost $197, his hotel cost $612, and his food cost $140. Find his total expenses.

Solution

To find the total population, we must find the sum of the populations of the individual colonies.

```
   2
    510
    425
    900
 +2,500
  4,335
```

The total population was 4,335.

Answer: $1,364

Finding the perimeter of a rectangle and a square

A **rectangle** is a four-sided figure (like a dollar bill) whose opposite sides are of equal length. Either of the longer sides is called its **length,** and either of the shorter sides is called its **width.** A rectangle with all four sides of equal length is called a **square.** (See Figure 1-3.)

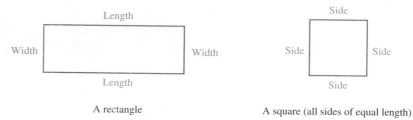

A rectangle   A square (all sides of equal length)

FIGURE 1-3

The distance around a rectangle or a square is called a **perimeter.** To find the perimeter of a rectangle, we add the lengths of its four sides.

$$\begin{array}{c}\text{The perimeter}\\\text{of a rectangle}\end{array} \quad = \quad \text{width} \quad + \quad \text{width} \quad + \quad \text{length} \quad + \quad \text{length.}$$

To find the perimeter of a square, we add the lengths of its four sides.

$$\begin{array}{c}\text{The perimeter}\\\text{of a square}\end{array} \quad = \quad \text{side} \quad + \quad \text{side} \quad + \quad \text{side} \quad + \quad \text{side.}$$

EXAMPLE 5   *Perimeter of a rectangle.* Find the perimeter of the rectangle.

Solution
To find the perimeter of the rectangle, we add the lengths of its four sides.

Length = 92 inches

Width = 47 inches

$$\begin{array}{r}1\\47\\47\\92\\+\ 92\\\hline 278\end{array}$$

The perimeter is 278 inches.
   To see whether this answer is reasonable, we estimate the answer. Because the rectangle is about 50 inches by 100 inches, its perimeter is approximately 50 + 50 + 100 + 100, or 300 inches. An answer of 278 inches is reasonable.

Self Check
Find the perimeter of a square with sides 63 centimeters long.

Answer: 252 cm

Subtracting whole numbers

Subtraction of whole numbers determines how many objects remain when some objects are removed from a set. For example, if we start with a set of nine objects and take four away, we are left with a set of five objects.

A set of nine objects.

★★★★★ ★★★★

We take away four objects

A set of five objects.

★★★★★

to get this set.

This corresponds to the subtraction fact

$9 - 4 = 5$ Read as "9 minus 4 equals 5."

In this subtraction fact, 9 is called the **minuend,** 4 is called the **subtrahend,** and 5 is called the **difference.**

With whole numbers, we cannot subtract in the opposite order and find the difference $4 - 9$, because we cannot take away 9 objects from 4 objects. Since subtraction of whole numbers cannot be done in either order, subtraction is not commutative.

Subtraction is not associative either, because if we group in different ways, we get different answers.

$$(9 - 5) - 1 = 4 - 1 \qquad \text{but} \qquad 9 - (5 - 1) = 9 - 4$$
$$= 3 \qquad\qquad\qquad\qquad = 5$$

EXAMPLE 6   *Subtracting whole numbers less than 10.*   Evaluate:

a. $9 - 3$ **b.** $8 - 5$ **c.** $9 - (6 - 3)$ **d.** $(9 - 6) - 3$

Solution

a. $9 - 3 = 6$ **b.** $8 - 5 = 3$

c. $9 - (6 - 3) = 9 - 3$ Do the subtraction in parentheses first: $6 - 3 = 3$.
$\qquad\qquad\quad = 6$

d. $(9 - 6) - 3 = 3 - 3$ Do the subtraction in parentheses first: $9 - 6 = 3$.
$\qquad\qquad\quad = 0$

Self Check
Evaluate:

a. $7 - 3$ **b.** $9 - 2$

c. $8 - (5 - 2)$ **d.** $(8 - 5) - 2$

Answer: **a.** 4,   **b.** 7,   **c.** 5,   **d.** 1

Whole numbers can be subtracted using a vertical format. Because subtractions often require subtracting a larger digit from a smaller digit, we may need to *borrow*. For example, to subtract 15 from 32, we write the minuend, 32, and the subtrahend, 15, in a vertical format, aligning the digits with the same place value.

$$\begin{array}{r} 3\,2 \\ -\,1\,5 \end{array}$$ Write the numerals in a column, with corresponding digits aligned vertically.

Since 5 can't be subtracted from 2, we borrow from the tens column of the minuend.

$$\begin{array}{r} 2 \\ \not{3}\,12 \\ -\,1\,\,\,5 \\ \hline 7 \end{array}$$ To subtract in the ones column, borrow 1 ten from the tens column and add 10 to the ones column of the minuend, giving 12. Then subtract: $12 - 5 = 7$.

$$\begin{array}{r} 2 \\ \not{3}\,12 \\ -\,1\,\,\,5 \\ \hline 1\,\,\,7 \end{array}$$ Subtract in the tens column: $2 - 1 = 1$.

Thus, $32 - 15 = 17$.

EXAMPLE 7 *Subtracting whole numbers greater than 10.* Subtract 576 from 2,021.

Solution

$$
\begin{array}{r}
2{,}0\ 2\ 1 \\
-\ \ \ 5\ 7\ 6 \\
\hline
\end{array}
$$
Write the numerals in a column, with the digits of the same place value aligned vertically.

$$
\begin{array}{r}
\ \ \ \ \ \ \ 1 \\
2{,}0\ \cancel{2}\ 11 \\
-\ \ \ 5\ 7\ 6 \\
\hline
5
\end{array}
$$
To subtract in the ones column, borrow 1 ten from the tens column and add it to the ones column. Then subtract: $11 - 6 = 5$.

Since we can't subtract 7 from 1 in the tens column, we borrow. Because there is a 0 in the hundreds column of the minuend, we must borrow from the thousands column. We can take 1 thousand from the thousands column (leaving 1 thousand behind) and write it as 10 hundreds, placing a 10 in the hundreds column. From these 10 hundreds, we take 1 hundred (leaving 9 hundreds behind) and write it as 10 tens. We add these 10 tens to the 1 ten that is already in the tens column to get 11 tens. From these 11 tens, we subtract 7 tens: $11 - 7 = 4$.

$$
\begin{array}{r}
1\ \ 9\ \ 11 \\
\cancel{2}{,}\cancel{10}\ \cancel{2}\ 11 \\
-\ \ \ 5\ 7\ 6 \\
\hline
4\ 5
\end{array}
$$
To subtract in the tens column, borrow 10 hundreds from the thousands digit and add it to the hundreds digit. Borrow 10 tens from the hundreds digit and add it to the tens digit. Then subtract: $11 - 7 = 4$.

$$
\begin{array}{r}
1\ \ 9\ \ 11 \\
\cancel{2}{,}\cancel{10}\ \cancel{2}\ 11 \\
-\ \ \ 5\ 7\ 6 \\
\hline
4\ 4\ 5
\end{array}
$$
Subtract in the hundreds column: $9 - 5 = 4$.

$$
\begin{array}{r}
1\ \ 9\ \ 11 \\
\cancel{2}{,}\cancel{10}\ \cancel{2}\ 11 \\
-\ \ \ 5\ 7\ 6 \\
\hline
1{,}4\ 4\ 5
\end{array}
$$
Subtract in the thousands column: $1 - 0 = 1$.

Thus, $2{,}021 - 576 = 1{,}445$.

Accent on Technology   *Recruiting athletes*

One year, 930,000 students participated in high school football, and 516,000 participated in high school basketball. To find how many more basketball players must be recruited to make basketball as popular as football, we must subtract 516,000 from 930,000. We can do this subtraction using a calculator by pressing these keys.

Evaluate: $930{,}000 - 516{,}000$

Keystrokes:
| 9 | 3 | 0 | 0 | 0 | 0 | − | 5 | 1 | 6 | 0 | 0 | 0 | = |

$\boxed{414000}$

414,000 more students must play basketball.

Words such as *minus, decrease, loss, debit, down, backward, fall, reduce, in the past,* and *to the left* indicate subtraction.

EXAMPLE 8 *Buying on sale.* A television set that regularly sells for $975 is reduced by $100. During a weekend sale, the price is reduced another $40. Find the sale price.

Solution We can do two subtractions. From the regular price of $975, we first subtract $100.

$$975 - 100 = 875$$

We then subtract $40 from $875:

$$875 - 40 = 835$$

The television set sells for $835.

The calculations in this example could be written as one expression:

$$975 - 100 - 40$$

The two subtractions are done from left to right:

$$(975 - 100) - 40 = 875 - 40$$ The parentheses emphasize that subtractions are performed from left to right.

$$= 835$$

We could also find the sale price by adding the two price reductions and subtracting that sum from the regular price. Since the total reduction was $100 + $40, or $140, the sale price was $835.

$$975 - 140 = 835$$

Additions and subtractions often appear in the same problem. It is important to read the problem carefully, extract the useful information, and organize it correctly.

EXAMPLE 9 *Bus passengers.* Twenty-seven people were riding a bus on Route 47. At Seventh Street, 16 riders got off and 5 got on. How many people were then on the bus?

Self Check

One share of ABC Corporation stock cost $75. The price fell $7 per share. However, it recovered and rose $13 per share. What is its current price?

Solution

The route and street number are not important. When 16 of the original 27 people got off, there were

$$27 - 16 = 11$$

people left. After 5 got on, there were

$$11 + 5 = 16$$

people on the bus. The number of riders on the bus can also be found by calculating the expression $27 - 16 + 5$.

$$(27 - 16) + 5 = 11 + 5$$ The parentheses emphasize that we do the calculations from left to right.

$$= 16$$

Answer: $81

STUDY SET   Section 1.2

VOCABULARY *Fill in the blanks to make a true statement.*

1. Numbers that are to be added are called _____ .

2. When two numbers are added, the result is called a _____ .

3. A _____ is a four-sided figure (like a dollar bill) whose opposite sides are of equal length.

4. Either of the longer sides of a rectangle is called its _____ . Either of its shorter sides is called its _____ .

5. A _____ is a rectangle with all sides of equal length.

6. When two numbers are subtracted, the result is called a _____ .

7. The property that guarantees that we can add two numbers in either order and get the same sum is called the _____ property of addition.

8. The property that allows us to group numbers in an addition in any way we want is called the _____ property of addition.

9. The distance around a rectangle is called its _____.

10. In a subtraction problem, the _____ is always subtracted from the _____.

CONCEPTS   *In Exercises 11–14, tell which property of addition guarantees that the quantities are equal.*

11. $3 + 4 = 4 + 3$

12. $(3 + 4) + 5 = 3 + (4 + 5)$

13. $7 + (8 + 2) = (7 + 8) + 2$

14. $(8 + 5) + 1 = 1 + (8 + 5)$

15. $(3 + 5) + 2 = (5 + 3) + 2$

16. $(6 + 5) + 3 = 6 + (5 + 3)$

17. Any number added to ___ stays the same.

18. In evaluating $(12 + 8) - 5$, what operation should be performed first?

NOTATION   *Fill in the blanks to make a true statement.*

19. The symbols (   ) are called _____.

20. The minus sign $(-)$ means _____.

Express the following facts in words.

21. $33 + 12 = 45$

22. $28 - 22 = 6$

In Exercises 23–24, complete each solution.

23. $(36 - 11) + 5 = $ ___ $+ 5$
$= 30$

24. $12 + (15 + 2) = 12 + $ ___
$= 29$

PRACTICE   *In Exercises 25–40, do each addition.*

25. $25 + 13 = $ ___

26. $47 + 12 = $ ___

27. $156 + 305 = $ ___

28. $647 + 38 = $ ___

29. $(95 + 16) + 39 = $ ___

30. $832 + (97 + 27) = $ ___

31. $25 + (321 + 17) = $ ___

32. $(4,231 + 213) + 5,234 = $ ___

33.   632
    $+347$

34.   423
    $+570$

35.   1,372
    $+\ \ 613$

36.   2,477
    $+\ \ 693$

37.   6,427
    $+3,573$

38.   3,567
    $+8,778$

39.   8,539
    $+7,368$

40.   5,799
    $+6,879$

In Exercises 41–44, use a calculator to find each sum.

41.   1,246
     578
    $+\ \ \ 37$

42.   4,689
     3,422
    $+\ \ \ 26$

43.   3,156
     1,578
    $+\ \ 578$

44.   2,379
     4,779
    $+2,339$

In Exercises 45–48, find the perimeter of each rectangle or square.

45.

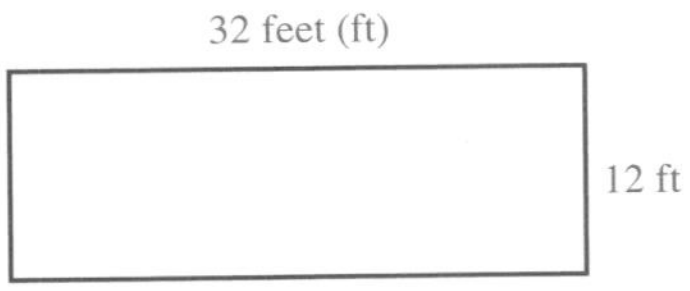

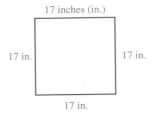

32 feet (ft)

12 ft

46.

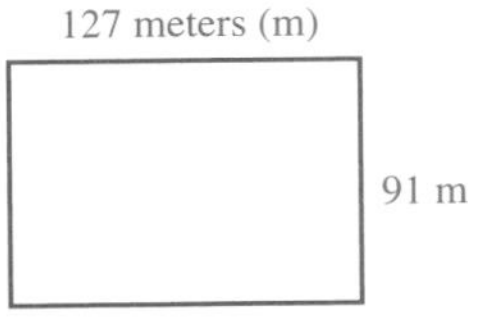

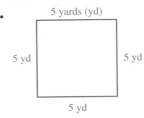

127 meters (m)

91 m

47.

17 inches (in.)

17 in.   17 in.

17 in.

48.

5 yards (yd)

5 yd   5 yd

5 yd

In Exercises 49–64, do each subtraction.

49. 17 − 14 = ___

50. 42 − 31 = ___

51. 39 − 14 = ___

52. 45 − 32 = ___

53. 174 − 71 = ___

54. 257 − 155 = ___

55. 633 − 598 − 30 = ___

56. 600 − (497 − 60) = ___

57. 367
       −343

58. 224
       −122

59. 423
       −305

60. 330
       −270

61. 1,537
       − 579

62. 2,470
       − 863

63. 4,267
       −2,578

64. 7,356
       −3,578

 In Exercises 65–68, use a calculator to find each difference.

65.   17,246
        − 6,789

66.   34,510
        −27,593

67.   15,700
        −15,397

68.   35,021
        −23,999

APPLICATIONS   *In Exercises 69–86, solve each problem.*

69. GARDENING   Beverly planted 27 seedlings, but only 21 of them survived. How many plants died?

70. RETAILING   Phil sold 42 toasters on Tuesday, 7 more than he sold on Monday. How many did he sell on Monday?

71. TRANSPORTATION   For a 17-mile trip, Wanda paid the taxi driver $23. If $5 was a tip, how much was the fare?

72. MAGAZINE CIRCULATION   *TV Guide* had a recent annual circulation of 16,969,260. By what amount did this exceed the circulation of *Reader's Digest,* with 16,566,650 readers?

73. BANKING   A savings account contained $370. After a deposit of $40 and a withdrawal of $197, how much is in the account?

74. TRAVEL   A student wants to make a 2,221-mile trip in three days. If she drives 751 miles on the first day and 875 miles on the second day, how far must she travel on the third day?

75. TAX DEDUCTIONS   For tax purposes, a woman kept the mileage records shown in Illustration 1. Find the total number of miles that she drove.

76. COMPANY BUDGET   A department head in a company prepared an annual budget with the line items shown in Illustration 2. Find the projected number of dollars to be spent.

Month	Miles driven
January	2,345
February	1,712
March	1,778
April	445
May	1,003
June	2,774

ILLUSTRATION 1

Line item	Amount
Equipment	17,242
Contractual	5,443
Travel	2,775
Supplies	10,553
Development	3,225
Maintenance	1,075

ILLUSTRATION 2

77. DIMENSIONS OF A HOUSE   Find the length of the house shown in Illustration 3.

18 ft   20 ft   24 ft

ILLUSTRATION 3

78. CONTROLLING CAR EMISSIONS   Illustration 4 shows the number of tons of hydrocarbons and nitrogen oxides that have been removed from the air because of antismog legislation in California. How many tons have been removed daily because of this legislation?

Step taken / Year	Tons removed daily
State gasoline reformulation / 1996	215
Federal gasoline reformulation / 1995	85
Auto nitrogen oxide standard / 1993 models	117
Auto hydrocarbon standard / 1993 models	35
Diesel fuel reformulation / 1993	70
Smog check / 1984	150
Gas pump nozzles / 1976	120

ILLUSTRATION 4

79. CITY FLAG   To decorate a city flag, yellow fringe is to be sewn around its outside edges, as shown in Illustration 5. The fringe comes on long spools and is sold by the inch. How many inches of fringe must be purchased to complete the project?

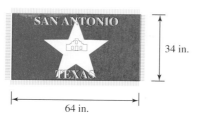

SAN ANTONIO

TEXAS

34 in.

64 in.

ILLUSTRATION 5

80. BOXING RING   How much padded rope is needed to create this square boxing ring, 24 feet on each side? (See Illustration 6.)

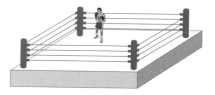

ILLUSTRATION 6

81. NUMBER OF GRADUATES   During the first year of this century, 18,549 boys and 25,182 girls graduated from American high schools. Find the total number of graduates.

82. CIVIL WAR CASUALTIES   In the Civil War, the Union army suffered 140,410 battle casualties and 224,100 deaths from other causes. Find the total number of casualties.

83. SCHIRRA IN SPACE   Astronaut Walter Schirra's first space flight orbited the earth 6 times and lasted 9 hours. His second flight orbited the earth 16 times and lasted 26 hours. How long was Schirra in space?

84. COLLEGE DEGREES AWARDED   The chart in Illustration 7 shows the number of each type of college degree that was granted in the years 1899 and 1900. How many degrees were granted?

Type of degree	Number
Bachelor's	15,539
Master's	1,015
Doctorate	149

ILLUSTRATION 7

85. DOW JONES AVERAGE   How much did the Dow rise on the day described by Illustration 8?

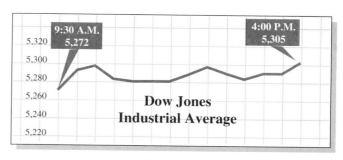

9:30 A.M. 5,272

4:00 P.M. 5,305

Dow Jones
Industrial Average

ILLUSTRATION 8

86. LENGTH OF A MOTOR Find the length of the motor on the machine shown in Illustration 9.

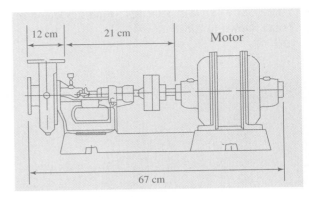

ILLUSTRATION 9

WRITING *Write a paragraph using your own words.*

87. Explain why the operation of addition is commutative.

88. Explain why the operation of subtraction is not commutative.

89. Explain why the operation of addition is associative.

90. Explain why the operation of subtraction is not associative.

REVIEW *In Exercises 91–92, write each numeral in expanded notation.*

91. 3,125

92. 60,037

In Exercises 93–96, round 6,354,784 to the specified place.

93. nearest ten

94. nearest hundred

95. nearest ten thousand

96. nearest hundred thousand

1.3 *Multiplying and Dividing Whole Numbers*

In this section, you will learn about

- **Multiplying whole numbers less than 10**
- **Multiplying whole numbers greater than 10**
- **Finding the area of a rectangle**
- **Dividing whole numbers**

INTRODUCTION. The importance of whole numbers increases even more when we learn to multiply and divide. Mastering these operations will enable us to find areas of geometric figures and solve business problems. For example, to find the area of a rectangle, we need to multiply its length by its width. To figure a paycheck, we need to multiply the number of hours worked by the hourly rate of pay.

We will need to divide to calculate how many miles a car can go on one gallon of gasoline or to calculate average costs of doing business.

Multiplying whole numbers less than 10

There are several notations that we can use to indicate multiplication.

Symbols that are used for multiplication

Symbol		*Example*
$\times$	times sign	4×5 or $\begin{array}{r} 234 \\ \times\ 12 \\ \hline \end{array}$
$\cdot$	raised dot	$4 \cdot 5$ or $117 \cdot 225$
( )	parentheses	(4)(5) or 4(5) or (4)5

Multiplication is repeated addition. For example, $4 \cdot 5$ means the sum of four 5's:

$$4 \cdot 5 = \overbrace{5 + 5 + 5 + 5}^{\text{The sum of four 5's.}}$$
$$= 20$$

In the above multiplication, the result of 20 is called a **product.** The numbers that were multiplied (4 and 5) are called **factors.**

$$\underset{\underset{4}{\downarrow}}{\text{Factor}} \quad \underset{\underset{5}{\downarrow}}{\text{Factor}} \quad = \quad \underset{\underset{20}{\downarrow}}{\text{Product}}$$

The multiplication $5 \cdot 4$ means the sum of five 4's:

$$5 \cdot 4 = \overbrace{4 + 4 + 4 + 4 + 4}^{\text{The sum of five 4's.}}$$
$$= 20$$

From these examples, we see that the sum of four 5's is equal to the sum of five 4's. The order in which we multiply makes no difference. This property is called the **commutative property of multiplication.**

Commutative property of multiplication

The order in which whole numbers are multiplied does not affect their product.

Table 1-1 summarizes the basic multiplication facts.

- To find the product of 6 and 8 using the table, we find the intersection of the **6** row and the **8** column. The product is 48.
- To find the product of 8 and 6, we find the intersection of the 8th row and the 6th column. Once again, the product is 48.

The symmetry of the table further illustrates that multiplication is commutative.

·	0	1	2	3	4	5	6	7	8	9
0	0	0	0	0	0	0	0	0	0	0
1	0	1	2	3	4	5	6	7	8	9
2	0	2	4	6	8	10	12	14	16	18
3	0	3	6	9	12	15	18	21	24	27
4	0	4	8	12	16	20	24	28	32	36
5	0	5	10	15	20	25	30	35	40	45
6	0	6	12	18	24	30	36	42	48	54
7	0	7	14	21	28	35	42	49	56	63
8	0	8	16	24	32	40	48	56	64	72
9	0	9	18	27	36	45	54	63	72	81

TABLE 1-1

From the table, we see that whenever we multiply a number by 0, the product is 0. For example,

$$0 \cdot 5 = 0, \qquad 0 \cdot 8 = 0, \qquad \text{and} \qquad 9 \cdot 0 = 0$$

We also see that whenever we multiply a number by 1, the number remains the same. For example,

$$3 \cdot 1 = 3, \qquad 7 \cdot 1 = 7, \qquad \text{and} \qquad 1 \cdot 9 = 9$$

These examples suggest the multiplicative properties of 0 and 1.

Multiplicative properties of 0 and 1	The product of any whole number and 0 is 0.
	The product of any whole number and 1 is that whole number.

EXAMPLE 1   *Computing daily wages.*   Raul worked an 8-hour day at an hourly rate of $9. How much money did he earn?

Solution

For each of the 8 hours, Raul earned $9. His total pay for the day is the sum of eight 9's. This can be calculated by multiplication.

$$\text{Total wages} = 9 + 9 + 9 + 9 + 9 + 9 + 9 + 9$$
$$= 8 \cdot 9$$
$$= 72 \qquad \text{See the multiplication table.}$$

Raul earned $72.

Self Check

At a rate of $8 per hour, how much will Wendy earn in 7 hours?

Answer: $56

To multiply three numbers, we first multiply two of them and then multiply that product by the third number. We can multiply $3 \cdot 2 \cdot 4$ in two ways. The parentheses show us what to do first.

Method 1: Group 3 · 2

$(3 \cdot 2) \cdot 4 = 6 \cdot 4$ Multiply 3 and 2 to get 6.
$= 24$ Then multiply 6 and 4 to get 24.

Method 2: Group 2 · 4

$3 \cdot (2 \cdot 4) = 3 \cdot 8$ Multiply 2 and 4 to get 8.
$= 24$ Then multiply 3 and 8 to get 24.

The products are the same. This illustrates that it doesn't matter how we group or associate numbers in multiplication. This property is called the **associative property of multiplication.**

Associative property of multiplication	The way in which whole numbers are grouped does not affect their product.

Multiplying whole numbers greater than 10

To find the product $8 \cdot 47$, it is inconvenient to add up eight 47's. Instead, we find the product by a multiplication process.

$$\begin{array}{r} 4\,7 \\ \times \quad 8 \\ \hline \end{array}$$ Write the factors in a column, with the corresponding digits aligned vertically.

$$\begin{array}{r} 5 \\ 4\,7 \\ \times \quad 8 \\ \hline 6 \end{array}$$ Multiply 7 by 8. The product is 56. Place 6 in the ones column and carry 5 to the tens column.

$$\begin{array}{r} 5 \\ 4\,7 \\ \times \quad 8 \\ \hline 3\,7\,6 \end{array}$$ Multiply 4 by 8. The product is 32. To the 32, add the carried 5 to get 37. Place the 7 in the tens column and the 3 in the hundreds column.

The product is 376.

EXAMPLE 2   ***Calculating distance.***   A car can travel 32 miles on 1 gallon of gasoline. How many miles can the car travel on 3 gallons of gasoline?

Self Check
How far can the car travel on 8 gallons of gasoline?

Solution
Each of the 3 gallons of gasoline enables the car to go 32 miles. The total distance the car can travel is the sum of three 32's. This can be calculated by multiplication:

$3 \cdot 32 = 96$

On 3 gallons of gasoline, the car can travel 96 miles.

Answer: 256 mi

To find the product $23 \cdot 435$, we use the multiplication process. Because $23 = 20 + 3$, we multiply 435 by 20 and by 3, and add the products.

$$\begin{array}{r} 4\,3\,5 \\ \times \quad 2\,3 \\ \hline \end{array}$$ Write the factors in a column, with the corresponding digits aligned vertically.

We begin by multiplying 435 by 3:

$$\begin{array}{r} \overset{1}{4\,3\,5} \\ \times\quad 2\,3 \\ \hline 5 \end{array}$$

Multiply 5 by 3. The product is 15. Place 5 in the ones column and carry 1 to the tens column.

$$\begin{array}{r} \overset{1\,1}{4\,3\,5} \\ \times\quad 2\,3 \\ \hline 0\,5 \end{array}$$

Multiply 3 by 3. The product is 9. To the 9, add the carried 1 to get 10. Place the 0 in the tens column and carry the 1 to the hundreds column.

$$\begin{array}{r} \overset{1\,1}{4\,3\,5} \\ \times\quad 2\,3 \\ \hline 1\,3\,0\,5 \end{array}$$

Multiply 4 by 3. The product is 12. Add the 12 to the carried 1 to get 13. Write 13.

We continue by multiplying 435 by 2 tens, or 20:

$$\begin{array}{r} \overset{1}{4\,3\,5} \\ \times\quad 2\,3 \\ \hline 1\,3\,0\,5 \\ 0 \end{array}$$

Multiply 5 by 2. The product is 10. Write 0 in the tens column and carry 1.

$$\begin{array}{r} \overset{1}{4\,3\,5} \\ \times\quad 2\,3 \\ \hline 1\,3\,0\,5 \\ 7\,0 \end{array}$$

Multiply 3 by 2. The product is 6. Add 6 to the carried 1 to get 7. Write the 7. There is no carry.

$$\begin{array}{r} \overset{1}{4\,3\,5} \\ \times\quad 2\,3 \\ \hline 1\,3\,0\,5 \\ 8\,7\,0 \end{array}$$

Multiply 4 by 2. The product is 8. There is no carry to add. Write the 8.

$$\begin{array}{r} 4\,3\,5 \\ \times\quad 2\,3 \\ \hline 1\,3\,0\,5 \\ 8\,7\,0 \\ \hline 1\,0\,0\,0\,5 \end{array}$$

Draw another line beneath the two completed rows. Add the two rows. This sum gives the product of 435 and 23.

Thus, $435 \cdot 23 = 10{,}005$.

Accent on Technology   *Calculating production*

The labor force of an electronics firm works two 8-hour shifts each day and manufactures 53 television sets each hour. To find how many sets will be manufactured in 5 days, we must find the following product:

2 shifts per day   8 hours per shift
↓   ↓

$$2 \cdot 8 \cdot 53 \cdot 5$$

↑   ↑
53 sets per hour   5 days

We can use a calculator to find this product by pressing these keys.

Evaluate: $2 \cdot 8 \cdot 53 \cdot 5$.

Keystrokes: $\boxed{2}\ \boxed{\times}\ \boxed{8}\ \boxed{\times}\ \boxed{5}\ \boxed{3}\ \boxed{\times}\ \boxed{5}\ \boxed{=}$ $\boxed{4240}$

4,240 television sets will be manufactured.

We can use multiplication to count objects arranged in rectangular patterns. For example, the display below shows a rectangular array consisting of 5 rows of 7 stars. The product 5 · 7, or 35, indicates the total number of stars.

Because multiplication is commutative, the array consisting of 7 rows of 5 stars contains the same number of stars:

EXAMPLE 3 *Computer science.* To draw graphics on a computer screen, a computer controls each *pixel* (one dot on the screen). A high-resolution computer graphics image is 800 pixels wide and 600 pixels high. How many pixels does the computer control?

Solution

The graphics image is a rectangular array of pixels. Each of its 600 rows consists of 800 pixels. The total number of pixels is the product of 600 and 800:

$$600 \cdot 800 = 480,000$$

The computer controls 480,000 pixels.

Self Check

On a color monitor, each of the pixels can be red, blue, or green. How many colored pixels does the computer control?

Answer: 1,440,000

Finding the area of a rectangle

One important application of multiplication is finding the area of a rectangle. The **area of a rectangle** is the measure of the amount of surface it encloses. Area is measured in square units, such as square inches (denoted as in.^{2}) or square centimeters (denoted as cm^2). (See Figure 1-4.)

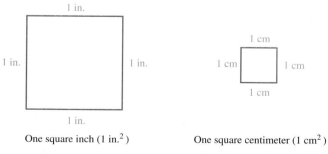

One square inch (1 in.^{2})   One square centimeter (1 cm^2)

FIGURE 1-4

The rectangle in Figure 1-5 has a length of 5 centimeters and a width of 3 centimeters. Since each small square covers an area of one square centimeter, each small square measures 1 cm². The small squares form a rectangular pattern, with 3 rows of 5 squares.

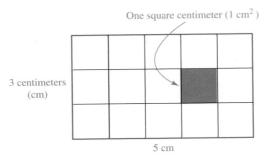

One square centimeter (1 cm²)

3 centimeters (cm)

5 cm

FIGURE 1-5

Because there are 5 · 3, or 15, small squares, the area of the rectangle is 15 cm². This suggests that the area of any rectangle is the product of its length and its width.

$$\text{Area of a rectangle} = \text{length} \cdot \text{width}.$$

By using the letter l to represent the length and the letter w to represent the width, we can write this formula in simpler form.

Area of a rectangle	The area, A, of a rectangle is the product of the rectangle's length, l, and its width, w:
	Area = (length)(width) or $A = l \cdot w$

EXAMPLE 4   *Area of a rectangle.*   Find the area of the rectangle.

13 m

21 m

Solution
We substitute 21 for l and 13 for w in the formula for the area of a rectangle.

$A = lw$ Note that lw means l times w.

$A = 21(13)$ Replace l with 21 and w with 13.

$ = 273$ Do the multiplication.

The area of the rectangle is 273 square meters (m²).

Self Check
A package of 9-inch-by-12-inch paper contains 500 sheets. How many square inches of paper is in one package?

Answer: 54,000 in.²

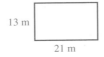 **WARNING!** Remember that the perimeter of a rectangle is the distance around it. The area of a rectangle is a measure of the surface it encloses.

Dividing whole numbers

If \$12 is distributed equally among 4 people, we must divide to see that each person would receive \$3.

$$4\overline{)12} \quad \to \quad 3$$

There are several notations that we can use to indicate division.

	Symbol		**Example**
Symbols that are used for division	$\div$	division sign	$12 \div 4$ or $1{,}242 \div 23$
	$)$	long division	$4)\overline{12}$ or $23)\overline{1{,}242}$
	$\overline{}$	fraction bar	$\dfrac{12}{4}$ or $\dfrac{1{,}242}{23}$

The answer to a division problem is called a **quotient.** The number that we are dividing *by* is called the **divisor.** The number we are dividing *into* is called the **dividend.** In the division problems

$$4)\overline{\overset{3}{12}}, \qquad 12 \div 4 = 3, \qquad \text{and} \qquad \frac{12}{4} = 3$$

the quotient is 3, the divisor is 4, and the dividend is 12.

Division can be thought of as repeated subtraction. To divide 12 by 4 is to ask, "How many 4's can be subtracted from 12?" Exactly three 4's can be subtracted from 12 to get 0:

$$12 - \overset{\text{Three 4's.}}{\overbrace{4 - 4 - 4}} = 0$$

Thus, $12 \div 4 = 3$.

Division is also related to multiplication.

$$\frac{12}{4} = 3 \quad \text{because} \quad 4 \cdot 3 = 12 \qquad \text{and} \qquad \frac{20}{5} = 4 \quad \text{because} \quad 5 \cdot 4 = 20$$

 WARNING! It is impossible to divide a number by 0. For example, consider $\frac{12}{0} = ?$. We know that $0 \cdot ?$ must give 12. However, there is no such number, because the product of 0 and any number is 0.

Division by zero	A number cannot be divided by 0.

The example $\frac{12}{1} = 12$ illustrates that *any number divided by 1 is the number itself.* The example $\frac{12}{12} = 1$ illustrates that *any number (except 0) divided by itself is 1.*

Division properties	A number divided by 1 is that number.
	Any nonzero number divided by itself equals 1.

We can use a process called **long division** to divide large numbers. To divide 832 by 23, for example, we proceed as follows:

$$\begin{array}{l} \text{Quotient} \rightarrow \\ \text{Divisor} \rightarrow 2\,3)\overline{8\,3\,2} \\ \qquad\qquad\quad \uparrow \\ \qquad\qquad \text{Dividend} \end{array}$$

Place the divisor and the dividend as indicated. The quotient will appear above the long division symbol.

We will find the quotient using the following division process.

$$\begin{array}{r} 4 \\ 2\,3\,\overline{)8\,3\,2} \end{array}$$

Ask: "How many times will 23 divide 83?" Because an estimate is 4, place 4 in the tens column of the quotient.

$$\begin{array}{r} 4 \\ 2\,3\,\overline{)8\,3\,2} \\ 9\,2 \end{array}$$

Multiply 23 · 4 and place the answer, 92, under the 83. Because 92 is larger than 83, our estimate of 4 for the tens column of the quotient was too large.

$$\begin{array}{r} 3 \\ 2\,3\,\overline{)8\,3\,2} \\ 6\,9\,\downarrow \\ \overline{1\,4\,2} \end{array}$$

Revise the estimate of the quotient to be 3. Multiply 23 · 3 to get 69, place 69 under the 83, draw a line, and subtract.

Bring down the 2 in the ones column.

$$\begin{array}{r} 3\,7 \\ 2\,3\,\overline{)8\,3\,2} \\ 6\,9 \\ \overline{1\,4\,2} \\ 1\,6\,1 \end{array}$$

Ask, "How many times 23 will divide 142?" The answer is approximately 7. Place 7 in the ones column of the quotient. Multiply 23 · 7 to get 161. Place 161 under 142. Because 161 is *larger* than 142, the estimate of 7 is too large.

$$\begin{array}{r} 3\,6 \\ 2\,3\,\overline{)8\,3\,2} \\ 6\,9 \\ \overline{1\,4\,2} \\ 1\,3\,8 \\ \overline{4} \end{array}$$

Revise the estimate of the quotient to be 6. Multiply: 23 · 6 = 138.

Place 138 under 142 and subtract.

The leftover 4 is the **remainder.** If $832 was distributed equally among 23 people, each person would receive $36, and there would be $4 left over.

To check the result of a division, we multiply the divisor by the quotient and then add the remainder. The result should be the dividend.

Quotient · divisor + remainder = dividend.

$$36 \quad \cdot \quad 23 \quad + \quad 4 \quad = \quad 832$$
$$828 + 4 = 832$$
$$832 = 832$$

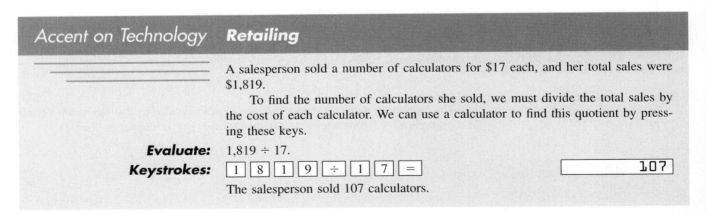

Accent on Technology   Retailing

A salesperson sold a number of calculators for $17 each, and her total sales were $1,819.

To find the number of calculators she sold, we must divide the total sales by the cost of each calculator. We can use a calculator to find this quotient by pressing these keys.

Evaluate: 1,819 ÷ 17.

Keystrokes: 1 8 1 9 ÷ 1 7 = | 107 |

The salesperson sold 107 calculators.

EXAMPLE 5 *Managing a soup kitchen.* A soup kitchen plans to feed 1,990 people. Because of space limitations, only 165 people can be served at one time. How many seatings will be necessary to feed everyone? How many will be served at the last seating?

Self Check

Each gram of fat in a meal provides 9 calories. A fast-food meal contains 243 calories from fat. How many grams of fat does the meal contain?

Solution

The 1,990 people can be fed 165 at a time. To find the number of seatings, we must divide.

```
        12
   165)1,990
       1 65↓
         340
         330
          10
```

The quotient is 12, and the remainder is 10. Thirteen seatings will be needed: 12 full-capacity seatings and one partial seating to serve the remaining 10 people.

Answer: 27

STUDY SET   Section 1.3

VOCABULARY   *Fill in the blanks to make a true statement.*

1. Multiplication means repeated _____.

2. Numbers that are to be multiplied are called _____.

3. A product is the result of a _____ problem.

4. The statement $6 \cdot 7 = 7 \cdot 6$ illustrates the _____ property of _____.

5. The statement $(5 \cdot 4) \cdot 7 = 5 \cdot (4 \cdot 7)$ illustrates the _____ property of _____.

6. If a square measures one inch on each side, its area is _____.

7. The result of a division problem is called a _____.

8. In a division, the _____ is divided by the _____.

CONCEPTS

9. Write $8 + 8 + 8 + 8$ as a multiplication.

10. Using the numbers 2, 3, and 4, write a statement that illustrates the associative property of multiplication.

11. Using the numbers 5 and 8, write a statement that illustrates the commutative property of multiplication.

12. How do we find the amount of surface enclosed by a rectangle?

13. A number can never be divided by ___.

14. Find the number of squares without counting all of them.

NOTATION

15. Write three symbols that are used for multiplication.

16. Write three symbols that are used for division.

PRACTICE   *In Exercises 17–32, do each multiplication.*

17. $4 \cdot 7 =$ ___

18. $7 \cdot 9 =$ ___

19. $12 \cdot 7 =$ ___

20. $15 \cdot 8 =$ ___

21. $27 \cdot 12 =$ ___

22. $35 \cdot 17 =$ ___

23. $9 \cdot (4 \cdot 5) =$ ___

24. $(3 \cdot 5) \cdot 12 =$ ___

25.   99
         ×77

26.   73
         ×59

27.   20
         ×53

28.   78
         ×20

29. 112
×   23

30. 232
×   53

31. 207
×   97

32. 768
×   70

 In Exercises 33–36, use a calculator to do each multiplication.

33. 13,456 · 217

34. 17,456 · 1,257

35. 3,302 · 15,358

36. 123,112 · 467

In Exercises 37–46, solve each problem.

37. FIGURING WAGES   Jennifer worked 12 hours at $11 per hour. How much did she earn?

38. FIGURING WAGES   Manuel worked 23 hours at $9 per hour. How much did he earn?

39. FINDING DISTANCE   A car with a tank that holds 14 gallons of gasoline goes 29 miles on one gallon. How far can the car go on a full tank?

40. RENTING APARTMENTS   Mia owns an apartment building with 18 units. Each unit generates a monthly income of $450. Find her total monthly income.

41. ATTENDING CONCERTS   A jazz quartet gave two concerts in each of 37 cities. Approximately 1,700 fans attended each concert. How many persons heard the group?

42. RAISINS IN CEREAL   A cereal maker advertises "Two cups of raisins in every box." Find the number of cups of raisins in a case of 36 boxes of cereal.

43. ORANGES IN JUICE   It takes 13 oranges to make one can of orange juice. Find the number of oranges used to make a case of 24 cans.

44. ROOM CAPACITY   A college lecture hall has 17 rows of 33 seats. A sign on the wall reads, "Occupancy by more than 570 persons is prohibited." If the seats are filled and there is one instructor, is the college breaking the rule?

45. CAPACITY OF AN ELEVATOR   There are 14 people in an elevator with a capacity of 2,000 pounds. If the average weight of a person on the elevator is 150 pounds, is the elevator overloaded?

46. CHANGING UNITS   There are 12 inches in 1 foot. How many inches are in 80 feet?

In Exercises 47–50, find the area of each rectangle or square.

47.

6 in.

14 in.

48.

50 m

22 m

49.

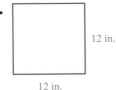

12 in.

12 in.

50.

20 cm

20 cm

In Exercises 51–62, do each division.

51. 40 ÷ 5 = ___

52. 40 ÷ 8 = ___

53. 42 ÷ 14 = ___

54. 65 ÷ 13 = ___

55. 132 ÷ 11 = ___

56. 132 ÷ 12 = ___

57. $\frac{221}{17}$ = ___

58. $\frac{221}{13}$ = ___

59. 13)‾949

60. 73)‾949

61. 33)‾1,353

62. 41)‾1,353

In Exercises 63–66, use a calculator to do each division.

63. $39\overline{)7,995}$ **64.** $71\overline{)7,313}$ **65.** $29\overline{)6,090}$ **66.** $13\overline{)7,410}$

In Exercises 67–74, do each division and give the quotient and the remainder.

67. $31\overline{)273}$ **68.** $25\overline{)290}$ **69.** $37\overline{)743}$ **70.** $79\overline{)931}$

71. $42\overline{)1,273}$ **72.** $83\overline{)3,280}$ **73.** $57\overline{)1,795}$ **74.** $99\overline{)9,876}$

APPLICATIONS   In Exercises 75–88, solve each problem.

75. DISTRIBUTING MILK   Juan's first grade class received 73 half-pint cartons of milk to distribute evenly to his 23 students. How many cartons were left over?

76. WAITING TABLES   A waitress earned $96 in tips. If each customer tipped $3, how many customers did she serve?

77. DRAINING A POOL   A 950,000-gallon pool empties in 16 hours. How many gallons are drained each hour?

78. RUNNING   Brian runs 7 miles each day. In how many days will Brian run 371 miles?

79. AREA OF WYOMING   The state of Wyoming is a rectangle 360 miles long and 270 miles wide. Find its perimeter and its area.

80. COMPARING ROOMS   Which has the greater area, a rectangular room that is 14 feet by 17 feet or a square room that is 16 feet on each side? Which has the greater perimeter?

81. SQUARES ON A CHESSBOARD   A chessboard consists of 8 rows, with 8 squares in each row. How many squares are there on a chessboard?

82. COMPARING MATTRESSES   A queen-size mattress measures 60 inches by 80 inches, and a full-size mattress measures 54 inches by 75 inches. How much more sleeping surface is there on a queen-size mattress?

83. GARDENING   A rectangular garden is 27 feet long and 19 feet wide. A path in the garden uses 125 square feet of space. How many square feet are left for planting?

84. AREA OF A POSTER BOARD   A poster board has dimensions of 24 inches by 36 inches. Find its area.

85. How many feet more than two miles is 11,000 feet? (*Hint:* 5,280 feet = 1 mile.)

86. ORDERING DOUGHNUTS   How many dozen doughnuts must be ordered for a meeting if 156 people are expected to attend, and each person will be served one doughnut?

87. TENNIS   See Illustration 1. First, find the number of square feet of court area a singles tennis player must defend. Then do the same for a doubles player. What is the difference between the two results?

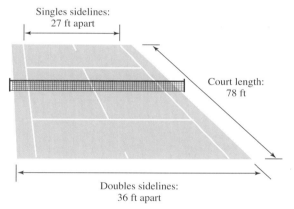

Singles sidelines: 27 ft apart

Court length: 78 ft

Doubles sidelines: 36 ft apart

ILLUSTRATION 1

88. VOLLEYBALL LEAGUE   A total of 216 girls tried out for a city volleyball program. How many girls should be put on each team roster if the following requirements must be met?

All the teams are to have the same number of players.

A reasonable number of players on a team is 7 to 10.

For scheduling purposes, there must be an even number of teams.

 In Exercises 89–90, use a calculator to solve each problem.

89. PRICE OF A TEXTBOOK   An author knows that her publisher received $954,193 on the sale of 23,273 textbooks. What is the cost of each book?

90. WATER DISCHARGE   The Susquehanna River discharges 38,200 cubic feet of water per second into the Chesapeake Bay. How long will it take for the river to discharge 1,719,000 cubic feet?

WRITING   *Write a paragraph using your own words.*

91. Explain why the division of two numbers is not commutative.

92. Explain why the division of two numbers is not associative.

93. Explain the difference between what perimeter measures and what area measures.

94. Give two examples each of how the concepts of perimeter and area are used when remodeling or redecorating a home.

REVIEW   *In Exercises 95–96, consider the number 372,856.*

95. What digit is in the hundreds column?

96. What digit is in the thousands column?

97. Round 45,995 to the nearest thousand.

98. Round 45,995 to the nearest hundred.

99. Add: 357, 39, and 476.

100. Subtract 987 from 1,010.

101. DISCOUNT SHOPPING   A radio, originally priced at $97, has been marked down to $75. By how many dollars has the radio been discounted?

102. DISCOUNTING CARS   A car, originally priced at $17,550, is being sold for $13,970. By how many dollars has the price been decreased?

ESTIMATION

Whole Numbers

In the previous two sections, we have used **estimation** as a means of checking the reasonableness of an answer. We now take a more in-depth look at the process of estimating.

Estimation is used to find an *approximate* answer to a problem. Estimates can be helpful in two ways. First, they serve as an accuracy check that can detect major computational errors. If an answer does not seem reasonable when compared to the estimate, the original problem should be reworked. Second, some situations call for only an approximate answer rather than the exact answer.

There are several ways to estimate, but there is one overriding theme of all the methods: The numbers in the problem are simplified so that the computation can be made easily and quickly. The first method we will study uses what is called **front-end rounding.** Each number is rounded to its largest place value, so that all but the first digit of each number is zero.

EXAMPLE 1 *Estimating sums, differences and products.*

a. Estimate the sum.

Solution: Use front-end rounding.

3,714 ──────→	4,000
2,489 ──────→	2,000
781 ──────→	800
5,500 ──────→	6,000
+ 303 ──────→	+ 300
	13,100

Each number is rounded to its largest place value. All but the first digit is zero.

13,100   This is the estimate.

b. Estimate the difference.

Solution: Use front-end rounding.

46,721 ──────→	50,000
−13,208 ──────→	−10,000
	40,000

Only the first digit is nonzero.

40,000   This is the estimate.

c. Estimate the product.

Solution: Use front-end rounding.

334 ──────→	300
× 59 ──────→	× 60
	18,000

334 rounds to 300, and 59 rounds to 60.

18,000   This is the estimate.

Self Check

a. Estimate the sum.

6,780
3,278
566
4,230
+1,923

b. Estimate the difference.

89,070
−15,331

c. Estimate the product.

707
×251

Answers: **a.** 16,600, **b.** 70,000, **c.** 210,000

To estimate quotients, we will use a method that approximates both the dividend and the divisor so that they will divide easily. With this method, some insight and intuition are needed. There is one rule of thumb for this method: If possible, round both numbers up or both numbers down.

EXAMPLE 2 *Estimating quotients.* Estimate the quotient: $170{,}715 \div 57$.

The dividend is approximately

$170{,}715 \div 57 \qquad 180{,}000 \div 60 = 3{,}000$ This is the estimate.

The divisor is approximately

This division can be done mentally.

Self Check
Estimate: $33{,}642 \div 42$.

Answer: 800

STUDY SET In Exercises 1–4, use front-end rounding to find an estimate to check the reasonableness of each answer. Write yes if it appears reasonable and no if it does not.

1.
```
    25,405
    11,222
     8,909
     1,076
    14,595
 + 33,999
    73,206
```

2.
```
   568,334
 −  31,225
   497,109
```

3.
```
      451
  ×    73
   39,923
```

4.
```
      616
  ×    98
   60,368
```

In Exercises 5–6, use estimation to check the reasonableness of each answer.

5. $57{,}238 \div 28 = 200$

6. $322\overline{)13{,}202}$ with quotient 41

In Exercises 7–10, use an estimation procedure to answer each problem.

7. CAMPAIGNING The number of miles flown each day by a politician on a campaign swing are shown here. Estimate the number of miles she flew in this period.

Day 1	3,546 miles
Day 2	567
Day 3	1,203
Day 4	342
Day 5	2,699

8. SHOPPING MALL The total sales income for a downtown mall in its first three years in operation are shown here.

1994	$5,234,301
1995	$2,898,655
1996	$6,343,433

Estimate the difference in income for 1995 and 1996 as compared to the first year, 1994.

9. GOLF COURSE Estimate the number of bags of grass seed needed to plant a fairway whose area is 86,625 square feet if the seed in each bag covers 2,850 square feet.

10. CENSUS FIGURES Estimate the total population of the seven largest counties in California as of July 1, 1995.

Los Angeles	9,138,788
San Diego	2,540,183
Orange	2,563,971
San Bernardino	1,569,586
Riverside	1,354,507
Alameda	1,323,312
Sacramento	1,096,697

1.4 Prime Factors and Exponents

In this section, you will learn about

- **Factoring whole numbers**
- **Prime numbers**
- **Composite numbers**
- **Even and odd whole numbers**
- **Finding prime factorizations with the tree method**
- **Exponents**
- **Finding prime factorizations with the division method**

INTRODUCTION. In this section, we will learn how to represent whole numbers in alternative forms. The procedures used to find these forms involve the operations of multiplication and division. We will then discuss exponents, a shortcut way to represent repeated multiplication.

Factoring whole numbers

The expression $3 \cdot 2 = 6$ has two parts: the numbers that are being multiplied, and the answer. The numbers that are being multiplied are *factors,* and the answer is the *product.* We say that 3 and 2 are factors of 6.

| **Factors** | Numbers that are multiplied together are called **factors.** |

EXAMPLE 1   *Finding factors of a whole number.*   Find the factors of 12.

Solution

We need to find the possible ways that we can multiply two whole numbers to get a product of 12.

$$1 \cdot 12 = 12, \qquad 2 \cdot 6 = 12, \qquad \text{and} \qquad 3 \cdot 4 = 12$$

In order, from least to greatest, the factors of 12 are 1, 2, 3, 4, 6, and 12.

Self Check

Find the factors of 20.

Answer: 1, 2, 4, 5, 10, and 20

Example 1 shows that 1, 2, 3, 4, 6, and 12 are the factors of 12. This observation was established by using multiplication facts. Each of these factors is related to 12 by division as well. Each of them divides 12, leaving a remainder of 0. Because of this fact, we say that 12 is **divisible** by each of its factors. When a division ends with a remainder of 0, we say that the division comes out even or that one of the numbers divides the other exactly.

| **Divisibility** | One number is **divisible** by another if, when dividing them, the remainder is 0. |

When we say that 3 is a factor of 6, we are using the word *factor* as a noun. Quite often, the word *factor* will be used as a verb.

Factoring a whole number	To **factor** a whole number means to express it as the product of other whole numbers.

EXAMPLE 2   *Factoring a whole number.*

a. Factor 40 using two factors.

b. Factor 40 using three factors.

Solution

a. There are several possibilities:

$$40 = 1 \cdot 40, \qquad 40 = 2 \cdot 20, \qquad 40 = 4 \cdot 10, \qquad \text{or} \qquad 40 = 5 \cdot 8$$

b. Again, there are several possibilities. Two of them are

$$40 = 5 \cdot 4 \cdot 2 \qquad \text{and} \qquad 40 = 2 \cdot 2 \cdot 10$$

Self Check

a. Factor 18 using two factors.

b. Factor 18 using three factors.

Answer: **a.** $1 \cdot 18$, $2 \cdot 9$, $3 \cdot 6$, **b.** $2 \cdot 3 \cdot 3$

Prime numbers

Prime numbers	A **prime number** is a whole number, greater than 1, that has only 1 and itself as factors.

EXAMPLE 3   *Finding the factors of a whole number.*   Find the factors of 17.

Solution $1 \cdot 17 = 17$

The only factors of 17 are 1 and 17.

In Example 3, we saw that 17 has exactly two factors: 1 and itself. Thus, 17 is a prime number. The prime numbers are the numbers

$$2, 3, 5, 7, 11, 13, 17, 19, 23, 29, 31, \ldots$$

The three dots at the end of the list indicate that there are infinitely many prime numbers.

Composite numbers

The set of whole numbers contains many prime numbers. It also contains many numbers that are not prime.

Composite numbers	The **composite numbers** are whole numbers, greater than 1, that are not prime.

The composite numbers are the numbers

$$4, 6, 8, 9, 10, 12, 14, 15, 16, 18, \ldots$$

The three dots at the end of the list indicate that there are infinitely many composite numbers.

EXAMPLE 4   *Prime and composite numbers.*

a. Is 37 a prime number?

b. Is 45 a prime number?

Solution

a. Since 37 is a whole number greater than 1 and its only factors are 1 and 37, it is prime.

b. The factors of 45 are 1, 3, 5, 9, 15, and 45. Since there are factors other than 1 and 45, 45 is not prime. It is a composite number.

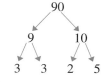

Self Check

a. Is 57 a prime number?

b. Is 39 a prime number?

Answer: **a.** no, **b.** no

> **WARNING!**   The numbers 0 and 1 are neither prime nor composite, because neither is a whole number greater than 1.

Even and odd whole numbers

Even and odd whole numbers	If a whole number is divisible by 2, it is called an **even** number.
	If a whole number is not divisible by 2, it is called an **odd** number.

The even whole numbers are the numbers

0, 2, 4, 6, 8, 10, 12, 14, 16, 18, . . .

The odd whole numbers are the numbers

1, 3, 5, 7, 9, 11, 13, 15, 17, 19, . . .

There are infinitely many even and infinitely many odd whole numbers.

Finding prime factorizations with the tree method

Every composite number can be formed by multiplying a specific combination of prime numbers. The process of finding that combination is called **prime factorization.**

Prime factorization	To find the **prime factorization** of a whole number means to write it as the product of only prime numbers.

Two methods can be used to find the prime factorization of a number. The first is called the **tree method.** We will use the tree method to find the prime factorization of 90 in two ways.

1. Factor 90 as $9 \cdot 10$.

2. Factor 9 and 10.

3. The process is complete when only prime numbers appear.

1. Factor 90 as $6 \cdot 15$.

2. Factor 6 and 15.

3. The process is complete when only prime numbers appear.

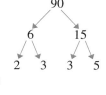

In either case, the prime factors are $2 \cdot 3 \cdot 3 \cdot 5$. Thus, $90 = 2 \cdot 3 \cdot 3 \cdot 5$.

The prime-factored form of 90 is $2 \cdot 3 \cdot 3 \cdot 5$. As we have seen, it does not matter how we factor the whole number. We will always get the same set of factors. No other combination of prime factors will multiply together and produce 90.

Fundamental Theorem of Arithmetic	Any composite number has exactly one set of prime factors.

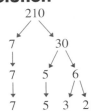

EXAMPLE 5 *Factoring whole numbers with factor trees.*
Use a factor tree to find the prime factorization of 210.

Solution

```
        210
       /   \
      7     30        Factor 210 as 7 · 30.
      |    /  \
      7   5    6      Factor 30 as 5 · 6.
      |   |   / \
      7   5  3   2    Factor 6 as 3 · 2.
```

The prime factorization of 210 is $7 \cdot 5 \cdot 3 \cdot 2$. If we write the prime factors in order, from least to greatest, we have $210 = 2 \cdot 3 \cdot 5 \cdot 7$.

Self Check
Use a factor tree to find the prime factorization of 120.

Answer: $2 \cdot 2 \cdot 2 \cdot 3 \cdot 5$ ■

Exponents

In the Self Check of Example 5, we saw that the prime factorization of 120 is $2 \cdot 2 \cdot 2 \cdot 3 \cdot 5$. Because this factorization has three factors of 2, we call 2 a *repeated factor*. To express a repeated factor, we use an **exponent.**

Exponent and base	An **exponent** is used to indicate repeated multiplication. It tells how many times the **base** is used as a factor.

The exponent is 3.

$$\underbrace{2 \cdot 2 \cdot 2}_{\text{Repeated factors.}} = 2^3 \qquad \text{The expression } 2^3 \text{ is called a power of 2.}$$

The base is 2.

In the expression 5^4, the base is 5, the exponent is 4, and the expression is called a power of 5. Using exponential notation, we can write the prime factorization of 120 as $2^3 \cdot 3 \cdot 5$.

EXAMPLE 6 *Exponents.* Use exponents to write each prime factorization.

a. $5 \cdot 5 \cdot 5$

b. $7 \cdot 7 \cdot 11$

c. $2 \cdot 2 \cdot 2 \cdot 2 \cdot 3 \cdot 3 \cdot 3$

Solution
a. 5^3 b. $7^2 \cdot 11$ c. $2^4 \cdot 3^3$

Self Check
Use exponents to write each prime factorization.

a. $3 \cdot 3 \cdot 7$

b. $5 \cdot 5 \cdot 7 \cdot 7$

c. $2 \cdot 2 \cdot 2 \cdot 3 \cdot 3 \cdot 5$

Answer: a. $3^2 \cdot 7$,
b. $5^2 \cdot 7^2$, c. $2^3 \cdot 3^2 \cdot 5$ ■

EXAMPLE 7 *Evaluating exponential expressions.*

a. $7^2 = 7 \cdot 7 = 49$ Read 7^2 as "7 squared."

b. $5^2 = 5 \cdot 5 = 25$ Read 5^2 as "5 squared."

c. $2^5 = 2 \cdot 2 \cdot 2 \cdot 2 \cdot 2 = 32$ Read 2^5 as "2 to the fifth power."

d. $10^3 = 10 \cdot 10 \cdot 10 = 1,000$ Read 10^3 as "10 cubed."

e. $6^1 = 6$ Write the base, 6, one time.

Self Check

Which of the numbers 3^5, 4^4, and 5^3 is the largest?

Answer: 4^4 or 256

EXAMPLE 8 *Evaluating exponential expressions.* Evaluate: $2^3 \cdot 3^4 \cdot 5$.

Solution

We find the value of each power and multiply.

$2^3 \cdot 3^4 \cdot 5 = \mathbf{8 \cdot 81 \cdot 5}$ $2^3 = 8$ and $3^4 = 81$.

$\qquad\qquad\quad = 3,240$ $8 \cdot 81 \cdot 5 = 3,240$.

Self Check

Evaluate: $3^3 \cdot 4^2 \cdot 5^2$.

Answer: 10,800

 WARNING! Note that 5^3 means $5 \cdot 5 \cdot 5$. It does not mean $5 \cdot 3$.

Accent on Technology **Bacterial growth**

At the end of one hour, a culture containing two bacteria doubles every hour. Use exponents to determine how many bacteria the culture will contain after 24 hours.

We can use a chart to help model the situation (see Table 1-2). From the chart, we see a pattern developing: The number of bacteria in the culture after 24 hours will be 2^{24}. We can evaluate this exponential expression using a scientific calculator.

Number of bacteria	Time
$2 = 2^1$	1 hr
$4 = 2^2$	2 hr
$8 = 2^3$	3 hr
$16 = 2^4$	4 hr
$? = 2^{24}$	24 hr

TABLE 1-2

Evaluate: 2^{24}.

Keystrokes: `2` `y^x` `2` `4` `=`   `16777216`

(On some calculators, the y^x is labeled x^y.)

Thus, $2^{24} = 16,777,216$. There will be 16,777,216 bacteria in the culture after 24 hours.

Finding prime factorizations with the division method

We can also find the prime factorization of a whole number by division. For example, to find the prime factorization of 363, we begin the division method by choosing the *smallest* prime number that will divide the given number exactly. We continue this process until the result of the division is a prime number.

Step 1: 2 doesn't divide 363 exactly, but 3 does. The result is 121, which is not prime. We continue the process.

$$3 \underline{|\ 363}$$
$$121$$

Step 2: We choose the smallest prime number that will divide 121. The primes 2, 3, 5, and 7 don't divide 121 exactly, but 11 does. The result is 11, which is prime. We are done.

$$3 \underline{|\ 363}$$
$$11 \underline{|\ 121}$$
$$11$$
$$363 = 3 \cdot 11 \cdot 11$$

Using exponents, we can write the prime factorization of 363 as $3 \cdot 11^2$.

EXAMPLE 9   *Factoring with the division method.*   Use the division method to find the prime factorization of 100. Use exponents to express the result.

Self Check

Use the division method to find the prime factorization of 108. Use exponents to express the result.

Solution

2 divides 100 exactly. The result is 50, which is not prime. ⟶ $2 \underline{|\ 100}$
2 divides 50 exactly. The result is 25, which is not prime. ⟶ $2 \underline{|\ 50}$
5 divides 25 exactly. The result is 5, which is prime. We are done. ⟶ $5 \underline{|\ 25}$
5

The prime factorization of 100 is $2^2 \cdot 5^2$.

Answer: $3^3 \cdot 2^2$

STUDY SET   Section 1.4

VOCABULARY   *Fill in the blanks to make a true statement.*

1. Numbers that are multiplied together are called _____.

2. One number is _____ by another if the remainder is 0 when they are divided.

3. A division with a remainder of 0 is said to come out _____.

4. To _____ a number means to express it as the product of other whole numbers.

5. A _____ number is a whole number, greater than 1, that has only 1 and _____ as factors.

6. Whole numbers, greater than 1, that are not prime numbers are called _____ numbers.

7. An _____ whole number can be divided evenly by 2.

8. An _____ whole number cannot be evenly divided by 2.

9. To prime _____ a number means to write it as a product of only _____ numbers.

10. An _____ is used to represent repeated multiplication.

11. In the exponential expression 6^4, 6 is called the _____, and 4 is called the _____.

12. Another way to say "5 to the second power" is 5 _____. Another way to say "7 to the third power" is 7 _____.

CONCEPTS

13. Write 27 as the product of two factors.

14. Write 30 as the product of three factors.

In Exercises 15–16, the complete list of the factors of a whole number is given. What is the whole number?

15. 2, 4, 22, 44, 11, 1

16. 20, 1, 25, 100, 2, 4, 5, 50, 10

In Exercises 17–20, find all factors of each whole number.

17. 11   **18.** 23   **19.** 37   **20.** 13

21. From the results obtained in Exercises 17–20, what can you say about each of the numbers?

22. Suppose a number is divisible by 10. Is 10 a factor of the number?

23. If 4 is a factor of a whole number, will 4 divide the number exactly?

24. Give examples of whole numbers that have 11 as a factor.

In Exercises 25–28, the prime factorization of a whole number is given. Find the number.

25. $2 \cdot 3 \cdot 3 \cdot 5$ **26.** $3^3 \cdot 2$ **27.** $11^2 \cdot 5$ **28.** $2 \cdot 2 \cdot 2 \cdot 7$

29. Do you think that we can change the order of the base and the exponent in an exponential expression and obtain the same result? In other words, does $3^2 = 2^3$?

30. Find the prime factors of 30 and 165. What prime factors do they have in common?

31. Find the prime factors of 30 and 242. What prime factor do they have in common?

32. Find the prime factors of 20 and 35. What prime factor do they have in common?

33. Find the prime factors of 20 and 80. What prime factors do they have in common?

34. Find 1^2, 1^3, and 1^4. From the results, what conclusion can you make about any power of 1?

35. Finish the process of prime factoring 150. Compare the results in both cases.

36. Find three numbers, less than 10, that would fit at the top of this tree diagram.

Number

Prime number   Prime number

37. Consider the factors of 24.
  a. Which two have a sum of 10?
  b. Which two have a difference of 10?

38. Consider the factors of 18.
  a. Which two have a sum of 11?
  b. Which two have a difference of 3?

39. When using the division method to find the prime factorization of an even number, what would be an obvious choice with which to start the division process?

40. When using the division method to find the prime factorization of a number ending in 5, what would be an obvious choice with which to start the division process?

NOTATION   *In Exercises 41–44, write each expression without using exponents.*

41. 7^3 **42.** 8^4 **43.** 3^5 **44.** 4^6

In Exercises 45–48, write each expression as a base raised to a power.

45. $2 \cdot 2 \cdot 2 \cdot 2 \cdot 2$ **46.** $3 \cdot 3 \cdot 3 \cdot 3 \cdot 3 \cdot 3$ **47.** $5 \cdot 5 \cdot 5 \cdot 5$ **48.** $9 \cdot 9 \cdot 9$

PRACTICE   *In Exercises 49–58, find the factors of each whole number.*

49. 10   **50.** 6   **51.** 40   **52.** 75

53. 18   **54.** 32   **55.** 44   **56.** 65

57. 77   **58.** 441

In Exercises 59–66, write each number in prime-factored form.

59. 39 **60.** 20 **61.** 30 **62.** 150

63. 162 **64.** 400 **65.** 220 **66.** 126

In Exercises 67–78, evaluate each exponential expression.

67. 3^4 **68.** 5^3 **69.** 2^5 **70.** 10^5

71. 12^2 **72.** 7^3 **73.** 8^4 **74.** 9^5

75. $3^2 \cdot 2^3$ **76.** $3^3 \cdot 4^2$ **77.** $2^3 \cdot 3^3 \cdot 4^2$ **78.** $3^2 \cdot 4^3 \cdot 5^2$

 In Exercises 79–82, use a calculator to evaluate each exponential expression.

79. 234^3 **80.** 51^4 **81.** $23^2 \cdot 13^3$ **82.** $12^3 \cdot 15^2$

APPLICATIONS *In Exercises 83–86, solve each problem.*

83. PERFECT NUMBERS A whole number is called a **perfect number** when the sum of its factors that are less than the number equals the number. Show that 6 and 28 are perfect numbers.

84. CRYPTOGRAPHY Information is often transmitted in code. Many codes involve writing products of large primes, because they are difficult to factor. To see how difficult, try finding two prime factors of 7,663. (*Hint:* Both primes are greater than 70.)

85. RECYCLING The number of residents of a city taking part in a trash recycling program skyrocketed over a four-year period. This rise in participation is shown in Illustration 1. Express the number of participants each year as a power of 5.

86. CELL DIVISION After one hour, a cell has divided to form another cell. In another hour, these two cells have divided so that 4 cells exist. In another hour, these four cells divide so that 8 exist.
a. How many cells exist after the fourth hour?

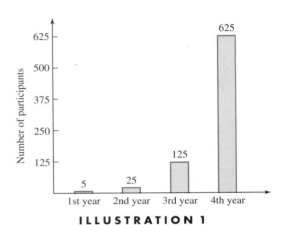

ILLUSTRATION 1

b. The number of cells that exist after each division can be found using an exponential expression. What would be its base?
c. Use a calculator to find the number of cells after 12 hours.

WRITING *Write a paragraph using your own words.*

87. Explain how to test a number to see whether it is prime.

88. Explain the difference between the *factors* of a number and the *prime factors* of the number.

89. Explain how to test a number to see whether it is even.

90. Explain how to test a number to see whether it is odd.

REVIEW *In Exercises 91–98, do the operations.*

91. $341 + 527$ **92.** $328 + 739$ **93.** $312 - 176$ **94.** $517 - 259$

95. $38 \cdot 42$ **96.** $234 \cdot 47$ **97.** $615 \div 15$ **98.** $1,073 \div 37$

99. BUYING STOCK An investor bought 100 shares of ABC stock at $44 per share and 100 shares of GFC at $32 per share. How much did he spend?

100. MARCHING BANDS When a university band lines up in eight rows of 15 musicians, there are five musicians left over. How many band members are there?

Order of Operations

In this section, you will learn about

- **Order of operations**
- **Grouping symbols**
- **The arithmetic mean (average)**

INTRODUCTION. Punctuation marks, such as commas, quotations, and periods, serve an important purpose when writing compositions. They determine the way in which sentences are to be read and interpreted. In this section, we will see that grouping symbols are the punctuation marks of mathematics. They determine how mathematical expressions are to be read and interpreted.

Order of operations

Suppose you are asked to contact a friend if you see a certain type of watch for sale while you are traveling in Europe. While in Switzerland, you spot the watch and send the following E-mail message.

E-Mail
WATCH $1500.   SHOULD I BUY IT FOR YOU?

The next day, you get this response from your friend.

E-Mail
NO PRICE TOO HIGH! REPEAT... NO! PRICE TOO HIGH.

Something is wrong. One statement says to buy the watch at any price. The other says not to buy it, because it's too expensive. The placement of the exclamation point makes us read these statements differently, resulting in different interpretations.

When reading a mathematical statement, the same kind of confusion is possible. For example, we consider the expression

$$3 + 2 \cdot 5$$

This expression contains two operations: addition and multiplication. We can calculate it in two different ways. We can do the addition first and then do the multiplication. Or we can do the multiplication first and then do the addition. However, we get different results.

Method 1: Add first

$3 + 2 \cdot 5 = 5 \cdot 5$ Add 3 and 2 first: $3 + 2 = 5$.

$= 25$ Multiply 5 and 5.

Method 2: Multiply first

$3 + 2 \cdot 5 = 3 + 10$ Multiply 2 and 5 first: $2 \cdot 5 = 10$.

$= 13$ Add 3 and 10.

———— Different results ————

If we don't establish an order of operations, the expression $3 + 2 \cdot 5$ has two different answers. To avoid this possibility, we will always do calculations in the following order.

Order of operations	1. Evaluate all powers.
	2. Do all multiplications and divisions as they occur from left to right.
	3. Do all additions and subtractions as they occur from left to right.

It may not be necessary to apply all of these steps in every problem. If, for example, an expression does not have any powers, we simply move to the next step in the list and look for multiplications and divisions.

To calculate $3 + 2 \cdot 5$ correctly, we apply the rules for the order of operations. Since there are no powers, we do the multiplication first and then do the addition.

Ignore the addition for now   and do the multiplication first: $2 \cdot 5 = 10$.

$$3 + 2 \cdot 5 = 3 + 10$$
$$= 13 \quad \text{Next, do the addition: } 3 + 10 = 13.$$

Using the rules for the order of operations, we see that the correct answer is 13.

EXAMPLE 1   *Order of operations.*   Evaluate: $2 \cdot 4^2 - 8$.

Solution

This expression contains multiplication, a power, and subtraction. We apply the rules for order of operations.

$2 \cdot 4^2 - 8 = 2 \cdot 16 - 8$ Find the power: $4^2 = 16$.
$\qquad\quad = 32 - 8$ Do the multiplication: $2 \cdot 16 = 32$.
$\qquad\quad = 24$ Do the subtraction: $32 - 8 = 24$.

Self Check
Evaluate: $4 \cdot 3^3 - 6$.

Answer: 102   ■

EXAMPLE 2   *Perimeter.*   The perimeter P of a rectangle with length l and width w is given by the formula

$$P = 2l + 2w$$

Find the perimeter of the rectangle shown.

Length: 65 yd
Width: 32 yd

Solution

$P = 2l + 2w$ The formula for the perimeter of a rectangle.
$P = 2(65) + 2(32)$ Replace l with 65 and w with 32.
$\quad = 130 + 64$ Do the multiplications first.
$\quad = 194$ Do the addition.

The perimeter of the rectangle is 194 yards.

Self Check
Find the perimeter of a rectangle having a length of 33 feet and a width of 15 feet.

Answer: 96 ft   ■

EXAMPLE 3   *Order of operations.*   Evaluate: $\dfrac{192}{6} - 5(3)2$.

Solution

$\dfrac{192}{6} - 5(3)2 = 32 - 5(3)2$ Working from left to right, do the division: $\frac{192}{6} = 32$.
$\qquad\qquad\quad = 32 - 15(2)$ Do the multiplication: $5(3) = 15$.
$\qquad\qquad\quad = 32 - 30$ Do the multiplication: $15(2) = 30$.
$\qquad\qquad\quad = 2$ Do the subtraction: $32 - 30 = 2$.

Self Check
Evaluate: $\dfrac{36}{9} + 4(2)3$.

Answer: 28   ■

EXAMPLE 4 *Order of operations.* Evaluate: $4^3 - 2(2^2)$.

Self Check
Evaluate: $5^2 + 3(3)^2$.

Solution

$\begin{aligned}
4^3 - 2(2^2) &= \mathbf{64} - 2(\mathbf{4}) & \text{Find the powers: } 4^3 = 64 \text{ and } 2^2 = 4. \\
&= 64 - 8 & \text{Do the multiplication.} \\
&= 56 & \text{Do the subtraction.}
\end{aligned}$

Answer: 52

 WARNING! One expression that is often evaluated incorrectly is $5 \cdot 2^2$. There are two operations involved here: multiplication and squaring 2. The order of operations rules tell us to find the power first and then do the multiplication.

$\begin{aligned}
5 \cdot 2^2 &= 5 \cdot \mathbf{4} & \text{Do the power first: } 2^2 = 4. \\
&= 20 & \text{Then do the multiplication: } 5 \cdot 4 = 20.
\end{aligned}$

Because of the parentheses, the expression $(5 \cdot 2)^2$ is evaluated differently.

$\begin{aligned}
(5 \cdot 2)^2 &= \mathbf{10}^2 & \text{Because of the parentheses, do the multiplication first: } 5 \cdot 2 = 10. \\
&= 100 & \text{Then find the power: } 10^2 = 100.
\end{aligned}$

Grouping symbols

Grouping symbols serve as mathematical punctuation marks. They help determine the order in which an expression is to be evaluated. Examples of grouping symbols are parentheses ( ), brackets [ ], braces { }, and the fraction bar ———.

Order of operations when grouping symbols are present	To evaluate an expression containing grouping symbols, use the rules for the order of operations within each pair of grouping symbols, working from the innermost pair to the outermost pair.
	In the case of a fraction bar, simplify the expression above the bar (called the **numerator**) and the expression below the bar (called the **denominator**) separately. Then do the division indicated by the fraction bar.

In the next example, we have two similar-looking expressions. However, because of the parentheses, we evaluate them in a different order.

EXAMPLE 5 *Working with grouping symbols.* Evaluate each expression: **a.** $12 - 3 + 5$ and **b.** $12 - (3 + 5)$.

Self Check
Evaluate each expression.

a. $20 - 7 + 6$

b. $20 - (7 + 6)$

Solution

a. This expression does not contain grouping symbols. We evaluate it by using the rules for the order of operations.

$\begin{aligned}
12 - 3 + 5 &= \mathbf{9} + 5 & \text{Working from left to right, do the subtraction:} \\
& & 12 - 3 = 9. \\
&= 14 & \text{Do the addition: } 9 + 5 = 14.
\end{aligned}$

b. This expression contains grouping symbols. We must do the operation inside the parentheses first.

$\begin{aligned}
12 - (3 + 5) &= 12 - \mathbf{8} & \text{Do the addition inside the parentheses first:} \\
& & 3 + 5 = 8. \\
&= 4 & \text{Do the subtraction: } 12 - 8 = 4.
\end{aligned}$

Answer: **a.** 19, **b.** 7

EXAMPLE 6 *Working with a fraction bar.* Evaluate: $\dfrac{2(13) - 2}{3(2^2)}$.

Self Check

Evaluate: $\dfrac{3(14) - 6}{2(3^2)}$.

Solution

A fraction bar is a grouping symbol. We will simplify the numerator and denominator separately by using the rules for the order of operations.

$$\dfrac{2(13) - 2}{3(2^2)} = \dfrac{26 - 2}{3(4)} \quad \text{In the numerator, do the multiplication.}$$

In the denominator, find the power.

$$= \dfrac{24}{12} \quad \text{In the numerator, do the subtraction.}$$

In the denominator, do the multiplication.

$$= 2 \quad \text{Do the division.}$$

Answer: 2

EXAMPLE 7 *Order of operations inside grouping symbols.*
Evaluate: $5 + 2(13 - 5 \cdot 2)$.

Self Check

Evaluate: $25 - 2(12 - 5 \cdot 2)$.

Solution

This expression contains grouping symbols. We will apply the rules for the order of operations *inside* the parentheses first.

$$5 + 2(13 - 5 \cdot 2) = 5 + 2(13 - 10) \quad \text{Do the multiplication inside the parentheses.}$$

$$= 5 + 2(3) \quad \text{Do the subtraction inside the parentheses.}$$

$$= 5 + 6 \quad \text{Do the multiplication.}$$

$$= 11 \quad \text{Do the addition.}$$

Answer: 21

If an expression contains more than one pair of grouping symbols, we always begin by working inside the innermost pair and then work to the outermost pair.

Innermost parentheses

$$16 + 2\,[\,14 - 3\,(\,5 - 2\,)\,]$$

Outermost brackets

EXAMPLE 8 *Grouping symbols inside grouping symbols.*
Evaluate: $16 + 2[14 - 3(5 - 2)]$.

Self Check

Evaluate: $46 - 2[4 + 3(6 - 2)]$.

Solution

$$16 + 2[14 - 3(5 - 2)] = 16 + 2[14 - 3(3)] \quad \text{Do the subtraction inside the parentheses.}$$

$$= 16 + 2(14 - 9) \quad \text{Do the multiplication inside the brackets.}$$

$$= 16 + 2(5) \quad \text{Do the subtraction inside the parentheses.}$$

$$= 16 + 10 \quad \text{Do the multiplication.}$$

$$= 26 \quad \text{Do the addition.}$$

Answer: 14

The arithmetic mean (average)

The **arithmetic mean,** or **average,** of several numbers is a value around which the values of the numbers are grouped. It gives you an indication of the "center" of the set of numbers. When finding the mean of a set of numbers, we usually need to apply the rules for the order of operations.

Finding an arithmetic mean	To find the mean of a set of scores, divide the sum of the scores by the number of scores.

EXAMPLE 9 **_The arithmetic mean._** The test scores of a student are shown in the table. Find the student's mean (average) test score.

Solution

To find the average score, we add the scores and divide by 6.

$$\text{Average} = \frac{70 + 80 + 80 + 100 + 80 + 70}{6}$$

$$= \frac{480}{6}$$

$$= 80$$

The student's average is 80.

Test	Score
#1	70
#2	80
#3	80
#4	100
#5	80
#6	70

Self Check

Jim earns scores of 72, 83, 66, 88, and 96 on five tests. Find his mean (average) score.

Answer: 81

Accent on Technology   _Order of operations and parentheses_

Scientific calculators have the rules for order of operations built in, because they all have a left parenthesis key (and a right parenthesis key). To use a calculator to evaluate $\frac{240}{20 - 15}$, we press the keys indicated below.

Evaluate: $\dfrac{240}{20 - 15}$.

Keystrokes: [2] [4] [0] [÷] [(] [2] [0] [−] [1] [5] [)] [=]   `48`

The answer is 48.

STUDY SET   Section 1.5

VOCABULARY   _Fill in the blanks to make a true statement._

1. Punctuation marks in mathematics are the _____ symbols.

2. The expression above a fraction bar is called the _____.

3. The expression below a fraction bar is called the _____.

4. To find the _____ of several values, we add the values and divide by the _____ of values.

CONCEPTS

5. Consider $5(2)^2 - 1$. How many operations will need to be performed to evaluate the expression? List them in the order in which they should be performed.

6. Consider $75 + 3 - (5 \cdot 2)^3$. How many operations will need to be performed to evaluate this expression? List them in the order in which they should be performed.

7. Consider $\frac{5 + 5(7)}{2 + (8 - 4)}$. In the numerator, what operation should be done first? In the denominator, what operation should be done first?

8. In the expression $\frac{3 - 5(2)}{5(2) + 4}$, the bar is a grouping symbol. What does it separate?

9. Explain the difference between $2 \cdot 3^2$ and $(2 \cdot 3)^2$.

10. In the expression $4 \cdot 3^2$, what operation should be done first?

NOTATION *In Exercises 11–14, fill in the blanks to make a true statement.*

11. The symbols ( ) are called _____.

12. The symbols [ ] are called _____.

13. The symbols { } are called _____.

14. A fraction bar separates the _____ from the denominator.

In Exercises 15–18, complete each solution.

15. $28 - 5(2)^2 = 28 - 5(\_\_\_)$ Find the power.
$ = 28 - \_\_\_$ Do the multiplication.
$ = 8$ Do the subtraction.

16. $2 + (5 + 6 \cdot 2) = 2 + (5 + \_\_\_)$ Do the multiplication.
$ = 2 + \_\_\_$ Do the addition.
$ = 19$ Do the addition.

17. $[4 \cdot (2 + 7)] - 6 = (4 \cdot \_\_\_) - 6$ Do the addition.
$ = \_\_\_ - 6$ Do the multiplication.
$ = 30$ Do the subtraction.

18. $\dfrac{5(3) + 12}{9 - 6} = \dfrac{\_\_\_ + 12}{}$ In the numerator, do the multiplication. In the denominator, do the subtraction.
$\phantom{\dfrac{5(3) + 12}{9 - 6}} = \dfrac{}{}$ In the numerator, do the addition.
$\phantom{\dfrac{5(3) + 12}{9 - 6}} = 9$ Do the division.

PRACTICE *In Exercises 19–66, simplify each expression by doing the operations.*

19. $7 + 4 \cdot 5$

20. $10 - 2 \cdot 2$

21. $10 + (8 + 5)$

22. $(10 + 8) + 5$

23. $20 - 10 + 5$

24. $80 - 5 - 4$

25. $15 - (8 - 3)$

26. $(15 - 8) - 3$

27. $7 \cdot 5 - 5 \cdot 6$

28. $13 - 3^2$

29. $4^2 + 3^2$

30. $12^2 - 5^2$

31. $2 \cdot 3^2$

32. $3^3 \cdot 5$

33. $3 + 2 \cdot 3^4 \cdot 5$

34. $3 \cdot 2^3 \cdot 4 - 12$

35. $6 + 2(5 + 4)$

36. $3(5 + 1) + 7$

37. $3 + 5(6 - 4)$

38. $7(9 - 2) - 1$

39. $(7 - 4)^2 + 1$

40. $(9 - 5)^3 + 8$

41. $6^3 - (10 - 8)$

42. $5^2 - (9 - 3)$

43. $50 - [2(4)^2 + 1]$

44. $30 + [2(3)^3 - 1]$

45. $3^4 - (3)(2)$

46. $2^2 + (4 - 2)^3$

47. $39 - 5(6) + 9 - 1$

48. $15 - 3(2) - 4 + 3$

49. $(18 - 12)^3 - 5^2$

50. $(9 - 2)^2 - 3^3$

51. $2(10 - 3^2) + 1$

52. $1 + 3(18 - 4^2)$

53. $6 + \dfrac{25}{5} + 6 \cdot 3$

54. $15 - \dfrac{24}{6} + 8 \cdot 2$

55. $3\left(\dfrac{18}{3}\right) - 2(2)$

56. $2\left(\dfrac{12}{3}\right) + 3(5)$

57. $(2 \cdot 6 - 4)^2$

58. $2(6 - 4)^2$

59. $11 - 2[3^2 - (4 + 1)]$

60. $5 + 3[2^2(5 - 5)]$

61. $\dfrac{5^2 + 17}{6 - 2^2}$

62. $\dfrac{3^2 - 2^2}{(3 - 2)^2}$

63. $\dfrac{(3 + 5)^2 + 2}{2(8 - 5)}$

64. $\dfrac{25 - (2 \cdot 3 - 1)}{2 \cdot 9 - 8}$

65. $\dfrac{(5 - 3)^2 + 2}{4^2 - (8 + 2)}$

66. $\dfrac{(4^3 - 2) + 7}{5(2 + 4) - 7}$

In Exercises 67–70, use the formula $P = 2l + 2w$ to find the perimeter of each rectangle.

67.

41 m

38 m

68.

65 in.

53 in.

69.

17 ft

44 ft

70.

79 cm

27 cm

In Exercises 71–74, simplify each expression by doing the operations.

71. $2{,}985 - (1{,}800 + 689)$ **72.** $\dfrac{897 - 655}{88 - 77}$ **73.** $3{,}245 - 25(16 - 12)^2$ **74.** $\dfrac{24^2 - 4^2}{22 + 58}$

APPLICATIONS   In Exercises 75–82, solve each problem.

75. BUYING MUSIC   Last week, Pat bought 3 audio tapes and 5 compact discs. If each tape cost $5 and each CD cost $12, how much did she pay?

76. BUYING GROCERIES   At the supermarket, Carlos has 3 cartons of coke, 5 bags of potato chips, and 2 cartons of dip in his cart. Each carton of coke costs $3, each bag of chips costs $2, and each carton of dip costs $1. If Carlos has $20, can he pay the bill?

77. JUDGING   The 6 scores received by a junior diver are as follows:

5	2	4	6	3	4

The formula for computing the score for the dive is as follows:

1. Throw out the lowest score.

2. Throw out the highest score.

3. Divide the sum of the remaining scores by 4.

Find the diver's score.

78. LABOR TOTALS   The hours needed to produce a television set like the one in Illustration 1 are as follows: 4 hours for the electronics and 5 hours for the cabinet. How many hours of labor are necessary to manufacture 50 television sets?

ILLUSTRATION 1

79. AVERAGE TEMPERATURE   Last week's temperatures were 75°, 80°, 83°, 80°, 77°, 72°, and 86°. Find the week's mean temperature.

80. HIGHWAY SPEEDS   Speeds of six randomly chosen expressway drivers were 65, 65, 72, 62, 60, and 66 mph. Find their mean speed.

81. TEST SCORES   On five tests, Andrew had scores of 56, 75, 82, 63, and 79. Find his average score.

82. AVERAGE SALARY   Six workers at an office earn $35,000, $25,000, $28,000, $32,000, $30,000, and $24,000. Find the mean wage.

WRITING   Write a paragraph using your own words.

83. Explain why a rule for order of operations is necessary.

84. Explain the difference between evaluating $5 \cdot 2^3$ and $(5 \cdot 2)^3$.

85. Explain the process of finding the mean of a large group of numbers. What does an average tell you?

86. Consider $10 - 2 \cdot 3$. Explain how inserting grouping symbols, $(10 - 2) \cdot 3$, causes an *override* of the rules for the order of operations.

REVIEW   In Exercises 87–94, do the operations.

87. $\begin{array}{r} 325 \\ +349 \\ \hline \end{array}$

88. $\begin{array}{r} 4{,}029 \\ +3{,}271 \\ \hline \end{array}$

89. $\begin{array}{r} 5{,}628 \\ -4{,}509 \\ \hline \end{array}$

90. $\begin{array}{r} 4{,}263 \\ -3{,}764 \\ \hline \end{array}$

91. $\begin{array}{r} 417 \\ \times\ 23 \\ \hline \end{array}$

92. $\begin{array}{r} 3{,}227 \\ \times\ \ 71 \\ \hline \end{array}$

93. $43\overline{)31{,}175}$

94. $82\overline{)50{,}430}$

Order of Operations

When asked to evaluate a numerical expression, you must perform the operations in the expression in the proper order. One of the major objectives of this course is that you be able to apply the rules for the order of operations.

Evaluate the expression $2 + 3 \cdot 5$ in two ways.

<u>Method 1:</u> Multiply first

$$2 + 3 \cdot 5$$

<u>Method 2:</u> Add first

$$2 + 3 \cdot 5$$

Why are the rules for the order of operations necessary?

Complete this list of the rules for the order of operations.

1. Evaluate all _____.
2. Do all _____ and divisions as they occur from _____ to right.
3. Do all additions and _____ as they occur from left to _____.

To apply the rules for the order of operations, we must identify the operations involved in an expression. List the operations that must be performed to evaluate each of the following expressions.

1. $10 + 4 - 3^2$

2. $\dfrac{180}{6} - (4)3$

3. $2(3) - 12 \div 6 \cdot 3$

4. $2(3)^3(4) + 6$

After identifying the operations, we must perform them in the proper order. Evaluate each expression.

5. $10 + 4 - 3^2$

6. $\dfrac{180}{6} - (4)3$

7. $2(3) - 12 \div 6 \cdot 3$

8. $2(3)^3(4) + 6$

When expressions involve grouping symbols, we perform the operations inside them first. Evaluate each of the following expressions.

9. $2(4 + 3 \cdot 2)^2 + 6$

10. $1 + 3[6 - (1 + 5)]$

| SECTION 1.1 | *Whole Numbers, Place Value, and Rounding* |

CONCEPT

A *set* is a collection of objects. The set of *natural numbers* is

1, 2, 3, 4, 5, . . .

The set of *whole numbers* is

0, 1, 2, 3, 4, 5, . . .

Whole numbers are often used in tables, bar graphs, and line graphs.

REVIEW EXERCISES

1. Consider the set 0, 2, $\frac{3}{2}$, 5, 7.2, 9.
  a. List each natural number in the set.

  b. List each whole number in the set.

2. Consider the data in the table, listing the number of building permits issued in the city of Springsville for the period 1993–1996.

Year	1993	1994	1995	1996
Building permits	12	15	10	7

  a. Construct a bar graph of the data.

  b. Construct a line graph of the data.

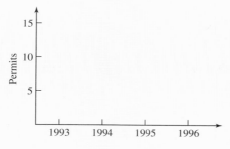

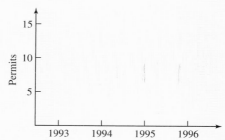

The digits in a *whole number* have *place value*.

3. Consider the number 2,365,720.
  a. Which digit is in the ten thousands column?

  b. Which digit is in the hundreds column?

A whole number is written in *expanded notation* when its digits are written with their place values.

4. Write each number in expanded notation.
  a. 570,302

  b. 37,309,054

5. Write each number in standard notation.
  a. 3 thousands + 2 hundreds + 7 ones
  b. twenty-three million, two hundred fifty-three thousand, four hundred twelve

The symbol < means "is less than." The symbol > means "is greater than."

6. Place an < or > symbol between the numerals to make a true statement.
  a. 9 ___ 7

  b. 3 ___ 5

To give approximate answers, we often use *rounded numbers*.

7. Round 2,507,348 to the specified place.
  a. nearest hundred
  c. nearest ten

  b. nearest ten thousand
  d. nearest hundred thousand

Adding and Subtracting Whole Numbers

Adding whole numbers corresponds to combining sets of objects. Do additions within parentheses first.

Commutative property of addition: The order in which whole numbers are added does not affect their sum.

Associative property of addition: The way whole numbers are grouped does not affect their sum.

The *perimeter* of a rectangle is the distance around it.

Subtracting whole numbers tells how many objects remain when some are removed from a set.

8. Find each sum.
  a. $7 + 6$
  b. $6 + 7$
  c. $4 + (7 + 3)$
  d. $(4 + 7) + 3$
  e. $5 + (6 + 9)$
  f. $(9 + 3) + 6$

9. Do each addition.
  a. $135 + 213$
  b. $4,447 + 7,478$
  c. $\begin{array}{r} 236 \\ +282 \\ \hline \end{array}$
  d. $\begin{array}{r} 5,345 \\ +\ 655 \\ \hline \end{array}$

10. Find the perimeter of the rectangle.

731 ft

642 ft

11. Do each subtraction.
  a. $8 - 5$
  b. $9 - (7 - 2)$
  c. $235 - 218$
  d. $5,231 - 5,177$
  e. $\begin{array}{r} 343 \\ -269 \\ \hline \end{array}$
  f. $\begin{array}{r} 7,800 \\ -5,725 \\ \hline \end{array}$

12. TRAVEL   A direct flight to San Francisco costs \$237. A flight with one stop in Reno costs \$192. How much can be saved by taking the inexpensive flight?

13. SAVINGS ACCOUNTS   A savings account contains \$931. If the owner deposits \$271 and makes withdrawals of \$37 and \$380, find the final balance.

14. FARMING   In a shipment of 350 animals, 124 were hogs, 79 were sheep, and the rest were cattle. Find the number of cattle in the shipment.

Multiplying and Dividing Whole Numbers

Multiplication is repeated addition. For example,

The sum of four 6's

$$4 \cdot 6 = \overbrace{6 + 6 + 6 + 6}$$
$$= 24$$

The result, 24, is called the *product,* and the 4 and 6 are called *factors.*

Commutative property of multiplication: The order in which whole numbers are multiplied does not affect their product.

Associative property of multiplication: The way whole numbers are grouped does not affect their product.

15. Do each multiplication.
  a. $8 \cdot 7$
  b. $7 \cdot 8$
  c. $8 \cdot 0$
  d. $7 \cdot 1$
  e. $(5 \cdot 7) \cdot 6$
  f. $5 \cdot (7 \cdot 6)$

16. Do each multiplication.
  a. $157 \cdot 21$
  b. $3,723 \cdot 48$
  c. $\begin{array}{r} 356 \\ \times\ 89 \\ \hline \end{array}$
  d. $\begin{array}{r} 5,624 \\ \times\ 81 \\ \hline \end{array}$

17. If a student worked for 38 hours and was paid \$9 per hour, how much did she earn?

The *area A of a rectangle* is the product of its length *l* and its width *w*.

$$A = lw$$

18. Find the area of each rectangle.

a.
8 cm
4 cm

b.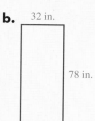
32 in.
78 in.

19. Sarah worked 12 hours at $7 per hour, and Santiago worked 15 hours at $6 per hour. Who earned the most money?

20. There are 12 eggs in one dozen, and 12 dozen in one gross. How many eggs are in a shipment of 5 gross?

Division is an operation that determines how many times a number (the *divisor*) is contained in another number (the *dividend*). *Remember that you can never divide by 0.*

21. Do each division.

a. $357 \div 17$

b. $1{,}443 \div 39$

c. $21\overline{)405}$

d. $54\overline{)1{,}269}$

22. If 745 candies are divided equally among 45 children, how many will each child receive?

23. In Problem 22, how many candies will be left over?

Prime Factors and Exponents

Numbers that are multiplied together are called *factors*.

A *prime number* is a whole number greater than 1 that has only 1 and itself as factors. Whole numbers greater than 1 that are not prime are called *composite numbers*.

Whole numbers that are divisible by 2 are *even* numbers. Whole numbers that are not divisible by 2 are *odd* numbers.

The *prime factorization* of a whole number is the product of its prime factors.

An *exponent* is used to indicate repeated multiplication. In the expression 6^3, 6 is the *base*, and 3 is the exponent.

24. Find all of the factors of each number.

a. 18

b. 25

25. Identify each number as a prime, composite, or neither.

a. 31

b. 100

c. 1

d. 0

e. 125

f. 47

26. Identify each number as an even or odd number.

a. 171

b. 214

c. 0

d. 1

27. Find the prime factorization of each number.

a. 42

b. 75

28. Write each expression using exponents.

a. $6 \cdot 6 \cdot 6 \cdot 6$

b. $5 \cdot 5 \cdot 5 \cdot 13 \cdot 13$

29. Evaluate each expression.

a. 5^3

b. 11^2

c. $2^3 \cdot 5^2$

d. $2^2 \cdot 3^3 \cdot 5^2$

Do mathematical operations in the following order:

1. Evaluate all powers.
2. Do all multiplications and divisions in order from left to right.
3. Do all additions and subtractions in order from left to right.

To evaluate an expression containing grouping symbols, use the rules for the order of operations within each pair of grouping symbols, working from the innermost pair to the outermost pair.

The *arithmetic mean* (average) is a value around which number values are grouped.

The *perimeter* of a rectangle having length *l* and width *w* is given by the formula
$$P = 2l + 2w$$

30. Evaluate each expression.
- **a.** $13 + 12 \cdot 3$
- **b.** $35 - 15 \div 5$
- **c.** $(13 + 12) \cdot 3$
- **d.** $(35 - 15) \div 5$
- **e.** $8 \cdot 5 - 4 \div 2$
- **f.** $8 \cdot (5 - 4 \div 2)$
- **g.** $2 + 3(10 - 4 \cdot 2)$
- **h.** $4(20 - 5 \cdot 3 + 2) - 4$
- **i.** $\dfrac{4(6) - 6}{2(3^2)}$
- **j.** $\dfrac{12 + 3 \cdot 7}{5^2 - 14}$
- **k.** $7 + 3[10 - 3(4 - 2)]$
- **l.** $5 + 2[(15 - 3 \cdot 4) - 2]$

31. Find the arithmetic mean of each set of scores.

a.

Test	1	2	3	4
Score	80	74	66	88

b.

Test	1	2	3	4	5
Score	73	77	81	69	90

32. Find the perimeter of a basketball court that is 50 feet wide and 94 feet long.

1. List the whole numbers less than 5.

2. Write "five thousand two hundred sixty-six" in expanded notation.

3. Write "7 thousands + 5 hundreds + 7 ones" in standard notation.

4. Round 34,752,341 to the nearest million.

In Problems 5–6, refer to the data in the table.

Lot number	1	2	3	4
Defective bolts	7	10	5	15

5. Use the data to make a bar graph.

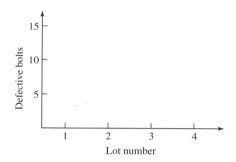

6. Use the data to make a line graph.

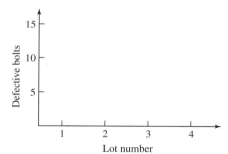

In Problems 7–8, place one of the symbols < or > between the numbers to make a true statement.

7. 15 ___ 10

8. 12 ___ 17

In Problems 9–12, do each operation.

9. Add: 327 + 435.

10. Subtract 287 from 535.

11. Add: 4,521
         +3,579

12. Subtract: 4,521
            −3,579

13. A rectangle is 327 inches wide and 757 inches long. Find its perimeter.

14. On Tuesday, a share of KBJ Company was selling at $73. The price rose $12 on Wednesday and fell $9 on Thursday. Find its price on Thursday.

In Problems 15–18, do each operation.

15. Multiply: 53
           $\times\ 8$

16. Multiply: 367
           $\times\ 73$

17. Divide: $63\overline{)4,536}$

18. Divide: $73\overline{)8,379}$

19. Find the perimeter and the area of the rectangle.

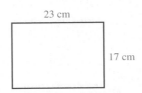

23 cm

17 cm

20. If 3,451 students are placed in groups of 74, how many will be left over?

21. Find the prime factorization of 1,260.

22. Evaluate $3 \cdot 4^2 - 2^2$.

23. Evaluate $9 + 4 \cdot 5$.

24. Evaluate $10 + 2[12 - 2(6 - 4)]$.

25. A student scored 73, 52, and 70 on three exams and received 0 on two missed exams. Find his mean score.

26. What information does the arithmetic mean (or average) give us about a set of values?

27. Explain the difference between what the perimeter and the area of a rectangle measure.

28. What are parentheses, and how have they been used in this chapter?

The Integers

2

IN THIS CHAPTER, THE CONCEPT OF A NEGATIVE NUMBER IS INTRODUCED AS WE EXPLORE AN EXTENSION OF THE SET OF WHOLE NUMBERS CALLED THE INTEGERS.

2.1 An Introduction to the Integers

In this section, you will learn about

- **Graphing on the number line**
- **Inequality symbols**
- **The integers**
- **Extending the number line**
- **Absolute value**
- **The opposite of a number**
- **The − symbol**

INTRODUCTION.   Whole numbers are not adequate to describe many situations that arise in everyday life. For example, 10 degrees below zero, $50 overdrawn, or 300 feet below sea level cannot be represented by whole numbers. To describe these situations, we need to use negative numbers. An excellent way to learn about negative numbers is with the help of a number line.

Graphing on the number line

A **number line** is a horizontal or vertical line that is used to represent numbers graphically. Like a ruler, a number line is straight and has uniform markings. (See Figure 2-1.) To construct a number line, we begin on the left with a point on the line representing the number 0. This point is called the **origin.** We then proceed to the right, drawing equally spaced marks and labeling them with whole numbers that increase in size. The arrowhead at the right is used to indicate that the number line continues forever.

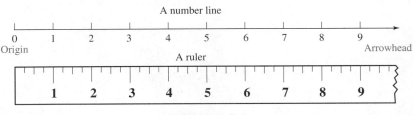

FIGURE 2-1

Using a process known as **graphing,** a single number or a set of numbers can be represented on a number line. *The graph of a number* is the point on the number line that corresponds to that number. *To graph a number* means to locate its position on the number line and then to highlight it using a heavy dot. The graph of the set of whole numbers less than 6 is shown in Figure 2-2.

FIGURE 2-2

Inequality symbols

As you move to the right on a number line, the values of the numbers increase. In Figure 2-3, we know that 8 is greater than 2 because the graph of 8 is to the right of the graph of 2.

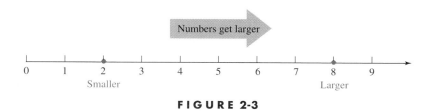

FIGURE 2-3

Two symbols, called **inequality symbols,** can be used to express the fact that 8 is greater than 2. They are the "is greater than" symbol ($>$) and the "is less than" symbol ($<$).

$8 > 2$ Read as "8 is greater than 2."

$2 < 8$ Read as "2 is less than 8."

To distinguish between these two symbols, remember that they always point to the smaller of the two numbers involved.

$10 > 3$ $4 < 9$

└──── Points to the ────┘
smaller number.

EXAMPLE 1   *Using inequality symbols.*   Use inequality symbols to write
a. Ten is greater than two   and   **b.** Two is less than ten.

Solution
a. $10 > 2$ Since 2 is smaller, the inequality symbol points to it.

b. $2 < 10$

Self Check

Use inequality symbols to write:

a. Eighty is less than eighty-one.

b. Eighty-one is greater than eighty.

Answer:
a. $80 < 81$,   **b.** $81 > 80$

Three other commonly used inequality symbols are the "is not equal to" symbol ($\neq$), the "is less than or equal to" symbol ($\leq$), and the "is greater than or equal to" symbol ($\geq$).

$5 \neq 2$ Read as "5 is not equal to 2."

$6 \leq 10$ Read as "6 is less than or equal to 10." This statement is true, because $6 < 10$.

$12 \leq 12$ Read as "12 is less than or equal to 12." This statement is true, because $12 = 12$.

$17 \geq 15$ Read as "17 is greater than or equal to 15." This statement is true, because $17 > 15$.

$20 \geq 20$ Read as "20 is greater than or equal to 20." This statement is true, because $20 = 20$.

EXAMPLE 2   *Inequality symbols.* Use inequality symbols to write **a.** "Eight is not equal to 5" and **b.** "50 is greater than or equal to 40."

Solution
a. $8 \neq 5$

b. $50 \geq 40$

Self Check
Use inequality symbols to write "30 is less than or equal to 35."

Answer: $30 \leq 35$

The integers

To describe a temperature of 10 degrees below zero, or $50 overdrawn, or an elevation of 300 feet below sea level, we need to use negative numbers. **Negative numbers** are numbers less than 0, and they are written using a **negative sign, −.**

In words	In symbols	Read as
10 degrees below zero	-10	"negative ten"
$50 overdrawn	-50	"negative fifty"
300 feet below sea level	-300	"negative three hundred"

Positive and negative numbers

Positive numbers are greater than 0. **Negative numbers** are less than 0.

 WARNING! Zero is neither positive nor negative.

Together, the set of positive numbers, the set of negative numbers, and zero make up the set of **signed numbers.** Negative numbers must always be written with a negative sign. However, positive numbers are not always written with a positive sign. For example, we can write 1,500 feet above sea level as $+1,500$ or 1,500.

The collection of all positive and negative whole numbers, along with zero, is called the set of **integers.**

The integers

. . . , $-5, -4, -3, -2, -1, 0, 1, 2, 3, 4, 5, \ldots$.

Since every natural number is also an integer, the natural numbers form a subset of the integers. Likewise, the whole numbers form a subset of the integers.

 WARNING! Note that not all integers are natural numbers, and that not all integers are whole numbers.

Extending the number line

Negative numbers can be represented on a number line by extending the line to the left. Beginning with zero, we move left, marking equally spaced points and then labeling them with progressively smaller negative whole numbers. (See Figure 2-4.) As you move to the right on a number line, the values of the numbers increase. As you move to the left, the values decrease.

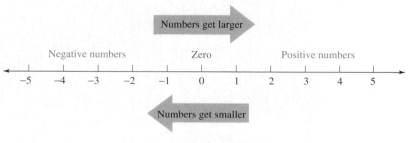

FIGURE 2-4

A thermometer is an example of a number line. The thermometer on the left is scaled in degrees, and it shows a temperature of $-10°$. In the study set for this section, you will see examples of number lines used to illustrate historical and scientific facts.

Negative numbers are graphed in the same way we graphed positive numbers.

EXAMPLE 3 *Graphing integers.* Graph 2, -3, 4, -1.

Solution

Self Check
Graph 3, -4, 1, -2.

Answer:

EXAMPLE 4 *Using inequality symbols.* Use one of the symbols $>$ or $<$ to make each statement true: **a.** 4 ___ -5 and **b.** -4 ___ 5.

Solution

a. Since 4 is to the right of -5 on the number line, $4 > -5$.

b. Since -4 is to the left of 5 on the number line, $-4 < 5$.

Self Check
Use one of the symbols $>$ or $<$ to make each statement true:

a. 6 ___ -6

b. -6 ___ -5

Answer: **a.** $6 > -6$, **b.** $-6 < -5$

By extending the number line to include negative numbers, we can represent more situations using bar graphs and line graphs. For example, the bar graph shown in Figure 2-5 illustrates the annual profits *and losses* of a car rental agency. Note that the profit in 1995 was $30 million and that the loss in 1994 was $40 million.

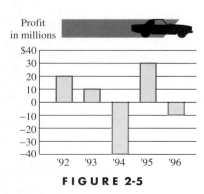

FIGURE 2-5

Absolute value

Using a number line, we can see that the numbers 3 and -3 are both a distance of 3 units away from 0, as shown in Figure 2-6.

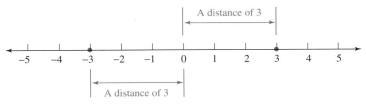

FIGURE 2-6

The **absolute value** of a number gives the distance between the number and 0 on a number line. To indicate absolute value, the number is inserted between two vertical bars. For the example above, we would write $|-3| = 3$. This is read as "The absolute value of negative 3 is 3," and it tells us that the distance between -3 and 0 is 3 units. In the example, we also see that $|3| = 3$.

Absolute value	The **absolute value** of a number is the distance on a number line between the number and 0.

> **WARNING!** Absolute value expresses distance. The absolute value of a number is always positive or zero, but never negative!

EXAMPLE 5   *Evaluating an absolute value.* Evaluate each absolute value: **a.** $|8|$ and **b.** $|-5|$.

Solution

a. On a number line, the distance between 8 and 0 is 8. Therefore,

$$|8| = 8$$

b. On a number line, the distance between -5 and 0 is 5. Therefore,

$$|-5| = 5$$

Self Check

Evaluate each absolute value:

a. $|-9|$

b. $|4|$

c. $|0|$

Answer: **a.** 9, **b.** 4, **c.** 0 ■

The opposite of a number

Opposites or negatives	Two numbers represented by points on a number line that are the same distance away from the origin, but on opposite sides of it, are called **opposites** or **negatives.**

The numbers 4 and -4 are opposites because they are the same distance from the origin. (See Figure 2-7.)

To write the opposite of a number, a $-$ symbol is used. For example, the opposite, or negative, of 5 can be written as -5. Parentheses are needed to express the opposite of a negative number. The opposite of -5 is written as $-(-5)$. Since 5 and

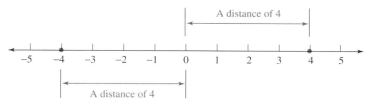

FIGURE 2-7

-5 are the same distance from zero, the opposite of -5 is 5. Therefore, in symbols, we can write $-(-5) = 5$. This leads us to the following conclusion.

The double negative rule	The **double negative rule** states: The opposite of the opposite of a number is that number.

	Number	*Opposite*	
	57	-57	Read -57 as "negative fifty-seven."
	-8	$-(-8) = 8$	$-(-8)$ is read as "the opposite of negative eight." Apply the double negative rule.
	0	$-0 = 0$	The opposite of 0 is 0.

The concept of opposite can also be applied to an absolute value. For example, the opposite of the absolute value of -8 can be written $-|-8|$. Think of this as a two-step process. Find the absolute value first, and then attach the $-$ to that result.

First, find the absolute value.

$$-|-8| = -8$$

Then attach a $-$ sign.

The $-$ symbol

The symbol used to indicate the opposite of a number, the symbol used to represent the operation of subtraction, and the symbol used to write a negative number are all the same. The key to interpreting the $-$ symbol correctly is to examine the context in which it is used.

Using the negative, the opposite, and the minus sign	-12	Negative twelve	A $-$ symbol directly in front of a single number is read as "negative."
	$-(-12)$	The opposite of negative twelve	The first $-$ symbol is read as "the opposite of" and the second as "negative."
	$12 - 5$	Twelve minus five	Notice the space used before and after the $-$ sign. This indicates subtraction and is read as "minus."

VOCABULARY *In Exercises 1–10, fill in the blanks to make a true statement.*

1. Numbers can be represented by points equally spaced on a number _____.

2. The point on a number line representing 0 is called the _____.

3. To _____ a number means to locate it on a number line and highlight it with a _____.

4. The _____ of a number is the point on the number line that represents that number.

5. The symbols > and < are called _____ symbols.

6. _____ numbers are less than 0.

7. The _____ value of a number is the distance between the number and zero on a number line.

8. Two numbers on a number line that are the same distance from zero, but on opposite sides of the origin, are called _____.

9. The collection of all whole numbers and their opposites is called the _____.

10. The _____ negative rule states that the opposite of the _____ of a number is that number.

CONCEPTS

11. Refer to each graph and use an inequality symbol (> or <) to make a true statement.

a.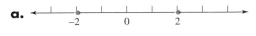

−2 ____ 2

b.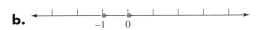

0 ____ −1

c.

−1 ____ 0   and   1 ____ 0

12. Tell what is wrong with each number line.

a.

b.

c.

d.

13. Does every number on the number line have an opposite?

14. Is the absolute value of a number always positive?

15. Which of the following contains a minus sign: 15 − 8, −(−15), or −15?

16. Is there a number that is both greater than 10 and less than 10 at the same time?

17. Express the fact 12 < 15 using the > symbol.

18. Express the fact 5 > 4 using the < symbol.

19. Represent each of these situations using a signed number, and then describe its opposite in words.
  a. $225 overdrawn
  b. 10 seconds before liftoff
  c. 3 degrees below normal
  d. A deficit of $12,000
  e. A race horse finished 2 lengths behind the leader.

20. Represent each of these situations using a signed number, and then describe its opposite in words.
  a. A trade surplus of $3 million
  b. A bacteria count 70 more than the standard
  c. A profit of $67
  d. A business $1 million in the "black"
  e. 20 units over their quota

21. If a number is less than 0, describe what type of number it must be.

22. If a number is greater than 0, describe what type of number it must be.

23. On a number line, what number is 3 units to the right of −7?

24. On a number line, what number is 4 units to the left of 2?

25. Name two numbers on a number line that are a distance of 5 away from −3.

26. Name two numbers on a number line that are a distance of 4 away from 3.

27. Which number is closer to −3 on the number line, 2 or −7?

28. Which number is farther from 1 on the number line, −5 or 8?

29. Give examples of the − symbol used in three different ways.

30. What is the opposite of 0?

PRACTICE *In Exercises 31–42, evaluate each expression.*

31. $|9|$

32. $|12|$

33. $|-8|$

34. $|-1|$

35. $|-14|$

36. $|-85|$

37. $-|20|$

38. $-|110|$

39. $-|-6|$

40. $|0|$

41. $|203|$

42. $-|-11|$

In Exercises 43–46, simplify each expression.

43. $-(-4)$

44. $-(-9)$

45. $-(-12)$

46. $-(-25)$

In Exercises 47–50, graph each set of numbers on a number line labeled from −5 to 5.

47. −3, 0, 3, 4, −1

48. −4, −1, 2, 5, 1

49. The opposite of −3, the opposite of 5, and the absolute value of −2

50. The absolute value of 3, the opposite of 3, and the number 1 less than −3

In Exercises 51–58, insert one of the symbols >, <, or = in the blank to make a true statement.

51. −5 ___ 5

52. 0 ___ −1

53. −12 ___ −6

54. −6 ___ −7

55. −10 ___ −11

56. −11 ___ −20

57. $|-2|$ ___ 0

58. $|-30|$ ___ −40

In Exercises 59–62, insert one of the symbols ≥ or ≤ in the blank to make a true statement.

59. −1,255 ___ −(−1,254)

60. 0 ___ −3

61. $-|-3|$ ___ 4

62. $-|-163|$ ___ −150

In Exercises 63–70, use the information in each problem to write an inequality.

63. This year's newspaper recycling drive total of 540 pounds exceeded the 391 pounds collected last year.

64. This year's bake sale proceeds of $617 were less than last year's total of $723.

65. The commander of a submarine notified the navigator to keep the vessel at a depth of 500 feet below sea level.

66. 475 people attended a banquet, which was more than the 250 people who attended the dance.

67. The 400-meter relay team's time of 49 seconds beat the old record of 50 seconds.

68. The weatherman said that the temperature tomorrow will be 5° warmer than today's temperature of 67°.

69. The altitude of Beach Town is 475 feet above sea level.

70. In a golf tournament, the winning score was 72, and the second-place score was 73.

APPLICATIONS

71. FLIGHT OF A BALL A boy throws a ball from the top of a building, as shown in Illustration 1. At the instant he does this, his friend starts a stop watch and keeps track of the time as the ball rises to a peak and then falls to the ground. Use the vertical number line to complete the table by finding the position of the ball at the specified times.

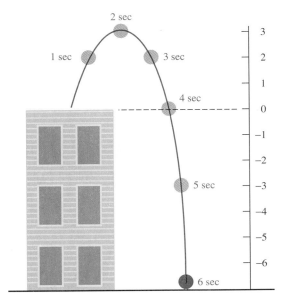

ILLUSTRATION 1

Time	Position of ball
1 sec	
2 sec	
3 sec	
4 sec	
5 sec	
6 sec	

72. SHOOTING GALLERY At an amusement park, a shooting gallery contains moving ducks. The path of one duck is shown in Illustration 2, along with the time it takes the duck to reach certain positions on the gallery wall. Complete the table at the top of the page using the horizontal number line in the illustration.

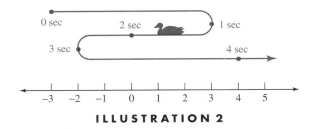

ILLUSTRATION 2

Time	Position of duck
0 sec	
1 sec	
2 sec	
3 sec	
4 sec	

73. TECHNOLOGY The readout from a testing device is shown in Illustration 3. It is important to know the value of each of the "peaks" and each of the "valleys" shown on the screen. Use the vertical number line to find each of these numbers.

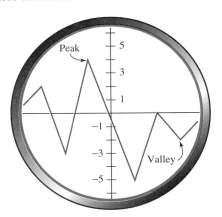

ILLUSTRATION 3

74. FLOODING A week of daily reports listing the height of a river in comparison to flood stage is given in the table. Complete the bar graph in Illustration 4. It should show a day-by-day account of the height of the river compared to flood stage.

Flood stage report	
Sun.	2 ft below
Mon.	3 ft over
Tue.	4 ft over
Wed.	2 ft over
Thu.	1 ft below
Fri.	3 ft below
Sat.	4 ft below

ILLUSTRATION 4

75. GOLF STATISTICS   In Illustration 5, each golf ball represents the score (in relation to par) of a professional golfer on the last hole of a tournament.

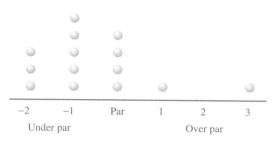

ILLUSTRATION 5

a. What score was shot most often on this hole?

b. What was the best score on this hole?

c. Explain why this hole appears to be too easy for a professional golfer.

76. PAYCHECK   Examine the items listed on the paycheck stub in Illustration 6. Then write two columns on your paper—one headed "positive" and the other "negative." List each item under the appropriate heading.

Tom Dryden Dec. 96	Christmas bonus	$100
Gross pay $2,000	**Reductions**	
Overtime $300	Retirement	$200
Deductions	**Taxes**	
Union dues $30	Federal withholding	$160
U.S. Bonds $100	State withholding	$35

ILLUSTRATION 6

77. WEATHER MAP   Illustration 7 shows the predicted Fahrenheit temperatures for a day in mid-January.

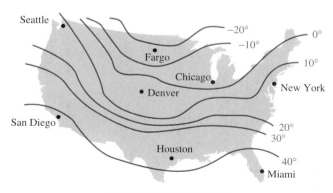

ILLUSTRATION 7

a. What is the temperature range for the region including Fargo, North Dakota?

b. Approximately how much colder will it be in Chicago than in Denver?

c. According to this prediction, what is the coldest it should get in Seattle?

78. ENDANGERED SPECIES   Illustration 8 tracks a 3-year effort by zoologists to restore the population of an endangered species. Their goal was to bring the population back to 400 animals worldwide.

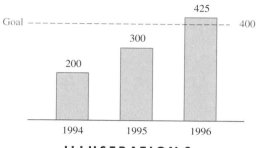

ILLUSTRATION 8

a. Use signed numbers to express how far below (or above) the goal the population was in the years 1994, 1995, and 1996.

b. If the trends indicated by this chart continue, estimate the population in 1997.

79. HISTORICAL TIME LINE   Number lines can be used to display historical data. Some important world events are shown on the time line in Illustration 9.

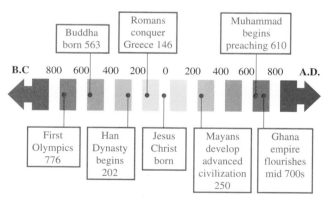

ILLUSTRATION 9

a. What basic unit was used to scale this time line?

b. On this time line, what represents the origin?

c. What could be thought of as positive numbers?

d. What could be thought of as negative numbers?

80. LINE GRAPH Each thermometer in Illustration 10 gives the daily high temperature in degrees Fahrenheit. Plot each daily high temperature on the grid and then construct a line graph.

ILLUSTRATION 10

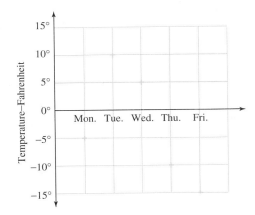

81. SOUND The loudness of sound is measured in decibels (abbreviated dB). The decibel scale can be represented by a number line. Use the number line provided and label it from 30 to 170 in increments of 20. On the number line, graph each loudness measurement in Illustration 11 and label it.

Vacuum cleaner	70 dB
Motorcycle	90 dB
Thunder	115 dB
Rocket launch	155 dB
Human speech	40 dB
Jet plane takeoff	130 dB

ILLUSTRATION 11

82. THERMOMETER The thermometer in Illustration 12 is designed to record the overnight low and the daytime high temperatures. Use the number line provided to mathematically model this thermometer. Use a basic unit of 10 to scale your model. Graph the high and low temperatures for that day.

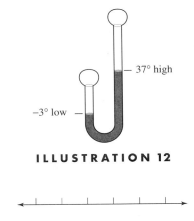

ILLUSTRATION 12

WRITING *Write a paragraph using your own words.*

83. Explain the concept of the opposite of a number.

85. Explain why the absolute value of a number is never negative.

84. What real-life situation do you think gave rise to the concept of a negative number?

86. Give an example of the use of a number line that you have seen in another course.

REVIEW

87. Round 23,456 to the nearest hundred.

89. Is this statement true or false? $3 \le 3$

91. Give the name of the property illustrated here:
$(13 \cdot 2) \cdot 5 = 13 \cdot (2 \cdot 5)$

88. Evaluate: $19 - 2 \cdot 3$.

90. Is this statement true or false? $345 < 354$

92. Write four times five using three different notations.

Addition of Integers

In this section, you will learn about

- **Adding integers**
- **Adding two integers with the same sign**
- **Adding two integers with different signs**
- **The addition property of zero**
- **The additive inverse of a number**

INTRODUCTION. We have seen that integers are used in various settings to describe a temperature (4 degrees above zero), an altitude (20 feet below sea level), or a debt (a loss of $10). After learning how to use integers to describe a quantity, the next step is to learn how to add, subtract, multiply, and divide them. We begin with the operation of addition.

Adding integers

When adding two integers, they either have the same sign or they have different signs.

Observations	**Same sign**	**Different signs**
	If the two integers to be added have the same sign, then	If the two numbers to be added have different signs, then
	both are positive: $\quad 4 + 3$ or both are negative: $\quad -4 + (-3)$	one is positive and the other negative: $4 + (-3) \quad$ or $\quad -4 + 3$

We will now examine each of the specific problems listed above.

Adding two integers with the same sign

$4 + 3$
both positive

To explain addition of signed numbers, we can use a number line. (See Figure 2-8.) To compute $4 + 3$, we begin at the origin and draw an arrow 4 units long, pointing to the right. This represents positive 4. From that point, we draw an arrow 3 units long, pointing to the right, to represent positive 3. The second arrow points to 7. Therefore, $4 + 3 = 7$.

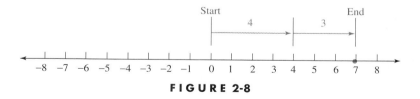

FIGURE 2-8

As a check, let's think of the problem in terms of money. If you had $4 and earned $3 more, you would have a total of $7.

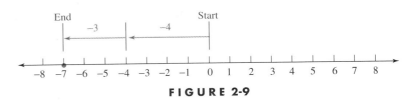

$$-4 + (-3)$$
both negative

To compute $-4 + (-3)$ on a number line, we begin at the origin and draw an arrow 4 units long, pointing to the left. This represents -4. From there, we draw an arrow 3 units long, pointing in the negative direction to represent -3. We end up at -7. (See Figure 2-9.) The end of the second arrow points to the sum: $-4 + (-3) = -7$.

End   Start

-3 -4

-8 -7 -6 -5 -4 -3 -2 -1 0 1 2 3 4 5 6 7 8

FIGURE 2-9

Let's think of this problem in terms of money. If you had a debt of $4 (negative 4) and then incurred $3 more debt (negative 3), you would be in debt $7 (negative 7). Here are some observations about adding two numbers that have the same sign.

Observations
1. Both arrows point in the same direction and "build" on each other.
2. The result has the *same sign* as the two numbers being added.

$$4 \ + \ 3 \ = \ 7 \qquad\qquad -4 \ + \ (-3) \ = \ -7$$

positive + positive = positive   negative + negative = negative
answer   answer

These observations suggest the following rule.

Adding two integers with the same sign

To add two integers with the **same sign,** add their absolute values and attach their common sign to the sum. If both integers are positive, their sum is positive. If both integers are negative, their sum is negative.

EXAMPLE 1 *Adding two negative integers.* Find the sum: $-9 + (-4)$.

Solution These integers are both negative. The rule for adding two integers with the same sign tells us to first add the absolute values of each of the integers.

$$|-9| + |-4| = 9 + 4 = 13$$

The rule then says to attach the common sign (which is negative) to this result. Therefore,

Attach the common negative sign.

$$-9 + (-4) = -13$$

After some practice, you will be able to do this kind of problem in your head. It will not be necessary to show all the steps as we have done here. ■

EXAMPLE 2 *Adding two integers.* Find the sum: **a.** $6 + 4$ and **b.** $-80 + (-60)$.

Solution

a. Since both integers are positive, the answer will be positive.

$$6 + 4 = 10$$

b. Since both integers are negative, the answer will be negative.

$$-80 + (-60) = -140$$

Self Check
Find the sum:

a. $7 + 5$

b. $-300 + (-100)$

Answer: **a.** 12, **b.** -400 ■

Adding two integers with different signs

$$4 + (-3)$$
one positive, one negative

To compute $4 + (-3)$ on a number line, we start at the origin and draw an arrow 4 units long, pointing to the right. This represents positive 4. From there, we draw an arrow 3 units long, pointing to the left, to represent -3. We end up at 1. (See Figure 2-10.) The end of the second arrow points to the answer: $4 + (-3) = 1$.

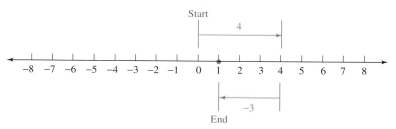

FIGURE 2-10

In terms of money, if you had \$4 (positive 4) and then incurred a debt of \$3 (negative 3), you would have \$1 (positive 1) left.

$$-4 + 3$$
one positive, one negative

The problem $-4 + 3$ can be illustrated by drawing an arrow 4 units long from the origin, pointing to the left. This represents -4. From there, we draw an arrow 3 units long, pointing to the right, to represent positive 3. (See Figure 2-11.) The end of the second arrow points to the answer: $-4 + 3 = -1$.

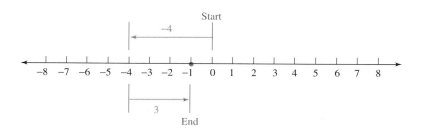

FIGURE 2-11

This problem can be thought of as owing \$4 (negative 4) and then paying back \$3 (positive 3). You will still owe \$1 (negative 1).

The last two examples lead us to some observations about adding two integers with different signs.

Observations

1. The arrows representing the integers point in *opposite* directions.

2. The *longer* of the two arrows determines the sign of the result.

These observations suggest the following rule.

Adding two integers with different signs

To add two integers with **different signs,** subtract their absolute values, the smaller from the larger. Then attach to that result the sign of the integer with the larger absolute value.

EXAMPLE 3 *Adding a positive and a negative integer.* Find the sum: $5 + (-7)$.

Solution One integer is positive and the other is negative—they have different signs. The rule tells us to first subtract the smaller absolute value from the larger.

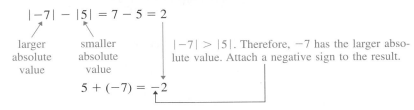

$$|-7| - |5| = 7 - 5 = 2$$

larger absolute value   smaller absolute value $|-7| > |5|$. Therefore, -7 has the larger absolute value. Attach a negative sign to the result.

$$5 + (-7) = -2$$

EXAMPLE 4 *Adding a positive and a negative integer.* Find the sum:
a. $-8 + 5$ and **b.** $11 + (-5)$.

Solution

a. Since -8 has the larger absolute value, the answer will be negative.

$-8 + 5 = -3$ Subtract 5 from 8 and use a $-$ sign.

b. Since 11 has the larger absolute value, the answer will be positive.

$11 + (-5) = +6$ Subtract 5 from 11 and use a $+$ sign.
$ = 6$

Self Check
Find the sum:

a. $-2 + 7$

b. $6 + (-9)$

Answer: **a.** 5, **b.** -3

EXAMPLE 5 *Weather predictions.* A television weather person announced that because of a warming trend, a low of $-6°$ on Friday would give way to a $35°$ rise in temperature for Saturday. What high temperature can be expected on Saturday?

Solution The words *rise in temperature* indicate addition. This situation can be represented mathematically by $-6 + 35$.

$$-6 + 35 = 29$$

The high temperature that can be expected on Saturday is $29°$.

EXAMPLE 6 *Adding several integers.* Add: $-3 + 5 + (-12) + 2$.

Solution
This expression contains four integers. We will add them, working from left to right.

$-3 + 5 + (-12) + 2 = 2 + (-12) + 2$ Add: $-3 + 5 = 2$.
$ = -10 + 2$ Add: $2 + (-12) = -10$.
$ = -8$ Do the addition.

Self Check
Add: $-12 + 8 + (-6) + 1$.

Answer: -9

EXAMPLE 7 *Order of operations.* Evaluate: $(-1 + 5) + [-6 + (-5)]$.

Solution
To evaluate this expression, we will need to apply the rules for order of operations.

$(-1 + 5) + [-6 + (-5)] = 4 + (-11)$ Do the operations inside the grouping symbols first: $-1 + 5 = 4$ and $-6 + (-5) = -11$.
$ = -7$ Do the addition.

Self Check
Evaluate:
$[-1 + (-9)] + (-8 + 4)$.

Answer: -14

Your calculator can add positive and negative numbers.

- You do not have to do anything special to enter positive numbers. When you press 3, for example, a positive 3 is entered.
- To enter a negative 3 on a scientific calculator, the ± key must be pressed *after* entering 3. It is called the *opposite,* or *change of sign* key.

Evaluate: -238 $+$ $(-1,097)$

Keystrokes: | 2 | 3 | 8 | +/− | + | 1 | 0 | 9 | 7 | +/− | = | | −1335 |

The sum is −1,335.

The addition property of zero

When 0 is added to a number, the number remains the same. For example, $5 + 0 = 5$, and $0 + (-4) = -4$. Because of this, we call 0 the **additive identity.**

Addition property of zero

The sum of any number and 0 is that number.

The additive inverse of a number

A second fact concerning 0 and the operation of addition can be demonstrated by considering the sum of a number and its opposite. To illustrate this, we use the number line in Figure 2-12 to add 6 and its opposite, -6. We see that $6 + (-6) = 0$.

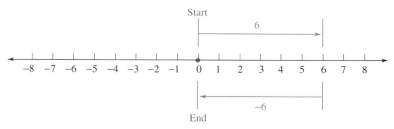

FIGURE 2-12

If the sum of two numbers is 0, the numbers are said to be **additive inverses** of each other. Since $6 + (-6) = 0$, we say that 6 and -6 are additive inverses.

We can now classify a pair of numbers such as 6 and -6 in three ways: as opposites, negatives, or additive inverses.

The additive inverse of a number

Numbers are said to be **additive inverses** if their sum is 0.

EXAMPLE 8   *The additive inverse.*   What is the additive inverse of -3? Justify your result.

Solution

The additive inverse of -3 is its opposite, 3. To justify the result, we add and show that the sum is 0.

$$-3 + 3 = 0$$

Self Check

What is the additive inverse of 12? Justify your result.

Answer: -12;
$12 + (-12) = 0$

VOCABULARY   *In Exercises 1–2, fill in the blanks to make a true statement.*

1. When 0 is added to a number, the number remains the same. We call 0 the additive _____.

2. Since $-5 + 5 = 0$, 5 is the additive _____ of -5.

CONCEPTS   *In Exercises 3–6, use the number line in Illustration 1 to find each answer.*

3. $-3 + 6$

4. $-3 + (-2)$

5. $-5 + 3$

6. $-1 + (-3)$

$$\overset{\displaystyle\longleftarrow\;|\;|\;|\;|\;|\;|\;|\;|\;|\;|\;|\;\longrightarrow}{\;\;-5\;-4\;-3\;-2\;-1\;\;\;0\;\;\;1\;\;\;2\;\;\;3\;\;\;4\;\;\;5}$$

ILLUSTRATION 1

7. Is the sum of two positive integers always positive?

8. Is the sum of two negative integers always negative?

9. What is the sum of a number and its additive inverse?

10. What is the sum of a number and its opposite?

11. What property states that $-7 + 8 = 8 + (-7)$?

12. What number, when added to 15, gives 0?

In Exercises 13–14, fill in the blanks to make a true statement.

13. To add two integers with unlike signs, _____ their absolute values, the _____ from the larger. Then attach to that result the sign of the number with the _____ absolute value.

14. To add two integers with like signs, add their _____ values and attach their common _____ to the sum.

NOTATION   *In Exercises 15–18, complete each solution.*

15. Evaluate: $-16 + (-2) + (-1)$.

$-16 + (-2) + (-1) =$ ___ $+ (-1)$ Work from left to right. Add -16 and -2.

$\qquad\qquad\qquad = -19$ Do the addition.

16. Evaluate: $-8 + (-2) + 6$.

$-8 + (-2) + 6 =$ ___ $+ 6$ Work from left to right. Add -8 and -2.

$\qquad\qquad\quad = -4$ Do the addition.

17. Evaluate: $(-3 + 8) + (-3)$.

$(-3 + 8) + (-3) =$ ___ $+ (-3)$ Do the addition inside the parentheses.

$\qquad\qquad\quad = 2$ Do the addition.

18. Evaluate: $-5 + [2 + (-9)]$.

$-5 + [2 + (-9)] = -5 +$ ___   Do the addition inside the brackets.

$\qquad\qquad\quad = -12$ Do the addition.

PRACTICE   *In Exercises 19–26, find the additive inverse of each number.*

19. -11

20. 9

21. -23

22. $-(-43)$

23. 0

24. $|4|$

25. $|-14|$

26. 250

In Exercises 27–50, find each sum.

27. $6 + (+3)$

28. $2 + (+2)$

29. $-5 + (-5)$

30. $-8 + (-8)$

31. $-6 + 7$

32. $-2 + 4$

33. $-15 + 8$

34. $-18 + 10$

35. $20 + (-40)$

36. $25 + (-10)$

37. $30 + (-15)$

38. $8 + (-20)$

39. $-1 + 9$

40. $-2 + 7$

41. $-7 + 9$

42. $-3 + 6$

43. $5 + (-15)$

44. $16 + (-26)$

45. $24 + (-15)$

46. $-4 + 14$

47. $35 + (-27)$

48. $46 + (-73)$

49. $24 + (-45)$

50. $-65 + 31$

In Exercises 51–58, evaluate each expression.

51. $-2 + 6 + (-1)$ **52.** $4 + (-3) + (-2)$ **53.** $-9 + 1 + (-2)$ **54.** $5 + 4 + (-6)$

55. $6 + (-4) + (-13) + 7$ **56.** $8 + (-5) + (-10) + 6$ **57.** $9 + (-3) + 5 + (-4)$ **58.** $-3 + 7 + 1 + (-4)$

59. Find the sum of -6, -7, and -8.

60. Find the sum of -11, -12, and -13.

In Exercises 61–68, find each sum.

61. $-7 + 0$ **62.** $6 + 0$ **63.** $9 + 0$ **64.** $0 + (-15)$

65. $-4 + 4$ **66.** $18 + (-18)$ **67.** $2 + (-2)$ **68.** $-10 + 10$

69. What number must be added to -5 to obtain 0?

70. What number must be added to 8 to obtain 0?

In Exercises 71–80, evaluate each expression.

71. $2 + (-10 + 8)$

72. $(-9 + 12) + (-4)$

73. $(-4 + 8) + (-11 + 4)$

74. $(-12 + 6) + (-6 + 8)$

75. $[-3 + (-4)] + (-5 + 2)$

76. $[9 + (-10)] + (-7 + 9)$

77. $[6 + (-4)] + [8 + (-11)]$

78. $[5 + (-8)] + [9 + (-15)]$

79. $-2 + [-8 + (-7)]$

80. $-8 + [-5 + (-2)]$

APPLICATIONS   *In Exercises 81–90, use signed numbers to help answer each question.*

81. G FORCES   As a fighter pilot dives and loops, different forces are exerted on the body, just like the forces you experience when riding on a roller coaster. Some of the forces, called G's, are positive and some are negative. The force of gravity, 1G, is constant. Complete the diagram in Illustration 2.

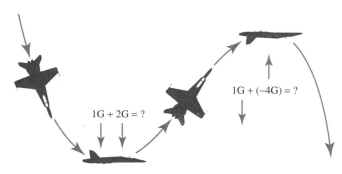

$1G + 2G = ?$

$1G + (-4G) = ?$

ILLUSTRATION 2

82. CHEMISTRY   The first several steps of a chemistry lab experiment are listed here. The experiment begins with a compound that is stored at $-40°$ F.

    Step 1: Raise the temperature of the compound $200°$.

    Step 2: Add sulfur dioxide and then raise the temperature $10°$.

    Step 3: Add 10 milliliters of water, stir, and raise the temperature $25°$.

What is the resulting temperature of the mixture after step 3?

83. CASH FLOW   The maintenance costs, utilities, and taxes on a duplex are \$900 per month. The owner of the apartments receives monthly rental payments of \$450 and \$380. Does this investment produce a positive cash flow each month?

84. JOGGING   A businessman's lunchtime workout includes jogging up 10 stories of stairs in his high-rise office building. If he starts on the fourth level below ground in the underground parking garage, on what story of the building will he finish his workout?

85. OCEANOGRAPHY   In an experiment to measure ocean water temperatures, a probe at the end of a cable was lowered from a Sea Institute ship, and readings were taken at various depths. The results of the test are shown in Illustration 3.

Depth (ft)	Temperature (F)
-5	$68°$
-25	$64°$
-45	$58°$
-65	$46°$
-85	$42°$

ILLUSTRATION 3

a. What does -85 represent?

b. How much cable was reeled in after each temperature reading was taken?

c. Between what two depths did the greatest temperature change take place?

86. SPREADSHEET   Monthly rain totals for four counties are listed in the spreadsheet shown in Illustration 4. The −1 entered in cell B1 means that the rain total for Suffolk County for a certain month was one inch below average. We can analyze this data by asking the computer to perform various operations.

	A	B	C	D	E	F
1	Suffolk	−1	−1	0	+1	+1
2	Marin	0	−2	+1	+1	−1
3	Logan	−1	+1	+2	+1	+1
4	Tipton	−2	−2	+1	−1	−3

ILLUSTRATION 4

a. To ask the computer to add the numbers in cells C1, C2, C3, and C4, we type SUM(C1:C4). Find this sum.

b. Find SUM(B4:F4).

87. ATOMS   An atom is composed of protons, neutrons, and electrons. A proton has a positive charge (represented by +1), a neutron has no charge, and an electron has a negative charge (−1). Two simple models of atoms are shown in Illustration 5 at the top of the page. What is the net charge of each atom?

88. POLITICAL POLLS   Six months before a general election, the incumbent senator found himself trailing the challenger by 18 points. To overtake his opponent, the campaign staff decided to use a four-part strategy.

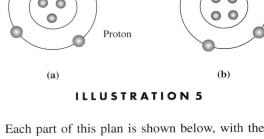

Electron

Proton

(a)   (b)

ILLUSTRATION 5

Each part of this plan is shown below, with the anticipated point gain.

1. Intense TV ad blitz   +10
2. Ask for union endorsement   +2
3. Voter mailing   +3
4. Get-out-the vote campaign   +1

With these gains, will the incumbent overtake the challenger on election day?

89. FLOODING   After a heavy rainstorm, a river that had been 4 feet under flood stage rose 11 feet in a 48-hour period. Find the height of the river after the storm in comparison to flood stage.

90. INSURANCE   An auto body shop gives Marta an estimate of $600 to repair her car, damaged in a traffic collision. If her insurance coverage includes a $450 deductible, what will the insurance company pay?

In Exercises 91–94, use a calculator to answer each problem.

91. 789 + (−9,135)

92. 2,701 + (−4,089)

93. −675 + (−456) + 99

94. −9,750 + (−780) + 2,345

95. FILM PROFITS   A movie studio produced four films, two financial successes and two failures. The profits and losses of the films are shown in Illustration 6. Find the studio's profit, if any, for the year.

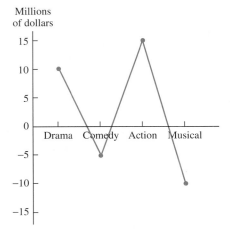

Millions of dollars

ILLUSTRATION 6

96. MONTHLY SALES   A car dealership has a quota of 200 new cars to sell each month. At the end of October, the dealership was 270 cars under the quota for that point in the year. The owners decided to slash prices in the last two months of the year in an effort to make up for the shortfall. Auto sales for November and December are listed in Illustration 7.

a. Complete the chart.

Month	Cars sold	Over quota
November	273	
December	399	

ILLUSTRATION 7

b. Did the dealership reach its year-end sales quota?

97. Explain why we cannot say with certainty what type of number will result when adding a positive and a negative number.

98. How do you explain the fact that when asked to *add* $-4 + 8$, we must actually *subtract* to obtain the result?

99. Why is the sum of two negative numbers a negative number?

100. Write an application problem that will require adding -50 and -60.

REVIEW

101. Find the area of the rectangle in Illustration 8.

5 ft

3 ft

ILLUSTRATION 8

102. A car with a tank that holds 15 gallons of gasoline goes 25 miles on one gallon. How far can it go on a full tank?

103. Find the perimeter of the rectangle in Illustration 8.

104. What property does the statment $5 \cdot 15 = 15 \cdot 5$ illustrate?

105. Prime factor 125. Use exponents to express the result.

106. Do the division: $\dfrac{144}{12}$.

2.3   *Subtraction of Integers*

In this section, you will learn about

- **Adding the opposite**
- **Order of operations**
- **Applications of subtraction**

INTRODUCTION.   In this section, we will study another way to think about subtraction. This new procedure will be helpful when subtraction problems involve negative numbers.

Adding the opposite

The subtraction problem $6 - 4$ can be thought of as taking away 4 from 6. We can use a number line to illustrate this. (See Figure 2-13.) Beginning at the origin, we draw an arrow of length 6 units in the positive direction. From that point, we move back 4 units to the left. The answer, called the **difference,** is 2.

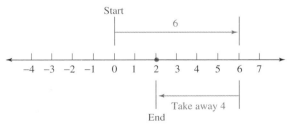

FIGURE 2-13

The work shown in Figure 2-13 looks a lot like the illustration for the *addition* problem $6 + (-4) = 2$, shown in Figure 2-14.

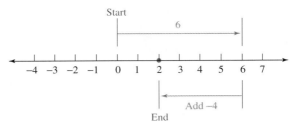

FIGURE 2-14

In the first problem, $6 - 4$, we subtracted 4 from 6. In the second, $6 + (-4)$, we added -4 (which is the opposite of 4) to 6. In each case, the result was 2.

Subtracting 4.   Adding the opposite of 4.

$$6 - 4 = 2 \qquad\qquad 6 + (-4) = 2$$

The same result.

This observation helps to justify the following rule for subtraction.

| **Rule for subtraction** | To subtract two integers, add the opposite of the second integer to the first integer. |

In words, this rule says that subtraction is the same as adding the opposite of the number to be subtracted. Then we apply the rules for addition.

You won't need to use this rule for every subtraction problem. For example, $6 - 4$ is obviously 2; it does not need to be rewritten as adding the opposite. But for more complicated problems such as $-6 - 4$ or $-6 - (-4)$, where the result is not obvious, the subtraction rule will be quite helpful.

EXAMPLE 1   *Adding the opposite.*   Find $-6 - 4$.

Solution
We are asked to subtract 4 from -6. We apply the subtraction rule.

$$-6 - 4 = -6 + (-4) \qquad \text{Write the subtraction as an addition of the opposite of 4.}$$
$$\text{The opposite of 4 is } -4.$$
$$ = -10 \qquad \text{To add } -6 \text{ and } -4, \text{ we apply the rule for adding two negative numbers.}$$

Self Check
Find $-2 - 3$.

Answer: -5

EXAMPLE 2   *Adding the opposite.*   Find $3 - (-5)$.

Solution
We apply the subtraction rule.

$$3 - (-5) = 3 + [-(-5)] \qquad \text{Add the opposite of } -5. \text{ Attach a } - \text{ sign in front of } -5.$$
$$\text{Use brackets to do this.}$$
$$ = 3 + 5 \qquad \text{Use the double negative rule: } -(-5) = 5.$$
$$ = 8 \qquad \text{Do the addition.}$$

Self Check
Find $3 - (-2)$.

Answer: 5

| EXAMPLE 3 | *Adding the opposite.* Find $-8 - (-5)$. |

Self Check

Find $-2 - (-6)$.

Solution

$$\begin{aligned} -8 - (-5) &= -8 + [-(-5)] && \text{Add the opposite of } -5. \\ &= -8 + 5 && \text{Apply the double negative rule.} \\ &= -3 && \text{Do the addition.} \end{aligned}$$

Answer: 4

Remember that any subtraction problem can be rewritten as an equivalent addition. We just add the opposite of the number that is to be subtracted.

Subtraction can be written as addition . . .

$$\begin{aligned} 4 - 8 &= 4 + (-8) = -4 \\ 4 - (-8) &= 4 + 8 = 12 \\ -4 - 8 &= -4 + (-8) = -12 \\ -4 - (-8) &= -4 + 8 = 4 \end{aligned}$$

of the opposite of the number to be subtracted.

Accent on Technology   *Subtraction with negative numbers*

We can subtract positive and negative numbers using a calculator.

- We will need to use the opposite (or sign change) key $\boxed{+/-}$ when applying the subtraction rule of this section.

Evaluate: $-1{,}021 \quad - \quad (-504)$

Keystrokes: $\boxed{1}\,\boxed{0}\,\boxed{2}\,\boxed{1}\,\boxed{+/-}\,\boxed{-}\,\boxed{5}\,\boxed{0}\,\boxed{4}\,\boxed{+/-}\,\boxed{=}$ $\boxed{\qquad\qquad -517}$

The difference is -517. We could have written this subtraction as an addition of the opposite before using the calculator. That is, $-1{,}021 + 504$.

Order of operations

Expressions can contain repeated subtraction or subtraction in combination with grouping symbols. To work these problems, we will apply the rule for order of operations.

| EXAMPLE 4 | *Repeated subtraction.* Evaluate: $-1 - (-2) - 10$. |

Self Check

Evaluate: $-3 - 5 - (-1)$.

Solution

This subtraction problem involves three numbers. We work from left to right, rewriting each subtraction as an addition of the opposite.

$$\begin{aligned} -1 - (-2) - 10 &= -1 + [-(-2)] + (-10) && \text{Add the opposite of } -2 \text{ and } 10. \\ &= -1 + 2 + (-10) && \text{Apply the double negative rule:} \\ & && -(-2) = 2. \\ &= 1 + (-10) && \text{Work from left to right. Add} \\ & && -1 + 2. \\ &= -9 && \text{Add 1 and } -10. \end{aligned}$$

Answer: -7

EXAMPLE 5   *Order of operations.*   Evaluate: $-8 - (-2 - 2)$.

Solution

We must do the subtraction inside the parentheses first.

$-8 - (-2 - 2) = -8 - [-2 + (-2)]$ Do the subtraction inside the parentheses first. Add the opposite of 2, which is -2.

$\qquad\qquad = -8 - (-4)$ Add -2 and -2.

$\qquad\qquad = -8 + 4$ Add the opposite of -4, which is 4.

$\qquad\qquad = -4$ Add -8 and 4.

Applications of subtraction

Things are constantly changing in our daily lives. The temperature, the amount of money we have in the bank, and our ages are a few examples. In mathematics, the operation of subtraction is used to measure change.

In general, to find the change in a quantity, subtract the earlier value from the later value. If you subtract the amount of money you had in the bank last week from the amount of money you have in the bank this week, the result will tell you how much your bank balance has changed in a week.

EXAMPLE 6   *Change of water level.*   On Monday, the water level in a city storage tank was 6 feet above normal. By Friday, the level had fallen to a mark 4 feet below normal. Find the change in the water level from Monday to Friday. (See Figure 2-15.)

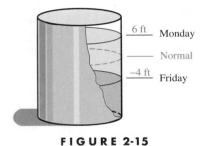

FIGURE 2-15

Solution   We will use subtraction to find the amount of change. The water levels of 4 feet below normal (the later value) and 6 feet above normal (the earlier value) can be represented by -4 and 6, respectively.

Water level Friday	minus	water level Monday	equals	change of water level.

$-4 - 6 = -4 + (-6)$ Add the opposite of 6.

$\qquad\quad = -10$ Do the addition. The negative result indicates that the water level fell.

The water level fell 10 feet from Monday to Friday.

In the next example, a number line will serve as a mathematical model of a real-life situation. You will see how the operation of subtraction can be used to find the distance between two points on a number line.

EXAMPLE 7   *Artillery accuracy.*   In a practice session, an artillery group fired two rounds at a target. The first landed 65 yards short of the target, and the second landed 50 yards past it. (See Figure 2-16.) How far apart were the two impact points?

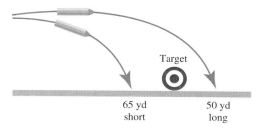

Target

65 yd
short

50 yd
long

FIGURE 2-16

Solution   We can use a number line to model this situation. The target is the origin. The words *short of the target* indicate a negative number, and the words *past it* indicate a positive number. Therefore, we graph the impact points at −65 and 50 in Figure 2-17.

Short   Target   Long

−65   0   50

FIGURE 2-17

The phrase *how far apart* tells us to subtract.

Position of long	minus	position of short	equals	distance between impact points.

$$50 - (-65) = 50 + 65 \quad \text{Add the opposite of } -65.$$
$$= 115 \quad \text{Do the addition.}$$

The impact points are 115 yards apart.

STUDY SET   Section 2.3

VOCABULARY   *In Exercises 1–4, fill in the blanks to make a true statement.*

1. The answer to a subtraction problem is called the _____.

2. [ ] are grouping symbols called _____.

3. Two numbers represented by points on a number line that are the same distance away from the origin, but on opposite sides of it, are called _____.

4. The double _____ rule states that the opposite of the opposite of a number is that number.

CONCEPTS   *In Exercises 5–12, fill in the blanks to make a true statement.*

5. _____ is the same as adding the opposite of the number to be subtracted.

6. Subtracting 3 is the same as adding ___.

7. Subtracting −6 is the same as adding ___.

8. The opposite of −8 is written as _____.

9. Every subtraction problem can be written as an equivalent _____ problem.

10. $12 - 7 = 12 +$ _____

11. After using parentheses as grouping symbols, if another set of grouping symbols is needed, we use _____.

12. We can find the _____ in a quantity by subtracting the earlier value from the later value.

13. Write this problem using mathematical symbols: negative eight minus negative four.

14. Write this problem using mathematical symbols: negative eight subtracted from negative four.

15. Find the distance between -4 and 3 on a number line.

16. Find the distance between -10 and 1 on a number line.

NOTATION *In Exercises 17–20, complete each solution.*

17. Evaluate: $1 - 3 - (-2)$.

$1 - 3 - (-2) = 1 + (\underline{}) + 2$ Add the opposite of 3 and -2.

$\qquad\qquad = -2 + \underline{}$ Add 1 and -3.

$\qquad\qquad = 0$ Do the addition.

18. Evaluate: $-6 + 5 - (-5)$.

$-6 + 5 - (-5) = -6 + 5 + \underline{}$ Add the opposite of -5.

$\qquad\qquad = \underline{} + 5$ Add -6 and 5.

$\qquad\qquad = 4$ Do the addition.

19. Evaluate: $(-8 - 2) - (-6)$.

$(-8 - 2) - (-6) = [-8 + (\underline{})] - (-6)$ Add the opposite of 2.

$\qquad\qquad = \underline{} - (-6)$ Do the addition in the brackets.

$\qquad\qquad = -10 + \underline{}$ Add the opposite of -6.

$\qquad\qquad = -4$ Do the addition.

20. Evaluate: $-5 - (-1 - 4)$.

$-5 - (-1 - 4) = -5 - [-1 + (\underline{})]$ Add the opposite of 4.

$\qquad\qquad = -5 - (\underline{})$ Do the addition in the brackets.

$\qquad\qquad = -5 + \underline{}$ Add the opposite of -5.

$\qquad\qquad = 0$ Do the addition.

PRACTICE *In Exercises 21–44, find the difference.*

21. $8 - (-1)$

22. $3 - (-8)$

23. $-4 - 9$

24. $-7 - 6$

25. $-5 - 5$

26. $-7 - 7$

27. $-5 - (-4)$

28. $-9 - (-1)$

29. $-1 - (-1)$

30. $-4 - (-3)$

31. $-2 - (-10)$

32. $-6 - (-12)$

33. $0 - (-5)$

34. $0 - 8$

35. $0 - 4$

36. $0 - (-6)$

37. $-2 - 2$

38. $-3 - 3$

39. $-10 - 10$

40. $4 - 4$

41. $9 - 9$

42. $4 - (-4)$

43. $-3 - (-3)$

44. $-5 - (-5)$

In Exercises 45–56, evaluate each expression.

45. $-4 - (-4) - 15$

46. $-3 - (-3) - 10$

47. $-3 - 3 - 3$

48. $-1 - 1 - 1$

49. $5 - 9 - (-7)$

50. $6 - 8 - (-4)$

51. $10 - 9 - (-8)$

52. $16 - 14 - (-9)$

53. $-1 - (-3) - 4$

54. $-2 - 4 - (-1)$

55. $-5 - 8 - (-3)$

56. $-6 - 5 - (-1)$

In Exercises 57–66, evaluate each expression.

57. $(-6 - 5) - 3$

58. $(-2 - 1) - 5$

59. $(6 - 4) - (1 - 2)$

60. $(5 - 3) - (4 - 6)$

61. $-9 - (6 - 7)$

62. $-3 - (6 - 12)$

63. $-8 - [4 - (-6)]$

64. $-1 - [5 - (-2)]$

65. $[-4 + (-8)] - (-6)$

66. $[-5 + (-4)] - (-2)$

67. Subtract -3 from 7.

68. Subtract 8 from -2.

69. Subtract -6 from -10.

70. Subtract -4 from -9.

71. SCUBA DIVING   After descending 50 feet, a scuba diver paused to check his equipment before descending an additional 70 feet. Use a signed number to represent the diver's final depth.

72. TEMPERATURE CHANGE   Rashawn flew from his New York home to Hawaii for a week of vacation. He left blizzard conditions and a temperature of −6°, and stepped off the airplane into 85° weather. What temperature change did he experience?

73. READING PROGRAM   In a state reading test administered at the start of a school year, an elementary school's performance was 23 points below the county average. The principal immediately began a special tutorial program. At the end of the school year, retesting showed the students to be only 7 points below the average. How many points did the school's reading score improve over the year?

74. SUBMARINE   A submarine was traveling 2,000 feet below the ocean's surface when the radar system warned of an impending collision with another sub. The captain ordered the navigator to dive an additional 200 feet and then level off. Find the depth of the submarine after the dive.

75. AMPERAGE   During normal operation, the ammeter on a car reads +5. (See Illustration 1.) If the headlights, which draw a current of 7 amps, and the radio, which draws a current of 6 amps, are both turned on, what number will the ammeter register?

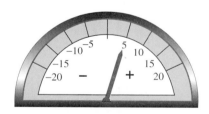

I L L U S T R A T I O N 1

76. TRADE DEFICIT   For the first half of the year, figures released by the government showed a trade deficit with one European country of $6 million. The situation worsened in the second half of the year, with an imbalance of $9 million occurring. On the other hand, there was a trade surplus of $5 million for the year with a South American country. Supply the missing information in the bar graph.

77. GEOGRAPHY   Death Valley, California, is the lowest land point in the United States, at 283 feet below sea level. The lowest land point on the earth is the Dead Sea, which is 1,290 feet below sea level. How much lower is the Dead Sea than Death Valley?

78. WEATHER   On Monday, the temperature reached a low of −4° in eastern Canada. Weather service reports that evening warned of colder temperatures, perhaps 5° to 10° colder, by the end of the week. If the low temperature Friday was −16°, was the prediction correct?

79. FOOTBALL   A college football team records the outcome of each of its plays during a game on a stat sheet. (See Illustration 2.) Find the net gain (or loss) after the 3rd play.

Down	Play	Result
1st	run	lost 1 yd
2nd	pass—sack!	lost 6 yd
penalty	delay of game	lost 5 yd
3rd	pass	gained 8 yd
punt		

I L L U S T R A T I O N 2

80. TRUCKING   A scheduler has designed a route for pickups and deliveries by a trucker, as shown on the route slip in Illustration 3. By law, the entire truck's weight load must never exceed vehicle code standards. Before stop #1, the truck is 1 ton under this standard. As the day progresses, is the weight limit ever violated? If so, do you have any suggestions for a pickup–delivery route that will not violate the vehicle code?

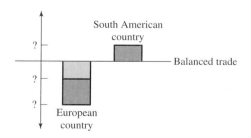

Stop			
#1	Delivery	Sammuel's	3 tons
#2	Pickup	Liu Tran Co.	5 tons
#3	Delivery	Carpet Town	2 tons

I L L U S T R A T I O N 3

81. DIVING A diver jumps from a platform. After she hits the water, her momentum takes her to the bottom of the pool. (See Illustration 4.)

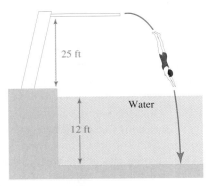

ILLUSTRATION 4

a. Use a number line and signed numbers to model this situation. Show the top of the platform, the water line, and the bottom of the pool.

b. Find the total length of the dive from the top of the platform to the bottom of the pool.

82. SNOW PACK The Department of Water measures the snow pack in the Sierras every winter to help predict the spring runoff. The figures are reported with respect to an average snowfall. In the past 50 years, annual figures have ranged from 12 feet over this average mark to 8 feet below the average mark. (See Illustration 5.)

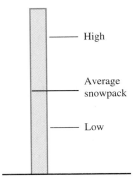

ILLUSTRATION 5

a. Use a number line and signed numbers to model this situation. What can be represented by the origin?

b. What is the range between these high and low marks?

 In Exercises 83–90, use a calculator.

83. $-1,557 - 890$

84. $-345 - (-789)$

85. $20,007 - (-496)$

86. $-979 - (-44,879)$

87. $-162 - (-789) - 2,303$

88. $-787 - 1,654 - (-232)$

89. CHECKING ACCOUNT Michael has $1,303 in his checking account. Can he pay his car insurance premium of $676, his utility bills of $121, and his rent of $750 without having to make another deposit? Explain your answer.

90. HISTORY Two of the greatest Greek mathematicians were Archimedes (287–212 B.C.) and Pythagoras (569–500 B.C.). How many years apart were they born?

WRITING *Write a paragraph using your own words.*

91. Explain what is meant when we say that every subtraction can be written as an equivalent addition.

92. Give an example showing that it is possible to subtract something from nothing.

93. Subtraction can undo addition. Explain this concept using an example.

94. Explain why algebra students don't need to change every subtraction they encounter into an equivalent addition problem. Give some examples.

REVIEW

95. Round 5,989 to the nearest ten.

96. Round 5,999 to the nearest hundred.

97. List the factors of 20.

98. How far can a car go on 8 gallons of gas if it gets 24 miles per gallon?

99. It takes 13 oranges to make one can of orange juice. Find the number of oranges used to make 12 cans.

100. Evaluate: $12 - (5 - 4)$.

101. Write 4,502 in expanded notation.

102. What property does the following illustrate? $15 + 12 = 12 + 15$

2.4   *Multiplication of Integers*

In this section, you will learn about

- **The product of two positive integers**
- **The product of a positive and a negative integer**
- **The product of a negative and a positive integer**
- **Multiplication by zero**
- **The product of two negative integers**
- **Powers of integers**

INTRODUCTION. We now turn our attention to multiplication of integers. When multiplying two nonzero integers, the first factor can be positive or negative, and the same is true for the second factor. This means that there are four possible combinations to consider.

Positive · positive

Positive · negative

Negative · positive

Negative · negative

In this section, we will discuss these four combinations and use our observations to establish rules to apply when multiplying integers.

The product of two positive integers

> **4(3)**
> like signs: both positive

We begin by considering the product of two positive integers, 4(3). Since both factors are positive, we can say that their signs are *like*. In Chapter 1, we learned that multiplication is repeated addition. Therefore, 4(3) represents the sum of four 3's.

$4(3) = 3 + 3 + 3 + 3$ Multiplication is repeated addition. Write 3 four times.

$4(3) = 12$ The result is a positive number.

This result, a positive 12, suggests that *the product of two positive integers is positive.*

The product of a positive and a negative integer

> **4(−3)**
> unlike signs:
> one positive, one negative

Next, we will consider 4(−3). This is the product of a positive and a negative integer. The signs of these factors are *unlike*. According to the definition of multiplication, 4(−3) means that we are to add −3 four times.

$$4(-3) = (-3) + (-3) + (-3) + (-3)$$ Use the definition of multiplication. Write -3 four times.

$$4(-3) = \quad\; (-6) + (-3) + (-3)$$ Work from left to right. Apply the rule for adding two negative numbers.

$$4(-3) = \qquad\qquad (-9) + (-3)$$ Work from left to right. Apply the rule for adding two negative numbers.

$$4(-3) = \qquad\qquad\qquad -12$$ Do the addition.

This result is -12, which suggests that *the product of a positive integer and a negative integer is negative.*

The product of a negative and a positive integer

$$-3(4)$$
unlike signs:
one negative, one positive

To develop a rule for multiplying a negative and a positive integer, we will consider $-3(4)$. Notice that the factors have *unlike* signs. Because of the commutative property of multiplication, the answer to $-3(4)$ will be the same as the answer to $4(-3)$. We know that $4(-3) = -12$ from the previous discussion, so $-3(4) = -12$. This suggests that *the product of a negative integer and a positive integer is negative.*

Putting the results of the last two cases together leads us to the rule for multiplying two integers with unlike signs.

Multiplying two integers with unlike signs	To multiply a positive integer and a negative integer, or a negative integer and a positive integer, multiply their absolute values. Then make the answer negative.

EXAMPLE 1   *Multiplying two integers with unlike signs.*   Find each product: **a.** $7(-5)$ and   **b.** $20(-8)$.

Solution

To multiply integers with unlike signs, we multiply their absolute values and make the product negative.

a. $7(-5) = -35$ Multiply 7 and 5. Then make the result negative.

b. $20(-8) = -160$ Multiply 20 and 8. Then make the result negative.

Self Check

Find each product:

a. $2(-6)$

b. $30(-2)$

Answer: a. -12, **b.** -60 ■

EXAMPLE 2   *Multiplying two integers with unlike signs.*   Multiply: **a.** $-8 \cdot 5$ and   **b.** $-5(30)$.

Solution

To multiply integers with unlike signs, we multiply their absolute values and make the product negative.

a. $-8 \cdot 5 = -40$ Multiply 8 and 5. Then make the result negative.

b. $-5(30) = -150$ Multiply 5 and 30. Then make the result negative.

Self Check

Multiply:

a. $-7(6)$

b. $-15 \cdot 2$

Answer: a. -42, **b.** -30 ■

Multiplication by zero

Before we can proceed, we need to examine multiplication by zero. If 4(3) means that we are to find the sum of four 3's, then $0(-3)$ means that we are to find the sum of zero -3's. Obviously, the sum would be 0. Thus, $0(-3) = 0$.

The commutative property of multiplication guarantees that we can change the order of the factors in the multiplication problem without affecting the result.

$$(-3)(0) = 0(-3) \;\; = \;\; 0$$

Change the order   The result is
of the factors.   still 0.

We can see that the order in which you write the factors 0 and -3 doesn't matter— their product is 0.

Multiplication by zero	The product of any integer and 0 is 0.

The product of two negative integers

$$-3(-4)$$

like signs: both negative

To develop a rule for multiplying two negative integers, we consider the pattern displayed below. There, we multiply -4 by a series of factors that decrease by 1. After determining each product, we then graph it on a number line (Figure 2-18). See if you can determine the answers to the last three multiplication problems by examining the pattern of answers leading up to them.

This factor decreases by 1 as you read down the column.

Look for a pattern here.

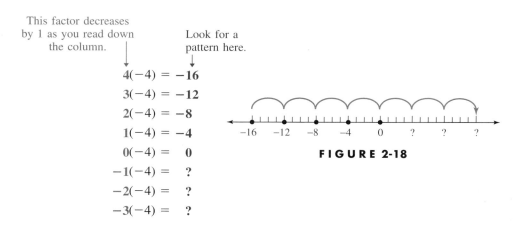

$$4(-4) = -16$$
$$3(-4) = -12$$
$$2(-4) = -8$$
$$1(-4) = -4$$
$$0(-4) = 0$$
$$-1(-4) = ?$$
$$-2(-4) = ?$$
$$-3(-4) = ?$$

FIGURE 2-18

The products are graphed in Figure 2-18. From the pattern, we see that

$$-1(-4) \;\; = \;\; 4$$
$$-2(-4) \;\; = \;\; 8$$
$$-3(-4) \;\; = \;\; 12$$

For two negative factors,   the product is a positive.

These results suggest that *the product of two negative integers is positive.* Earlier in this section, we saw that the product of two positive integers is also positive. This leads to the following conclusion.

Multiplying two integers with like signs	To multiply two positive integers, or two negative integers, multiply their absolute values. The answer is positive.

EXAMPLE 3   *Multiplying two negative integers.*   Find each product:
a. $-5(-9)$ and   **b.** $(-8)(-10)$.

Solution
To multiply two negative integers, we multiply their absolute values and make the result positive.

a. $-5(-9) = 45$ Multiply 5 and 9. The answer is positive.

b. $(-8)(-10) = 80$ Multiply 8 and 10. The answer is positive.

Self Check
Find each product:

a. $-9(-7)$

b. $(-12)(-2)$

Answer: **a.** 63,   **b.** 24 ∎

We now summarize the rules for multiplying two integers.

Multiplying two integers	To multiply two integers, multiply their absolute values. 1. The product of two integers with *like* signs is positive. 2. The product of two integers with *unlike* signs is negative.

EXAMPLE 4   *Supermarket giveaway.*   At Thanksgiving time, a supermarket gives a free turkey with every grocery purchase of $100 or more. If the turkey costs the store $6, and 200 people take advantage of the offer, what is the loss suffered by the store?

Solution   Each of the 200 turkeys given away is a loss of $6, which can be expressed as -6.

$$-6(200) = -1,200$$ The two factors have unlike signs, so the product is negative.

The supermarket will lose $1,200. ∎

EXAMPLE 5   *Multiplying three factors.*   Multiply $-4(5)(-3)$ in two ways.

Solution
First, we work from left to right.

$$-4(5)(-3) = -20(-3)$$ Multiply -4 and 5 first.

$$= 60$$ Do the multiplication.

An alternate approach would be to multiply the two negative numbers first.

$$-4(5)(-3) = -4(-3)(5)$$ Apply the commutative property of multiplication. Change the order of the factors.

$$= 12(5)$$ Do the multiplication: $-4(-3) = 12$.

$$= 60$$ Do the multiplication.

After some practice, you will be able to do this work in your head.

Self Check
Multiply $-5(5)(-2)$ in two ways.

Answer: 50 ∎

Powers of integers

Recall that exponential expressions are used to represent repeated multiplication. For example, 2 to the third power, or 2^3, is a shorthand way of writing $2 \cdot 2 \cdot 2$. In this expression, 3 is the exponent, and the base is positive 2. In the next example, we evaluate exponential expressions with bases that are negative numbers.

EXAMPLE 6 *Evaluating powers of integers.* Find each power: **a.** $(-2)^4$ and **b.** $(-5)^3$.

Solution

a. $(-2)^4 = (-2)(-2)(-2)(-2)$ Write -2 as a factor 4 times.

$\quad\quad\quad = 4(-2)(-2)$ Work from left to right. Multiply -2 and -2 to get 4.

$\quad\quad\quad = -8(-2)$ Work from left to right. Multiply 4 and -2 to get -8.

$\quad\quad\quad = 16$ Do the multiplication.

b. $(-5)^3 = (-5)(-5)(-5)$ Write -5 as a factor 3 times.

$\quad\quad\quad = 25(-5)$ Work from left to right. Multiply -5 and -5 to get 25.

$\quad\quad\quad = -125$ Do the multiplication.

Self Check

Find each power:

a. $(-3)^4$

b. $(-4)^3$

Answer: a. 81, **b.** -64 ■

In Example 6, part a, -2 was raised to an even power, and the answer was positive. In part b, another negative number, -5, was raised to an odd power, and the answer was negative. These results suggest a general rule.

Even and odd powers of a negative integer	When a negative integer is raised to an even power, the result is positive.
	When a negative integer is raised to an odd power, the result is negative.

EXAMPLE 7 *Evaluating a power of an integer.* Find $(-1)^5$.

Solution

We have a negative integer raised to an odd power. The result will be negative.

$(-1)^5 = (-1)(-1)(-1)(-1)(-1)$

$\quad\quad = -1$

Self Check

Find the power $(-1)^8$.

Answer: 1 ■

Although the expressions -3^2 and $(-3)^2$ look somewhat alike, they are not. In -3^2, the base is 3 and the exponent 2. The $-$ sign in front of 3^2 means the opposite of 3^2. In $(-3)^2$, the base is -3 and the exponent is 2. When we evaluate them, it becomes clear that they are not equivalent.

-3^2 represents *the opposite of* 3^2.

$\mathbf{-3^2 = -(3 \cdot 3)}$ Write 3 as a factor 2 times.

$\quad\quad = -9$ Work inside the parentheses first.

$(-3)^2$ represents $(-3)(-3)$.

$\mathbf{(-3)^2 = (-3)(-3)}$ Write -3 as a factor 2 times.

$\quad\quad = 9$ The product of two negative numbers is positive.

Notice that the results are different.

EXAMPLE 8   *The opposite of a power.*   Evaluate -2^2.

Solution

$-2^2 = -(2 \cdot 2)$ Since 2 is the base, write 2 as a factor two times.

$ = -4$ Do the multiplication inside the parentheses.

Self Check

Evaluate **a.** -4^2 and
b. $(-5)^2$.

Answer: a. -16, **b.** 25 ■

Accent on Technology   **Raising a negative number to a power**

Negative numbers can be raised to a power using a calculator. You will need to use the sign change key $\boxed{+/-}$ and the power key $\boxed{y^x}$.

Evaluate: $(-5)^6$

Keystrokes: $\boxed{5}$ $\boxed{+/-}$ $\boxed{y^x}$ $\boxed{6}$ $\boxed{=}$

The result is 15,625.

$\boxed{15625}$

STUDY SET   Section 2.4

VOCABULARY   *In Exercises 1–6, fill in the blanks to make a true statement.*

1. In the multiplication $-5(-4)$, the numbers -5 and -4 are called _____. The answer, 20, is called the _____.

2. The definition of multiplication tells us that $5(-4)$ represents repeated _____.

3. The numbers . . . $-4, -3, -2, -1, 0, 1, 2, 3, 4, . . .$ are called _____.

4. The definition of an exponent tells us that 5^4 represents repeated _____.

5. In the expression -3^5, ___ is the base and 5 is the _____.

6. In the expression $(-3)^5$, ___ is the base and ___ is the exponent.

CONCEPTS   *In Exercises 7–10, fill in the blanks to make a true statement.*

7. The product of two integers with _____ signs is negative.

8. The product of two integers with like signs is _____.

9. The _____ property of multiplication implies that $-2(-3) = -3(-2)$.

10. The product of zero and any number is ___.

11. Find $-1(9)$. In general, what is the result when multiplying a positive number by -1?

12. Find $-1(-9)$. In general, what is the result when multiplying a negative number by -1?

13. When multiplying two integers, there are four possible combinations of signs. List each of them.

14. When multiplying two integers, there are four possible combinations of signs. How can they be grouped into two categories?

15. If each of the following powers were evaluated, what would be the *sign* of the result?
  a. $(-5)^{13}$ **b.** $(-3)^{20}$

16. An algebra student claimed, "A positive and a negative is negative." What is wrong with this statement?

17. Find each absolute value.
   a. $|-3|$ **b.** $|12|$
   c. $|-5|$ **d.** $|9|$
   e. $|10|$ **f.** $|-25|$

18. Draw a number line from -5 to 5. Graph each of these products on the number line. What is the distance between each product?
$$-2(2), \ -2(1), \ -2(0), \ -2(-1), \ -2(-2)$$

19. Complete the table in Illustration 1.

20. Complete the table in Illustration 2.

Problem	Number of negative factors	Answer
$-2(-2)$		
$-2(-2)(-2)(-2)$		
$-2(-2)(-2)(-2)(-2)(-2)$		

ILLUSTRATION 1

Problem	Number of negative factors	Answer
$-2(-2)(-2)$		
$-2(-2)(-2)(-2)(-2)$		
$-2(-2)(-2)(-2)(-2)(-2)(-2)$		

ILLUSTRATION 2

The answers entered in the table help to justify the following rule: The product of an _____ number of negative integers is positive.

The answers entered in the table help to justify the following rule: The product of an _____ number of negative integers is negative.

NOTATION   *In Exercises 21–22, complete each solution.*

21. Find $-3(-2)(-4)$.

$-3(-2)(-4) = $ ___ (-4) Work from left to right. Multiply -3 and -2.

$\qquad\qquad = -24$ Do the multiplication.

22. Find $(-3)^4$.

$(-3)^4 = (-3)(-3)(-3)(-3)$ Write -3 as a factor 4 times.

$\qquad = $ ___ $(-3)(-3)$ Work from left to right. Multiply -3 and -3.

$\qquad = $ ___ (-3) Work from left to right. Multiply 9 and -3.

$\qquad = 81$ Do the multiplication.

PRACTICE   *In Exercises 23–42, find each product.*

23. $-9(-6)$ **24.** $-5(-5)$ **25.** $-3 \cdot 5$ **26.** $-6 \cdot 4$

27. $12(-3)$ **28.** $11(-4)$ **29.** $(-8)(-7)$ **30.** $(-9)(-3)$

31. $(-2)10$ **32.** $(-3)8$ **33.** $-40(3)$ **34.** $-50(2)$

35. $-8(0)$ **36.** $0(-27)$ **37.** $-1(-6)$ **38.** $-1(-8)$

39. $-7(-1)$ **40.** $-5(-1)$ **41.** $1(-23)$ **42.** $-35(1)$

In Exercises 43–58, evaluate each expression.

43. $-6(-4)(-2)$ **44.** $-3(-2)(-3)$ **45.** $5(-2)(-4)$ **46.** $3(-3)(3)$

47. $2(3)(-5)$ **48.** $6(2)(-2)$ **49.** $6(-5)(2)$ **50.** $4(-2)(2)$

51. $(-1)(-1)(-1)$ **52.** $(-1)(-1)(-1)(-1)$ **53.** $-2(-3)(3)(-1)$ **54.** $5(-2)(3)(-1)$

55. $3(-4)(0)$ **56.** $-7(-9)(0)$ **57.** $-2(0)(-10)$ **58.** $-6(0)(-12)$

59. Find the product of -6 and the opposite of 10.

60. Find the product of the opposite of 9 and the opposite of 8.

In Exercises 61–72, find each power.

61. $(-4)^2$ **62.** $(-6)^2$ **63.** $(-5)^3$ **64.** $(-6)^3$

65. $(-2)^3$ **66.** $(-4)^3$ **67.** $(-9)^2$ **68.** $(-10)^2$

69. $(-1)^5$ **70.** $(-1)^6$ **71.** $(-1)^8$ **72.** $(-1)^9$

In Exercises 73–76, evaluate each pair of expressions.

73. $(-7)^2$ and -7^2 **74.** $(-5)^2$ and -5^2

75. -12^2 and $(-12)^2$ **76.** -11^2 and $(-11)^2$

APPLICATIONS *In Exercises 77–84, use signed numbers to help answer each problem.*

77. DIETING After giving a patient a physical exam, a physician felt that the patient should begin a diet. Two options were discussed. (See Illustration 3.)

	Plan #1	Plan #2
Length	10 weeks	14 weeks
Daily exercise	1 hour	30 min
Weight loss per week	3 lb	2 lb

ILLUSTRATION 3

a. Find the expected weight loss from each diet plan. Express the answer as a signed number.

b. With which plan should the patient expect to lose the most weight? Explain why the patient might not choose it.

78. INVENTORY A spreadsheet is used to record inventory losses at a warehouse. The items, their cost, and the number missing are listed in Illustration 4.

	A	B	C	D
1	Item	Cost	Number of units	$ losses
2	CD	$5	−11	
3	TV	$200	−2	
4	Radio	$20	−4	

ILLUSTRATION 4

a. What instruction should be given to find the total losses for each type of item? Find each of those losses and fill in column D.

b. What instruction should be given to find the *total* inventory losses for the warehouse? Find this number.

79. MAGNIFICATION Using an electronic testing device, a mechanic can check the emissions of a car. The results of the test are displayed on a screen. (See Illustration 5.)

a. Find the high and low values for this test as shown on the screen.

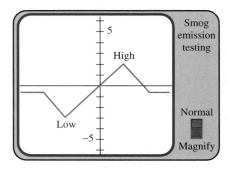

ILLUSTRATION 5

b. By switching a setting on the monitor, the picture on the screen can be magnified. Draw a picture of the new screen display if every value is doubled.

80. BUSINESS PROFITS Use signed numbers to express a company's profits or losses, as shown in Illustration 6, for the first three months of the year.

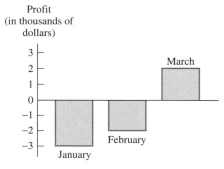

ILLUSTRATION 6

81. TEMPERATURE CHANGE A farmer, worried about his fruit trees suffering frost damage, calls the weather service for temperature information. He is told that temperatures will be decreasing approximately 4° every hour for the next 5 hours. What signed number represents the total change in temperature expected over the next five hours?

82. DEPRECIATION For each of the last four years, a businesswoman has filed a $200 depreciation allowance on her income tax return, for an office computer system. Use a signed number to represent the total amount of depreciation written off over the four-year period.

83. EROSION A levy protects a town in a low-lying area from flooding. According to geologists, the banks of the levy are eroding at a rate of 2 feet per year. If something isn't done to correct the problem, what signed number indicates how much of the levy will erode during the next decade?

84. DECK SUPPORT After a winter storm, a homeowner has an engineering firm inspect the damaged deck of his beach house. Their recommendation is to install new pilings, sunk three times deeper than the originals. (See Illustration 7.) Use a signed number to represent this proposed new depth.

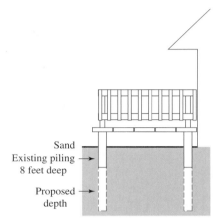

ILLUSTRATION 7

 In Exercises 85–94, use a calculator.

85. −76(787)

86. 407(−32)

87. $(-81)^4$

88. $(-6)^5$

89. (−32)(−12)(−67)

90. (−56)(−9)(−23)

91. $(-25)^4$

92. $(-41)^5$

93. PROMOTIONAL GIVEAWAY On hat night, the L.A. Lakers give away a team cap to everyone in attendance. This promotion always attracts a sellout crowd of 14,505. If the hat costs $3, use a signed number to express the financial loss from this giveaway.

94. HEALTH CARE A health care provider for a company estimates that 75 hours per week are lost by employees suffering from stress-related or preventable illness. In a 52-week year, how many hours are lost? Use a signed number to answer.

WRITING *Write a paragraph using your own words.*

95. If a product contains an even number of negative factors, how do we know that the result will be positive?

96. Explain why the product of a positive number and a negative number is negative, using 5(−3) as an example.

97. Explain why the result is the opposite of the original number when a number is multiplied by −1.

98. Can you think of any number that yields a nonzero result when multiplied by 0? Explain your response.

REVIEW

99. The prime factorization of a number is $3^2 \cdot 5$. What is the number?

100. Round, 10,345 to the nearest hundred.

101. The enrollment at a college went from 10,200 to 12,300 in one year. What was the increase in enrollment?

102. Name four types of grouping symbols used in mathematics.

103. Find the perimeter of a square with a side of length 6 yards.

104. What does the symbol $<$ mean?

105. List the first ten prime numbers.

106. Find the area of a rectangle with a width of 7 meters and a length of 9 meters.

Division of Integers

In this section, you will learn about

- **The relationship between multiplication and division**
- **Rules for dividing integers**
- **Division and zero**

INTRODUCTION. In this section, we will develop rules for division of integers, just as we did for multiplication of integers. We will also consider two types of division involving zero.

The relationship between multiplication and division

When we solved equations in Chapter 1, multiplication was used to undo division, and division was used to undo multiplication. Because one operation undoes the other, multiplication and division are said to be *inverse operations*. Every division fact containing three numbers can be written as an equivalent multiplication fact involving the same three numbers. For example,

$$\frac{6}{3} = 2 \quad \text{because} \quad 3(2) = 6$$

Rules for dividing integers

We will now use the relationship between multiplication and division to help develop rules for dividing integers. There are four cases to consider.

In the first case, a positive integer is divided by a positive integer. From years of experience, we already know that the result is positive. Therefore, *the quotient of two positive integers is positive.*

Next, we consider the quotient of two negative integers. As an example, consider the division $\frac{-12}{-2} = ?$ We can do this division by examining its related multiplication statement, $-2(?) = -12$. Our objective is to find the number that should replace the question mark. To do this, we need to use the rules for multiplying integers introduced in the previous section.

Multiplication statement

$$-2(?) = -12$$

└─ This must be *positive* 6 if the product is to be *negative* 12.

Division statement

$$\frac{-12}{-2} = ?$$

So the quotient is *positive* 6.

Therefore, $\frac{-12}{-2} = 6$. From this example, we can see that *the quotient of two negative integers is positive.*

The third case we will examine is the quotient of a positive integer and a negative integer. Let's consider $\frac{12}{-2} = ?$ Its equivalent multiplication statement is $-2(?) = 12$.

Multiplication statement	Division statement
$-2(?) = 12$	$\dfrac{12}{-2} = ?$

This must be −6 if the product is to be *positive* 12.

So the quotient is −6.

Therefore, $\frac{12}{-2} = -6$. This result shows that *the quotient of a positive integer and a negative integer is negative.*

Finally, to find the quotient of a negative integer and a positive integer, let's consider $\frac{-12}{2} = ?$ Its equivalent multiplication statement is $2(?) = -12$.

Multiplication statement	Division statement
$2(?) = -12$	$\dfrac{-12}{2} = ?$

This must be −6 if the product is to be −12.

So the quotient is −6.

Therefore, $\frac{-12}{2} = -6$. From this example, we can see that *the quotient of a negative integer and a positive integer is negative.*

We now summarize the results from the previous discussion.

Dividing integers	To divide two integers, divide their absolute values.
	1. The quotient of two integers with *like* signs is positive.
	2. The quotient of two integers with *unlike* signs is negative.

The rules for division of integers are similar to those for multiplication of integers.

EXAMPLE 1   *Dividing integers.*   Find each quotient:

a. $\dfrac{-35}{7}$ and   **b.** $\dfrac{20}{-5}$.

Solution

To divide integers with unlike signs, we find the quotient of their absolute values and make the quotient negative.

a. $\dfrac{-35}{7} = -5$ Dividing 35 by 7, we get 5. The quotient is negative.

b. $\dfrac{20}{-5} = -4$ Dividing 20 by 5, we get 4. The quotient is negative.

Self Check

Find each quotient:

a. $\dfrac{-45}{5}$

b. $\dfrac{60}{-20}$

Answer: a. −9,   b. −3 ■

EXAMPLE 2   *Dividing integers.*   Divide: $\dfrac{-12}{-3}$.

Solution

The integers have like signs. The quotient will be positive.

$$\dfrac{-12}{-3} = 4$$

Self Check

Divide: $\dfrac{-21}{-3}$.

Answer: 7 ■

EXAMPLE 3 **Price reduction.** Over the course of a year, the price of a television set was uniformly reduced each month by a retailer, because it was not selling. By the end of the year, the cost was $132 less than it was at the beginning of the year. How much did the price fall each month?

Solution We will label the drop in price of $132 for the year as -132. It occurred in 12 equal (uniform) reductions. This indicates division.

$$\frac{-132}{12} = -11 \quad \text{The quotient of a negative number and a positive number is negative.}$$

The drop in price each month was $11.

Division and zero

We will now consider two types of division that involve zero. In the first case, we will examine division *of* zero; in the second, division *by* zero.

To help explain the concept of division of zero, we will look at $\frac{0}{2} = ?$ The equivalent multiplication statement is $2(?) = 0$.

Multiplication statement

$2(?) = 0$

└── This must be 0 if the product is to be 0.

Division statement

$\frac{0}{2} = ?$

So the quotient ──┘ is 0.

Therefore, $\frac{0}{2} = 0$. This example suggests that *the quotient of zero divided by any nonzero number is zero.*

To illustrate division by zero, let's look at $\frac{2}{0} = ?$ The equivalent multiplication statement is $0(?) = 2$.

Multiplication statement

$0(?) = 2$

└── There is no number that gives 2 when multiplied by 0.

Division statement

$\frac{2}{0} = ?$

There is no ──┘ quotient.

Therefore, $\frac{2}{0}$ does not have an answer. We say that division by zero is **undefined.** This example suggests that *the quotient of any number divided by zero is undefined.*

| **Division with zero** | 1. The quotient of 0 divided by any nonzero number is 0. |
| | 2. Division by 0 is undefined. |

EXAMPLE 4 **Division with zero.** Find $\frac{-4}{0}$, if possible.

Solution

Since $\frac{-4}{0}$ is division by 0, the division is undefined.

Self Check

Find $\frac{0}{12}$.

Answer: 0

You can divide positive and negative numbers using your calculator. You will need to use the opposite (or change of sign) key $\boxed{+/-}$ when entering negative numbers.

Evaluate: $\dfrac{-7{,}163}{551}$

Keystrokes: $\boxed{7}\ \boxed{1}\ \boxed{6}\ \boxed{3}\ \boxed{+/-}\ \boxed{\div}\ \boxed{5}\ \boxed{5}\ \boxed{1}\ \boxed{=}$ $\boxed{\qquad\qquad -13}$
The quotient is -13.

STUDY SET   Section 2.5

VOCABULARY   *In Exercises 1–4, fill in the blanks to make a true statement.*

1. In $\dfrac{-27}{3} = -9$, the number -9 is called the
_____, and the number 3 is the
_____.

2. Division by zero is _____. Division of zero by a number that is not zero is ___.

3. The _____ value of a number is the distance between it and 0 on the number line.

4. The numbers . . . , $-4, -3, -2, -1, 0, 1, 2, 3, 4, \ldots$ are the _____.

CONCEPTS

5. Write the related multiplication statement for $\frac{-25}{5} = -5$.

6. Write the related multiplication statement for $\frac{0}{-15} = 0$.

7. Show that there is no answer for $\frac{-6}{0}$ by writing the related multiplication statement.

8. Find the value of $\frac{0}{8}$.

9. Write a related division statement for $5(-4) = -20$.

10. How do the rules for multiplying integers compare with the rules for dividing integers?

11. The quotient of two negative integers is
_____.

12. The quotient of a negative integer and a positive integer is _____.

13. Tell whether each statement is always true, sometimes true, or never true.
a. The product of a positive integer and a negative integer is negative.
b. The sum of a positive integer and a negative integer is negative.
c. The quotient of a positive integer and a negative integer is negative.

14. Tell whether each statement is always true, sometimes true, or never true.
a. The product of two negative integers is positive.
b. The sum of two negative integers is negative.
c. The quotient of two negative integers is negative.

PRACTICE   *In Exercises 15–38, find each quotient, if possible.*

15. $\dfrac{-14}{2}$

16. $\dfrac{-10}{5}$

17. $\dfrac{-8}{-4}$

18. $\dfrac{-12}{-3}$

19. $\dfrac{-25}{-5}$

20. $\dfrac{-36}{-12}$

21. $\dfrac{-45}{-15}$

22. $\dfrac{-81}{-9}$

23. $\dfrac{40}{-2}$

24. $\dfrac{35}{-7}$

25. $\dfrac{50}{-25}$

26. $\dfrac{80}{-40}$

27. $\dfrac{0}{-16}$

28. $\dfrac{0}{-6}$

29. $\dfrac{-6}{0}$

30. $\dfrac{-8}{0}$

31. $\dfrac{-5}{1}$

32. $\dfrac{-9}{1}$

33. $-5 \div (-5)$

34. $-11 \div (-11)$

35. $\dfrac{-9}{9}$

36. $\dfrac{-15}{15}$

37. $\dfrac{-10}{-1}$

38. $\dfrac{-12}{-1}$

In Exercises 39–46, divide.

39. $\dfrac{-100}{25}$

40. $\dfrac{-100}{50}$

41. $\dfrac{75}{-25}$

42. $\dfrac{300}{-100}$

43. $\dfrac{-500}{-100}$

44. $\dfrac{-60}{-30}$

45. $\dfrac{-200}{50}$

46. $\dfrac{-500}{100}$

47. Find the quotient of -45 and 9.

48. Find the quotient of -36 and -4.

49. Divide 8 by -2.

50. Divide -16 by -8.

APPLICATIONS *In Exercises 51–58, use signed numbers to solve each problem.*

51. TEMPERATURE DROP During a five-hour period, the temperature steadily dropped. (See Illustration 1.) What was the average change per hour over this five-hour time span?

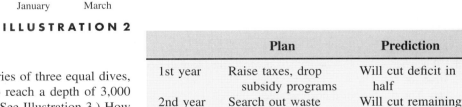

ILLUSTRATION 1

52. PRICE DROP Over a three-month period, the cost of a VCR steadily fell. (See Illustration 2.) What was the average monthly change in the price of the VCR over this period?

ILLUSTRATION 2

53. SUBMARINE DIVE In a series of three equal dives, a submarine is programmed to reach a depth of 3,000 feet below the ocean surface. (See Illustration 3.) How deep will each of the three dives be?

Surface

Dive 1

Dive 2

Dive 3

3,000 ft

ILLUSTRATION 3

54. GRAND CANYON TRIP A mule train is to travel from a stable on the rim of the Grand Canyon to a camp on the canyon floor, approximately 5,000 feet below the rim. If the guide wants the mules to be rested after every 1,000 feet of descent, how many stops will be made on the trip to the bottom of the canyon?

55. BASEBALL TRADE At the midway point of the season, a baseball team finds itself 12 games behind the league leader. Team management decides to trade for a talented hitter, in hopes of making up at least half of the deficit in the standings by the end of the year. How many games behind the leader does management expect to finish at season's end?

56. BUDGET DEFICIT A politician proposed a two-year plan for cutting a county's $20 million budget deficit, as shown in Illustration 4. If this plan is put into effect, what is the predicted budget deficit in two years?

	Plan	Prediction
1st year	Raise taxes, drop subsidy programs	Will cut deficit in half
2nd year	Search out waste and fraud	Will cut remaining deficit in half

ILLUSTRATION 4

57. PRICE MARKDOWN The owner of a clothing store decides to reduce the price on a line of jeans that are not selling. She feels she can afford to lose $300 of projected income on these pants. By how much can she mark down each of the 20 pairs of jeans?

58. WATER RESERVOIR Over a week's time, engineers at a city water reservoir released enough water to lower the water level 35 feet. On average, how much was the water level lowered each day during this period?

59. $\dfrac{-13,550}{25}$ **60.** $\dfrac{-3,876}{-19}$ **61.** $\dfrac{272}{-17}$ **62.** $\dfrac{-6,776}{-77}$

63. PAY CUT   In a cost-cutting effort, a business decides to lower expenditures on salaries by $9,135,000. To do this, all of the 5,250 employees will have their salaries reduced by an equal dollar amount. What size pay cut will each employee experience?

64. STOCK MARKET   On Monday, the value of Maria's 255 shares of stock was at an all-time high. By Friday, the value had fallen $4,335. What was her per-share loss that week?

WRITING   *Write a paragraph using your own words.*

65. Explain why the quotient of two negative numbers is positive.

66. Think of a real-life situation that could be represented by $\frac{0}{4}$. Explain why the answer would be zero.

67. Using a specific example, explain how multiplication can be used as a check for division.

68. Explain what it means when we say that division by zero is undefined.

REVIEW

69. Evaluate: $3\left(\dfrac{18}{3}\right) - 2(2)$.

70. List the set of whole numbers.

71. Find the prime factorization of 210.

72. The statement $(4 + 8) + 10 = 4 + (8 + 10)$ illustrates what property?

73. True or false: $17 \geq 17$.

74. Use the expression $8 - 2$ to show that subtraction is not commutative.

75. Evaluate: 3^4.

76. Sharif has scores of 55, 70, 80, and 75 on four mathematics tests. What is his mean (average) score?

2.6   *Order of Operations and Estimation*

In this section, you will learn about

- **Order of operations**
- **Grouping symbols**
- **Absolute value**
- **Estimation**

INTRODUCTION.   In this section, we will evaluate expressions which involve more than one operation. To do this, we will apply the rules for order of operations discussed in Chapter 1, as well as the rules for working with integers. We will also continue the discussion of estimating an answer. Estimation can be used when you need a quick indication of the size of the actual answer to a calculation.

Order of operations

In Section 1.5, the rule for order of operations was introduced. You will recall that it is an agreed-upon list of steps in which the operations of arithmetic are to be completed.

The rule for order of operations	1. Evaluate all the powers.
	2. Do all multiplications and divisions as they occur from left to right.
	3. Do all additions and subtractions as they occur from left to right.

EXAMPLE 1 *Order of operations.* Evaluate: $-4(-3)^2 - (-2)$.

Solution

This expression contains the operations of multiplication, raising to a power, and sub-traction. The rule for order of operations tells us to find the power first.

$$\begin{aligned} -4(-3)^2 - (-2) &= -4(9) - (-2) & \text{Evaluate the power: } (-3)^2 = 9. \\ &= -36 - (-2) & \text{Do the multiplication: } -4(9) = -36. \\ &= -36 + 2 & \text{To do the subtraction, add the opposite of } -2. \\ &= -34 & \text{Do the addition.} \end{aligned}$$

Self Check
Evaluate: $-5(-2)^2 - (-6)$.

Answer: -14 ■

EXAMPLE 2 *Order of operations.* Evaluate: $2(3) + (-5)(-3)(-2)$.

Solution

This expression contains the operations of multiplication and addition. By the rule for order of operations, we do the multiplications first.

$$\begin{aligned} 2(3) + (-5)(-3)(-2) &= 6 + (-30) & \text{Working left to right, do the multiplications.} \\ &= -24 & \text{Do the addition.} \end{aligned}$$

Self Check
Evaluate: $4(2) + (-4)(-3)(-2)$.

Answer: -16 ■

EXAMPLE 3 *Order of operations.* Evaluate: $40 \div (-4)5$.

Solution

This expression contains the operations of division and multiplication. The rule says to do the divisions and multiplications as they occur from left to right.

$$\begin{aligned} 40 \div (-4)5 &= -10 \cdot 5 & \text{Do the division first: } 40 \div (-4) = -10. \\ &= -50 & \text{Do the multiplication.} \end{aligned}$$

Self Check
Evaluate: $45 \div (-5)3$.

Answer: -27 ■

oo

EXAMPLE 4 *Order of operations.* Evaluate: $-2^2 - (-2)^2$.

Solution

This expression contains the operations of raising to a power and subtraction. We are to find the powers first. (Recall that -2^2 means the *opposite* of 2^2.)

$$\begin{aligned} -2^2 - (-2)^2 &= -4 - 4 & \text{Find the powers: } -2^2 = -4 \text{ and } (-2)^2 = 4. \\ &= -8 & \text{Do the subtraction.} \end{aligned}$$

Self Check
Evaluate: $-3^2 - (-3)^2$.

Answer: -18 ■

Grouping symbols

Recall that grouping symbols are mathematical punctuation marks. They help to de-termine the order in which an expression is to be evaluated.

Order of operations when grouping symbols are present	To evaluate an expression containing grouping symbols, use the rule for order of operations within the grouping symbols, working from the innermost set to the outermost set.
	In the case of a fraction bar, simplify the numerator and denominator separately. Then simplify the resulting fraction, if possible.

EXAMPLE 5   *Working with grouping symbols.*
Evaluate: $-15 + 3(-4 + 7)$.

Self Check
Evaluate: $-18 + 4(-7 + 9)$.

Solution

$$-15 + 3(-4 + 7) = -15 + 3(3)$$ Do the addition inside the parentheses: $-4 + 7 = 3$.

$$= -15 + 9$$ Do the multiplication: $3(3) = 9$.

$$= -6$$ Do the addition.

Answer: -10

EXAMPLE 6   *Working with a fraction bar.*   Evaluate: $\dfrac{-20 + 3(-5)}{(-4)^2 - 21}$.

Self Check
Evaluate: $\dfrac{-9 + 6(-4)}{(-5)^2 - 28}$.

Solution

We first simplify the expressions in the numerator and the denominator, separately.

$$\frac{-20 + 3(-5)}{(-4)^2 - 21} = \frac{-20 + (-15)}{16 - 21}$$ In the numerator, do the multiplication: $3(-5) = -15$.
In the denominator, evaluate the power: $(-4)^2 = 16$.

$$= \frac{-35}{-5}$$ In the numerator, add: $-20 + (-15) = -35$.
In the denominator, subtract: $16 - 21 = -5$.

$$= 7$$ Do the division.

Answer: 11

EXAMPLE 7   *Grouping symbols within grouping symbols.*
Evaluate: $-5[1 + (2 - 8)]$.

Self Check
Evaluate: $-4[2 + (5 - 9)]$.

Solution

We begin by working within the innermost pair of grouping symbols and work to the outermost pair.

$$-5[1 + (2 - 8)] = -5[1 + (-6)]$$ Do the subtraction inside the parentheses.

$$= -5(-5)$$ Do the addition inside the brackets.

$$= 25$$ Do the multiplication.

Answer: 8

Absolute value

You will recall that the absolute value of a number is the distance between the number and 0 on a number line. Earlier in this chapter, we evaluated simple absolute value expressions such as $|-3|$ and $|10|$. Absolute value symbols are also used in combination with more complicated expressions, such as $|-4 - 6|$ and $|6(-5)| - 7$. When applying the rule for order of operations to evaluate these expressions, *the absolute value symbols are considered to be grouping symbols,* and any operations inside them are to be completed first.

EXAMPLE 8 *Absolute value.* Find each absolute value:
  a. $|-4(3)|$ and **b.** $|-6 + 1|$.

Solution

We do the operations inside the absolute value symbols first.

a. $|-4(3)| = |-12|$ Do the multiplication inside the absolute value symbols: $-4(3) = -12$.

        $= 12$ Find the absolute value of -12.

b. $|-6 + 1| = |-5|$ Do the addition inside the absolute value symbols: $-6 + 1 = -5$.

        $= 5$ Find the absolute value of -5.

EXAMPLE 9 *Grouping symbols.* Evaluate: $8 - 4|-6 - 2|$.

Solution

We do the operation inside the absolute value symbols first.

$$8 - 4|-6 - 2| = 8 - 4|-8|$$ Do the subtraction inside the absolute value symbols: $-6 - 2 = -8$.

$$= 8 - 4(8)$$ Find the absolute value: $|-8| = 8$.

$$= 8 - 32$$ Do the multiplication: $4(8) = 32$.

$$= -24$$ Do the subtraction.

Estimation

Recall that the idea behind estimation is to simplify calculations by using rounded numbers that are close to the actual values in the problem. Estimating can be used when an exact answer is not necessary and a quick approximation will do.

As an example, consider the problem $-78 + 51$. We can estimate the sum quickly by approximating the numbers involved. The number -80 is close to the given number -78, and the number 50 is close to the given number 51. Therefore, we compute $-80 + 50$ for our estimate. The result is -30.

EXAMPLE 10 *Dow Jones Average.* The Dow Jones Industrial Average is announced at the end of each trading day to give investors an indication of how the New York Stock Exchange performed. A positive number indicates good performance, while a negative number indicates poor performance. Estimate the net gain or loss of points in the Dow for the week shown in Figure 2-19.

Monday	Tuesday	Wednesday	Thursday	Friday
+37	+21	−47	+12	−39

FIGURE 2-19

Solution We will approximate each of these numbers. For example, 37 is close to 40, and -47 is close to -50. To find an estimate of the net gain or loss, we add the approximations.

$$40 + 20 + (-50) + 10 + (-40) = 70 + (-90)$$ Add positive and negative numbers separately to get subtotals.

$$= -20$$ Do the addition.

This estimate tells us that there was a loss of approximately 20 points in the Dow.

VOCABULARY *In Exercises 1–4, fill in the blanks to make a true statement.*

1. When asked to evaluate expressions containing more than one operation, we should apply the rules for _____ of _____.

2. In situations where an exact answer is not needed, an approximation or _____ can be a quick way of obtaining an idea of what the size of the actual answer would be.

3. Absolute value symbols, parentheses, and brackets are types of _____ symbols.

4. If an expression involves two sets of grouping symbols, always begin working within the _____ symbols and then work to the _____.

CONCEPTS

5. Consider $5(-2)^2 - 1$. How many operations will need to be performed to evaluate this expression? List them in the order in which they should be performed.

6. Consider $15 - 3 + (5 \cdot 2)^3$. How many operations will need to be performed to evaluate this expression? List them in the order in which they should be performed.

7. Consider $\dfrac{5 + 5(7)}{2 + (4 - 8)}$. In the numerator, what operation should be performed first? In the denominator, what operation should be performed first?

8. In the expression $4 + 2(-7 - 1)$, how many operations need to be performed? List them in the order in which they should be performed.

9. Explain the difference between -3^2 and $(-3)^2$.

10. In the expression $-2 \cdot 3^2$, what operation should be performed first?

NOTATION *In Exercises 11–14, complete each solution.*

11. Evaluate: $-8 - 5(-2)^2$.

$$\begin{aligned}
-8 - 5(-2)^2 &= -8 - 5(\underline{}) &&\text{Evaluate the power.}\\
&= -8 - \underline{} &&\text{Do the multiplication.}\\
&= -8 + (\underline{}) &&\text{Add the opposite.}\\
&= -28 &&\text{Do the addition.}
\end{aligned}$$

12. Evaluate: $2 + (5 - 6 \cdot 2)$.

$$\begin{aligned}
2 + (5 - 6 \cdot 2) &= 2 + (5 - \underline{}) &&\text{Work inside the parentheses first. Do the multiplication.}\\
&= 2 + [5 + (\underline{})] &&\text{Add the opposite.}\\
&= 2 + (\underline{}) &&\text{Do the addition.}\\
&= -5 &&\text{Do the addition.}
\end{aligned}$$

13. Evaluate: $[-4(2 + 7)] - 6$

$$\begin{aligned}
[-4(2 + 7)] - 6 &= [-4(\underline{})] - 6 &&\text{Do the addition within the innermost grouping symbols.}\\
&= \underline{} - 6 &&\text{Do the multiplication within the brackets.}\\
&= -42 &&\text{Do the subtraction.}
\end{aligned}$$

14. Evaluate: $\dfrac{|-9 + (-3)|}{9 - 6}$.

$$\begin{aligned}
\frac{|-9 + (-3)|}{9 - 6} &= \frac{|\underline{}|}{3} &&\text{In the numerator, do the addition inside the absolute value symbols. In the denominator, do the subtraction.}\\
&= \frac{\underline{}}{3} &&\text{In the numerator, evaluate the absolute value.}\\
&= 4 &&\text{Do the division.}
\end{aligned}$$

PRACTICE *In Exercises 15–34, evaluate each expression.*

15. $(-3)^2 - 4^2$

16. $-7 + 4 \cdot 5$

17. $3^2 - 4(-2)(-1)$

18. $2^3 - 3^3$

19. $(2 - 5)(5 + 2)$

20. $-3(2)^2 4$

21. $-10 - 2^2$

22. $-5|-7|$

23. $\dfrac{-6 - 8}{2}$

24. $\dfrac{-6 - 6}{-2 - 2}$

25. $\dfrac{-5 - 5}{2}$

26. $\dfrac{-7 - (-3)}{2 - 4}$

27. $-12 \div (-2) \cdot 2$　　**28.** $-60 \cdot (-2) \div 3$　　**29.** $-16 - 4 \div (-2)$　　**30.** $-24 + 4 \div (-2)$

31. $|-5(-6)|$　　**32.** $|-7 - 9|$　　**33.** $|-4 - (-6)|$　　**34.** $|-2 + 6 - 5|$

In Exercises 35–46, evaluate each expression.

35. $(7 - 5)^2 - (1 - 4)^2$　　**36.** $5^2 - (-9 - 3)$　　**37.** $-1(2^2 - 2 + 1^2)$　　**38.** $(-7 - 4)^2 - (-1)$

39. $-50 - 2(-3)^3$　　**40.** $(-2)^3 - (-3)(-2)$　　**41.** $-6^2 + 6^2$　　**42.** $-9^2 + 9^2$

43. $3\left(\dfrac{-18}{3}\right) - 2(-2)$　　**44.** $2\left(\dfrac{-12}{3}\right) + 3(-5)$　　**45.** $6 + \dfrac{25}{-5} + 6 \cdot 3$　　**46.** $-5 - \dfrac{24}{6} + 8(-2)$

In Exercises 47–66, evaluate each expression.

47. $\dfrac{1 - 3^2}{-2}$　　**48.** $\dfrac{-3 - (-7)}{2^2 - 3}$　　**49.** $\dfrac{-4(-5) - 2}{-6}$　　**50.** $\dfrac{(-6)^2 - 1}{-4 - 3}$

51. $-3\left(\dfrac{32}{-4}\right) - (-1)^5$　　**52.** $-5\left(\dfrac{16}{-4}\right) - (-1)^4$　　**53.** $6(2^3)(-1)$　　**54.** $2(3^3)(-2)$

55. $2 + 3[5 - (1 - 10)]$　　**56.** $12 - 2[1 - (-8 + 2)]$　　**57.** $-7(2 - 3 \cdot 5)$　　**58.** $-4(1 + 3 \cdot 5)$

59. $-[6 - (1 - 4)]$　　**60.** $-[9 - (9 - 12)]$　　**61.** $15 + (-3 \cdot 4 - 8)$　　**62.** $11 + (-2 \cdot 2 + 3)$

63. $|-3 \cdot 4 + (-5)|$　　**64.** $|-8 \cdot 5 - 2 \cdot 5|$　　**65.** $|(-5)^2 - 2 \cdot 7|$　　**66.** $|8 \div (-2) - 5|$

In Exercises 67–74, make a mental estimate.

67. $-379 + (-103) + 287$　　　　**68.** $\dfrac{-67 - 9}{-18}$

69. $-39 \cdot 8$　　　　**70.** $-568 - (-227)$

71. $-3{,}887 + (-5{,}106)$　　　　**72.** $-333(-4)$

73. $\dfrac{6{,}267}{-5}$　　　　**74.** $-36 + (-78) + 59 + (-4)$

APPLICATIONS　*In Exercises 75–78, solve each problem.*

75. TEST GRADING　In an effort to discourage her students from guessing on multiple-choice tests, a professor uses the grading scale shown in Illustration 1. If unsure of an answer, a student does best to skip the question, because incorrect responses are penalized very heavily. Find the test score of a student who gets 12 correct and 3 wrong and leaves 5 questions blank.

76. QUARTERLY REPORT　The financial performance of a company in each of the four quarters of the year is shown in Illustration 2. Find the company's quarterly average for the year. (The figures are in millions of dollars.)

Response	Value
Correct	+3
Incorrect	−4
Left blank	−1

ILLUSTRATION 1

Quarterly Profits/Losses

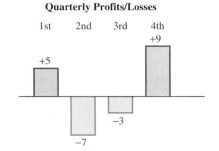

ILLUSTRATION 2

77. FOOTBALL   The results of each carry that a running back had during a football game are recorded in Illustration 3. Find his average yards gained per carry for the game.

Play	Result
Sweep	Gain: 16 yd
Pitch	Gain: 10 yd
Dive	Gain: 4 yd
Dive	Loss: 2 yd
Dive	No gain
Sweep	Loss: 4 yd

ILLUSTRATION 3

78. SPREADSHEET   Illustration 4 shows the data from a chemistry experiment in spreadsheet form. To obtain a result, the chemist needs to add the values in row 1, double that sum, and then divide that number by the smallest value in column C. What is the final result of these calculations?

	A	B	C	D
1	12	−5	6	−2
2	15	4	5	−4
3	6	4	−2	8

ILLUSTRATION 4

In Exercises 79–80, use estimation to answer each question.

79. OIL PRICES   The price per barrel of crude oil fluctuates with supply and demand. It can rise and fall quickly. The line graph in Illustration 5 shows how many cents the price per barrel *rose or fell each day* for a week. Estimate the net gain or loss in the value of a barrel of crude oil for the week.

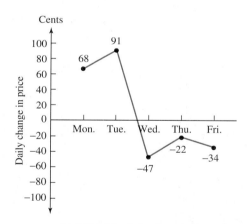

ILLUSTRATION 5

80. ESTIMATION   Quickly determine a reasonable estimate of the exact answer in each of the following situations.
  a. A diver, swimming at a depth of 34 feet below sea level, spots a sunken ship beneath him. He dives down another 57 feet to reach it. What is the depth of the sunken ship?
  b. A dental hygiene company offers a money-back guarantee on its tooth whitener kit, which is sold by mail order. When the kit is returned by a dissatisfied customer, the company loses the $11 it cost to produce it, because it cannot be resold. How much money has the company lost because of this return policy if 56 kits have been mailed back by customers?
  c. A tram line makes a 7,561-foot descent from a mountaintop in 18 equal stages. How much does it descend in each stage?

In Exercises 81–84, use a calculator to evaluate each expression.

81. $-2(-34)^2 - (-605)$ **82.** $11 - (-15)(24)^2$ **83.** $-60 - \dfrac{1{,}620}{-36}$ **84.** $\dfrac{2^5 - 4^6}{-42 + 58}$

WRITING   *Write a paragraph using your own words.*

85. When evaluating expressions, why is a rule for order of operations necessary?

86. In the rule for order of operations, what does the phrase *as they occur from left to right* mean?

87. Name a situation in daily life where you use estimation.

88. List some advantages and some disadvantages of the process of estimation.

Signed Numbers

In algebra, we work with both positive and negative numbers. We study negative numbers because they are necessary to describe many situations in daily life.

Represent each of these situations using a signed number.

1. Stocks fell 5 points.

2. The river was 12 feet over flood stage.

3. 30 seconds before going on the air

4. A business $6 million in the red

5. 10 degrees above normal

6. The year 2000 B.C.

7. $205 overdrawn

8. 14 units under their quota

A number line can be used to illustrate positive and negative numbers.

9. On this number line, label the location of the positive numbers and the negative numbers.

10. On this number line, graph 2 and its opposite.

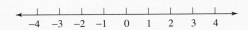

11. Two numbers, -8 and -4, are graphed on this number line. What can you say about their relative sizes?

12. On a number line, the distance between -3 and 0, called the absolute value of -3, is written $|-3|$. Show this distance on the number line.

In the space provided, summarize how addition, multiplication, and division are performed with two integers having the same sign and with two integers having different signs. Explain the method that is used to subtract signed numbers.

13. Addition
Same sign:

Different signs:

14. Division
Same sign:

Different signs:

15. Multiplication
Same sign:

Different signs:

16. Subtraction with signed numbers

CHAPTER REVIEW

An Introduction to the Integers

CONCEPTS

A *number line* is a horizontal or vertical line used to represent numbers graphically.

Integers: . . . , −3, −2, −1, 0, 1, 2, 3, . . .

Inequality symbols:
 > is greater than
 < is less than
 ≠ is not equal to
 ≥ is greater than or equal to
 ≤ is less than or equal to

A *negative* number is less than 0. A *positive* number is greater than 0.

The *absolute value* of a number is the distance between it and 0 on the number line.

Double negative rule: The opposite of the opposite of a number is that number.

On a number line, two numbers the same distance away from the origin but on different sides of it are called *opposites*.

REVIEW EXERCISES

1. Graph each set of numbers.
 a. −3, −1, 0, 4
 b. The integers greater than −3 but less than 4.

2. Insert one of the symbols >, <, or = in the blank to make a true statement.
 a. −7 ___ 0
 b. −20 ___ −19
 c. |−16| ___ −16
 d. 56 ___ 60

3. Tell whether each statement is true or false.
 a. 3 ≠ 7
 b. 19 ≤ 19
 c. 100 ≥ 50
 d. 25 + 25 ≥ 40 + 20

4. Represent each of these situations using a signed number.
 a. A deficit of $1,200
 b. 10 seconds before going on the air

5. Evaluate each expression.
 a. |−4|
 b. |0|
 c. |−43|
 d. −|12|

6. Explain the meaning of each red − sign.
 a. −5
 b. −(−5)
 c. −(−5)
 d. 5 − (−5)

7. Find each of the following.
 a. −(−12)
 b. The opposite of 8
 c. The opposite of −8
 d. −0

Addition of Integers

8. Use a number line to find each sum.
 a. 4 + (−2)
 b. −1 + (−3)

To add two integers with *like signs*, add their absolute values and attach their common sign to that sum.

To add two integers with *unlike signs*, subtract their absolute values, the smaller from the larger. Attach the sign of the number with the larger absolute value to that result.

Addition property of zero: 0 added to any integer is that integer.

If the sum of two integers is 0, the integers are called *additive inverses*.

9. Add.
   a. $-6 + (-4)$
   b. $-2 + (-3)$
   c. $-23 + (-60)$
   d. $-1 + (-4) + (-3)$

10. Add.
   a. $-4 + 3$
   b. $-7 + 9$
   c. $-12 + 3$
   d. $-28 + 40$
   e. $15 + (-4)$
   f. $9 + (-20)$
   g. $3 + (-2) + (-4)$
   h. $14 + [(-5) + 4]$

11. Add.
   a. $-4 + 0$
   b. $0 + (-20)$
   c. $-8 + 8$
   d. $3 + (-3)$

12. Give the additive inverse of each number.
   a. -11
   b. 4
   c. $|-7|$
   d. $-(-10)$

13. DROUGHT   During a drought, the water level in a reservoir fell to a point 100 feet below normal. After two rainy months, it rose 35 feet. How far below normal was the water level after the rain?

SECTION 2.3	*Subtraction of Integers*

Rule for subtraction: To subtract two integers, add the opposite of the second integer to the first integer.

14. Subtract.
   a. $5 - 8$
   b. $-9 - 12$
   c. $-4 - (-8)$
   d. $-6 - 10$
   e. $-8 - (-2)$
   f. $7 - 1$
   g. $0 - 3$
   h. $0 - (-30)$

15. Fill in the blanks to make a true statement: Subtracting a number is the same as _____ the _____ of that number.

16. Evaluate each expression.
   a. $-9 - 7 + 12$
   b. $7 - [(-6) - 2]$
   c. $1 - (2 - 7)$
   d. $-12 - (6 - 10)$

17. GOLD MINING   Some miners discovered a small vein of gold at a depth of 150 feet. This prompted them to continue their exploration. After descending another 75 feet, they came upon a much larger find. Use a signed number to represent the depth of the second discovery.

18. Evaluate: $2 - [-(-3)]$.

To find the *change* in a quantity, subtract the earlier value from the later value.

19. RECORD TEMPERATURES   The lowest and highest recorded temperatures for Alaska and Virginia are shown here. For each state, find the difference in temperature between the record high and low.

Alaska: Low $-80°$ Jan. 23, 1971   Virginia: Low $-30°$ Jan. 22, 1985
         High $100°$ June 27, 1915   High $110°$ July 15, 1954

Multiplication of Integers

The product of two integers with *like signs* is positive. The product of two integers with *unlike signs* is negative.

20. Multiply.
   a. $-9 \cdot 5$ **b.** $-3(-6)$ **c.** $7(-2)$ **d.** $(-8)(-4)$

   e. $-20 \cdot 5$ **f.** $-1(-1)$ **g.** $-1(25)$ **h.** $(5)(-30)$

21. Multiply.
   a. $(-6)(-2)(-3)$ **b.** $4(-3)3$
   c. $0(-7)$ **d.** $(-1)(-1)(-1)(-1)$

22. TAX DEFICIT   A state agency's prediction of a tax shortfall proved to be two times worse than the actual deficit of \$3 million. The federal prediction of the same shortfall was even more inaccurate—three times the amount of the actual deficit. Complete Illustration 1, which summarizes these incorrect forecasts.

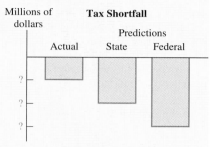

ILLUSTRATION 1

An *exponent* is used to represent repeated multiplication.

23. Find each power.
   a. $(-5)^2$ **b.** $(-2)^5$ **c.** $(-8)^2$ **d.** $(-4)^3$

When a negative integer is raised to an *even* power, the result is positive. When it is raised to an *odd* power, the result is negative.

24. When $(-5)^9$ is evaluated, will the result be positive, or will it be negative?

25. Explain the difference between -2^2 and $(-2)^2$ and then evaluate each.

Division of Integers

26. Fill in the blank to make a true statement: We know that $\dfrac{-15}{5} = -3$ because _____ $= -15$.

The quotient of two integers with *unlike* signs is negative.

The quotient of two numbers with *like* signs is positive.

27. Divide.
   a. $\dfrac{-14}{7}$ **b.** $\dfrac{25}{-5}$ **c.** $\dfrac{-64}{8}$ **d.** $\dfrac{-20}{-2}$

The quotient of 0 divided by any nonzero number is 0.

Division by 0 is undefined.

28. Find each quotient, if possible.
   a. $\dfrac{0}{-5}$ **b.** $\dfrac{-4}{0}$ **c.** $\dfrac{-6}{-6}$ **d.** $\dfrac{-10}{-1}$

29. PRODUCTION TIME   Because of improved production procedures, the time needed to produce an electronic component dropped by 12 minutes over the past six months. If the drop in production time was uniform, how much did it drop each month over this period?

Order of Operations and Estimation

The rule for order of operations:
1. Evaluate all powers.
2. Do all the multiplications and divisions, working from left to right.
3. Do all the additions and subtractions, working from left to right.

To evaluate an expression containing grouping symbols, use the rule for order of operations within them, working from the innermost set to the outermost set.

A fraction bar is a grouping symbol.

An *estimation* is an approximation that gives a quick idea of what the actual answer would be.

30. Evaluate each expression.

  a. $2 + 4(-6)$

  b. $7 - (-2)^2 + 1$

  c. $2 - 5(4) + (-25)$

  d. $-3(-2)^3 - 16$

  e. $-2(5)(-4) + \dfrac{|-9|}{3^2}$

  f. $-4^2 + (-4)^2$

  g. $-12 - (8 - 9)$

  h. $7|-8| - 2(3)(4)$

31. Evaluate each expression.

  a. $-4\left(\dfrac{15}{-3}\right) - 2^3$

  b. $-20 + 2(12 - 5 \cdot 2)$

  c. $20 + 2[12 - (-7 + 5)]$

  d. $8 - |-3 \cdot 4 + 5|$

32. Evaluate each expression.

  a. $\dfrac{10 + (-6)}{-3 - 1}$

  b. $\dfrac{3(-6) - 11 + 1}{4^2 - 3^2}$

33. Estimate each answer.

  a. $-89 + 57 + (-42)$

  b. $\dfrac{-507}{-24}$

  c. $(-681)(9)$

  d. $317 - (-775)$

1. Insert one of the symbols $>$, $<$, or $=$ in the blank to make a true statement.

    a. -8 ___ -9 **b.** -8 ___ $|-8|$

    c. The opposite of 5 ___ 0

2. List the integers.

3. Write "3° below zero" as a signed number.

4. Use a number line to find the sum: $-3 + (-2)$.

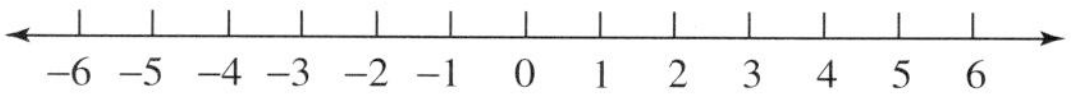

5. Subtract.

    a. $-7 - 6$ **b.** $-7 - (-6)$

6. Find each product.

    a. $(-10)7$ **b.** $-4(-2)(-6)$

    c. $(-2)(-2)(-2)(-2)$

7. Write the related multiplication statement for
$$\frac{-20}{-4} = 5.$$

8. Find each quotient, if possible.

    a. $\dfrac{-32}{4}$ **b.** $\dfrac{8}{6-6}$ **c.** $\dfrac{-5}{1}$

9. BUSINESS TAKEOVER   Six businessmen are contemplating taking over a company that has potential, but they must retire the debt incurred by the company over the past three quarters. (See Illustration 1.) If they plan equal ownership, how much will each have to contribute to retire the debt?

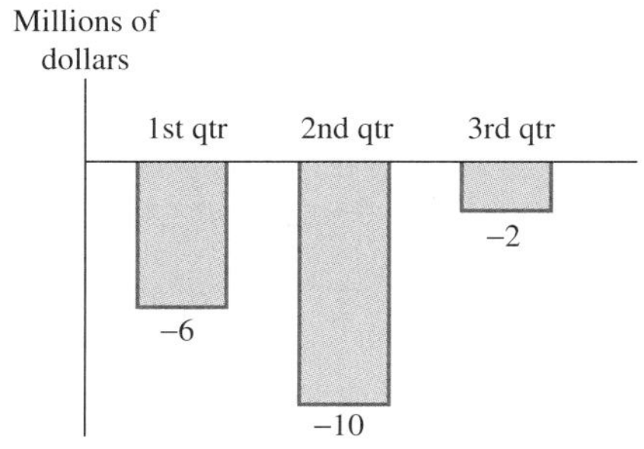

10. Evaluate each expression.

    a. $-(-6)$ **b.** $|-7|$ **c.** $|-9 + 3|$

ILLUSTRATION 1

11. Find each power.
   a. $(-4)^2$ **b.** -4^2
   c. $(-4 - 3)^2$

12. Evaluate: $-18 \div 2 \cdot 3$.

13. Evaluate: $4 - (-3)^2 + 6$.

14. Evaluate: $-3 + \dfrac{-16}{4} - 3^3$.

15. Evaluate: $10 + 2[6 - (-2)(-5)]$.

16. Evaluate: $\dfrac{4(-6) - 2^2}{-3 - 4}$.

17. HIKING   On an outing in the desert, a boy scout troop set up base camp on the desert floor, 650 feet below sea level. As part of their training, they hiked to a hill-top, at an elevation of 400 feet above sea level. What was the change in elevation they experienced on this hike?

18. HOSPITAL CAPACITY   One morning, the number of beds occupied by patients in a hospital was 3 under capacity. By afternoon, the number of unoccupied beds was 21. If no new patients were admitted, how many patients were released to go home?

19. Explain what is meant by the statement "Subtraction is the same as adding the opposite." Give an example.

20. Explain why a rule for order of operations is necessary. Give an example.

In Exercises 1–4, consider the numbers $-2, -1, 0, 1, 2, \frac{3}{2}, 5, 9$.

1. List each natural number.

2. List each whole number.

3. List each negative number.

4. List each integer.

In Exercises 5–8, consider the number 7,326,549.

5. Which digit is in the thousands column?

6. Which digit is in the hundred thousands column?

7. Round to the nearest hundred.

8. Round to the nearest ten thousand.

In Exercises 9–12, do each operation.

9. $237 + 549$

10. $6,375 - 2,569$

11.
$$\begin{array}{r} 5,369 \\ -\ \ 685 \\ \hline \end{array}$$

12.
$$\begin{array}{r} 7,899 \\ +5,237 \\ \hline \end{array}$$

13. Find the perimeter of the rectangular garden in Illustration 1.

14. In a shipment of 147 pieces of furniture, 27 pieces were sofas, 55 were leather chairs, and the rest were wooden chairs. Find the number of wooden chairs.

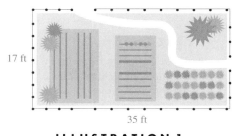

17 ft

35 ft

ILLUSTRATION 1

In Exercises 15–18, do each operation.

15. $435 \cdot 27$

16. $1,261 \div 97$

17.
$$\begin{array}{r} 4,587 \\ \times\ \ \ 67 \\ \hline \end{array}$$

18. $38\overline{)17,746}$

19. Find the area of the garden in Exercise 13.

20. Find all of the factors of 18.

In Exercises 21–24, identify each number as a prime, a composite, an even, or an odd.

21. 17

22. 18

23. 0

24. 1

25. Find the prime factorization of 504.

26. Write the expression $11 \cdot 11 \cdot 11 \cdot 11$ using one exponent.

In Exercises 27–30, evaluate each expression.

27. $5^2 \cdot 7$

28. $2^3 \cdot 5 \cdot 7^2$

29. $25 + 5 \cdot 5$

30. $\dfrac{16 - 2 \cdot 3}{2 + (9 - 6)}$

31. A traffic officer monitored several cars on a city street. She found that the speeds of the cars were as follows:

38, 42, 36, 38, 48, 44

On the average, were the drivers obeying the 40-mph speed limit?

32. A golfer shot rounds of 84, 86, 92, and 94. Is her average score less than 90?

In Exercises 33–34, graph each set of numbers on a number line.

33. $-2, -1, 0, 2$

34. The integers greater than -4 but less than 2.

In Exercises 35–46, evaluate each expression.

35. $-2 + (-3)$

36. $-15 + 10$

37. $|-3| - 5$

38. $15 - |-3|$

39. $(-8)(-3)$

40. $5(-7)$

41. $\dfrac{-14}{-7}$

42. $\dfrac{45}{-9}$

43. $5 + (-3)(-7)$

44. $-6(5) - 5$

45. $\dfrac{10 - (-5)}{1 - 2 \cdot 3}$

46. $\dfrac{3(-6) - 10}{3^2 - 4^2}$

47. BUYING A BUSINESS   When 12 investors decided to buy a bankrupt company, they agreed to assume equal shares of the company's debt of $1,512,444. How much was each person's share?

48. CAVITIES   The following table gives the numbers of cavities for six 7-year-old boys. Make a bar graph of the data.

Joe	John	Jim	Juan	Josh	Jeb
5	2	3	1	2	4

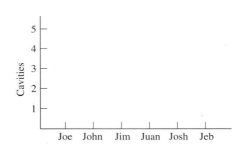

Fractions and Mixed Numbers

3

FRACTIONS PROVIDE THE BASIS FOR SOLVING MANY PROBLEMS THAT COULD NOT BE SOLVED USING JUST WHOLE NUMBERS.

3.1 | The Fundamental Property of Fractions

In this section, you will learn about
- **Basic facts about fractions**
- **Equivalent fractions**
- **Simplifying a fraction**
- **Expressing a fraction in higher terms**

INTRODUCTION. There is no better place to start a study of fractions than with **the fundamental property of fractions.** This property is the foundation for two fundamental procedures that are used when working with fractions. But first, we review some basic facts about fractions.

Basic facts about fractions

1. A Fraction Can Be Used to Indicate Equal Parts of a Whole.

In our everyday lives, we often deal with parts of a whole. For example, we talk about parts of an hour, parts of an inch, and parts of a pound.

2. A Fraction Has Three Parts.

A fraction is composed of a **numerator,** a **denominator,** and a **fraction bar.**

$$\text{Fraction bar} \longrightarrow \frac{3 \longleftarrow \text{Numerator}}{4 \longleftarrow \text{Denominator}}$$

The denominator (in this case, 4) tells us that a whole was divided into four equal parts. The numerator tells us that we are considering three of those equal parts.

3. Fractions Can be Proper or Improper.

If the numerator of a fraction is less than its denominator, the fraction is called a **proper fraction.** Fractions whose numerators are greater than or equal to their denominators are called **improper fractions.**

Proper fractions	*Improper fractions*
$\frac{1}{3}$ and $\frac{98}{99}$	$\frac{3}{2}, \frac{98}{97}, \frac{98}{98},$ and $\frac{8}{1}$

EXAMPLE 1

Fractional parts of a whole. **a.** In Figure 3-1, what fractional part of the barrel is full? **b.** What fractional part is empty?

FIGURE 3-1

Solution

The barrel has been divided into three equal parts.

a. Two of the three parts are full. Therefore, the barrel is $\frac{2}{3}$ full.

b. One of the three equal parts is not filled. The barrel is $\frac{1}{3}$ empty.

The fractions $\frac{2}{3}$ and $\frac{1}{3}$ are both proper fractions.

Self Check

a. According to the calendar below, what fractional part of the month has passed? **b.** What fractional part remains?

X	X	X	X	X	X	X
X	X	X	X	12	13	14
15	16	17	18	19	20	21
22	23	24	25	26	27	28
29	30	31				

Answer: **a.** $\dfrac{11}{31}$, **b.** $\dfrac{20}{31}$

4. The Denominator of a Fraction Cannot Be 0.

$\dfrac{7}{0}$, $\dfrac{23}{0}$, and $\dfrac{0}{0}$ are meaningless expressions. However, $\dfrac{0}{5} = 0$ and $\dfrac{0}{12} = 0$.

5. Fractions Can Be Negative.

There are times when a negative fraction is needed to describe a quantity. For example, a drop of seven-eighths of a point in the price of a stock can be denoted as $-\frac{7}{8}$.

Negative fractions can be written in three ways. The negative sign can appear in the numerator, in the denominator, or in front of the fraction.

$$\frac{-7}{8} = \frac{7}{-8} = -\frac{7}{8}$$

Fractions are often called **rational numbers.** All integers are rational numbers, because every integer can be written as a fraction with a denominator of 1. For example,

$$2 = \frac{2}{1} \qquad -5 = \frac{-5}{1} \qquad \text{and} \qquad 0 = \frac{0}{1}$$

Since every integer is also a rational number, the integers form a subset of the rational numbers.

 WARNING! Note that not all rational numbers (fractions) are integers. For example, the rational number $\frac{7}{8}$ is not an integer.

Equivalent fractions

Fractions can look different but still represent the same number. To show this, let's divide the rectangle in Figure 3-2(a) in two ways. In Figure 3-2(b), we divide it into halves (2 equal-sized parts). In Figure 3-2(c), we divide it into fourths (4 equal-sized parts). Notice that one-half of the figure is exactly the same size as two-fourths of the figure.

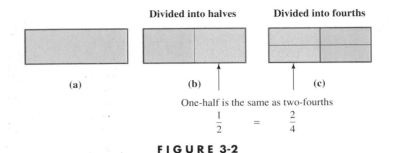

Divided into halves

Divided into fourths

(a)

(b)

(c)

One-half is the same as two-fourths

$$\frac{1}{2} = \frac{2}{4}$$

FIGURE 3-2

The fractions $\frac{1}{2}$ and $\frac{2}{4}$ look different, but Figure 3-2 shows that they represent the same amount. We say that they are **equivalent fractions.**

Equivalent fractions	Two fractions are **equivalent** if they represent the same number.

Simplifying a fraction

If we replace a fraction with an equivalent fraction that contains smaller numbers, we are **simplifying** or **reducing the fraction.**

The fundamental property of fractions	Multiplying or dividing the numerator and the denominator of a fraction by the same nonzero number does not change the value of the fraction.

As an example, we consider $\frac{24}{28}$. It is apparent that 24 and 28 have a common factor of 4. By the fundamental property of fractions, we can divide the numerator and denominator of this fraction by 4.

Divide the numerator by 4.

$$\frac{24}{28} = \frac{24 \div 4}{28 \div 4}$$

Divide the denominator by 4.

$$= \frac{6}{7} \qquad 24 \div 4 = 6 \text{ and } 28 \div 4 = 7.$$

Thus, $\frac{24}{28} = \frac{6}{7}$. Since $\frac{24}{28}$ and $\frac{6}{7}$ are equal, they are equivalent fractions and represent the same number.

In practice, we show the previous simplification in a slightly different way.

$$\frac{24}{28} = \frac{4 \cdot 6}{4 \cdot 7}$$ Once you see a common factor of the numerator and the denominator, factor each of them so that it shows. In this case, they share a common factor of 4.

$$= \frac{\overset{1}{\cancel{4}} \cdot 6}{\underset{1}{\cancel{4}} \cdot 7}$$ Apply the fundamental property of fractions. Divide the numerator and the denominator by 4 by drawing slashes through the common factors. Use small 1's to represent the result of each division of 4 by 4.

$$= \frac{6}{7}$$ Multiply in the numerator and in the denominator: $1 \cdot 6 = 6$ and $1 \cdot 7 = 7$.

In the second step of the previous simplification, we say that we divided out the common factor of 4.

Simplifying a fraction	We can **simplify** a fraction by factoring its numerator and denominator and then dividing out all common factors in the numerator and denominator.

When a fraction can be simplified no further, we say that it is written in **lowest terms.**

Lowest terms	A fraction is in **lowest terms** if the only factor common to the numerator and denominator is 1.

EXAMPLE 2   *Simplifying a fraction.*   Simplify to lowest terms: $-\dfrac{25}{75}$.

Solution

By inspection, the numerator and the denominator have a common factor of 25.

$$-\frac{25}{75} = -\frac{25 \cdot 1}{25 \cdot 3}$$ Factor 25 as $25 \cdot 1$ and 75 as $25 \cdot 3$.

$$= -\frac{\overset{1}{25} \cdot 1}{\underset{1}{25} \cdot 3}$$ Divide out the common factor of 25.

$$= -\frac{1}{3}$$ Multiply in the numerator and in the denominator.

Self Check

Simplify to lowest terms: $-\dfrac{20}{80}$.

Answer: $-\dfrac{1}{4}$

EXAMPLE 3   *Simplifying a fraction using prime factorization.*
Simplify $\dfrac{90}{126}$.

Solution

To find the common factors that will divide out, we find the prime factorizations of 90 and 126.

$$\frac{90}{126} = \frac{5 \cdot 3 \cdot 3 \cdot 2}{7 \cdot 3 \cdot 3 \cdot 2}$$ Factor 90 as $5 \cdot 3 \cdot 3 \cdot 2$ and 126 as $7 \cdot 3 \cdot 3 \cdot 2$.

$$= \frac{5 \cdot \overset{1}{3} \cdot \overset{1}{3} \cdot \overset{1}{2}}{7 \cdot \underset{1}{3} \cdot \underset{1}{3} \cdot \underset{1}{2}}$$ Divide out the common factors of 3, 3, and 2.

$$= \frac{5}{7}$$ Multiply in the numerator and in the denominator.

Self Check

Simplify $\dfrac{42}{150}$.

Answer: $\dfrac{7}{25}$

EXAMPLE 4   *Simplifying a fraction using prime factorization.*
Simplify $\dfrac{225}{150}$.

Solution

To find the common factors that will divide out, we find the prime factorizations of 225 and 150.

$$\frac{225}{150} = \frac{3 \cdot 3 \cdot 5 \cdot 5}{2 \cdot 3 \cdot 5 \cdot 5}$$ Factor 225 as $3 \cdot 3 \cdot 5 \cdot 5$ and 150 as $2 \cdot 3 \cdot 5 \cdot 5$.

$$= \frac{3 \cdot \overset{1}{3} \cdot \overset{1}{5} \cdot \overset{1}{5}}{2 \cdot \underset{1}{3} \cdot \underset{1}{5} \cdot \underset{1}{5}}$$ Divide out the common factors of 3, 5, and 5.

$$= \frac{3}{2}$$ Multiply in the numerator and in the denominator.

Self Check

Simplify $\dfrac{88}{12}$.

Answer: $\dfrac{22}{3}$

Expressing a fraction in higher terms

It is sometimes necessary to replace a fraction with an equivalent fraction that involves larger numbers or more complex terms. This is called **expressing the fraction in higher terms** or **building up** the fraction.

For example, to write $\frac{3}{8}$ as an equivalent fraction with a denominator of 40, we can use the fundamental property of fractions and multiply the numerator and denominator by 5.

Multiply the numerator by 5.

$$\frac{3}{8} = \frac{3 \cdot 5}{8 \cdot 5}$$

Multiply the denominator by 5.

$$= \frac{15}{40} \qquad \text{Do the multiplications in the numerator and denominator.}$$

Therefore, $\frac{3}{8} = \frac{15}{40}$.

EXAMPLE 5 *Expressing a fraction in higher terms.* Write $\frac{5}{7}$ as an equivalent fraction with denominator 28.

Solution

We need to multiply the denominator by 4 to obtain 28. By the fundamental property of fractions, we must multiply the numerator by 4 as well.

$$\frac{5}{7} = \frac{5 \cdot 4}{7 \cdot 4} \qquad \text{Multiply the numerator and denominator by 4.}$$

$$= \frac{20}{28} \qquad \text{Do the multiplication in the numerator and in the denominator.}$$

Self Check

Write $\frac{2}{3}$ as an equivalent fraction with denominator 24.

Answer: $\dfrac{16}{24}$

EXAMPLE 6 *Expressing a whole number as a fraction.* Write 4 as a fraction with denominator 6.

Solution

First, express 4 as a fraction: $4 = \frac{4}{1}$. To obtain a denominator of 6, we need to multiply the numerator and denominator by 6.

$$\frac{4}{1} = \frac{4 \cdot 6}{1 \cdot 6}$$

$$= \frac{24}{6} \qquad 4 \cdot 6 = 24 \text{ and } 1 \cdot 6 = 6.$$

Self Check

Write 5 as a fraction with denominator 3.

Answer: $\dfrac{15}{3}$

STUDY SET   Section 3.1

VOCABULARY   *In Exercises 1–8, fill in the blanks to make a true statement.*

1. For the fraction $\frac{7}{8}$, 7 is the _____ and 8 is the _____.

2. Fractions are often called _____ numbers.

3. We can simplify a fraction that is not in lowest terms by _____ the numerator and the denominator by the _____ number.

4. A fraction is said to be in _____ terms if 1 is the only number that will divide numerator and denominator evenly.

5. Two fractions are _____ if they have the same value.

7. Multiplying the numerator and denominator of a fraction by a number to obtain an equivalent fraction that involves larger numbers or more complex terms is called expressing the fraction in higher _____.

6. A _____ can be used to indicate the number of equal parts of a whole.

8. We can _____ a fraction that is not in lowest terms by applying the fundamental property of _____. We divide the numerator and the denominator by the _____ number.

CONCEPTS

9. Given:

$$\frac{15}{35} = \frac{\overset{1}{\cancel{5} \cdot 3}}{7 \cdot \cancel{5}}$$

In this work, what do the slashes and small 1's mean?

10. Given:

$$\frac{14}{35}$$

a. What common factor do the numerator and denominator share?

b. By what number can the numerator and denominator be divided to simplify the fraction?

11. What concept studied in this section is shown by Illustration 1?

12. Why can't we say that $\frac{2}{5}$ of the figure in Illustration 2 is shaded?

ILLUSTRATION 1

ILLUSTRATION 2

13. a. Explain the difference in the two approaches used to simplify $\frac{20}{28}$.

$$\frac{\overset{1}{\cancel{4}} \cdot 5}{\cancel{4} \cdot 7} \quad \text{and} \quad \frac{\overset{1}{\cancel{2}} \cdot \overset{1}{\cancel{2}} \cdot 5}{\cancel{2} \cdot \cancel{2} \cdot 7}$$

b. Are the results the same?

14. What concept studied in this section does this statement illustrate?

$$\frac{1}{2} = \frac{2}{4} = \frac{3}{6} = \frac{4}{8} = \frac{5}{10}$$

15. Why isn't this a valid application of the fundamental property of fractions?

$$\frac{10}{11} = \frac{2+8}{2+9} = \frac{\overset{1}{\cancel{2}}+8}{\cancel{2}+9} = \frac{9}{10}$$

16. Write the fraction $\dfrac{7}{-8}$ in two other ways.

17. Using the fraction $\frac{3}{8}$ and the number 4, apply the fundamental property of fractions to write an equivalent fraction.

18. Fill in the missing numbers in the following statement:

$$\frac{5 \cdot \underline{}}{9 \cdot \underline{}} = \frac{15}{27}$$

NOTATION *In Exercises 19–20, complete each solution.*

19. Simplify $\dfrac{18}{24}$.

$$\frac{18}{24} = \frac{3 \cdot \cdot 2}{3 \cdot 2 \cdot \cdot 2} \qquad \text{Prime factor 18 and 24.}$$

$$= \frac{\cancel{} \cdot 3 \cdot \cancel{}}{\cancel{} \cdot 2 \cdot 2 \cdot \cancel{}} \qquad \text{Divide out the common factors of 3 and 2.}$$

$$= \frac{3}{4} \qquad \text{Multiply in the numerator and denominator.}$$

20. Simplify $\dfrac{60}{90}$.

$$\frac{60}{90} = \frac{ \cdot 2}{ \cdot 3} \qquad \text{Factor 60 and 90.}$$

$$= \frac{\overset{1}{\cancel{30}} \cdot }{\cancel{30} \cdot } \qquad \text{Divide out the common factor of 30.}$$

$$= \frac{2}{3} \qquad \text{Multiply in the numerator and in the denominator.}$$

21. $\frac{3}{9}$

22. $\frac{5}{20}$

23. $-\frac{7}{21}$

24. $-\frac{6}{30}$

25. $\frac{20}{30}$

26. $\frac{12}{30}$

27. $\frac{15}{6}$

28. $\frac{24}{16}$

29. $-\frac{28}{56}$

30. $\frac{-45}{54}$

31. $\frac{90}{105}$

32. $\frac{26}{78}$

33. $\frac{60}{108}$

34. $\frac{75}{125}$

35. $\frac{180}{210}$

36. $\frac{76}{28}$

37. $\frac{55}{67}$

38. $\frac{41}{51}$

39. $\frac{36}{96}$

40. $\frac{48}{120}$

41. $\frac{25}{35}$

42. $\frac{16}{20}$

43. $\frac{12}{15}$

44. $\frac{10}{15}$

45. $\frac{12}{14}$

46. $\frac{20}{25}$

47. $\frac{21}{24}$

48. $\frac{20}{42}$

49. $-\frac{10}{30}$

50. $-\frac{14}{28}$

51. $\frac{150}{250}$

52. $\frac{160}{240}$

53. $\frac{35}{28}$

54. $\frac{35}{25}$

55. $\frac{56}{28}$

56. $\frac{32}{8}$

In Exercises 57–72, write each fraction as an equivalent fraction with the indicated denominator.

57. $\frac{7}{8}$, 40

58. $\frac{3}{4}$, 24

59. $\frac{4}{5}$, 35

60. $\frac{5}{7}$, 49

61. $\frac{-5}{6}$, 54

62. $\frac{-11}{16}$, 32

63. $-\frac{1}{2}$, 30

64. $-\frac{1}{3}$, 60

65. $\frac{2}{7}$, 14

66. $\frac{3}{10}$, 50

67. $\frac{9}{10}$, 60

68. $\frac{2}{3}$, 27

69. $-\frac{5}{4}$, 20

70. $-\frac{9}{4}$, 44

71. $\frac{-2}{15}$, 45

72. $\frac{-5}{12}$, 36

In Exercises 73–80, write each number as a fraction with the indicated denominator.

73. 3 as fifths

74. 4 as thirds

75. −6 as eighths

76. −3 as sixths

77. 4 as ninths

78. 7 as fourths

79. −2 as halves

80. −10 as ninths

APPLICATIONS In Exercises 81–92, use the concept of fraction in answering each question.

81. COMMUTING TO WORK How much of the commute from home to work has the motorist in Illustration 3 made?

ILLUSTRATION 3

82. TIME CLOCK How much of the hour has passed?

a.

b.

c.

d.

83. CHEMICAL MIXTURE   Three identical beakers are shown in Illustration 4. If beakers 1 and 2 are poured into the empty beaker 3, how full will it be?

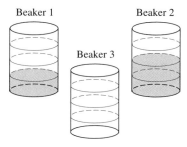

ILLUSTRATION 4

84. POLITICAL PARTIES   Some civic leaders were polled as to their political preference. The results are shown in Illustration 5.

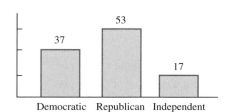

ILLUSTRATION 5

a. What fraction are Democrats?
b. What fraction are independent?
c. What fraction are Republicans?

85. PERSONNEL RECORDS   Complete the chart in Illustration 6 by finding the amount of the job that will be completed by each person working alone for the given number of hours.

Name	Total time to complete the job alone	Time worked alone	Amount of job completed
Bob	10 hours	7 hours	
Ali	8 hours	1 hour	

ILLUSTRATION 6

86. GAS TANK   See Illustration 7. How full does the gauge indicate the gas tank is? How much of the tank has been used?

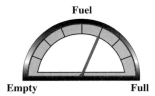

ILLUSTRATION 7

87. ALMANAC   The diagram in Illustration 8 appeared in an almanac. What information does it give us?

ILLUSTRATION 8

88. USING A RULER   Illustration 9 shows a ruler. First, tell how many spaces there are between the numbers 0 and 1. Then tell to what number the arrow is pointing.

ILLUSTRATION 9

89. MACHINERY   The operator of a machine is to turn this dial from setting A to setting B. Express this in two different ways, using fractions.

90. EARTH'S ROTATION   The earth rotates about its vertical axis once every 24 hours.
a. What is the significance of $\frac{1}{24}$ of a rotation to us on earth?
b. What significance does $\frac{24}{24}$ of a revolution have?

91. SUPERMARKET DISPLAY   The amount of space to be given each type of snack food in a supermarket display case is expressed as a fraction. Complete the model of the display, showing where the adjustable shelves should be located, and label where each snack food should be stocked.

$\frac{3}{8}$: potato chips

$\frac{2}{8}$: peanuts

$\frac{1}{8}$: pretzels

$\frac{2}{8}$: tortilla chips

Snacks

92. MEDICAL CENTER Hospital designers have located a nurse's station at the center of a circular building. Show how to divide up the surrounding office space so that each medical department has the proper fractional amount allocated to it. Label each department.

$\dfrac{2}{12}$: Radiology

$\dfrac{5}{12}$: Pediatrics

$\dfrac{1}{12}$: Laboratory

$\dfrac{3}{12}$: Orthopedics

$\dfrac{1}{12}$: Pharmacy

WRITING *Write a paragraph using your own words.*

93. Explain the concept of equivalent fractions.

94. What does it mean for a fraction to be in lowest terms?

95. Explain the difference between three-fourths and three-fifths of a pizza.

96. Explain both parts of the fundamental property of fractions.

REVIEW

97. List the first five natural numbers.

98. List the first five whole numbers.

99. Round 564,112 to the nearest thousand.

100. Give the definition of a prime number.

101. Simplify: $3 + 4 \cdot 2$.

102. Add: $2 + (-3) + (-5)$.

103. Evaluate the power: $(-10)^2$.

104. Multiply: $(-3)(-4)$.

3.2 Multiplying Fractions

In this section, you will learn about

- **Multiplying fractions**
- **Simplifying when multiplying fractions**
- **Powers of a fraction**
- **Applications involving multiplication of fractions**

INTRODUCTION. In the next three sections, we will discuss how to add, subtract, multiply, and divide fractions. We begin with the operation of multiplication.

Multiplying fractions

Suppose that a television network is going to take out a full-page ad to publicize its fall lineup of shows. The prime-time shows are to get $\frac{3}{5}$ of the ad space and daytime programming the remainder. Of the space devoted to prime time, $\frac{1}{2}$ is to be used to promote weekend programs. How much of the newspaper page will be used to advertise weekend prime-time programs?

The ad for the weekend prime-time shows will occupy $\frac{1}{2}$ of $\frac{3}{5}$ of the page. This can be expressed as $\frac{1}{2} \cdot \frac{3}{5}$. We can simplify this expression by using a series of illustrations.

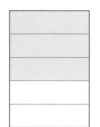

Step 1: Divide the page into fifths and shade three of them. This represents the fraction $\frac{3}{5}$, the amount of the page needed to advertise prime-time shows.

Step 2: Next, find $\frac{1}{2}$ of the shaded part of the page by first dividing the page into halves, using a vertical line.

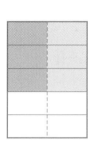

Step 3: Then, highlight $\frac{1}{2}$ of the shaded parts determined in step 2. This highlighted part (shown at left in purple) represents $\frac{3}{10}$ of the page, the amount of the page needed to advertise the weekend prime-time shows. This leads us to the conclusion that $\frac{1}{2} \cdot \frac{3}{5} = \frac{3}{10}$.

Two observations can be made from this result.

Observations

1. The numerator of the answer is the product of the numerators of the original fractions.

$$\overbrace{\frac{1}{2} \cdot \frac{3}{5} = \frac{3}{10}}^{1 \cdot 3 = 3} \quad \text{Answer}$$
$$\underbrace{\phantom{\frac{1}{2} \cdot \frac{3}{5} = \frac{3}{10}}}_{2 \cdot 5 = 10}$$

2. The denominator of the answer is the product of the denominators of the original fractions.

These observations suggest the following rule for multiplying two fractions.

Multiplying fractions To multiply two fractions, multiply the numerators and multiply the denominators. Then simplify the result, if possible.

EXAMPLE 1 *Multiplying fractions.* Multiply: $\dfrac{7}{8} \cdot \dfrac{3}{5}$.

Solution

$$\frac{7}{8} \cdot \frac{3}{5} = \frac{7 \cdot 3}{8 \cdot 5}$$ Multiply the numerators and multiply the denominators.

$$= \frac{21}{40}$$ $7 \cdot 3 = 21$ and $8 \cdot 5 = 40$.

Self Check

Multiply: $\dfrac{5}{9} \cdot \dfrac{2}{3}$.

Answer: $\dfrac{10}{27}$

The rules for multiplying integers also hold for multiplying fractions. When multiplying two fractions with *like* signs, the product is positive. When multiplying two fractions with *unlike* signs, the product is negative.

EXAMPLE 2 *Multiplication with a negative fraction.*
Multiply: $-\dfrac{3}{4}\left(\dfrac{1}{8}\right)$.

Self Check
Multiply: $\dfrac{5}{6}\left(-\dfrac{1}{3}\right)$.

Solution

$-\dfrac{3}{4}\left(\dfrac{1}{8}\right) = -\dfrac{3 \cdot 1}{4 \cdot 8}$ Multiply the numerators and the denominators. Since the fractions have unlike signs, the product is negative.

$= -\dfrac{3}{32}$ $3 \cdot 1 = 3$ and $4 \cdot 8 = 32$.

Answer: $-\dfrac{5}{18}$

Simplifying when multiplying fractions

After multiplying two fractions, we should simplify the result, if possible.

EXAMPLE 3 *Simplifying when multiplying fractions.* Find $\dfrac{5}{8} \cdot \dfrac{4}{5}$.

Self Check
Find $\dfrac{6}{25} \cdot \dfrac{5}{6}$.

Solution

$\dfrac{5}{8} \cdot \dfrac{4}{5} = \dfrac{5 \cdot 4}{8 \cdot 5}$ Write the products in the numerator and denominator.

$= \dfrac{5 \cdot 4}{4 \cdot 2 \cdot 5}$ In the denominator, factor 8 as $4 \cdot 2$.

$= \dfrac{\overset{1}{\cancel{5}} \cdot \overset{1}{\cancel{4}}}{\underset{1}{\cancel{4}} \cdot 2 \cdot \underset{1}{\cancel{5}}}$ Divide out the common factors of 4 and 5.

$= \dfrac{1}{2}$ $1 \cdot 1 = 1$ and $1 \cdot 2 \cdot 1 = 2$.

Answer: $\dfrac{1}{5}$

When multiplying a fraction and an integer, we express the integer as a fraction with a denominator of 1.

EXAMPLE 4 *Multiplying three fractions.* Find $45\left(-\dfrac{1}{14}\right)\left(-\dfrac{7}{10}\right)$.

Solution This expression involves three factors, and one of them is an integer.

$45\left(-\dfrac{1}{14}\right)\left(-\dfrac{7}{10}\right) = \dfrac{45}{1}\left(\dfrac{1}{14}\right)\left(\dfrac{7}{10}\right)$ Write 45 as a fraction: $45 = \frac{45}{1}$. The product of two negative numbers is positive.

$= \dfrac{45 \cdot 1 \cdot 7}{1 \cdot 14 \cdot 10}$ Multiply the numerators and multiply the denominators.

$= \dfrac{5 \cdot 3 \cdot 3 \cdot 1 \cdot 7}{1 \cdot 7 \cdot 2 \cdot 5 \cdot 2}$ Prime factor 45 as $5 \cdot 3 \cdot 3$, 14 as $7 \cdot 2$, and 10 as $5 \cdot 2$.

$= \dfrac{\overset{1}{\cancel{5}} \cdot 3 \cdot 3 \cdot 1 \cdot \overset{1}{\cancel{7}}}{1 \cdot \underset{1}{\cancel{7}} \cdot 2 \cdot \underset{1}{\cancel{5}} \cdot 2}$ Divide out the common factors of 5 and 7.

$= \dfrac{9}{4}$ Multiply in the numerator and denominator.

EXAMPLE 5 *Multiplication of fractions and whole numbers.*

Multiply: $\frac{1}{4}(8)$.

Solution

$\frac{1}{4}(8) = \frac{1}{4} \cdot \frac{8}{1}$ Write 8 as a fraction: $8 = \frac{8}{1}$.

$= \frac{1 \cdot 8}{4 \cdot 1}$ Multiply the numerators and multiply the denominators.

$= \frac{1 \cdot 2 \cdot \overset{1}{\cancel{4}}}{\underset{1}{\cancel{4}} \cdot 1}$ Factor 8 as $2 \cdot 4$ and divide out the common factor of 4.

$= \frac{2}{1}$ Multiply in the numerator and denominator.

$= 2$ $\frac{2}{1} = 2$.

Self Check

Multiply: $\frac{1}{5}(5)$.

Answer: 1

Powers of a fraction

If the base of an exponential expression is a fraction, the exponent tells us how many times to write that fraction as a factor. For example,

$$\left(\frac{2}{3}\right)^2 = \frac{2}{3} \cdot \frac{2}{3} = \frac{4}{9}$$

EXAMPLE 6 *Power of a fraction.* Find $\left(-\frac{2}{3}\right)^2$.

Solution

Exponents are used to indicate repeated multiplication.

$\left(-\frac{2}{3}\right)^2 = \left(-\frac{2}{3}\right)\left(-\frac{2}{3}\right)$ Write $-\frac{2}{3}$ as a factor 2 times.

$= \frac{2 \cdot 2}{3 \cdot 3}$ The product of two fractions with like signs is positive. Multiply the numerators and denominators.

$= \frac{4}{9}$ Multiply in the numerator and denominator.

Self Check

Find $\left(-\frac{3}{4}\right)^3$.

Answer: $-\frac{27}{64}$

Applications involving multiplication of fractions

EXAMPLE 7 *House of Representatives.* In the United States House of Representatives, a bill was introduced that would require a $\frac{3}{5}$ vote of the 435 members to authorize any tax increase. Under this requirement, how many representatives would have to vote for a tax increase before it could become law?

Solution

$$\frac{3}{5} \text{ of } 435 = \frac{3}{5} \cdot \frac{435}{1}$$

The word *of* means to multiply. $435 = \frac{435}{1}$.

$$= \frac{3 \cdot 435}{5 \cdot 1}$$

Multiply the numerators and multiply the denominators.

$$= \frac{3 \cdot 5 \cdot 3 \cdot 29}{5 \cdot 1}$$

Prime factor 435 as $5 \cdot 3 \cdot 29$.

$$= \frac{3 \cdot \overset{1}{\cancel{5}} \cdot 3 \cdot 29}{\underset{1}{\cancel{5}} \cdot 1}$$

Divide out the common factor of 5.

$$= \frac{261}{1}$$

Multiply in the numerator: $3 \cdot 1 \cdot 3 \cdot 29 = 261$.

Multiply in the denominator: $1 \cdot 1 = 1$.

$$= 261$$

$\frac{261}{1} = 261$.

It would take 261 representatives voting in favor to pass a tax increase. ■

As Figure 3-3 shows, a triangle has three sides. The length of the base of the triangle can be represented by the letter b and the height by the letter h.

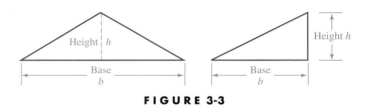

FIGURE 3-3

The area of the triangle can be found by using the following formula.

Area of a triangle	The area, A, of a triangle is one-half the product of its base, b, and its height, h: $$\text{Area} = \frac{1}{2}(\text{base})(\text{height}) \qquad \text{or} \qquad A = \frac{1}{2}bh$$

EXAMPLE 8 **Geography.** Estimate the area of the state of Virginia using the triangle in Figure 3-4.

Solution We will estimate the area of the state by finding the area of the triangle.

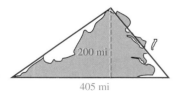

FIGURE 3-4

$$A = \frac{1}{2}bh$$

The formula for the area of a triangle.

$$= \frac{1}{2}(405)(200)$$

Substitute 405 for b and 200 for h.

$$= \frac{1}{2}\left(\frac{405}{1}\right)\left(\frac{200}{1}\right)$$

Write 405 and 200 as fractions.

$$= \frac{1 \cdot 405 \cdot 200}{2}$$

Multiply the numerators. Multiply the denominators.

$$= \frac{1 \cdot 405 \cdot 2 \cdot 100}{2}$$

Factor 200: $200 = 2 \cdot 100$.

$$= \frac{1 \cdot 405 \cdot \overset{1}{\cancel{2}} \cdot 100}{\underset{1}{\cancel{2}}}$$ Divide out the common factor of 2.

$$= 40,500$$ $405 \cdot 100 = 40,500.$

The estimated area of the state of Virginia is 40,500 square miles.

STUDY SET   Section 3.2

VOCABULARY   *In Exercises 1–6, fill in the blanks to make a true statement.*

1. The word *of* in mathematics usually means _____.

2. The _____ of a figure is the amount of surface that it encloses.

3. The result of a multiplication problem is called the _____.

4. To _____ a fraction means to divide the numerator and denominator by the same number.

5. In a triangle, *b* stands for the length of the _____ and *h* stands for the _____.

6. The _____ for the area of a triangle is $A = \frac{1}{2}bh$.

CONCEPTS

7. Fill in the blank: To multiply fractions, we multiply their _____ and multiply their denominators.

8. Write each of the following as fractions:
   a. 4   **b.** -3

9. Use the following rectangle to find $\frac{1}{3} \cdot \frac{1}{4}$.

   a. Using vertical lines, divide the rectangle into four equal parts and lightly shade one of them. What fractional part of the rectangle did you shade?
   b. To find $\frac{1}{3}$ of the shaded portion, use two horizontal lines to divide the rectangle into three equal parts and lightly shade one of them. Into how many equal parts is the rectangle now divided? How many parts have been shaded twice? What is $\frac{1}{3} \cdot \frac{1}{4}$?

10. Use the following rectangle to find $\frac{2}{3} \cdot \frac{1}{5}$.

   a. Using vertical lines, divide the rectangle into five equal parts and lightly shade one of them. What fractional part of the rectangle did you shade?
   b. To find $\frac{2}{3}$ of the shaded portion, use two horizontal lines to divide the rectangle into three equal parts and lightly shade two of them. Into how many equal parts is the rectangle now divided? How many parts have been shaded twice? What is $\frac{2}{3} \cdot \frac{1}{5}$?

11. If we evaluated $\left(-\frac{4}{5}\right)^{10}$, would the result be positive or negative?

12. a. Multiply $\frac{9}{10}$ and 20.
   b. When we multiply two numbers, is the product always larger than both those numbers?

13. Tell whether each statement is true or false.
   a. $\left(\frac{1}{2}\right)^2 = \frac{1}{2} \cdot \frac{1}{2}$ **b.** $\left(\frac{1}{2}\right)^3 = \frac{1}{2} + \frac{1}{2} + \frac{1}{2}$

14. What is the numerator of the result for the multiplication problem shown here?

$$\frac{4}{15} \cdot \frac{3}{4} = \frac{\overset{1}{\cancel{4}} \cdot \overset{1}{\cancel{3}}}{5 \cdot \underset{1}{\cancel{3}} \cdot \underset{1}{\cancel{4}}}$$

15. Multiply: $\dfrac{5}{8} \cdot \dfrac{7}{15}$.

$\dfrac{5}{8} \cdot \dfrac{7}{15} = \dfrac{5 \cdot }{8 \cdot }$ Multiply the numerators.
Multiply the denominators.

$= \dfrac{5 \cdot 7}{8 \cdot 5 \cdot }$ Factor 15.

$= \dfrac{\overset{1}{\cancel{5}} \cdot 7}{8 \cdot \underset{1}{\cancel{5}} \cdot}$ Divide out the common factor of 5.

$= \dfrac{7}{24}$ Multiply in the numerator and denominator.

16. Multiply: $\dfrac{7}{12} \cdot \dfrac{4}{21}$.

$\dfrac{7}{12} \cdot \dfrac{4}{21} = \dfrac{7 \cdot 4}{ \cdot }$ Multiply the numerators.
Multiply the denominators.

$= \dfrac{7 \cdot 4}{4 \cdot \cdot \cdot 3}$ Factor 12 and 21.

$= \dfrac{\overset{1}{\cancel{7}} \cdot \overset{1}{\cancel{4}}}{\cancel{4} \cdot 3 \cdot \cancel{7} \cdot 3}$ Divide out the common factors of 4 and 7.

$= \dfrac{1}{9}$ Multiply in the numerator and denominator.

PRACTICE *In Exercises 17–50, multiply. Write all answers in lowest terms.*

17. $\dfrac{1}{4} \cdot \dfrac{1}{2}$

18. $\dfrac{1}{3} \cdot \dfrac{1}{5}$

19. $\dfrac{3}{8} \cdot \dfrac{7}{16}$

20. $\dfrac{5}{9} \cdot \dfrac{2}{7}$

21. $\dfrac{2}{3} \cdot \dfrac{6}{7}$

22. $\dfrac{5}{12} \cdot \dfrac{3}{4}$

23. $\dfrac{14}{15} \cdot \dfrac{11}{8}$

24. $\dfrac{5}{16} \cdot \dfrac{8}{3}$

25. $-\dfrac{15}{24} \cdot \dfrac{8}{25}$

26. $-\dfrac{20}{21} \cdot \dfrac{7}{16}$

27. $\left(-\dfrac{11}{21}\right)\left(-\dfrac{14}{33}\right)$

28. $\left(-\dfrac{16}{35}\right)\left(-\dfrac{25}{48}\right)$

29. $\dfrac{7}{10}\left(\dfrac{20}{21}\right)$

30. $\left(\dfrac{7}{6}\right)\dfrac{9}{49}$

31. $\dfrac{3}{4} \cdot \dfrac{4}{3}$

32. $\dfrac{4}{5} \cdot \dfrac{5}{4}$

33. $\dfrac{1}{3} \cdot \dfrac{15}{16} \cdot \dfrac{4}{25}$

34. $\dfrac{3}{15} \cdot \dfrac{15}{7} \cdot \dfrac{14}{27}$

35. $\left(\dfrac{2}{3}\right)\left(-\dfrac{1}{16}\right)\left(-\dfrac{4}{5}\right)$

36. $\left(\dfrac{3}{8}\right)\left(-\dfrac{2}{3}\right)\left(-\dfrac{12}{27}\right)$

37. $\dfrac{5}{6} \cdot 18$

38. $6\left(-\dfrac{2}{3}\right)$

39. $15\left(-\dfrac{4}{5}\right)$

40. $-2\left(-\dfrac{7}{8}\right)$

41. $\dfrac{3}{16} \cdot 4 \cdot \dfrac{2}{3}$

42. $5 \cdot \dfrac{7}{5} \cdot \dfrac{3}{14}$

43. $\dfrac{5}{3}\left(-\dfrac{6}{15}\right)(-4)$

44. $\dfrac{5}{6}\left(-\dfrac{2}{3}\right)(-12)$

45. $-\dfrac{2}{3}\left(-\dfrac{6}{5}\right)\left(-\dfrac{20}{3}\right)$

46. $-\dfrac{3}{2}\left(-\dfrac{5}{6}\right)\left(\dfrac{12}{5}\right)$

47. $-\dfrac{11}{12} \cdot \dfrac{18}{55} \cdot 5$

48. $-\dfrac{24}{5} \cdot \dfrac{7}{12} \cdot \dfrac{1}{14}$

49. $\dfrac{3}{4}\left(\dfrac{5}{7}\right)\left(\dfrac{2}{3}\right)\left(\dfrac{7}{3}\right)$

50. $-\dfrac{5}{4}\left(\dfrac{8}{15}\right)\left(\dfrac{2}{3}\right)\left(\dfrac{7}{2}\right)$

In Exercises 51–58, find each power.

51. $\left(\dfrac{2}{3}\right)^2$

52. $\left(\dfrac{3}{5}\right)^2$

53. $\left(-\dfrac{5}{9}\right)^2$

54. $\left(-\dfrac{5}{6}\right)^2$

55. $\left(\dfrac{4}{3}\right)^2$

56. $\left(\dfrac{3}{2}\right)^2$

57. $\left(-\dfrac{1}{4}\right)^3$

58. $\left(-\dfrac{2}{5}\right)^3$

59. Complete the multiplication table of fractions in Illustration 1.

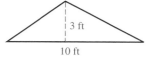

	$\frac{1}{2}$	$\frac{1}{3}$	$\frac{1}{4}$	$\frac{1}{5}$	$\frac{1}{6}$
$\frac{1}{2}$					
$\frac{1}{3}$					
$\frac{1}{4}$					
$\frac{1}{5}$					
$\frac{1}{6}$					

ILLUSTRATION 1

60. Complete Illustration 2 by finding the original fraction, given its square.

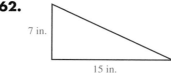

Original fraction squared	Original fraction
$\frac{1}{9}$	
$\frac{1}{100}$	
$\frac{4}{25}$	
$\frac{16}{49}$	
$\frac{81}{36}$	
$\frac{9}{121}$	

ILLUSTRATION 2

In Exercises 61–64, find the area of each triangle.

61.

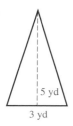

3 ft
10 ft

62.

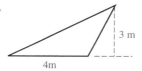

7 in.
15 in.

63.

5 yd
3 yd

64.

3 m
4m

APPLICATIONS *In Exercises 65–72, solve each problem.*

65. THE CONSTITUTION Article V of the United States Constitution requires a two-thirds vote of the House of Representatives to propose a constitutional amendment. The House has 435 members. Find the number of votes needed to meet this requirement.

66. EMPLOYMENT Three-fifths of the students who attend Citrus College are also employed. Of those employed, one-fourth of them work more than 40 hours per week. What fraction of the student body works more than 40 hours per week?

67. STAMPS The best designs in a contest to create a wildlife stamp are shown in Illustration 3. To save on

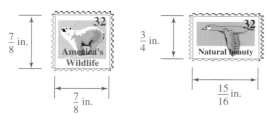

$\frac{7}{8}$ in.
$\frac{7}{8}$ in.
$\frac{3}{4}$ in.
$\frac{15}{16}$ in.

ILLUSTRATION 3

paper costs, the postal service has decided to choose the stamp that has the smaller area. Which one is that?

68. TENNIS BALL A tennis ball is dropped from a height of 54 inches. Each time it hits the ground, it rebounds one-third of the previous height it fell. See Illustration 4 and find the three missing rebound heights.

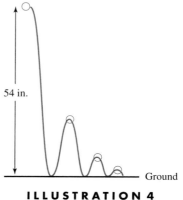

54 in.
Ground

ILLUSTRATION 4

69. PLANT GROWTH RATES In an experiment, monthly growth rates of three types of plants doubled when nitrogen was added to the soil. Complete Illustration 5 by charting the improved growth rate next to each normal growth rate.

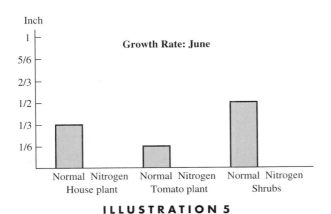

ILLUSTRATION 5

70. COOKING Shani plans to make a batch of cookies. Use the recipe below, along with the concept of multiplication with fractions, to find how much sugar and molasses he needs to make one dozen cookies.

Gingerbread cookies

$\frac{3}{4}$ cup sugar $\frac{1}{2}$ cup water

2 cups flour $\frac{2}{3}$ cup shortening

$\frac{1}{8}$ teaspoon allspice $\frac{1}{4}$ teaspoon salt

$\frac{1}{3}$ cup dark molasses $\frac{3}{4}$ teaspoon ginger

Makes two dozen gingerbread men.

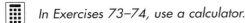 *In Exercises 73–74, use a calculator.*

73. THE EARTH'S SURFACE The surface of the earth covers an area of approximately 196,800,000 square miles. About $\frac{3}{4}$ of that area is covered by water. Find the number of square miles of the surface covered by water.

71. WINDSURFING Estimate the area of the sail on the windsurfing board in Illustration 6. The total height of the mast is 12 feet.

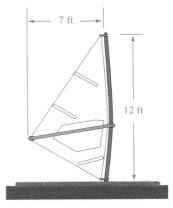

ILLUSTRATION 6

72. TILE DESIGN A design for bathroom tile is shown in Illustration 7. Find the amount of area on a tile that is blue.

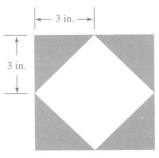

ILLUSTRATION 7

74. GEOGRAPHY Estimate the area of the state of New Hampshire, using the triangle in Illustration 8.

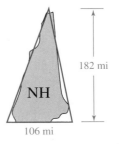

ILLUSTRATION 8

WRITING *Write a paragraph using your own words.*

75. In mathematics, the word *of* usually means multiply. Give three real-life examples of this usage.

77. Explain how to multiply two fractions.

76. Explain how you could multiply the number 5 and another number and obtain an answer that is less than 5.

78. Explain how to raise a fraction to the second power.

79. Round 6,794 to the nearest hundred.

80. Find the distance covered by a motorist traveling at 60 miles per hour for 3 hours.

81. Evaluate $5 + 7 \cdot 3$.

82. Multiply: 324
　　　　　$\underline{45}$

83. Find the prime factorization of 125.

84. Evaluate: $-2(-3)(-4)$.

3.3　*Dividing Fractions*

In this section, you will learn about
- **Division with fractions**
- **Reciprocals**
- **A rule for dividing fractions**

INTRODUCTION. In this section, we will discuss how to divide fractions. We will examine problems involving positive and negative fractions as well as algebraic fractions. The skills you learned in Section 3.2 will be useful in this section.

Division with fractions

Suppose that the manager of a candy store purchases large bars of chocolate and divides each one into four equal parts. How many fourths can be obtained from 5 bars?

We are asking, "How many $\frac{1}{4}$'s are there in 5?" To answer the question, we need to use the operation of division. We can represent this division as $5 \div \frac{1}{4}$. See Figure 3-5.

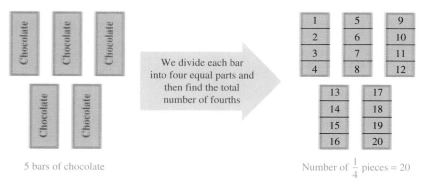

5 bars of chocolate　　　　　　　　　　　Number of $\frac{1}{4}$ pieces = 20

FIGURE 3-5

There are 20 fourths in the 5 bars.

Observations　The numbers 5 and $\frac{1}{4}$ are involved in the original problem, but to obtain the result of 20, the numbers 5 and 4 seem to be more important. This problem can be solved using *multiplication* instead of division: $5 \cdot 4 = 20$.

Reciprocals

Division of fractions involves working with **reciprocals.** To present the concept of reciprocal, we will consider the problem $\frac{7}{8} \cdot \frac{8}{7}$.

$$\frac{7}{8} \cdot \frac{8}{7} = \frac{7 \cdot 8}{8 \cdot 7}$$ Multiply the numerators and multiply the denominators.

$$= \frac{\overset{1}{7} \cdot \overset{1}{8}}{\underset{1}{8} \cdot \underset{1}{7}}$$ Divide out the common factors of 7 and 8.

$$= \frac{1}{1}$$ Multiply in the numerator and multiply in the denominator.

$$= 1$$ $\frac{1}{1} = 1$.

The product of $\frac{7}{8}$ and $\frac{8}{7}$ is 1.

Whenever the product of two numbers is 1, we say that those numbers are *reciprocals.* Therefore, $\frac{7}{8}$ and $\frac{8}{7}$ are reciprocals. To find the reciprocal of a fraction, *we invert the numerator and the denominator.*

Reciprocals	Two numbers are called **reciprocals** if their product is 1.

 WARNING! Zero does not have a reciprocal, because the product of 0 and a number can never be 1.

EXAMPLE 1 *Reciprocals.* For each number, find its reciprocal and show that their product is 1: **a.** $\frac{2}{3}$, **b.** $-\frac{3}{4}$, and **c.** 5.

Solution

a. The reciprocal of $\frac{2}{3}$ is $\frac{3}{2}$.

$$\frac{2}{3} \cdot \frac{3}{2} = \frac{\overset{1}{2} \cdot \overset{1}{3}}{\underset{1}{3} \cdot \underset{1}{2}} = 1$$

b. The reciprocal of $-\frac{3}{4}$ is $-\frac{4}{3}$.

$$-\frac{3}{4}\left(-\frac{4}{3}\right) = \frac{\overset{1}{3} \cdot \overset{1}{4}}{\underset{1}{4} \cdot \underset{1}{3}} = 1$$ The product of two fractions with like signs is positive.

c. $5 = \frac{5}{1}$, so the reciprocal of 5 is $\frac{1}{5}$.

$$5 \cdot \frac{1}{5} = \frac{5}{1} \cdot \frac{1}{5} = \frac{\overset{1}{5} \cdot 1}{1 \cdot \underset{1}{5}} = 1$$

Self Check

For each number, find its reciprocal and show that their product is 1:

a. $\frac{3}{5}$

b. $-\frac{5}{6}$

c. 8

Answer: **a.** $\frac{5}{3}$, **b.** $-\frac{6}{5}$,

c. $\frac{1}{8}$

A rule for dividing fractions

As the candy store example showed, division by a fraction is the same as multiplication by its reciprocal.

Dividing fractions	To divide fractions, multiply the first fraction by the reciprocal of the second fraction.

For example, to work the problem $\frac{5}{7} \div \frac{3}{4}$, we multiply the first fraction by the reciprocal of the second.

Change the division to a multiplication.

$$\frac{5}{7} \div \frac{3}{4} = \frac{5}{7} \cdot \frac{4}{3}$$

The reciprocal of $\frac{3}{4}$ is $\frac{4}{3}$.

$$= \frac{5 \cdot 4}{7 \cdot 3}$$

$$= \frac{20}{21} \qquad \text{Multiply in the numerator and denominator:} \\ 5 \cdot 4 = 20 \text{ and } 7 \cdot 3 = 21.$$

Therefore, $\frac{5}{7} \div \frac{3}{4} = \frac{20}{21}$.

EXAMPLE 2   *Dividing fractions.*   Divide: $\frac{1}{3} \div \frac{4}{5}$.

Solution

$$\frac{1}{3} \div \frac{4}{5} = \frac{1}{3} \cdot \frac{5}{4} \qquad \text{Multiply } \tfrac{1}{3} \text{ by the reciprocal of } \tfrac{4}{5}, \text{ which is } \tfrac{5}{4}.$$

$$= \frac{1 \cdot 5}{3 \cdot 4} \qquad \text{Multiply the numerators and multiply the denominators.}$$

$$= \frac{5}{12} \qquad \text{Multiply in the numerator and denominator.}$$

Self Check

Divide: $\frac{2}{3} \div \frac{7}{8}$.

Answer: $\frac{16}{21}$

EXAMPLE 3   *Dividing fractions.*   Divide: $\frac{9}{16} \div \frac{3}{20}$.

Solution

$$\frac{9}{16} \div \frac{3}{20} = \frac{9}{16} \cdot \frac{20}{3} \qquad \text{Multiply } \tfrac{9}{16} \text{ by the reciprocal of } \tfrac{3}{20}, \text{ which is } \tfrac{20}{3}.$$

$$= \frac{9 \cdot 20}{16 \cdot 3} \qquad \text{Multiply the numerators and multiply the denominators.}$$

$$= \frac{3 \cdot 3 \cdot 4 \cdot 5}{4 \cdot 4 \cdot 3} \qquad \text{Factor 9 as } 3 \cdot 3, 20 \text{ as } 4 \cdot 5, \text{ and } 16 \text{ as } 4 \cdot 4.$$

$$= \frac{\overset{1}{\cancel{3}} \cdot 3 \cdot \overset{1}{\cancel{4}} \cdot 5}{\underset{1}{\cancel{4}} \cdot 4 \cdot \underset{1}{\cancel{3}}} \qquad \text{Divide out the common factors of 3 and 4.}$$

$$= \frac{15}{4} \qquad \text{Multiply in the numerator and denominator.}$$

Self Check

Divide: $\frac{4}{5} \div \frac{8}{25}$.

Answer: $\frac{5}{2}$

EXAMPLE 4   *Surfboard design.*   Most surfboards are made of polyurethane foam plastic covered with several layers of fiberglas to keep them water-tight. How many layers are needed to build up a finish three-eighths of an inch thick if each layer of fiberglas has a thickness of one-sixteenth of an inch?

Solution   We need to know how many one-sixteenths there are in three-eighths. To answer this question, we will use division and find $\frac{3}{8} \div \frac{1}{16}$.

$$\frac{3}{8} \div \frac{1}{16} = \frac{3}{8} \cdot \frac{16}{1}$$ Multiply $\frac{3}{8}$ by the reciprocal of $\frac{1}{16}$, which is $\frac{16}{1}$.

$$= \frac{3 \cdot 16}{8 \cdot 1}$$ Multiply the numerators and multiply the denominators.

$$= \frac{3 \cdot 8 \cdot 2}{8 \cdot 1}$$ Factor 16 as $8 \cdot 2$.

$$= \frac{3 \cdot \overset{1}{\cancel{8}} \cdot 2}{\underset{1}{\cancel{8}} \cdot 1}$$ Divide out the common factor of 8.

$$= \frac{6}{1}$$ Multiply in the numerator and denominator.

$$= 6$$ $\frac{6}{1} = 6$.

The number of layers of fiberglas to be applied is 6.

When working with divisions involving negative fractions, we use the rules for multiplying two numbers with *like* or *unlike* signs.

EXAMPLE 5   *Division with a negative fraction.*
Divide: $\dfrac{1}{6} \div \left(-\dfrac{1}{18} \right)$.

Solution

$$\frac{1}{6} \div \left(-\frac{1}{18} \right) = \frac{1}{6}\left(-\frac{18}{1} \right)$$ Multiply $\frac{1}{6}$ by the reciprocal of $-\frac{1}{18}$, which is $-\frac{18}{1}$.

$$= -\frac{1 \cdot 18}{6 \cdot 1}$$ The product of two fractions with unlike signs is negative. Multiply the numerators and multiply the denominators.

$$= -\frac{1 \cdot 6 \cdot 3}{6 \cdot 1}$$ Factor 18 as $6 \cdot 3$.

$$= -\frac{1 \cdot \overset{1}{\cancel{6}} \cdot 3}{\underset{1}{\cancel{6}} \cdot 1}$$ Divide out the common factor of 6.

$$= -\frac{3}{1}$$ Multiply in the numerator and denominator.

$$= -3$$ $\frac{3}{1} = 3$.

Self Check
Divide: $\dfrac{2}{3} \div \left(-\dfrac{7}{6} \right)$.

Answer: $-\dfrac{4}{7}$

EXAMPLE 6 *Division with a fraction and an integer.*

Divide: $-\dfrac{21}{36} \div (-3)$.

Solution

$$-\dfrac{21}{36} \div (-3) = -\dfrac{21}{36}\left(-\dfrac{1}{3}\right)$$ Multiply $-\frac{21}{36}$ by the reciprocal of -3, which is $-\frac{1}{3}$.

$$= \dfrac{21 \cdot 1}{36 \cdot 3}$$ The product of two fractions with like signs is positive. Multiply the numerators and multiply the denominators.

$$= \dfrac{3 \cdot 7 \cdot 1}{36 \cdot 3}$$ Factor 21 as $3 \cdot 7$.

$$= \dfrac{\overset{1}{\cancel{3}} \cdot 7 \cdot 1}{36 \cdot \underset{1}{\cancel{3}}}$$ Divide out the common factor of 3.

$$= \dfrac{7}{36}$$ Multiply in the numerator and denominator.

Self Check

Divide: $-\dfrac{24}{25} \div (-8)$.

Answer: $\dfrac{3}{25}$

STUDY SET   Section 3.3

VOCABULARY   *In Exercises 1–2, fill in the blanks to make a true statement.*

1. Two numbers are called _____ if their product is 1.

2. The result of a division problem is called the _____.

CONCEPTS

3. Complete this statement:

$\dfrac{2}{3} \div \dfrac{3}{5} = $ ___ $\cdot$ ___

4. Find the reciprocal of each number.

a. $\dfrac{2}{5}$ **b.** -3 **c.** $\dfrac{1}{5}$

5. Using horizontal lines, divide each rectangle in Illustration 1 into thirds. What division problem does this illustrate? What is the quotient of that problem?

6. Using horizontal lines, divide each rectangle in Illustration 2 into fifths. What division problem does this illustrate? What is the quotient of that problem?

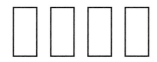

ILLUSTRATION 1

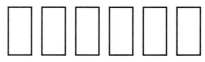

ILLUSTRATION 2

7. Multiply $\frac{4}{5}$ and its reciprocal. What is the result?

9. a. Find $15 \div 3$.
   b. Rewrite $15 \div 3$ as multiplication by the reciprocal of 3 and find the result.
   c. Complete this statement: Division by 3 is the same as multiplication by ___.

8. Multiply $-\frac{3}{5}$ and its reciprocal. What is the result?

10. a. Find $10 \div \frac{1}{5}$.
   b. Rewrite $10 \div \frac{1}{5}$ as multiplication by the reciprocal of $\frac{1}{5}$ and find the result.
   c. Complete this statement: Division by $\frac{1}{5}$ is the same as multiplication by ___.

11. Divide: $\dfrac{25}{36} \div \dfrac{10}{9}$.

$\dfrac{25}{36} \div \dfrac{10}{9} = \dfrac{25}{36} \cdot \text{——}$ Multiply by the reciprocal.

$= \dfrac{25 \cdot \text{——}}{36 \cdot \text{——}}$ Multiply the numerators and denominators.

$= \dfrac{5 \cdot \text{ } \cdot 9}{4 \cdot 9 \cdot 2 \cdot 5}$ Factor 25, 36, and 10.

$= \dfrac{\overset{1}{\diagup} \cdot 5 \cdot \overset{1}{\diagup}}{4 \cdot \underset{1}{\diagup} \cdot 2 \cdot \underset{1}{\diagup}}$ Divide out the common factors of 5 and 9.

$= \dfrac{5}{8}$ Multiply in the numerator and denominator.

12. Divide: $\dfrac{4}{9} \div \dfrac{8}{27}$.

$\dfrac{4}{9} \div \dfrac{8}{27} = \dfrac{4}{9} \cdot \text{——}$ Multiply by the reciprocal.

$= \dfrac{4 \cdot \text{——}}{9 \cdot \text{——}}$ Multiply the numerators and denominators.

$= \dfrac{4 \cdot 3 \cdot 9}{9 \cdot 4 \cdot 2}$ Factor 27 and 8.

$= \dfrac{\overset{1}{\diagup} \cdot 3 \cdot \overset{1}{\diagup}}{\underset{1}{\diagup} \cdot \underset{1}{\diagup} \cdot 2}$ Divide out the common factors of 4 and 9.

$= \dfrac{3}{2}$ Multiply in the numerator and denominator.

PRACTICE *In Exercises 13–40, do the operations.*

13. $\dfrac{1}{2} \div \dfrac{3}{5}$

14. $\dfrac{5}{7} \div \dfrac{5}{6}$

15. $\dfrac{3}{16} \div \dfrac{1}{9}$

16. $\dfrac{5}{8} \div \dfrac{2}{9}$

17. $\dfrac{4}{5} \div \dfrac{4}{5}$

18. $\dfrac{2}{3} \div \dfrac{2}{3}$

19. $\left(-\dfrac{7}{4}\right) \div \left(-\dfrac{21}{8}\right)$

20. $\left(-\dfrac{15}{16}\right) \div \left(-\dfrac{5}{8}\right)$

21. $3 \div \dfrac{1}{12}$

22. $9 \div \dfrac{3}{4}$

23. $-\dfrac{9}{10} \div \dfrac{4}{15}$

24. $-\dfrac{3}{4} \div \dfrac{3}{2}$

25. $-\dfrac{4}{5} \div (-6)$

26. $-\dfrac{7}{8} \div (-14)$

27. $\dfrac{9}{10} \div \left(-\dfrac{3}{25}\right)$

28. $\dfrac{11}{16} \div \left(-\dfrac{9}{16}\right)$

29. $\dfrac{5}{49} \div \dfrac{10}{7}$

30. $\dfrac{8}{9} \div \dfrac{28}{3}$

31. $\dfrac{11}{12} \div \dfrac{55}{24}$

32. $\dfrac{5}{39} \div \dfrac{10}{13}$

33. $\dfrac{6}{7} \div 12$

34. $\dfrac{5}{8} \div (-10)$

35. $30 \div \dfrac{3}{5}$

36. $44 \div \dfrac{4}{7}$

37. $-\dfrac{15}{32} \div \dfrac{5}{64}$

38. $-\dfrac{28}{15} \div \dfrac{21}{10}$

39. $\dfrac{25}{7} \div \left(-\dfrac{30}{21}\right)$

40. $\dfrac{39}{25} \div \left(-\dfrac{13}{10}\right)$

APPLICATIONS *In Exercises 41–46, solve each problem.*

41. MARATHON   Each lap around a stadium track is $\frac{1}{4}$ of a mile. How many laps would a runner have to complete to get a 26-mile workout?

42. COOKING   A recipe calls for $\frac{3}{4}$ cup of flour, and the only measuring container you have holds $\frac{1}{8}$ of a cup. How many $\frac{1}{8}$ cups of flour would you need to add to follow the recipe?

43. LASER TECHNOLOGY   Using a laser, a technician slices thin pieces of aluminum off the end of a rod that is $\frac{7}{8}$ of an inch long. How many $\frac{1}{64}$-inch-wide slices can be cut from this rod?

44. FURNITURE   A production process applies several layers of a clear acrylic coat to outdoor furniture to help protect it from the weather. If each protective coat is $\frac{3}{32}$ of an inch thick, how many applications will be needed to build up $\frac{3}{8}$ of an inch of clear finish?

45. UNDERGROUND CABLE   In Illustration 3, which construction proposal will require the fewest days to install underground TV cable from the broadcasting station to the subdivision?

Proposal	Amount of cable installed per day	Comments
Route 1	$\frac{3}{5}$ of a mile	Longer than Route 2
Route 2	$\frac{2}{5}$ of a mile	Terrain very rocky

ILLUSTRATION 3

46. PRODUCTION PLANNING The materials used to make a pillow are shown in Illustration 4. Examine the inventory list to decide how many pillows can be manufactured in one production run with the materials in stock.

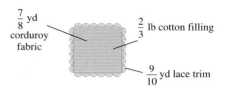

$\frac{7}{8}$ yd corduroy fabric

$\frac{2}{3}$ lb cotton filling

$\frac{9}{10}$ yd lace trim

Factory Inventory List

Materials	*Amount in stock*
Lace trim	135 yd
Corduroy fabric	154 yd
Cotton filling	98 lb

ILLUSTRATION 4

In Exercises 47–48, use a calculator.

47. FORESTRY A set of forestry maps divides the 6,284 acres of an old-growth forest into $\frac{4}{5}$-acre sections. How many sections do the maps contain?

48. HARDWARE A hardware chain purchases large amounts of nails and packages them in $\frac{9}{16}$-pound bags for sale. How many of these bags of nails can be obtained from 2,871 pounds of nails?

WRITING *Write a paragraph using your own words.*

49. Explain how to divide two fractions.

50. Explain why 0 does not have a reciprocal.

51. Write an application problem that could be solved by finding $10 \div \frac{1}{5}$.

52. Explain why dividing a fraction by 2 is the same as finding $\frac{1}{2}$ of it.

REVIEW *In Exercises 53–56, fill in the blanks to make a true statement.*

53. The symbol $<$ means _____.

54. The statement $9 \cdot 8 = 8 \cdot 9$ illustrates the _____ property of multiplication.

55. Zero is neither _____ nor negative.

56. The sum of two negative numbers is _____.

57. True or false: If equal amounts are subtracted from the numerator and the denominator of a fraction, the result will be an equivalent fraction.

58. Graph each of these numbers on a number line: -2, 0, $|-4|$, and the opposite of 1.

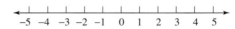

59. Evaluate: $-5 + (-3)(8)$.

60. Find each power:
a. 3^5 **b.** $(-2)^5$

3.4 *Adding and Subtracting Fractions*

In this section, you will learn about

- **Fractions with the same denominator**
- **Fractions with different denominators**
- **Finding the LCD**
- **Comparing fractions**

INTRODUCTION. In this section, we will discuss how to add and subtract fractions.

Fractions with the same denominator

In arithmetic and algebra, *we can only add or subtract objects that are similar.*

For example, we can add dollars to dollars, but we cannot add dollars to oranges. This concept is important when adding or subtracting fractions. Consider the problem $\frac{3}{5} + \frac{1}{5}$. When this problem is written in words, it is apparent that we are adding similar objects.

three-**fifths**   +   one-**fifth**

└── Similar objects ──┘

Because the denominators of $\frac{3}{5}$ and $\frac{1}{5}$ are the same, we say that they have a **common denominator.** Since the fractions have a common denominator, we can add them. Using the illustration in Figure 3-6, we can draw some conclusions about how this is done.

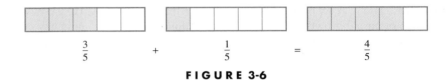

$$\frac{3}{5} \qquad + \qquad \frac{1}{5} \qquad = \qquad \frac{4}{5}$$

FIGURE 3-6

Observations

1. The *sum* of the numerators is the numerator of the answer.

$$\frac{3}{5} \quad + \quad \frac{1}{5} \quad = \quad \frac{4}{5}$$

2. The answer is a fraction whose denominator is the *same* as the two fractions that were added.

These observations suggest the following rule.

Adding or subtracting fractions with the same denominators	To add (or subtract) fractions with the same denominators, add (or subtract) their numerators and write that result over the common denominator. Simplify the result, if possible.

EXAMPLE 1   *Adding fractions with the same denominator.*

Add: $\dfrac{1}{18} + \dfrac{5}{18}$.

Solution

$$\frac{1}{18} + \frac{5}{18} = \frac{1+5}{18} \qquad \text{Add the numerators. Write the sum over the common denominator 18.}$$

$$= \frac{6}{18} \qquad \text{Do the addition: } 1 + 5 = 6. \text{ Note that this fraction can be simplified.}$$

$$= \frac{\overset{1}{\cancel{6}} \cdot 1}{\underset{1}{\cancel{6}} \cdot 3} \qquad \text{Factor 6 as } 6 \cdot 1 \text{ and 18 as } 6 \cdot 3. \text{ Divide out the common factor of 6.}$$

$$= \frac{1}{3} \qquad \text{Multiply in the numerator and in the denominator.}$$

Self Check

Add: $\dfrac{9}{12} + \dfrac{1}{12}$.

Answer: $\dfrac{5}{6}$

■

EXAMPLE 2 *Subtracting negative fractions.* Subtract: $-\dfrac{7}{3} - \left(-\dfrac{2}{3}\right)$.

Solution

$$-\frac{7}{3} - \left(-\frac{2}{3}\right) = -\frac{7}{3} + \frac{2}{3} \qquad \text{Add the opposite of } -\frac{2}{3}.$$

$$= \frac{-7}{3} + \frac{2}{3} \qquad -\frac{7}{3} = \frac{-7}{3}.$$

$$= \frac{-7 + 2}{3} \qquad \text{Add the numerators. Write the sum over the common denominator 3.}$$

$$= \frac{-5}{3} \qquad \text{Do the addition: } -7 + 2 = -5.$$

$$= -\frac{5}{3} \qquad \frac{-5}{3} = -\frac{5}{3}.$$

Self Check

Subtract: $-\dfrac{9}{11} - \left(-\dfrac{3}{11}\right)$.

Answer: $-\dfrac{6}{11}$

EXAMPLE 3 *Combining three fractions.* Do the operations:
$$\frac{3}{8} + \frac{2}{8} - \left(-\frac{1}{8}\right).$$

Solution

$$\frac{3}{8} + \frac{2}{8} - \left(-\frac{1}{8}\right) = \frac{5}{8} - \left(-\frac{1}{8}\right) \qquad \text{Work left to right. Do the addition: } \frac{3}{8} + \frac{2}{8} = \frac{5}{8}.$$

$$= \frac{5}{8} + \frac{1}{8} \qquad \text{Change the subtraction to addition of the opposite of } -\frac{1}{8}. \; -\left(-\frac{1}{8}\right) = \frac{1}{8}.$$

$$= \frac{6}{8} \qquad \text{Add the numerators and keep the common denominator 8.}$$

$$= \frac{3}{4} \qquad \text{Simplify the fraction: } \frac{6}{8} = \frac{\overset{1}{\cancel{2}} \cdot 3}{\underset{1}{\cancel{2}} \cdot 4} = \frac{3}{4}.$$

Self Check

Do the operations: $\frac{2}{9} + \frac{2}{9} - \left(-\frac{2}{9}\right)$.

Answer: $\frac{2}{3}$

Fractions with different denominators

Let's consider the problem $\frac{3}{5} + \frac{1}{3}$. Since the denominators are different, we cannot add these fractions in their present form.

three-fifths   +   one-third
└ Not similar objects ┘

To add these fractions, we need to find a common denominator. The smallest common denominator (called the **least** or **lowest common denominator**) usually is the easiest common denominator to work with.

Least common denominator	The **least common denominator (LCD)** for a set of fractions is the smallest number each denominator will divide exactly.

In the problem $\frac{3}{5} + \frac{1}{3}$, the denominators are 5 and 3. The numbers 5 and 3 divide many numbers exactly; 30, 45, and 60, to name a few. But the smallest number that 5 and 3 divide exactly is 15. This is the LCD. We will now build each fraction into a fraction with a denominator of 15 by applying the fundamental property of fractions.

$$\frac{3}{5} + \frac{1}{3} = \frac{3 \cdot 3}{5 \cdot 3} + \frac{1 \cdot 5}{3 \cdot 5}$$ Apply the fundamental property of fractions to express each fraction in terms of 15ths.

We need to multiply this denominator by 3 to obtain 15.   We need to multiply this denominator by 5 to obtain 15.

$$= \frac{9}{15} + \frac{5}{15}$$ Do the multiplications in the numerators and denominators.

$$= \frac{9 + 5}{15}$$ Add the numerators and write the sum over the common denominator 15.

$$= \frac{14}{15}$$ Do the addition: $9 + 5 = 14$.

Figure 3-7 shows $\frac{3}{5}$ and $\frac{1}{3}$ expressed as equivalent fractions with a denominator of 15. Once the denominators are the same, the fractions can be added easily.

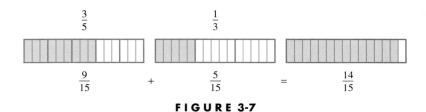

$$\frac{9}{15} \qquad + \qquad \frac{5}{15} \qquad = \qquad \frac{14}{15}$$

FIGURE 3-7

We can now summarize the procedure used to add or subtract fractions with different denominators.

Adding or subtracting fractions with different denominators	To add or subtract fractions with different denominators, 1. Find the LCD. 2. Express each fraction as a fraction with a denominator that is the LCD. 3. Add or subtract the resulting fractions. Simplify the result if possible.

EXAMPLE 4   *Adding fractions with different denominators.*
Add: $\frac{1}{8} + \frac{2}{3}$.

Self Check
Add: $\frac{1}{2} + \frac{2}{5}$.

Solution
Since the smallest number the denominators 8 and 3 divide evenly is 24, the LCD is 24.

$$\frac{1}{8} + \frac{2}{3} = \frac{1 \cdot 3}{8 \cdot 3} + \frac{2 \cdot 8}{3 \cdot 8}$$ Apply the fundamental property of fractions to express each fraction in terms of 24ths.

$$= \frac{3}{24} + \frac{16}{24}$$ Do the multiplications in the numerators and denominators.

$$= \frac{3 + 16}{24}$$ Add the numerators and write the sum over the common denominator 24.

$$= \frac{19}{24}$$ Do the addition in the numerator: $3 + 16 = 19$.

Answer: $\frac{9}{10}$

<blockquote>
EXAMPLE 5   *Adding an integer and a fraction.*   Add: $-5 + \dfrac{1}{4}$.
</blockquote>

<blockquote>
Self Check

Add: $-6 + \dfrac{2}{15}$.
</blockquote>

Solution

$$-5 + \frac{1}{4} = \frac{-5}{1} + \frac{1}{4} \qquad -5 = \frac{-5}{1}. \text{ The LCD for the two fractions is 4.}$$

$$= \frac{-5 \cdot 4}{1 \cdot 4} + \frac{1}{4} \qquad \text{Apply the fundamental property of fractions to express } \tfrac{5}{1} \text{ in terms of 4ths.}$$

$$= \frac{-20}{4} + \frac{1}{4} \qquad \text{Multiply in the numerator and denominator.}$$

$$= \frac{-20 + 1}{4} \qquad \text{Write the sum of the numerators over the common denominator 4.}$$

$$= \frac{-19}{4} \qquad \text{Do the addition: } -20 + 1 = -19.$$

$$= -\frac{19}{4} \qquad \tfrac{-19}{4} = -\tfrac{19}{4}.$$

Answer: $-\dfrac{88}{15}$

Finding the LCD

When adding or subtracting fractions with different denominators, the least common denominator is not always obvious. We will now develop a procedure that can be used to find the LCD.

Let's find the least common denominator of $\frac{3}{8}$ and $\frac{1}{10}$. We have learned that both 8 and 10 must divide the LCD exactly. Therefore, the LCD must include the prime factorization of 8 ($2 \cdot 2 \cdot 2$) and the prime factorization of 10 ($2 \cdot 5$). In Figure 3-8, we see that $2 \cdot 2 \cdot 2 \cdot 5$ is the smallest number that meets both of these requirements. Notice that the common factor of 2 is "shared" by both prime factorizations.

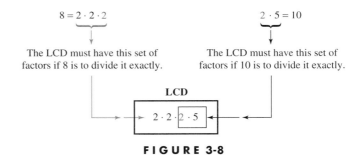

FIGURE 3-8

Finding the LCD	To find the LCD for a set of fractions,

1. Prime factor each of the denominators.

2. Select one of the prime factorizations and compare the others to it, one at a time. It must contain each of the other factorizations. If it does not, multiply it by any of the prime factors it lacks.

3. The LCD is the product of the prime factors that result from this comparison.

You will find that the product that gives the LCD contains each of the prime factors the greatest number of times they appear in any single prime factorization.

EXAMPLE 6 *Subtracting fractions with different denominators.*
Subtract: $\dfrac{20}{21} - \dfrac{5}{6}$.

Solution

To find the LCD, we factor each denominator: $21 = 7 \cdot 3$ and $6 = 3 \cdot 2$. The factorization $7 \cdot 3$ does not contain the factorization $3 \cdot 2$; it lacks the factor of 2. Therefore, we multiply $7 \cdot 3$ by 2 to obtain the LCD.

$$21 = 7 \cdot 3 \longrightarrow \boxed{7 \cdot \boxed{3} \cdot 2} \longleftarrow 3 \cdot 2 = 6$$

$$\text{LCD} = 42$$

$$\frac{20}{21} - \frac{5}{6} = \frac{20 \cdot 2}{21 \cdot 2} - \frac{5 \cdot 7}{6 \cdot 7} \qquad \text{Express each fraction in terms of 42nds.}$$

$$= \frac{40}{42} - \frac{35}{42} \qquad \text{Do the multiplications in the numerators and denominators.}$$

$$= \frac{40 - 35}{42} \qquad \text{Write the difference of the numerators over the common denominator 42.}$$

$$= \frac{5}{42} \qquad \text{Do the subtraction: } 40 - 35 = 5.$$

EXAMPLE 7 *Television viewing habits.* Students on a college campus were asked to estimate to the nearest hour how much television they watched each day. The results are given in the pie chart in Figure 3-9. For example, the chart tells us that $\frac{1}{4}$ of those responding watched 1 hour per day. Find the fraction of the student body watching from 0 to 2 hours daily.

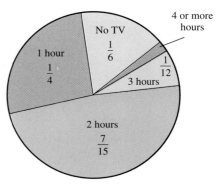

FIGURE 3-9

Solution To answer this question, we need to add three fractions: $\frac{1}{6}$, $\frac{1}{4}$, and $\frac{7}{15}$. To find the LCD of these fractions, we prime-factor each of the denominators.

$$4 = 2 \cdot 2 \longrightarrow \boxed{2 \cdot \boxed{2} \cdot \boxed{3} \cdot 5} \longleftarrow 3 \cdot 5 = 15$$
$$6 = 2 \cdot 3$$
$$\text{LCD} = 2 \cdot 2 \cdot 3 \cdot 5 = 60$$

$$\frac{1}{6} + \frac{1}{4} + \frac{7}{15} = \frac{1 \cdot 10}{6 \cdot 10} + \frac{1 \cdot 15}{4 \cdot 15} + \frac{7 \cdot 4}{15 \cdot 4} \qquad \text{Express each fraction in terms of 60ths.}$$

$$= \frac{10}{60} + \frac{15}{60} + \frac{28}{60} \qquad \text{Do the multiplication in the numerators and denominators.}$$

$$= \frac{10 + 15 + 28}{60} \qquad \text{Add the numerators. Write the sum over the common denominator 60.}$$

$$= \frac{53}{60} \qquad \text{Do the addition: } 10 + 15 + 28 = 53.$$

The fraction of the student body watching 0 to 2 hours of television daily is $\frac{53}{60}$. ∎

Comparing fractions

If fractions have the same denominator, the fraction with the larger numerator is the larger fraction. If their denominators are different, we need to write the fractions with a common denominator before we can make a comparison.

Comparing fractions	To compare unlike fractions, write the fractions as equivalent fractions with the same denominator—preferably the LCD. Then compare their numerators. The fraction with the larger numerator is the larger fraction.

EXAMPLE 8 *Comparing fractions.* Which fraction is larger: $\dfrac{5}{6}$ or $\dfrac{7}{8}$?

Solution

To compare these fractions, we express each with an LCD of 24.

$$\frac{5}{6} = \frac{5 \cdot 4}{6 \cdot 4} \qquad \frac{7}{8} = \frac{7 \cdot 3}{8 \cdot 3} \qquad \text{Express each fraction in terms of 24ths.}$$

$$= \frac{20}{24} \qquad\qquad = \frac{21}{24} \qquad \text{Do the multiplications in the numerators and denominators.}$$

Next, we compare the numerators. Since $21 > 20$, we conclude that $\frac{21}{24}$ is greater than $\frac{20}{24}$. Thus, $\frac{7}{8} > \frac{5}{6}$.

Self Check

Which fraction is larger: $\dfrac{5}{9}$ or $\dfrac{3}{5}$?

Answer: $\dfrac{3}{5}$

STUDY SET Section 3.4

VOCABULARY *In Exercises 1–4, fill in the blanks to make a true statement.*

1. The _____ common denominator for a set of fractions is the smallest number each denominator will divide exactly.

2. _____ fractions are fractions that represent the same amount.

3. To express a fraction in _____ terms, we multiply the numerator and denominator by the _____ number.

4. _____ up a fraction is the process of multiplying the numerator and the denominator of the fraction by the same number.

CONCEPTS

5. Fill in the blanks. To add two fractions with a common denominator, we add the _____ and then write that result over the _____ denominator.

6. a. Divide the figure on the left into fourths and shade one part. Divide the figure on the right into thirds and shade one part. Which shaded part is larger?

b. Express the shaded part of each figure in part a as a fraction. Show that one of those fractions is larger than the other by expressing both in terms of a common denominator and comparing.

7. Add the indicated fractions.

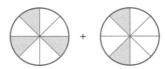

8. Subtract the indicated fractions.

9. Why must we do some preliminary work before doing the following addition?

$$\frac{2}{9} + \frac{2}{5}$$

10. Why must we do some preliminary work before doing the following subtraction?

$$\frac{5}{6} - \frac{5}{18}$$

11. By what number are the numerator and the denominator of the following fraction being multiplied?

$$\frac{5 \cdot 4}{6 \cdot 4}$$

12. By what number are the numerator and the denominator of the following fraction being multiplied?

$$\frac{8 \cdot 5}{3 \cdot 5}$$

13. The denominators of two fractions involved in a subtraction problem have the prime-factored forms $2 \cdot 2 \cdot 5$ and $2 \cdot 3 \cdot 5$. What is the LCD for the fractions?

14. The denominators of three fractions involved in a subtraction problem have the prime-factored forms $2 \cdot 2 \cdot 5$, $2 \cdot 3 \cdot 5$, $2 \cdot 3 \cdot 3 \cdot 5$. What is the LCD for the fractions?

15. Write 5 as a fraction.

16. Write $-\frac{12}{3}$ as an integer.

NOTATION *In Exercises 17–18, complete each solution.*

17. Add: $\frac{2}{5} + \frac{1}{3}$.

$$\frac{2}{5} + \frac{1}{3} = \frac{2 \cdot}{5 \cdot} + \frac{1 \cdot 5}{3 \cdot 5}$$ Express each fraction in terms of 15ths.

$$= \frac{}{15} + \frac{}{15}$$ Do the multiplications in the numerators and denominators.

$$= \frac{ + }{15}$$ Write the sum of the numerators over the common denominator 15.

$$= \frac{11}{15}$$ Do the addition.

18. Subtract: $\frac{7}{8} - \frac{2}{3}$.

$$\frac{7}{8} - \frac{2}{3} = \frac{7 \cdot 3}{\cdot 3} - \frac{2 \cdot 8}{\cdot 8}$$ Express each fraction in terms of 24ths.

$$= \frac{21}{} - \frac{16}{}$$ Do the multiplications in the numerators and denominators.

$$= \frac{21 - 16}{}$$ Write the difference of the numerators over the common denominator 24.

$$= \frac{5}{24}$$ Do the subtraction.

PRACTICE *In Exercises 19–26, write each fraction (or whole number) as an equivalent fraction with the indicated denominator.*

19. $\frac{3}{4}$, denominator 12

20. $\frac{1}{16}$, denominator 32

21. 5, denominator 2

22. 2, denominator 6

23. $\frac{7}{4}$, denominator 28

24. $\frac{2}{3}$, denominator 30

25. $\frac{5}{9}$, denominator 45

26. $\frac{8}{11}$, denominator 55

In Exercises 27–66, do each operation. Simplify when necessary.

27. $\frac{3}{7} + \frac{1}{7}$

28. $\frac{16}{25} - \frac{9}{25}$

29. $\frac{37}{103} - \frac{17}{103}$

30. $\frac{54}{53} - \frac{52}{53}$

31. $\frac{1}{4} + \frac{3}{8}$

32. $\frac{2}{3} + \frac{1}{6}$

33. $\frac{4}{5} - \frac{1}{2}$

34. $\frac{3}{4} - \frac{2}{3}$

35. $\frac{4}{5} + \frac{2}{3}$

36. $\frac{5}{16} + \frac{7}{8}$

37. $-\frac{5}{8} - \frac{1}{3}$

38. $-\frac{7}{20} - \frac{1}{5}$

39. $\dfrac{16}{25} - \left(-\dfrac{3}{10}\right)$ **40.** $\dfrac{3}{8} - \left(-\dfrac{1}{6}\right)$ **41.** $-\dfrac{7}{16} - \dfrac{1}{4}$ **42.** $-\dfrac{17}{20} + \dfrac{4}{5}$

43. $\dfrac{4}{7} - \dfrac{1}{3}$ **44.** $\dfrac{4}{5} + \dfrac{2}{7}$ **45.** $-\dfrac{5}{9} + \dfrac{1}{2}$ **46.** $-\dfrac{3}{5} + \dfrac{5}{3}$

47. $\dfrac{7}{8} - \dfrac{5}{7}$ **48.** $\dfrac{5}{6} + \dfrac{6}{7}$ **49.** $\dfrac{4}{5} - \dfrac{2}{9}$ **50.** $\dfrac{3}{16} + \dfrac{5}{8}$

51. $-3 + \dfrac{2}{5}$ **52.** $-6 + \dfrac{5}{8}$ **53.** $-\dfrac{3}{4} - 5$ **54.** $-2 - \dfrac{7}{8}$

55. $\dfrac{1}{3} + \dfrac{1}{4} + \dfrac{1}{5}$ **56.** $\dfrac{1}{10} + \dfrac{1}{8} + \dfrac{1}{5}$ **57.** $-\dfrac{2}{3} + \dfrac{5}{4} + \dfrac{1}{6}$ **58.** $-\dfrac{3}{4} + \dfrac{3}{8} + \dfrac{7}{6}$

59. $\dfrac{5}{24} + \dfrac{3}{16}$ **60.** $\dfrac{17}{20} - \dfrac{4}{15}$ **61.** $-\dfrac{11}{15} - \dfrac{2}{9}$ **62.** $-\dfrac{19}{18} - \dfrac{5}{12}$

63. $\dfrac{7}{25} + \dfrac{1}{15}$ **64.** $\dfrac{11}{20} - \dfrac{1}{8}$ **65.** $\dfrac{4}{27} + \dfrac{1}{6}$ **66.** $\dfrac{8}{9} - \dfrac{7}{12}$

67. Find the difference of $\dfrac{11}{60}$ and $\dfrac{2}{45}$.   **68.** Find the sum of $\dfrac{9}{48}$ and $\dfrac{7}{40}$.

APPLICATIONS *In Exercises 69–80, use addition or subtraction of fractions to solve each problem.*

69. BOTANY   To assess the effects of smog on tree development, botanists cut down a pine tree and measured the width of the growth rings for the last two years. (See Illustration 1.)

ILLUSTRATION 1

  a. What was the growth over this two-year period?

  b. What is the difference in the two ring widths?

70. MAGAZINE LAYOUT   The page design for a magazine cover includes a blank strip at the top, called a header, and a blank strip at the bottom of the page, called a footer. In Illustration 2, how much page length is lost because of the header and footer?

71. FAMILY DINNER   A family bought two large pizzas for dinner. Several pieces of each pizza were not eaten, as shown in Illustration 3. How much pizza was left? Could the family have been fed with just one pizza?

72. GASOLINE BARRELS   The contents of two identical-sized barrels are shown in Illustration 4. If they are dumped into an empty third barrel that is the same size, how much of the third barrel will they fill?

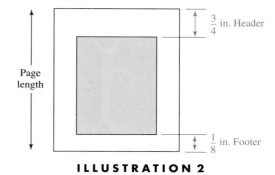

ILLUSTRATION 2

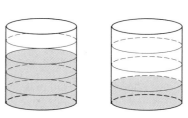

ILLUSTRATION 3

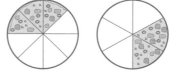

ILLUSTRATION 4

73. WEIGHTS AND MEASURES A consumer protection agency verifies the accuracy of butcher shop scales by placing a known three-quarter-pound weight on the scale and then comparing that to the scale's readout. According to Illustration 5, by how much is this scale off? Does it result in undercharging or overchanging customers on their meat purchases?

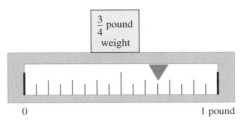

ILLUSTRATION 5

74. WRENCHES A mechanic likes to hang his wrenches above his tool bench in order of narrowest to widest. What is the proper order of the wrenches in Illustration 6?

ILLUSTRATION 6

75. HIKING Illustration 7 shows the length of each part of a three-part hike. Rank the lengths from longest to shortest.

ILLUSTRATION 7

76. FIGURE DRAWING As an aid in drawing the human body, artists divide the body into three parts. Each part is then expressed as a fraction of the total body height. (See Illustration 8.) For example, the torso is $\frac{4}{15}$ of the body height. What fraction of body height is the head?

77. STUDY HABITS College students taking a full load were asked to give the average number of hours they

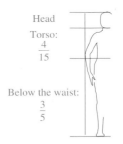

ILLUSTRATION 8

studied each day. The results are shown in the pie chart in Illustration 9. What fraction of the students study 2 hours or more daily?

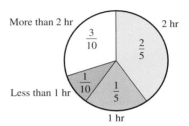

ILLUSTRATION 9

78. MUSICAL NOTES The notes used in music have fractional values. Their names and the symbols used to represent them are shown in Illustration 10(a). In common time, the values of the notes in each measure must add to 1. Is the measure in Illustration 10(b) complete?

(a)

(b)

ILLUSTRATION 10

79. GARAGE DOOR OPENER What is the difference in strength between a $\frac{1}{3}$-hp and a $\frac{1}{2}$-hp garage door opener?

80. DELIVERY TRUCK A truck can safely carry a one-ton load. Could it be used to deliver one-half ton of sand, one-third ton of gravel, and one-fifth ton of cement in one trip to a job site?

81. How are the procedures for expressing a fraction in higher terms and simplifying a fraction to lowest terms similar, and how are they different?

82. Given two fractions, how do we find their lowest common denominator?

83. How do we compare the relative sizes of two fractions with different denominators?

84. What is the difference between a common denominator and a lowest common denominator?

REVIEW

85. Evaluate: $|6 - 10|$.

86. Round 674 to the nearest ten.

87. Evaluate: $3^2 \cdot 4^3$.

88. Find the prime factors of 100.

89. Evaluate: $\dfrac{(2 + 5)^2 + 17}{2(9 - 6)}$.

90. Evaluate: $\dfrac{7 - 7}{7}$.

91. What is the formula for finding the perimeter of a rectangle?

92. What is the formula for finding the area of a rectangle?

3.5   *Multiplying and Dividing Mixed Numbers*

In this section, you will learn about

- **Mixed numbers**
- **Writing mixed numbers as improper fractions**
- **Writing improper fractions as mixed numbers**
- **Graphing fractions and mixed numbers**
- **Multiplying and dividing mixed numbers**

INTRODUCTION.   In the next two sections, we will show how to add, subtract, multiply, and divide *mixed numbers*. These numbers are widely used in daily life. Here are a few examples.

- The job took $2\frac{3}{4}$ hours to complete.
- A recipe calls for $4\frac{1}{3}$ cups flour.
- The next exit off the freeway is in $1\frac{1}{2}$ miles.

In this section, we will learn how to multiply and divide mixed numbers.

Mixed numbers

A **mixed number** is the *sum* of a whole number and a proper fraction. For example, $2\frac{3}{4}$ is a mixed number.

$$2\frac{3}{4} \quad = \quad 2 \quad + \quad \frac{3}{4}$$

Mixed number   Whole number   Proper fraction

WARNING!   Note that $2\frac{3}{4}$ means $2 + \frac{3}{4}$, even though the $+$ sign is not written. Do not confuse $2\frac{3}{4}$ with $2 \cdot \frac{3}{4}$ or $2\left(\frac{3}{4}\right)$, which indicate the multiplication of 2 and $\frac{3}{4}$.

In this section, we will work with negative as well as positive mixed numbers. For example, the negative mixed number $-4\frac{3}{4}$ could be used to represent $4\frac{3}{4}$ feet below sea level. We think of $-4\frac{3}{4}$ as $-4 - \frac{3}{4}$.

Writing mixed numbers as improper fractions

To see that mixed numbers are related to improper fractions, consider $2\frac{3}{4}$. To write $2\frac{3}{4}$ as an improper fraction, we need to find out how many *fourths* it represents. One way is to use the fundamental property of fractions.

$$2\frac{3}{4} = 2 + \frac{3}{4}$$ Write the mixed number $2\frac{3}{4}$ as a sum.

$$= \frac{2}{1} + \frac{3}{4}$$ Write 2 as a fraction: $2 = \frac{2}{1}$.

$$= \frac{2 \cdot 4}{1 \cdot 4} + \frac{3}{4}$$ Use the fundamental property of fractions to express $\frac{2}{1}$ as a fraction with denominator 4.

$$= \frac{8}{4} + \frac{3}{4}$$ Do the multiplications in the numerator and denominator.

$$= \frac{11}{4}$$ Add the numerators. Write the sum over the common denominator.

Thus, $2\frac{3}{4} = \frac{11}{4}$.

We can obtain the same result with far less work. To change $2\frac{3}{4}$ to an improper fraction, we simply multiply 2 by 4 and add 3 to get the numerator, and keep the denominator of 4.

$$2\frac{3}{4} = \frac{2(4) + 3}{4} = \frac{11}{4}$$

This example illustrates the following general rule.

Writing a mixed number as an improper fraction	To write a mixed number as an improper fraction, multiply the whole-number part by the denominator of the fraction and add the result to the numerator. Write this sum over the denominator.

EXAMPLE 1   *Writing a mixed number as an improper fraction.*
Write the mixed number $5\frac{1}{6}$ as an improper fraction.

Solution

$$5\frac{1}{6} = \frac{5(6) + 1}{6}$$ Multiply 5 by the denominator 6. Add the numerator 1. Write this sum over the denominator 6.

$$= \frac{30 + 1}{6}$$ Do the multiplication: $5(6) = 30$.

$$= \frac{31}{6}$$ Do the addition: $30 + 1 = 31$.

Self Check
Write the mixed number $3\frac{3}{8}$ as an improper fraction.

Answer: $\dfrac{27}{8}$

To write a negative mixed number in fractional form, ignore the $-$ sign and use the method shown in Example 1 on the positive mixed number. Once that procedure is completed, write a $-$ sign in front of the result. For example, $-3\frac{1}{4} = -\frac{13}{4}$.

Writing improper fractions as mixed numbers

To write an improper fraction as a mixed number, we must find two things: the *whole-number part* and the *fractional part* of the mixed number. To develop a procedure to do this, let's consider the improper fraction $\frac{7}{3}$. To find the number of groups of three in 7, we can divide 7 by 3. This will find the whole-number part of the mixed number. The remainder is the numerator of the fractional part of the mixed number.

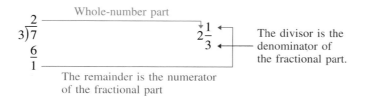

This example suggests the following general rule.

Writing an improper fraction as a mixed number	To write an improper fraction as a mixed number, divide the numerator by the denominator to obtain the whole-number part. The remainder over the divisor is the fractional part.

EXAMPLE 2   *Writing an improper fraction as a mixed number.* Write $\frac{29}{6}$ as a mixed number.

Solution

$$\begin{array}{r} 4 \\ 6\overline{)29} \\ \underline{24} \\ 5 \end{array}$$ Divide the numerator by the denominator.

The remainder is 5.

Thus, $\dfrac{29}{6} = 4\dfrac{5}{6}$.

Self Check

Write $\dfrac{43}{5}$ as a mixed number.

Answer: $8\dfrac{3}{5}$

Graphing fractions and mixed numbers

Earlier, we graphed whole numbers and integers on a number line. Fractions and mixed numbers can also be graphed on a number line.

EXAMPLE 3   *Graphing fractions and mixed numbers.*   Graph $-2\dfrac{3}{4}, -1\dfrac{1}{2}, -\dfrac{1}{8}, 2\dfrac{3}{5}$ on a number line.

Solution

Self Check

Graph $-1\dfrac{7}{8}, -\dfrac{2}{3}, 2\dfrac{1}{4}$ on a number line.
Answer:

Multiplying and dividing mixed numbers

Multiplying and dividing mixed numbers	To multiply or divide mixed numbers, first change the mixed numbers to improper fractions. Then do the multiplication or division of the fractions.

EXAMPLE 4 *Multiplying mixed numbers.* Multiply: $5\frac{1}{5} \cdot 1\frac{2}{13}$.

Self Check

Multiply: $9\frac{3}{5} \cdot 3\frac{3}{4}$.

Solution

$$5\frac{1}{5} \cdot 1\frac{2}{13} = \frac{26}{5} \cdot \frac{15}{13} \qquad \text{Write each mixed number as an improper fraction.}$$

$$= \frac{26 \cdot 15}{5 \cdot 13} \qquad \text{Multiply the numerators and multiply the denominators.}$$

$$= \frac{13 \cdot 2 \cdot 5 \cdot 3}{5 \cdot 13} \qquad \text{Prime factor 26 as } 13 \cdot 2 \text{ and 15 as } 5 \cdot 3.$$

$$= \frac{\overset{1}{13} \cdot 2 \cdot \overset{1}{5} \cdot 3}{\underset{1}{5} \cdot \underset{1}{13}} \qquad \text{Divide out the common factors of 13 and 5.}$$

$$= \frac{6}{1} \qquad \text{Multiply in the numerator and denominator.}$$

$$= 6 \qquad \tfrac{6}{1} = 6.$$

Answer: 36

EXAMPLE 5 *Dividing mixed numbers.* Divide: $-3\frac{3}{8} \div 2\frac{1}{4}$.

Self Check

Divide: $3\frac{4}{15} \div \left(-2\frac{1}{10}\right)$.

Solution

$$-3\frac{3}{8} \div 2\frac{1}{4} = -\frac{27}{8} \div \frac{9}{4} \qquad \text{Write each mixed number as an improper fraction.}$$

$$= -\frac{27}{8} \cdot \frac{4}{9} \qquad \text{Multiply by the reciprocal of } \tfrac{9}{4}.$$

$$= -\frac{27 \cdot 4}{8 \cdot 9} \qquad \text{The product of two fractions with unlike signs is negative. Multiply the numerators and multiply the denominators.}$$

$$= -\frac{\overset{1}{9} \cdot 3 \cdot \overset{1}{4}}{\underset{1}{4} \cdot 2 \cdot \underset{1}{9}} \qquad \text{Factor 27 as } 9 \cdot 3 \text{ and 8 as } 4 \cdot 2. \text{ Divide out the common factors of 9 and 4.}$$

$$= -\frac{3}{2} \qquad \text{Multiply in the numerator and denominator.}$$

$$= -1\frac{1}{2} \qquad \text{Write } -\tfrac{3}{2} \text{ as a mixed number.}$$

Answer: $-1\frac{5}{9}$

EXAMPLE 6   *Government grant.*   If \$12$\frac{1}{2}$ million is to be divided equally among five cities to fund recreation programs, how much will each city receive?

Solution   To find the amount received by each city, we divide the grant money by 5.

$$12\frac{1}{2} \div 5 = \frac{25}{2} \div \frac{5}{1}$$ Write $12\frac{1}{2}$ as an improper fraction, and write 5 as a fraction.

$$= \frac{25}{2} \cdot \frac{1}{5}$$ Multiply by the reciprocal of $\frac{5}{1}$.

$$= \frac{25 \cdot 1}{2 \cdot 5}$$ Multiply the numerators and multiply the denominators.

$$= \frac{\overset{1}{\cancel{5}} \cdot 5 \cdot 1}{2 \cdot \underset{1}{\cancel{5}}}$$ Factor 25 as $5 \cdot 5$. Divide out the common factor of 5.

$$= \frac{5}{2}$$ Multiply in the numerator and denominator.

$$= 2\frac{1}{2}$$ Write $\frac{5}{2}$ as a mixed number.

Each city will receive \$2$\frac{1}{2}$ million.

STUDY SET   Section 3.5

VOCABULARY   *In Exercises 1–4, fill in the blanks to make a true statement.*

1. A _____ number is the sum of a whole number and a proper fraction.

2. An _____ fraction is a fraction with a numerator that is greater than or equal to its denominator.

3. To _____ a number means to locate its position on a number line and highlight it using a heavy dot.

4. Multiplying or dividing the _____ and _____ of a fraction by the same nonzero number does not change the value of the fraction.

CONCEPTS

5. What signed number could be used to describe each situation?
 a. A temperature of five and one-half degrees below zero.
 b. A stock dropped $1\frac{7}{8}$ points.

6. What signed number could be used to describe each situation?
 a. A rain total two and three-tenths of an inch lower than the average.
 b. Three and one-half minutes before liftoff.

7.

ILLUSTRATION 1

 a. In Illustration 1, the divisions on the face of the meter are in terms of fractions. What value is the arrow registering?
 b. If the arrow moves two marks to the left, what value will it register?

8. a. In Illustration 2, the divisions on the face of the meter are in terms of fractions. What value is the arrow registering?

 b. If the arrow moves up one mark, what value will it register?

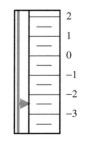

ILLUSTRATION 2

9. What fractions have been graphed on the number line in Illustration 3?

ILLUSTRATION 3

11. Graph $\frac{1}{8}$, $1\frac{1}{8}$, and $2\frac{1}{8}$.

13. Draw $\frac{17}{8}$ pizzas.

10. What mixed numbers have been graphed on the number line in Illustration 4?

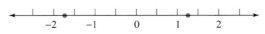

ILLUSTRATION 4

12. Graph $-1\frac{1}{7}$, $-2\frac{1}{8}$, and $-\frac{1}{9}$.

14. What mixed number and what improper fraction are depicted in Illustration 5?

ILLUSTRATION 5

NOTATION   *In Exercises 15–16, complete each solution.*

15. Multiply: $-5\frac{1}{4} \cdot 1\frac{1}{7}$.

$-5\frac{1}{4} \cdot 1\frac{1}{7} = -\frac{21}{4} \cdot \frac{}{7}$ Write each mixed number as an improper fraction.

$= -\frac{21 \cdot }{4 \cdot 7}$ The product of two fractions with unlike signs is negative. Multiply the numerators and multiply the denominators.

$= -\frac{\overset{1}{\cancel{7}} \cdot 3 \cdot \overset{1}{\cancel{4}} \cdot 2}{\underset{1}{\cancel{4}} \cdot \underset{1}{\cancel{7}}}$ Factor 21 and 8. Divide out the common factors of 7 and 4.

$= -\frac{}{1}$ Multiply in the numerator and denominator.

$= -6$ $\frac{6}{1} = 6$.

16. Divide: $-5\frac{5}{6} \div 2\frac{1}{12}$.

$-5\frac{5}{6} \div 2\frac{1}{12} = -\frac{}{6} \div \frac{25}{12}$ Write each mixed number as an improper fraction.

$= -\frac{35}{6} \cdot \frac{12}{}$ Multiply by the reciprocal.

$= -\frac{35 \cdot 12}{6 \cdot }$ The product of two fractions with unlike signs is negative. Multiply the numerators and multiply the denominators.

$= -\frac{\overset{1}{\cancel{5}} \cdot \cdot \overset{2}{\cancel{6}} \cdot 2}{\underset{1}{\cancel{6}} \cdot \underset{1}{\cancel{5}} \cdot }$ Factor 35, 12, and 25. Divide out the common factors of 5 and 6.

$= -\frac{}{5}$ Multiply in the numerator and denominator.

$= -2\frac{4}{5}$ Write the improper fraction as a mixed number.

PRACTICE   *In Exercises 17–24, write each improper fraction as a mixed number. Simplify the result, if possible.*

17. $\frac{15}{4}$

18. $\frac{41}{6}$

19. $\frac{29}{5}$

20. $\frac{29}{3}$

21. $-\frac{20}{6}$

22. $-\frac{28}{8}$

23. $\frac{127}{12}$

24. $\frac{197}{16}$

In Exercises 25–32, write each mixed number as an improper fraction.

25. $6\frac{1}{2}$

26. $8\frac{2}{3}$

27. $20\frac{4}{5}$

28. $15\frac{3}{8}$

29. $-6\frac{2}{9}$

30. $-7\frac{1}{12}$

31. $200\frac{2}{3}$

32. $90\frac{5}{6}$

In Exercises 33–36, graph the set of numbers on the number line.

33. $-2\frac{8}{9}, 1\frac{2}{3}$

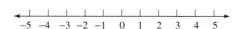

34. $-\frac{3}{4}, -1\frac{1}{4}$

35. $3\frac{1}{7}, -\frac{98}{99}$

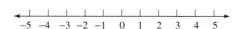

36. $-2\frac{1}{5}, \frac{4}{5}$

In Exercises 37–48, multiply.

37. $1\frac{2}{3} \cdot 2\frac{1}{7}$

38. $2\frac{3}{5} \cdot 1\frac{2}{3}$

39. $-7\frac{1}{2}\left(-1\frac{2}{5}\right)$

40. $-4\frac{1}{8}\left(-1\frac{7}{9}\right)$

41. $3\frac{1}{16} \cdot 4\frac{4}{7}$

42. $5\frac{3}{5} \cdot 1\frac{11}{14}$

43. $-6 \cdot 2\frac{7}{24}$

44. $-7 \cdot 1\frac{3}{28}$

45. $2\frac{1}{2}\left(-3\frac{1}{3}\right)$

46. $\left(-3\frac{1}{4}\right)\left(1\frac{1}{5}\right)$

47. $2\frac{5}{8} \cdot \frac{5}{27}$

48. $3\frac{1}{9} \cdot \frac{3}{32}$

49. Find the product of $1\frac{2}{3}$, 6, and $-\frac{1}{8}$.

50. Find the product of $-\frac{5}{6}$, -8, and $-2\frac{1}{10}$.

In Exercises 51–54, evaluate each power.

51. $\left(1\frac{2}{3}\right)^2$

52. $\left(3\frac{1}{2}\right)^2$

53. $\left(-1\frac{1}{3}\right)^3$

54. $\left(-1\frac{1}{5}\right)^3$

In Exercises 55–66, divide.

55. $3\frac{1}{3} \div 1\frac{5}{6}$

56. $3\frac{3}{4} \div 5\frac{1}{3}$

57. $-6\frac{3}{5} \div 7\frac{1}{3}$

58. $-4\frac{1}{4} \div 4\frac{1}{2}$

59. $-20\frac{1}{4} \div \left(-1\frac{11}{16}\right)$

60. $-2\frac{7}{10} \div \left(-1\frac{1}{14}\right)$

61. $6\frac{1}{4} \div 20$

62. $4\frac{2}{5} \div 11$

63. $1\frac{2}{3} \div \left(-2\frac{1}{2}\right)$

64. $2\frac{1}{2} \div \left(-1\frac{5}{8}\right)$

65. $8 \div 3\frac{1}{5}$

66. $15 \div 3\frac{1}{3}$

67. Find the quotient of $-4\frac{1}{2}$ and $2\frac{1}{4}$.

68. Find the quotient of 25 and $-10\frac{5}{7}$.

APPLICATIONS In Exercise 69–78, solve each problem.

69. CALORIES   A company advertises that its mints contain only $3\frac{1}{5}$ calories apiece. What is the calorie intake if you eat an entire package of 20 mints?

70. CEMENT MIXER   A cement mixer can carry $9\frac{1}{2}$ cubic yards of concrete. If it makes 8 trips to a job site, how much concrete will be delivered to the site?

71. SHOPPING   In Illustration 6, what is the cost of buying the fruit in the scale?

ILLUSTRATION 6

72. FRAMES   How many inches of molding is needed to produce the square picture frame in Illustration 7?

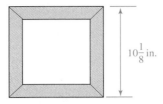

ILLUSTRATION 7

73. SUBDIVISION   A developer donated to the county 100 of the 1,000 acres of land she owned. She divided the remaining acreage into $1\frac{1}{3}$-acre lots. How many lots were created?

74. CATERING   How many people can be served $\frac{1}{3}$-pound hamburgers if a caterer purchases 200 pounds of ground beef?

75. GRAPH PAPER   Mathematicians use specially marked paper, called *graph paper,* when drawing figures. It is made up of $\frac{1}{4}$-inch squares. Find the length and width of the piece of graph paper in Illustration 8.

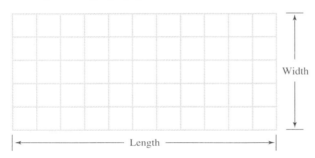

Width

Length

ILLUSTRATION 8

76. LUMBER   As Illustration 9 shows, 2-by-4's from the lumber yard do not really have dimensions of 2 inches

In Exercises 79–80, use a calculator.

79. FIRE ESCAPE   The fire escape stairway in an office building is shown in Illustration 11. Each riser is $7\frac{1}{2}$ inches high. If each floor is 105 inches high and the building is 43 stories tall, how many steps are there in the stairway?

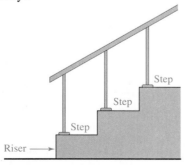

Step

Step

Step

Riser

ILLUSTRATION 11

by 4 inches. How wide and how high is the stack of 2-by-4s?

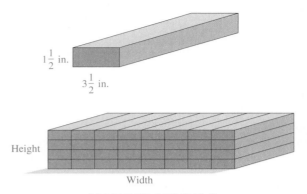

Height

Width

ILLUSTRATION 9

77. EMERGENCY EXIT   Illustration 10 shows a sign that marks the emergency exit on a school bus. Find the area.

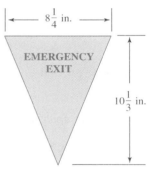

$8\frac{1}{4}$ in.

EMERGENCY EXIT

$10\frac{1}{3}$ in.

ILLUSTRATION 10

78. HORSE RACING   The race tracks on which thoroughbred horses run are marked off in $\frac{1}{8}$-mile-long segments called furlongs. How many furlongs are there in a $1\frac{1}{16}$-mile race?

80. LICENSE PLATE   Find the area of the license plate in Illustration 12.

$12\frac{1}{4}$ in.

WB COUNTY   UTAH   97

EJT852

$6\frac{1}{4}$ in.

ILLUSTRATION 12

81. Explain the difference between $2\frac{3}{4}$ and $2\left(\frac{3}{4}\right)$.

82. Give three examples of how you use mixed numbers in daily life.

83. Explain the procedure used to write an improper fraction as a mixed number.

84. Explain the procedure used to multiply two mixed numbers.

REVIEW

85. Evaluate: $3^2 \cdot 2^2$.

86. List the first eight natural numbers.

87. Write $8 + 8 + 8 + 8$ as a multiplication.

88. If a square measures 1 inch on each side, what is its area?

89. Simplify: $\dfrac{115}{25}$.

90. Multiply: $\left(-\dfrac{6}{5}\right)\left(\dfrac{35}{14}\right)$.

91. Evaluate: $(2 - 5)^2 - 2$.

92. The statement $(3 \cdot 4) \cdot 5 = 3 \cdot (4 \cdot 5)$ illustrates which property of multiplication?

3.6 Adding and Subtracting Mixed Numbers

In this section, you will learn about
- **Adding small mixed numbers**
- **Adding large mixed numbers**
- **Adding mixed numbers in vertical form**
- **Subtracting mixed numbers**

INTRODUCTION. In this section, we will discuss three methods for adding and subtracting mixed numbers. The first method works best when the whole-number parts of the mixed numbers are small. The second method works best when the whole-number parts of the mixed numbers are large. The third method uses columns as a way to organize the work.

Adding small mixed numbers

To add small mixed numbers, we follow these steps.

Adding small mixed numbers	1. Write each mixed number as an improper fraction.
	2. Write each improper fraction as an equivalent fraction with a denominator that is the LCD.
	3. Add the fractions.
	4. Change the result to a mixed number if desired.

EXAMPLE 1 *Adding small mixed numbers.* Add: $4\frac{1}{6} + 2\frac{3}{4}$.

Solution

$4\frac{1}{6} + 2\frac{3}{4} = \frac{25}{6} + \frac{11}{4}$ Write each mixed number as an improper fraction: $4\frac{1}{6} = \frac{25}{6}$ and $2\frac{3}{4} = \frac{11}{4}$.

By inspection, we see the common denominator is 12.

$= \frac{25 \cdot 2}{6 \cdot 2} + \frac{11 \cdot 3}{4 \cdot 3}$ Write each fraction as a fraction with a denominator of 12.

$= \frac{50}{12} + \frac{33}{12}$ Do the multiplications in the numerators and denominators.

$= \frac{83}{12}$ Add the numerators: $50 + 33 = 83$. Write the sum over the common denominator 12.

$= 6\frac{11}{12}$ Write the improper fraction as a mixed number: $\frac{83}{12} = 6\frac{11}{12}$.

Self Check

Add: $3\frac{2}{3} + 1\frac{1}{5}$.

Answer: $4\frac{13}{15}$

Adding large mixed numbers

To add large mixed numbers, we follow these steps.

Adding large mixed numbers	1. Write each mixed number as the sum of a whole number and a fraction.
	2. Use the commutative property of addition to group the whole numbers together and the fractions together.
	3. Add the whole numbers and the fractions separately.
	4. Write the result as a mixed number if necessary.

EXAMPLE 2 *Adding large mixed numbers.* Find the sum: $25\frac{3}{4} + 31\frac{1}{5}$.

Solution

$25\frac{3}{4} + 31\frac{1}{5} = 25 + \frac{3}{4} + 31 + \frac{1}{5}$ Write each mixed number as the sum of a whole number and a fraction.

$= 25 + 31 + \frac{3}{4} + \frac{1}{5}$ Use the commutative property of addition to change the order of the addition.

$= 56 + \frac{3}{4} + \frac{1}{5}$ Add the whole numbers: $25 + 31 = 56$.

$= 56 + \frac{3 \cdot 5}{4 \cdot 5} + \frac{1 \cdot 4}{5 \cdot 4}$ Write each fraction as a fraction with denominator 20.

$= 56 + \frac{15}{20} + \frac{4}{20}$ Multiply in the numerators and denominators.

$= 56 + \frac{19}{20}$ Add the numerators and write the sum over the common denominator 20.

$= 56\frac{19}{20}$ Write the sum as a mixed number.

Self Check

Find the sum: $75\frac{1}{6} + 81\frac{3}{5}$.

Answer: $156\frac{23}{30}$

Adding mixed numbers in vertical form

By working in columns, we can add mixed numbers quickly. The strategy is the same as in Example 2: Add whole numbers to whole numbers and fractions to fractions.

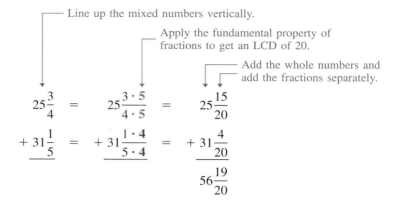

Line up the mixed numbers vertically.

Apply the fundamental property of fractions to get an LCD of 20.

Add the whole numbers and add the fractions separately.

$$
\begin{array}{r}
25\dfrac{3}{4} \\[2mm]
+\ 31\dfrac{1}{5}
\end{array}
\quad = \quad
\begin{array}{r}
25\dfrac{3\cdot 5}{4\cdot 5} \\[2mm]
+\ 31\dfrac{1\cdot 4}{5\cdot 4}
\end{array}
\quad = \quad
\begin{array}{r}
25\dfrac{15}{20} \\[2mm]
+\ 31\dfrac{4}{20} \\[1mm]
\hline
56\dfrac{19}{20}
\end{array}
$$

EXAMPLE 3

Suspension bridge. Find the total length of cable that must be ordered if cables a, d, and e of the suspension bridge in Figure 3-10 are to be replaced. (See the table below.)

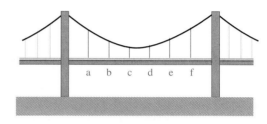

FIGURE 3-10

Bridge Specifications			
Cable	a	b	c
Length (feet)	$75\frac{1}{12}$	$54\frac{1}{6}$	$43\frac{1}{4}$

Solution To find the total length of cable to be ordered, we need to add the lengths of cables a, d, and e. Because of the symmetric design, cables e and b and cables d and c are the same length.

Length of cable a $+$ Length of cable d (or cable c) $+$ Length of cable e (or cable b) $=$ Total length needed.

$$
\begin{array}{r}
75\dfrac{1}{12} \\[2mm]
43\dfrac{1}{4} \\[2mm]
+\ 54\dfrac{1}{6}
\end{array}
\ = \
\begin{array}{r}
75\dfrac{1}{12} \\[2mm]
43\dfrac{1\cdot 3}{4\cdot 3} \\[2mm]
+\ 54\dfrac{1\cdot 2}{6\cdot 2}
\end{array}
\ = \
\begin{array}{r}
75\dfrac{1}{12} \\[2mm]
43\dfrac{3}{12} \\[2mm]
+\ 54\dfrac{2}{12} \\[1mm]
\hline
172\dfrac{6}{12}
\end{array}
\ = \ 172\dfrac{1}{2}
$$

Simplify

The total length of cable needed for the replacement is $172\frac{1}{2}$ feet.

When using vertical form to add mixed numbers, the sum of the fractions sometimes yields an improper fraction, as in the next example.

EXAMPLE 4 *Vertical form.* Add $45\frac{2}{3} + 96\frac{4}{5}$.

Solution

$$
\begin{array}{rcl}
45\dfrac{2}{3} &=& 45\dfrac{2\cdot 5}{3\cdot 5} \;=\; 45\dfrac{10}{15} \\[2ex]
+\,96\dfrac{4}{5} &=& +\,96\dfrac{4\cdot 3}{5\cdot 3} \;=\; +\,96\dfrac{12}{15} \\[2ex]
&& \hspace{2em} 141\dfrac{22}{15}
\end{array}
$$

The whole-number part of the answer. ⟶ | ⟵ The fractional part of the answer is an improper fraction.

Now write the improper fraction as a mixed number.

$$141\frac{22}{15} = 141 + \frac{22}{15} = 141 + 1\frac{7}{15} = 142\frac{7}{15}$$

Self Check

Add $76\frac{11}{12} + 49\frac{5}{8}$.

Answer: $126\dfrac{13}{24}$

Subtracting mixed numbers

Subtracting mixed numbers is similar to adding mixed numbers.

EXAMPLE 5 *Cooking.* How much butter is left in a 10-pound tub if $2\frac{2}{3}$ pounds are used for a wedding cake?

Solution The phrase "How much is left?" suggests subtraction. Since the numbers we are working with are small, we will write them as improper fractions and proceed as follows:

$$10 - 2\frac{2}{3} = \frac{10}{1} - \frac{8}{3} \qquad \text{Write 10 as a fraction: } 10 = \tfrac{10}{1}. \text{ Write } 2\tfrac{2}{3} \text{ as } \tfrac{8}{3}.$$

By inspection, we see that the LCD is 3.

$$\frac{10}{1} - 2\frac{2}{3} = \frac{10\cdot 3}{1\cdot 3} - \frac{8}{3} \qquad \text{Write the first fraction with a denominator of 3.}$$

$$= \frac{30}{3} - \frac{8}{3} \qquad \text{Do the multiplications in the first fraction.}$$

$$= \frac{30 - 8}{3} \qquad \text{Subtract the numerators and write the difference over the common denominator.}$$

$$= \frac{22}{3} \qquad \text{Do the subtraction: } 30 - 8 = 22.$$

$$= 7\frac{1}{3} \qquad \text{Write } \tfrac{22}{3} \text{ as a mixed number.}$$

There are $7\frac{1}{3}$ pounds of butter left in the tub.

In the next example, the fraction being subtracted *from* is smaller than the fraction being subtracted. Because of this, we will have to borrow.

EXAMPLE 6 *Borrowing.* Find the difference: $34\frac{1}{5} - 11\frac{2}{3}$.

Solution We will use the vertical form to subtract. The LCD is 15.

$$
\begin{array}{rcccl}
34\dfrac{1}{5} & = & 34\dfrac{1 \cdot 3}{5 \cdot 3} & = & 34\dfrac{3}{15} \\[2ex]
-\,11\dfrac{2}{3} & = & -\,11\dfrac{2 \cdot 5}{3 \cdot 5} & = & -\,11\dfrac{10}{15}
\end{array}
$$

Write each fraction as a fraction with a denominator of 15.

Since $\frac{10}{15}$ is larger than $\frac{3}{15}$, borrow 1 $\left(\text{in the form of } \frac{15}{15}\right)$ from 34 and add it to $\frac{3}{15}$. We obtain $33\frac{3}{15} + \frac{15}{15} = 33\frac{18}{15}$. Then we subtract the fractions and the whole numbers separately.

$$
\begin{array}{rcl}
33\dfrac{3}{15} + \dfrac{15}{15} & = & 33\dfrac{18}{15} \\[2ex]
-\,11\dfrac{10}{15} & = & -\,11\dfrac{10}{15} \\[2ex]
& & 22\dfrac{8}{15}
\end{array}
$$

STUDY SET Section 3.6

VOCABULARY *In Exercises 1–2, fill in the blanks to make a true statement.*

1. By the _____ property of addition, we can add numbers in any order.

2. A _____ number contains a whole-number part and a fractional part.

CONCEPTS

3. Write $4\frac{3}{5}$ as a sum.

4. For $76\frac{3}{4}$, list the whole-number part and the fractional part.

5. Use the commutative property of addition to get the whole numbers together.

$$14 + \frac{5}{6} + 53 + \frac{1}{6}$$

6. What property is being used here?

$$
\begin{array}{l}
25\dfrac{3 \cdot 5}{4 \cdot 5} \\[2ex]
+\,31\dfrac{1 \cdot 4}{5 \cdot 4}
\end{array}
$$

7. The denominators of two fractions, expressed in prime-factored form, are $5 \cdot 2$ and $5 \cdot 3$. Find the LCD for the fractions.

8. The denominators for three fractions, in prime-factored form, are $3 \cdot 5$, $2 \cdot 3$, and $3 \cdot 3$. Find the LCD for the fractions.

9. Simplify: $16\frac{12}{8}$.

10. Simplify: $45\frac{24}{20}$.

11. Add: $70\frac{3}{5} + 39\frac{2}{7}$.

$70\frac{3}{5} + 39\frac{2}{7} = \underline{} + \frac{3}{5} + \underline{} + \frac{2}{7}$ Write each mixed number as a sum.

$\phantom{70\frac{3}{5} + 39\frac{2}{7}} = \underline{} + \underline{} + \frac{3}{5} + \frac{2}{7}$ Apply the commutative property of addition.

$\phantom{70\frac{3}{5} + 39\frac{2}{7}} = 109 + \frac{3}{5} + \frac{2}{7}$ Add the whole numbers.

$\phantom{70\frac{3}{5} + 39\frac{2}{7}} = 109 + \frac{3 \cdot}{5 \cdot} + \frac{2 \cdot}{7 \cdot}$ Write each fraction as a fraction with denominator 35.

$\phantom{70\frac{3}{5} + 39\frac{2}{7}} = 109 + \frac{21}{} + \frac{10}{}$ Multiply the numerators. Multiply the denominators.

$\phantom{70\frac{3}{5} + 39\frac{2}{7}} = 109 + \frac{}{35}$ Add the fractions.

$\phantom{70\frac{3}{5} + 39\frac{2}{7}} = 109\frac{31}{35}$ Write the sum as a mixed number.

12. Subtract: $67\frac{3}{8} - 23\frac{2}{3}$.

$67\frac{3}{8} = 67\frac{3 \cdot}{8 \cdot}$

$-23\frac{2}{3} = -23\frac{2 \cdot}{3 \cdot}$ Write each fraction as a fraction with denominator 24.

$67\frac{9}{24} = \underline{}\frac{9}{24} + \underline{}$ Borrow 1 from 67 in the form of $\frac{24}{24}$. Add this to $\frac{9}{24}$.

$-23\frac{16}{24} = -23\frac{16}{24}$

$66\frac{}{24}$

$-23\frac{16}{24}$ Subtract the fractions. Subtract the whole numbers.

$43\frac{17}{24}$

PRACTICE In Exercises 13–40, find each sum or difference.

13. $2\frac{1}{5} + 2\frac{1}{5}$ **14.** $3\frac{1}{3} + 2\frac{1}{3}$ **15.** $8\frac{2}{7} - 3\frac{1}{7}$ **16.** $9\frac{5}{11} - 6\frac{2}{11}$

17. $3\frac{1}{4} + 4\frac{1}{4}$ **18.** $2\frac{1}{8} + 3\frac{3}{8}$ **19.** $4\frac{1}{6} + 1\frac{1}{5}$ **20.** $2\frac{2}{5} + 3\frac{1}{4}$

21. $2\frac{1}{2} - 1\frac{1}{4}$ **22.** $13\frac{5}{6} - 4\frac{2}{3}$ **23.** $2\frac{5}{6} - 1\frac{3}{8}$ **24.** $4\frac{5}{9} - 2\frac{1}{6}$

25. $5\frac{1}{2} + 3\frac{4}{5}$ **26.** $6\frac{1}{2} + 2\frac{2}{3}$ **27.** $7\frac{1}{2} - 4\frac{1}{7}$ **28.** $5\frac{3}{4} - 1\frac{3}{7}$

29. $56\frac{2}{5} + 73\frac{1}{3}$ **30.** $44\frac{3}{8} + 66\frac{1}{5}$ **31.** $380\frac{5}{6} + 17\frac{3}{4}$ **32.** $103\frac{1}{2} + 210\frac{4}{5}$

33. $28\frac{5}{9} + 44\frac{2}{3}$ **34.** $61\frac{7}{8} + 19\frac{1}{3}$ **35.** $78\frac{5}{7} - 55\frac{1}{3}$ **36.** $39\frac{1}{2} - 18\frac{3}{16}$

37. $40\frac{5}{6} - 29\frac{4}{5}$ **38.** $91\frac{1}{4} - 89\frac{1}{12}$ **39.** $22\frac{13}{16} - 21\frac{3}{8}$ **40.** $78\frac{3}{4} - 77\frac{5}{8}$

In Exercises 41–48, find each difference. You will need to borrow in each case.

41. $16\frac{1}{4} - 13\frac{3}{4}$ **42.** $40\frac{1}{7} - 19\frac{6}{7}$ **43.** $76\frac{1}{6} - 49\frac{7}{8}$ **44.** $101\frac{1}{4} - 70\frac{1}{2}$

45. $40\frac{3}{16} - 29\frac{3}{4}$ **46.** $11\frac{1}{3} - 8\frac{3}{4}$ **47.** $34\frac{1}{9} - 13\frac{5}{6}$ **48.** $42\frac{1}{8} - 29\frac{2}{3}$

In Exercises 49–60, find the sum or difference.

49. $7 - \frac{2}{3}$ **50.** $6 - \frac{1}{8}$ **51.** $9 - 8\frac{3}{4}$ **52.** $11 - 10\frac{4}{5}$

53. $4\dfrac{1}{7} - \dfrac{4}{5}$

54. $5\dfrac{1}{10} - \dfrac{4}{5}$

55. $6\dfrac{5}{8} - 3$

56. $10\dfrac{1}{2} - 6$

57. $\dfrac{7}{3} + 2$

58. $\dfrac{9}{7} + 3$

59. $2 + 1\dfrac{7}{8}$

60. $3\dfrac{3}{4} + 5$

In Exercises 61–64, find each sum.

61. $12\dfrac{1}{2} + 5\dfrac{3}{4} + 35\dfrac{1}{6}$

62. $31\dfrac{1}{3} + 20\dfrac{2}{5} + 10\dfrac{1}{15}$

63. $58\dfrac{7}{8} + 40 + 61\dfrac{1}{4}$

64. $91 + 33\dfrac{1}{16} + 16\dfrac{5}{8}$

In Exercises 65–68, find each sum or difference.

65. $-3\dfrac{3}{4} + \left(-1\dfrac{1}{2}\right)$

66. $-3\dfrac{2}{3} + \left(-1\dfrac{4}{5}\right)$

67. $-4\dfrac{5}{8} - 1\dfrac{1}{4}$

68. $-2\dfrac{1}{16} - 3\dfrac{7}{8}$

APPLICATIONS *In Exercises 69–80, solve each problem.*

69. FREEWAY TRAVEL   See Illustration 1. As you approach Citrus Avenue, your intended freeway exit, you learn from your car radio that traffic on Citrus is backed up due to an accident. How much farther will you have to travel on the freeway if you decide to use the Grand Avenue off-ramp instead?

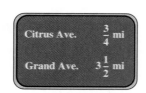

| Citrus Ave. | $\dfrac{3}{4}$ mi |
| Grand Ave. | $3\dfrac{1}{2}$ mi |

ILLUSTRATION 1

70. BASKETBALL   See Illustration 2. What is the difference in height between the tallest and the shortest of the starting players?

Heights of the Starting Five Players

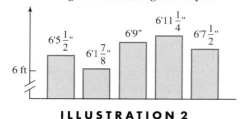

$6'5\dfrac{1}{2}''$ $6'1\dfrac{7}{8}''$ $6'9''$ $6'11\dfrac{1}{4}''$ $6'7\dfrac{1}{2}''$

6 ft

ILLUSTRATION 2

71. TRAIL MIX   A camper decides to prepare some trail mix according to the recipe in Illustration 3. She doubles up on the amount of sunflower seeds called for in the recipe and decides not to use any coconut. How much trail mix will the adjusted recipe yield?

Trail Mix	
Peanuts	$2\dfrac{3}{4}$ cups
Sunflower seeds	$\dfrac{1}{2}$ cup
Raisins	$\dfrac{2}{3}$ cup
Coconut	$\dfrac{1}{3}$ cup
Oat flakes	$1\dfrac{2}{3}$ cups

ILLUSTRATION 3

72. AIR TRAVEL   A businesswoman's flight leaves Los Angeles at 8 A.M. and arrives in Seattle at 9:45 A.M.
   a. Express the duration of the flight as a mixed number.
   b. Upon arrival, she boards a commuter plane at 11:15 A.M., arriving at her final destination at 11:45 A.M. Express the length of this flight as a fraction.
   c. Find the total time of these two flights.

73. HOSE REPAIR   To repair a bad connector, Ming Lin removes $1\frac{1}{2}$ feet from the end of a 50-foot garden hose. How long is the hose after the repair?

74. SEWING   To make some draperies, Liz needs $12\frac{1}{4}$ yards of material for the den and $8\frac{1}{2}$ yards for the living room. If the material only comes in 21-yard bolts, how much will be left over after completing both sets of drapes?

75. SHIPPING A passenger ship and a cargo ship leave San Diego harbor at midnight. During the first hour, the passenger ship travels south at $16\frac{1}{2}$ miles per hour while the cargo ship is traveling north at a rate of $5\frac{1}{5}$ miles per hour.

a. Complete the chart in Illustration 4.

	Speed (mph)	Time traveling (hr)	Distance traveled (mi)
Passenger ship		1	
Cargo ship		1	

ILLUSTRATION 4

b. How far apart are they at 1:00 A.M.?

76. HARDWARE See Illustration 5. To secure the bracket to the stock, a bolt and a nut are used. How long should the bolt be?

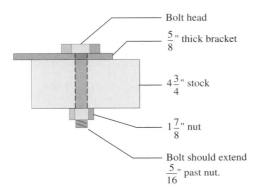

Bolt head

$\frac{5}{8}$" thick bracket

$4\frac{3}{4}$" stock

$1\frac{7}{8}$" nut

Bolt should extend $\frac{5}{16}$" past nut.

ILLUSTRATION 5

77. SERVICE STATION Use the service station sign in Illustration 6 to answer the following questions.

a. What is the difference in price between the least and most expensive types of gasoline at the self-service pump?

b. How much more is the cost per gallon for full service?

	Self Serve	Full Serve
Premium Unleaded	$149\frac{9}{10}$	$179\frac{9}{10}$
Unleaded	$139\frac{9}{10}$	$169\frac{9}{10}$
Premium Plus	$159\frac{9}{10}$	$189\frac{9}{10}$
	cents per gallon	

ILLUSTRATION 6

78. FUEL TANK In a small town, a 100-gallon gasoline tank is used to fill police vehicles. One week, the attendant failed to record the amount left in the tank after each fill-up. (See Illustration 7.) If the tank was full at the start of the week, complete the table.

Day of the the week	Amount of fill-up (gal)	Amount left in 100-gal tank
Monday	$11\frac{1}{4}$	
Tuesday	12	
Wednesday	$10\frac{1}{2}$	
Thursday	11	
Friday	$10\frac{3}{4}$	

ILLUSTRATION 7

79. WATER SLIDE An amusement park added a new section to a water slide to create a slide of length $341\frac{5}{12}$ feet. (See Illustration 8.) How long was the slide before the addition?

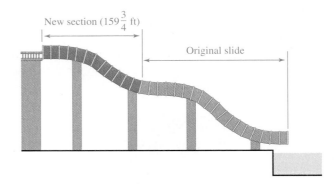

New section ($159\frac{3}{4}$ ft)

Original slide

ILLUSTRATION 8

80. JEWELRY A jeweler is to cut a 7-inch-long gold braid into three pieces. He aligns a 6-inch-long ruler directly below the braid and makes the proper cuts. (See Illustration 9.) Find the length of piece 2 of the braid.

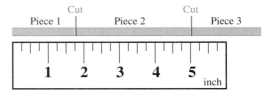

Cut

Piece 1   Piece 2   Cut   Piece 3

1 2 3 4 5 inch

ILLUSTRATION 9

WRITING *Write a paragraph using your own words.*

81. Of the two methods studied to add large mixed numbers, which do you like better and why?

82. When subtracting mixed numbers, when is borrowing necessary? How is it done?

83. Explain how to add $1\frac{3}{8}$ and $2\frac{1}{4}$ if we write each as an improper fraction.

84. Explain the process of simplifying $12\frac{16}{5}$.

85. Find the mean of 36, 48, and 72.

86. Multiply: $-3(-4)(-5)$.

87. Add: $\dfrac{2}{5} + \dfrac{1}{4}$.

88. Which fraction is larger: $\dfrac{11}{13}$ or $\dfrac{6}{7}$?

89. Evaluate: $-2 - (-8)$.

90. Find the area of a triangle with a base 6 inches long and a height of 8 inches.

91. What does area measure?

92. Find $|-12|$.

3.7   *Order of Operations and Complex Fractions*

In this section, you will learn about

- **Order of operations**
- **Evaluating formulas**
- **Complex fractions**
- **Simplifying complex fractions**

INTRODUCTION. In this section, we will evaluate expressions involving fractions and mixed numbers. We will also discuss complex fractions and the methods that are used to simplify them.

Order of operations

The rules for the order of operations are used to evaluate numerical expressions involving more than one operation.

EXAMPLE 1   *Order of operations.*   Evaluate: $\dfrac{3}{4} + \dfrac{5}{3}\left(-\dfrac{1}{2}\right)^3$.

Self Check

Evaluate: $\dfrac{7}{8} + \dfrac{3}{2}\left(-\dfrac{1}{4}\right)^2$.

Solution

The expression involves the operations of raising to a power, multiplication, and addition. By the rules for the order of operations, we must evaluate the power first, the multiplication second, and the addition last.

$$\dfrac{3}{4} + \dfrac{5}{3}\left(-\dfrac{1}{2}\right)^3 = \dfrac{3}{4} + \dfrac{5}{3}\left(-\dfrac{1}{8}\right) \qquad \text{Evaluate the power: } \left(-\tfrac{1}{2}\right)^3 = -\tfrac{1}{8}.$$

$$= \dfrac{3}{4} + \left(-\dfrac{5}{24}\right) \qquad \text{Do the multiplication: } \tfrac{5}{3}\left(-\tfrac{1}{8}\right) = -\tfrac{5}{24}.$$

$$= \dfrac{3 \cdot 6}{4 \cdot 6} + \left(-\dfrac{5}{24}\right) \qquad \text{The LCD is 24. Write the first fraction as a fraction with denominator 24.}$$

$$= \dfrac{18}{24} + \left(-\dfrac{5}{24}\right) \qquad \begin{array}{l}\text{Multiply in the numerator: } 3 \cdot 6 = 18.\\ \text{Multiply in the denominator: } 4 \cdot 6 = 24.\end{array}$$

$$= \dfrac{13}{24} \qquad \text{Add the numerators and write the sum over the common denominator.}$$

Answer: $\dfrac{31}{32}$

If an expression contains grouping symbols, we do the operations inside the grouping symbols first.

EXAMPLE 2 *Order of operations.* Evaluate: $\left(\dfrac{7}{8}-\dfrac{1}{4}\right)\div\left(-2\dfrac{3}{16}\right)$.

Solution $\left(\dfrac{7}{8}-\dfrac{1}{4}\right)\div\left(-2\dfrac{3}{16}\right)=\left(\dfrac{7}{8}-\dfrac{1\cdot 2}{4\cdot 2}\right)\div\left(-2\dfrac{3}{16}\right)$ Inside the first set of parentheses, write $\frac{1}{4}$ as a fraction with denominator 8.

$=\left(\dfrac{7}{8}-\dfrac{2}{8}\right)\div\left(-2\dfrac{3}{16}\right)$ Multiply in the numerator: $1\cdot 2=2$. Multiply in the denominator: $4\cdot 2=8$.

$=\dfrac{5}{8}\div\left(-2\dfrac{3}{16}\right)$ Subtract the numerators and write the difference over the common denominator: $7-2=5$.

$=\dfrac{5}{8}\div\left(-\dfrac{35}{16}\right)$ Write the mixed number as an improper fraction.

$=\dfrac{5}{8}\cdot\left(-\dfrac{16}{35}\right)$ Multiply by the reciprocal of $-\frac{35}{16}$.

$=-\dfrac{5\cdot 16}{8\cdot 35}$ The product of two fractions with unlike signs is negative. Multiply the numerators and multiply the denominators.

$=-\dfrac{\overset{1}{\cancel{5}}\cdot 2\cdot\overset{1}{\cancel{8}}}{\underset{1}{\cancel{8}}\cdot\underset{1}{\cancel{5}}\cdot 7}$ Factor 16 as $2\cdot 8$ and factor 35 as $5\cdot 7$. Divide out the common factors of 8 and 5.

$=-\dfrac{2}{7}$ Multiply in the numerator: $1\cdot 2\cdot 1=2$. Multiply in the denominator: $1\cdot 1\cdot 7=7$. ∎

Evaluating formulas

To evaluate a formula, we replace its letters with specific numbers and simplify by using the rules for the order of operations.

EXAMPLE 3 *Evaluating formulas.* The formula for the area of a trapezoid is $A=\frac{1}{2}h(a+b)$, where A is the area, h is the height, and a and b are the lengths of its bases. Find A when $h=1\frac{2}{3}$, $a=2\frac{1}{2}$, and $b=5\frac{1}{2}$.

Self Check

The formula for the area of a triangle is $A=\frac{1}{2}bh$. Find the area of a triangle with a base 12 meters long and a height of 15 meters.

Solution

$A=\dfrac{1}{2}h(a+b)$

$=\dfrac{1}{2}\left(1\dfrac{2}{3}\right)\left(2\dfrac{1}{2}+5\dfrac{1}{2}\right)$ Replace h, a, and b with the given values.

$=\dfrac{1}{2}\left(1\dfrac{2}{3}\right)(8)$ Do the addition inside the parentheses: $2\frac{1}{2}+5\frac{1}{2}=8$.

$=\dfrac{1}{2}\left(\dfrac{5}{3}\right)\left(\dfrac{8}{1}\right)$ Write $1\frac{2}{3}$ as an improper fraction and 8 as $\frac{8}{1}$.

$=\dfrac{1\cdot 5\cdot 8}{2\cdot 3\cdot 1}$ Multiply the numerators and multiply the denominators.

$=\dfrac{1\cdot 5\cdot\overset{1}{\cancel{2}}\cdot 4}{\underset{1}{\cancel{2}}\cdot 3\cdot 1}$ Factor 8 as $2\cdot 4$ and divide out the common factor of 2.

$$= \frac{20}{3}$$ Multiply in the numerator and the denominator.

$$A = 6\frac{2}{3}$$ Write $\frac{20}{3}$ as a mixed number.

Answer: 90 m²

EXAMPLE 4

Solution

Masonry. To build a wall, a mason will use blocks that are $5\frac{3}{4}$ inches high, held together with $\frac{3}{8}$-inch-thick layers of mortar. (See Figure 3-11.) If the plans call for 8 layers of blocks, what will be the height of the wall when completed?

To find the height, we must consider 8 layers of blocks and 8 layers of mortar. We will compute the height contributed by one block and one layer of mortar and then multiply that result by 8.

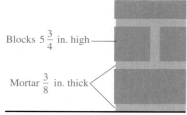

Blocks $5\frac{3}{4}$ in. high

Mortar $\frac{3}{8}$ in. thick

FIGURE 3-11

8 times	(	height of 1 block	plus	height of 1 layer of mortar	)	equals	height of block wall.

$$8 \quad \left(\quad 5\frac{3}{4} \quad + \quad \frac{3}{8} \quad \right) \quad = \quad \text{height of wall}$$

$$8\left(5\frac{3}{4} + \frac{3}{8}\right) = 8\left(\frac{23}{4} + \frac{3}{8}\right)$$ Write $5\frac{3}{4}$ as the improper fraction $\frac{23}{4}$.

$$= 8\left(\frac{23 \cdot 2}{4 \cdot 2} + \frac{3}{8}\right)$$ Express $\frac{23}{4}$ in terms of 8ths.

$$= 8\left(\frac{46}{8} + \frac{3}{8}\right)$$ Do the multiplication in the numerator and denominator.

$$= \frac{8}{1}\left(\frac{49}{8}\right)$$ Write the sum of the numerators over the common denominator 8. Write 8 as $\frac{8}{1}$.

$$= \frac{\overset{1}{\cancel{8}} \cdot 49}{\underset{1}{\cancel{8}}}$$ Multiply the numerators and the denominators. Divide out the common factor of 8.

$$= 49$$ $\frac{49}{1} = 49$.

The wall will be 49 inches high.

Complex fractions

Fractions whose numerators and/or denominators contain fractions are called *complex fractions*. Here is an example.

A fraction in the numerator ⟶ $\dfrac{\dfrac{3}{4}}{\dfrac{7}{8}}$ ⟵ The main fraction bar

A fraction in the denominator ⟶

Complex fraction

A **complex fraction** is a fraction whose numerator or denominator, or both, contain one or more fractions or mixed numbers.

Here are more examples of complex fractions.

$$\dfrac{-\dfrac{1}{4} - \dfrac{4}{5}}{2\dfrac{4}{5}}$$ ⟵ Numerator ⟶ $\dfrac{\dfrac{1}{3} + \dfrac{1}{4}}{\dfrac{1}{3} - \dfrac{1}{4}}$

⟵ Main fraction bar ⟶

⟵ Denominator ⟶

Simplifying complex fractions

To *simplify* complex fractions means to express them as fractions in simplified form.

Simplifying a complex fraction: method 1	Write the numerator and the denominator of the complex fraction as single fractions. Then do the indicated division of the two fractions and simplify.

Method 1 is based on the fact that the main fraction bar of the complex fraction indicates division.

$$\frac{\dfrac{1}{4}}{\dfrac{2}{5}} \longleftarrow \text{The main fraction bar means "divide the fraction in the numerator by the fraction in the denominator."} \longrightarrow \frac{1}{4} \div \frac{2}{5}$$

EXAMPLE 5 *Simplifying a complex fraction.* Simplify: $\dfrac{\frac{1}{4}}{\frac{2}{5}}$.

Solution

Since the numerator and the denominator of this complex fraction are single fractions, we can do the indicated division.

$$\frac{\dfrac{1}{4}}{\dfrac{2}{5}} = \frac{1}{4} \div \frac{2}{5} \qquad \text{Express the complex fraction as an equivalent division problem.}$$

$$= \frac{1}{4} \cdot \frac{5}{2} \qquad \text{Multiply by the reciprocal of } \tfrac{2}{5}.$$

$$= \frac{1 \cdot 5}{4 \cdot 2} \qquad \text{Multiply the numerators and multiply the denominators.}$$

$$= \frac{5}{8}$$

Self Check

Simplify: $\dfrac{\frac{1}{6}}{\frac{3}{8}}$.

Answer: $\dfrac{4}{9}$

A second method is based on the fundamental property of fractions.

Simplifying a complex fraction: method 2	Multiply both the numerator and the denominator of the complex fraction by the LCD of all the fractions that appear in its numerator and denominator. Then simplify.

EXAMPLE 6 *Simplifying a complex fraction.* Simplify: $\dfrac{-\frac{1}{4} + \frac{2}{5}}{\frac{1}{2} - \frac{4}{5}}$.

Solution Examine the numerator and the denominator of the complex fraction. The fractions involved have denominators of 4, 5, and 2. The LCD of these fractions is 20.

$$\frac{-\dfrac{1}{4} + \dfrac{2}{5}}{\dfrac{1}{2} - \dfrac{4}{5}} = \frac{20\left(-\dfrac{1}{4} + \dfrac{2}{5}\right)}{20\left(\dfrac{1}{2} - \dfrac{4}{5}\right)} \qquad \text{Apply the fundamental property of fractions. Multiply the numerator and the denominator of the complex fraction by 20.}$$

$$= \frac{20\left(-\dfrac{1}{4}\right) + 20\left(\dfrac{2}{5}\right)}{20\left(\dfrac{1}{2}\right) - 20\left(\dfrac{4}{5}\right)}$$

Apply the distributive property in the numerator and in the denominator.

$$= \frac{-5 + 8}{10 - 16}$$

Do the multiplications by 20.

$$= \frac{3}{-6}$$

Do the addition in the numerator and the subtraction in the denominator.

$$= -\frac{1}{2}$$

Simplify.

EXAMPLE 7   *Simplifying a complex fraction.*   Simplify: $\dfrac{7 - \dfrac{2}{3}}{4\dfrac{5}{6}}$.

Self Check

Simplify: $\dfrac{5 - \dfrac{3}{4}}{1\dfrac{7}{8}}$.

Solution

Examine the numerator and the denominator of the complex fraction. The fractions have denominators of 3 and 6. The LCD of these fractions is 6.

$$\frac{7 - \dfrac{2}{3}}{4\dfrac{5}{6}} = \frac{7 - \dfrac{2}{3}}{\dfrac{29}{6}}$$

Express $4\frac{5}{6}$ as an improper fraction.

$$= \frac{6\left(7 - \dfrac{2}{3}\right)}{6\left(\dfrac{29}{6}\right)}$$

Apply the fundamental property of fractions. Multiply the numerator and the denominator of the complex fraction by the LCD, 6.

$$= \frac{6(7) - 6\left(\dfrac{2}{3}\right)}{6\left(\dfrac{29}{6}\right)}$$

Apply the distributive property in the numerator. Distribute the 6.

$$= \frac{42 - 4}{29}$$

Do the multiplications by 6.

$$= \frac{38}{29}$$

Do the subtraction in the numerator.

$$= 1\frac{9}{29}$$

Write $\frac{38}{29}$ as a mixed number.

Answer: $2\dfrac{4}{15}$

STUDY SET   **Section 3.7**

VOCABULARY   *In Exercises 1–2, fill in the blanks to make a true statement.*

1. A _____ fraction is a fraction whose numerator or denominator, or both, contain one or more _____ or mixed numbers.

2. To _____ an algebraic expression, we substitute specific numbers for the letters in the expression and simplify.

CONCEPTS

3. What division is represented by this complex fraction?

$$\dfrac{\dfrac{2}{3}}{\dfrac{1}{5}}$$

4. Write this division as a complex fraction.

$$-\dfrac{7}{8} \div \dfrac{3}{4}$$

5. What is the common denominator of all the fractions in this complex fraction?

$$\dfrac{\dfrac{2}{3} - \dfrac{1}{5}}{\dfrac{1}{3} + \dfrac{4}{5}}$$

6. What is the common denominator of all the fractions in this complex fraction?

$$\dfrac{\dfrac{1}{8} - \dfrac{3}{16}}{-5\dfrac{3}{4}}.$$

7. When this complex fraction is simplified, will the result be positive or negative?

$$\dfrac{-\dfrac{2}{3}}{\dfrac{3}{4}}$$

8. What property is being applied?

$$\dfrac{1 + \dfrac{1}{11}}{\dfrac{1}{2}} = \dfrac{22\left(1 + \dfrac{1}{11}\right)}{22\left(\dfrac{1}{2}\right)}$$

9. What is the LCD of fractions with the denominators 6, 4, and 5?

10. What operations are involved in this numerical expression?

$$5\left(6\dfrac{1}{3}\right) + \left(-\dfrac{1}{4}\right)^2$$

NOTATION *In Exercises 11–12, complete each solution.*

11. Simplify: $\dfrac{\dfrac{1}{8}}{\dfrac{3}{4}}$.

$$\dfrac{\dfrac{1}{8}}{\dfrac{3}{4}} = \dfrac{1}{8} \div \underline{}$$
Write the complex fraction as an equivalent division problem.

$$= \dfrac{1}{8} \cdot \underline{}$$
Multiply by the reciprocal.

$$= \dfrac{1 \cdot }{8 \cdot 3}$$
Multiply the numerators and multiply the denominators.

$$= \dfrac{1 \cdot \overset{1}{\cancel{4}}}{2 \cdot \underset{1}{\cancel{4}} \cdot 3}$$
Factor the 8 as 2 · 4 and divide out the common factor of 4.

$$= \dfrac{1}{6}$$

12. Simplify: $\dfrac{\dfrac{1}{6} + \dfrac{1}{5}}{-\dfrac{1}{15}}$.

$$\dfrac{\dfrac{1}{6} + \dfrac{1}{5}}{-\dfrac{1}{15}} = \dfrac{30\left(\dfrac{1}{6} + \dfrac{1}{5}\right)}{\left(-\dfrac{1}{15}\right)}$$
Multiply the numerator and the denominator of the complex fraction by 30.

$$= \dfrac{\left(\dfrac{1}{6}\right) + \left(\dfrac{1}{5}\right)}{30\left(-\dfrac{1}{15}\right)}$$
Distribute 30.

$$= \dfrac{5 + 6}{}$$
Do the multiplications.

$$= \dfrac{}{-2}$$
Do the addition.

$$= -5\dfrac{1}{2}$$
Write the improper fraction as a mixed number.

PRACTICE *In Exercises 13–30, evaluate each expression.*

13. $\dfrac{2}{3}\left(-\dfrac{1}{4}\right) + \dfrac{1}{2}$

14. $-\dfrac{7}{8} - \left(\dfrac{1}{8}\right)\left(\dfrac{2}{3}\right)$

15. $\dfrac{4}{5} - \left(-\dfrac{1}{3}\right)^2$

16. $-\dfrac{3}{16} - \left(-\dfrac{1}{2}\right)^3$

17. $-4\left(-\dfrac{1}{5}\right) - \left(\dfrac{1}{4}\right)\left(-\dfrac{1}{2}\right)$

18. $(-3)\left(-\dfrac{2}{3}\right) - (-4)\left(-\dfrac{3}{4}\right)$

19. $1\dfrac{3}{5}\left(\dfrac{1}{2}\right)^2\left(\dfrac{3}{4}\right)$

20. $2\dfrac{3}{5}\left(-\dfrac{1}{3}\right)^2\dfrac{1}{2}$

21. $\dfrac{7}{8} - \left(\dfrac{4}{5} + 1\dfrac{3}{4}\right)$

22. $\left(\dfrac{5}{4}\right)^2 + \left(\dfrac{2}{3} - 2\dfrac{1}{6}\right)$

23. $\left(\dfrac{9}{20} \div 2\dfrac{2}{5}\right) + \left(\dfrac{3}{4}\right)^2$

24. $\left(1\dfrac{2}{3} \cdot 15\right) + \left(\dfrac{7}{9} \div \dfrac{7}{81}\right)$

25. $\left(-\dfrac{3}{4} \cdot \dfrac{9}{16}\right) + \left(\dfrac{1}{2} - \dfrac{1}{8}\right)$

26. $\left(\dfrac{8}{5} - 1\dfrac{1}{3}\right) - \left(-\dfrac{4}{5} \cdot 10\right)$

27. $\left(\dfrac{9}{10} - \dfrac{2}{3}\right) \div \left(-\dfrac{1}{5}\right)$

28. $\left(-\dfrac{3}{16} \div 2\dfrac{1}{4}\right) + \left(-2\dfrac{1}{8}\right)$

29. $\left(2 - \dfrac{1}{2}\right)^2 + \left(2 + \dfrac{1}{2}\right)^2$

30. $\left(1 - \dfrac{3}{4}\right)\left(1 + \dfrac{3}{4}\right)$

In Exercises 31–34, find $\frac{1}{2}$ of the given number and then square that result. Express your answer as an improper fraction.

31. -7 **32.** -5 **33.** $\dfrac{11}{2}$ **34.** $\dfrac{7}{3}$

In Exercises 35–38, evaluate each formula given that $l = 12$, $w = 8\frac{1}{2}$, $b = 10$, and $h = 7\frac{1}{5}$.

35. $A = lw$ **36.** $P = 2l + 2w$ **37.** $A = \dfrac{1}{2}bh$ **38.** $V = lwh$

In Exercises 39–40, find the perimeter of the figure.

39.

$2\dfrac{7}{8}$ in.

$1\dfrac{1}{4}$ in.

40.

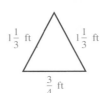

$1\dfrac{1}{3}$ ft $1\dfrac{1}{3}$ ft

$\dfrac{3}{4}$ ft

In Exercises 41–60, simplify each complex fraction.

41. $\dfrac{\dfrac{2}{3}}{\dfrac{4}{5}}$

42. $\dfrac{\dfrac{3}{5}}{\dfrac{9}{25}}$

43. $\dfrac{-\dfrac{14}{15}}{\dfrac{7}{10}}$

44. $\dfrac{\dfrac{5}{27}}{-\dfrac{5}{9}}$

45. $\dfrac{\dfrac{5}{10}}{21}$

46. $\dfrac{\dfrac{6}{3}}{8}$

47. $\dfrac{-\dfrac{5}{6}}{-1\dfrac{7}{8}}$

48. $\dfrac{-\dfrac{4}{3}}{-2\dfrac{5}{6}}$

49. $\dfrac{\dfrac{1}{2} + \dfrac{1}{4}}{\dfrac{1}{2} - \dfrac{1}{4}}$

50. $\dfrac{\dfrac{1}{3} + \dfrac{1}{4}}{\dfrac{1}{3} - \dfrac{1}{4}}$

51. $\dfrac{\dfrac{3}{8} + \dfrac{1}{4}}{\dfrac{3}{8} - \dfrac{1}{4}}$

52. $\dfrac{\dfrac{2}{5} + \dfrac{1}{4}}{\dfrac{2}{5} - \dfrac{1}{4}}$

53. $\dfrac{\dfrac{1}{5} + 3}{-\dfrac{4}{25}}$

54. $\dfrac{-5 - \dfrac{1}{3}}{\dfrac{1}{6} + \dfrac{2}{3}}$

55. $\dfrac{5\dfrac{1}{2}}{-\dfrac{1}{4} + \dfrac{3}{4}}$

56. $\dfrac{4\dfrac{1}{4}}{\dfrac{2}{3} + \left(-\dfrac{1}{6}\right)}$

57. $\dfrac{\dfrac{1}{5}-\left(-\dfrac{1}{4}\right)}{\dfrac{1}{4}+\dfrac{4}{5}}$ **58.** $\dfrac{\dfrac{1}{8}-\left(-\dfrac{1}{2}\right)}{\dfrac{1}{4}+\dfrac{3}{8}}$ **59.** $\dfrac{\dfrac{1}{3}+\left(-\dfrac{5}{6}\right)}{1\dfrac{1}{3}}$ **60.** $\dfrac{\dfrac{3}{7}+\left(-\dfrac{1}{2}\right)}{1\dfrac{3}{4}}$

APPLICATIONS *In Exercises 61–70, solve each problem.*

61. SANDWICH SHOP   A sandwich shop sells a $\frac{1}{2}$-pound club sandwich, made up of turkey meat and ham. The owner buys the turkey in $1\frac{3}{4}$-pound packages and the ham in $2\frac{1}{2}$-pound packages. If he mixes a package of each of the meats together, how many sandwiches can he make from the mixture?

62. SKIN CREAM   Using a formula of $\frac{1}{2}$ ounce of sun block, $\frac{2}{3}$ ounce of moisturizing cream, and $\frac{3}{4}$ ounce of lanolin, a beautician mixes her own brand of skin cream. She packages it in $\frac{1}{4}$-ounce tubes. How many tubes can be produced using this formula?

63. PHYSICAL FITNESS   Two people begin their workout from the same point on a bike path and travel in opposite directions, as shown in Illustration 1. How far apart are they in $1\frac{1}{2}$ hours? Use the chart to help organize your work.

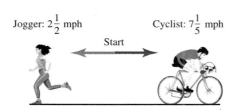

ILLUSTRATION 1

	Rate (mph)	Time (hr)	Distance (mi)
Jogger			
Cyclist			

64. STOCK MARKET   The value of a stock dropped for 5 consecutive days, as shown in Illustration 2. What was the total point loss for the week?

Mon Tues Wed Thurs Fri

$-\dfrac{3}{8}$ $-1\dfrac{3}{4}$ $-1\dfrac{3}{4}$ $-\dfrac{3}{8}$ $-\dfrac{3}{8}$

ILLUSTRATION 2

65. POSTAGE RATES   Can the advertising package in Illustration 3 be mailed for the one-ounce rate?

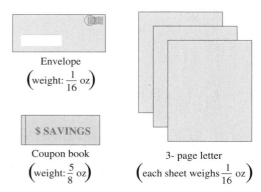

Envelope $\left(\text{weight: }\dfrac{1}{16}\text{ oz}\right)$

Coupon book $\left(\text{weight: }\dfrac{5}{8}\text{ oz}\right)$

3- page letter $\left(\text{each sheet weighs }\dfrac{1}{16}\text{ oz}\right)$

ILLUSTRATION 3

66. PLYWOOD   To manufacture a sheet of plywood, several layers of thin laminate are glued together, as shown in Illustration 4. Then an exterior finish is affixed to the top and bottom. How thick is the finished product?

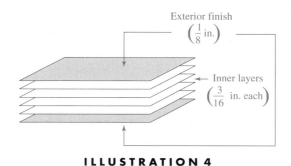

Exterior finish $\left(\dfrac{1}{8}\text{ in.}\right)$

Inner layers $\left(\dfrac{3}{16}\text{ in. each}\right)$

ILLUSTRATION 4

67. PHYSICAL THERAPY   After back surgery, a patient undertook a walking program to rehabilitate her back muscles, as specified in Illustration 5. What was the total distance she walked over this three-week period?

Week	Distance per day
#1	$\dfrac{1}{4}$ mile
#2	$\dfrac{1}{2}$ mile
#3	$\dfrac{3}{4}$ mile

ILLUSTRATION 5

68. READING PROGRAM To improve reading skills, elementary school children read silently at the end of the school day for $\frac{1}{4}$ hour on Mondays and for $\frac{1}{2}$ hour on Fridays. For the month of January, how many total hours did the children read silently in class? (See Illustration 6.)

S	M	T	W	T	F	S
	1	2	3	4	5	6
7	8	9	10	11	12	13
14	15	16	17	18	19	20
21	22	23	24	25	26	27
28	29	30	31			

ILLUSTRATION 6

69. AMUSEMENT PARK At the end of a ride at an amusement park, a boat splashes into a pool of water. The time in seconds that it takes for two pipes to refill the pool is given by

$$\frac{1}{\frac{1}{10} + \frac{1}{15}}$$

Find this time.

70. HIKING A scout troop plans to hike from the campground to Glenn Peak. (See Illustration 7.) Since the terrain is steep, they plan to stop and rest after every $\frac{2}{3}$ mile. With this plan, how many parts will there be to this hike?

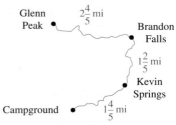

ILLUSTRATION 7

WRITING *Write a paragraph using your own words.*

71. What is a complex fraction?

72. Explain method 1 for simplifying complex fractions.

73. Explain why establishing rules for order of operations is important.

74. Explain method 2 for simplifying complex fractions.

REVIEW

75. Subtract 879 from 1,023.

76. Multiply 879 by 23.

77. Divide 1,665 by 45.

78. List the factors of 24.

79. Evaluate: $2 + 3[-3 - (-4 - 1)]$.

80. What is the sign of the quotient of two numbers with unlike signs?

81. Find the prime factorization of 288.

82. Subtract: $\frac{7}{8} - \frac{2}{3}$.

The Fundamental Property of Fractions

The **fundamental property of fractions** states that multiplying or dividing the numerator and the denominator of a fraction by the same nonzero number does not change the value of the fraction. This property is used to simplify fractions and to express fractions in higher terms. The following problems review both procedures. Complete each solution.

1. Simplify $\dfrac{15}{25}$.

Step 1: The numerator and denominator share a common factor of ___.

Step 2: Apply the fundamental property of fractions. Divide the numerator and denominator by the common factor.

$$\frac{15}{25} = \frac{15 \div}{25 \div}$$

Step 3: Do the divisions to simplify the fraction.

$$= \frac{3}{5}$$

2. In practice, we often show the simplifying process described in problem 1 in a different form.

Step 1: Factor 15 as ___ $\cdot$ 3 and 25 as ___ $\cdot$ 5.

$$\frac{15}{25} = \frac{\cdot\, 3}{\cdot\, 5}$$

Step 2: The slashes and small 1's indicate that the numerator and denominator have been divided by ___.

$$= \frac{\overset{1}{\cancel{5}} \cdot 3}{\underset{1}{\cancel{5}} \cdot 5}$$

Step 3: Multiply in the numerator and denominator.

$$= \frac{3}{5}$$

3. When adding or subtracting fractions, we often need to express a fraction in higher terms. Express $\frac{2}{3}$ as a fraction with a denominator of 18.

Step 1: We must multiply 3 by ___ to obtain 18.

Step 2: Apply the fundamental property of fractions. Multiply the numerator and denominator by ___.

$$\frac{2}{3} = \frac{2 \cdot}{3 \cdot}$$

Step 3: Do the multiplication in the numerator and the denominator.

$$= \frac{12}{18}$$

CHAPTER REVIEW

The Fundamental Property of Fractions

CONCEPTS

Fractions are used to indicate equal parts of a whole.

A fraction is composed of a *numerator*, a *denominator*, and a *fraction bar.*

The − sign of a negative fraction can be written in the numerator, in the denominator, or in front of the fraction.

Equivalent fractions represent the same number.

The *fundamental property of fractions:* Dividing the numerator and denominator of a fraction by the same nonzero number does not change the value of the fraction.

To *simplify* a fraction that is not in lowest terms, divide the numerator and denominator by the same number.

A fraction is in *lowest terms* if the only common factor the numerator and denominator share is 1.

The *fundamental property of fractions:* Multiplying the numerator and denominator of a fraction by a nonzero number does not change its value.

Expressing a fraction in higher terms results in an equivalent fraction that involves larger numbers or more complex terms.

REVIEW EXERCISES

1. If a woman gets seven hours of sleep each night, what part of a whole day does she spend sleeping?

2. In Illustration 1, why can't we say that $\frac{3}{4}$ of the figure is shaded?

ILLUSTRATION 1

3. Write the fraction $\frac{2}{-3}$ in two other ways.

4. What concept about fractions does Illustration 2 demonstrate?

ILLUSTRATION 2

5. Explain the procedure shown here.
$$\frac{4}{6} = \frac{4 \div 2}{6 \div 2} = \frac{2}{3}$$

6. Explain what the slashes and the 1's mean.
$$\frac{4}{6} = \frac{\overset{1}{\cancel{2} \cdot 2}}{\underset{1}{\cancel{2} \cdot 3}} = \frac{2}{3}$$

7. Simplify each fraction to lowest terms.
   a. $\frac{15}{45}$ **b.** $\frac{20}{48}$ **c.** $\frac{63}{84}$ **d.** $\frac{66}{108}$

8. Simplify each fraction to lowest terms.
   a. $\frac{20}{24}$ **b.** $\frac{25}{35}$ **c.** $\frac{18}{27}$ **d.** $\frac{21}{49}$

9. Explain what is being done and why it is valid.
$$\frac{5}{8} = \frac{5 \cdot 2}{8 \cdot 2} = \frac{10}{16}$$

10. Write each fraction or whole number with the indicated denominator.
   a. $\frac{2}{3}$, 18   **b.** $-\frac{3}{8}$, 16   **c.** $\frac{7}{15}$, 45   **d.** 4, 9

Multiplying Fractions

To *multiply two fractions,* multiply their numerators and multiply their denominators.

11. Multiply.

a. $\dfrac{1}{2} \cdot \dfrac{1}{3}$ **b.** $\dfrac{2}{5}\left(-\dfrac{7}{9}\right)$ **c.** $\dfrac{9}{16} \cdot \dfrac{20}{27}$ **d.** $\dfrac{5}{6} \cdot \dfrac{1}{3} \cdot \dfrac{18}{25}$

e. $\dfrac{3}{5} \cdot 7$ **f.** $(-4)\left(-\dfrac{9}{16}\right)$ **g.** $3\left(\dfrac{1}{3}\right)$ **h.** $-\dfrac{6}{7}\left(-\dfrac{7}{6}\right)$

12. Tell whether each statement is true or false.

a. $\dfrac{3}{4}(2) = \dfrac{3(2)}{4}$ **b.** $-\dfrac{5}{9}(3) = -\dfrac{5}{9(3)}$

13. Multiply.

a. $\dfrac{3}{5} \cdot \dfrac{10}{27}$ **b.** $-\dfrac{2}{3}\left(\dfrac{4}{7}\right)$ **c.** $\dfrac{4}{9} \cdot \dfrac{3}{28}$ **d.** $9\left(-\dfrac{5}{81}\right)$

An *exponent* indicates repeated multiplication.

14. Evaluate each power.

a. $\left(\dfrac{3}{4}\right)^{2}$ **b.** $\left(-\dfrac{5}{2}\right)^{3}$ **c.** $\left(\dfrac{2}{3}\right)^{2}$ **d.** $\left(-\dfrac{2}{5}\right)^{3}$

In mathematics, the word *of* usually means multiply.

15. GRAVITY ON THE MOON   Objects on the moon weigh only one-sixth as much as on earth. How much will an astronaut weigh on the moon if he weighs 180 pounds on earth?

The *area of a triangle:*

$A = \dfrac{1}{2}bh$

16. Find the area of the triangular sign in Illustration 3.

ILLUSTRATION 3

Dividing Fractions

Two numbers are called *reciprocals* if their product is 1.

17. Find the reciprocal of each number.

a. $\dfrac{1}{8}$ **b.** $-\dfrac{11}{12}$ **c.** 5 **d.** $\dfrac{8}{7}$

To *divide two fractions,* multiply the first by the reciprocal of the second.

18. Divide.

a. $\dfrac{1}{6} \div \dfrac{11}{25}$ **b.** $\dfrac{7}{8} \div \dfrac{1}{4}$

c. $\dfrac{15}{16} \div 10$ **d.** $8 \div \dfrac{16}{5}$

19. Divide.

a. $-\dfrac{3}{8} \div \dfrac{1}{4}$ **b.** $\dfrac{4}{5} \div \left(-\dfrac{1}{2}\right)$

c. $-\dfrac{2}{3} \div \left(-\dfrac{3}{2}\right)$ **d.** $\dfrac{2}{3} \div \left(-\dfrac{1}{9}\right)$

20. GOLD COINS How many $\frac{1}{16}$-ounce coins can be cast from a $\frac{3}{4}$-ounce bar of gold?

SECTION 3.4

Adding and Subtracting Fractions

To add (or subtract) fractions with like denominators, add (or subtract) their numerators and write the result over the common denominator.

21. Add or subtract.

a. $\dfrac{2}{7} + \dfrac{3}{7}$ b. $-\dfrac{3}{5} - \dfrac{3}{5}$ c. $\dfrac{3}{4} - \dfrac{1}{4}$ d. $\dfrac{7}{8} + \dfrac{3}{8}$

22. Explain why we cannot add $\frac{1}{2} + \frac{2}{3}$ without doing some preliminary work.

The *LCD* must include the set of prime factors of each of the denominators.

23. Use prime factorization to find the least common denominator for fractions with denominators of 45 and 30.

To add or subtract fractions with unlike denominators, we must first express them as equivalent fractions with the same denominator, preferably the LCD.

24. Add or subtract.

a. $\dfrac{1}{6} + \dfrac{2}{3}$ b. $\dfrac{2}{5} + \left(-\dfrac{3}{8}\right)$

c. $-\dfrac{3}{8} - \dfrac{5}{6}$ d. $3 - \dfrac{1}{7}$

e. $\dfrac{2}{25} - \dfrac{3}{10}$ f. $\dfrac{1}{3} + \dfrac{7}{4}$

g. $\dfrac{13}{6} - 6$ h. $\dfrac{1}{3} + \dfrac{1}{4} + \dfrac{1}{5}$

25. MACHINE SHOP See Illustration 4. How much must be milled off the $\frac{3}{4}$-inch-thick steel rod so that the collar will slip over the end of it?

$\dfrac{17}{32}$ in. $\dfrac{3}{4}$ in.

Steel rod

ILLUSTRATION 4

To *compare fractions,* write them as equivalent fractions with the same denominator. Then the fraction with the larger numerator will be the larger fraction.

26. TELEMARKETING In the first hour of work, a telemarketer made 2 sales out of 9 telephone calls. In the second hour, she made 3 sales out of 11 calls. During which hour was the rate of sales to calls better?

SECTION 3.5

Multiplying and Dividing Mixed Numbers

A mixed number is the sum of its whole-number part and its fractional part.

27. What fact about mixed numbers and improper fractions does Illustration 5 demonstrate?

ILLUSTRATION 5

To change an *improper fraction* to a mixed number, divide the numerator by the denominator to obtain the whole-number part. Write the remainder over the denominator for the fractional part.

28. Express each improper fraction as a mixed number or whole number.

a. $\dfrac{16}{5}$ **b.** $-\dfrac{47}{12}$ **c.** $\dfrac{6}{6}$ **d.** $\dfrac{14}{6}$

To change a *mixed number* to an improper fraction, multiply the whole number by the denominator and add the result to the numerator. Write this sum over the denominator.

29. Write each mixed number as an improper fraction.

a. $9\dfrac{3}{8}$ **b.** $-2\dfrac{1}{5}$ **c.** $100\dfrac{1}{2}$ **d.** $1\dfrac{99}{100}$

30. Graph $-2\frac{2}{3}$, $\frac{8}{9}$, $2\frac{11}{24}$.

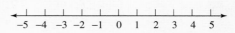

To *multiply* or *divide mixed numbers,* change the mixed numbers to improper fractions and then do the operations as usual.

31. Multiply or divide. Write answers as mixed numbers when appropriate.

a. $-5\dfrac{1}{4} \cdot \dfrac{2}{35}$ **b.** $\left(-3\dfrac{1}{2}\right) \div \left(-3\dfrac{2}{3}\right)$

c. $\left(-6\dfrac{2}{3}\right)(-6)$ **d.** $-8 \div 3\dfrac{1}{5}$

32. CAMERA TRIPOD   The three legs of a tripod can be extended to become $5\frac{1}{2}$ times their original length. If each leg is $8\frac{3}{4}$ inches long when collapsed, how long will a leg become when it is completely extended?

Adding and Subtracting Mixed Numbers

To add (or subtract) small mixed numbers, change each to an improper fraction and use the method of Section 3.4.

33. Add or subtract.

a. $1\dfrac{3}{8} + 2\dfrac{1}{5}$ **b.** $3\dfrac{1}{2} + 2\dfrac{2}{3}$

c. $2\dfrac{5}{6} - 1\dfrac{3}{4}$ **d.** $3\dfrac{7}{16} - 2\dfrac{1}{8}$

34. PAINTING SUPPLIES   In a project to restore a house, painters used $10\frac{3}{4}$ gallons of primer, $21\frac{1}{2}$ gallons of latex paint, and $7\frac{2}{3}$ gallons of enamel. Find the total number of gallons of paint used.

To add large mixed numbers, add the whole numbers and the fractions separately.

Vertical form can be used to add or subtract large mixed numbers.

35. Add or subtract.

a. $\begin{array}{r} 33\frac{1}{9} \\ + 49\frac{1}{6} \end{array}$ **b.** $\begin{array}{r} 98\frac{11}{20} \\ + 14\frac{3}{5} \end{array}$

c. $\begin{array}{r} 50\frac{5}{8} \\ - 19\frac{1}{6} \end{array}$ **d.** $\begin{array}{r} 75\frac{3}{4} \\ - 59 \end{array}$

If the fraction being subtracted is larger than the first fraction, we need to borrow from the whole number.

36. Subtract.

a. $23\dfrac{1}{3} - 2\dfrac{5}{6}$ **b.** $39 - 4\dfrac{5}{8}$

Order of Operations and Complex Fractions

A *complex fraction* is a fraction whose numerator or denominator, or both, contain one or more fractions or mixed numbers.

To simplify a complex fraction, *Method 1:* The main fraction bar of a complex fraction indicates division.

Method 2: Multiply the numerator and denominator of the complex fraction by the LCD of all the fractions that appear in it.

37. Evaluate each numerical expression.

a. $\dfrac{3}{4} + \left(-\dfrac{1}{3}\right)^2\left(\dfrac{5}{4}\right)$

b. $\left(\dfrac{2}{3} \div \dfrac{16}{9}\right) - \left(1\dfrac{2}{3} \cdot \dfrac{1}{15}\right)$

38. Simplify each complex fraction.

a. $\dfrac{\dfrac{3}{5}}{-\dfrac{17}{20}}$

b. $\dfrac{\dfrac{2}{3} - \dfrac{1}{6}}{-\dfrac{3}{4} - \dfrac{1}{2}}$

39. Evaluate the formula $P = 2l + 2w$ for the following values of l and w.

a. $l = \dfrac{3}{4}$ and $w = \dfrac{2}{5}$

b. $l = 2\dfrac{1}{3}$ and $w = 3\dfrac{1}{4}$

1. See Illustration 1.
 a. What fractional part of the plant is above ground?
 b. What fractional part of the plant is below ground?

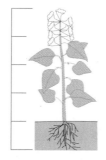

ILLUSTRATION 1

2. Simplify each fraction.
 a. $\dfrac{27}{36}$ **b.** $\dfrac{72}{180}$

3. Multiply: $-\dfrac{3}{4} \cdot \dfrac{1}{5}$.

4. COFFEE DRINKERS   Of 100 adults surveyed, $\frac{2}{5}$ said they started off their morning with a cup of coffee. Of the 100, how many would this be?

5. Divide: $\dfrac{4}{3} \div \dfrac{2}{9}$.

6. Subtract: $\dfrac{5}{6} - \dfrac{4}{5}$.

7. Express $\frac{7}{8}$ as an equivalent fraction with denominator 24.

8. Graph $2\dfrac{4}{5}$, $-1\dfrac{1}{7}$, and $\dfrac{7}{6}$.

```
 |----|----|----|----|----|----|
 -2   -1   0    1    2    3
```

9. SPORTS CONTRACT   A basketball player signed a nine-year contract for \$13$\frac{1}{2}$ million. How much is this per year?

10. Evaluate the formula $A = \frac{1}{2}bh$ when $b = \frac{7}{3}$ and $h = \frac{15}{4}$.

11. Add: $57\dfrac{5}{9} + 103\dfrac{3}{4}$.

12. Subtract: $67\dfrac{1}{4} - 29\dfrac{5}{6}$.

13. Add: $-\dfrac{3}{7} + 2$.

14. SEWING   When cutting material for a $10\frac{1}{2}$-inch-wide placemat, a seamstress allows $\frac{5}{8}$ inch on either end for a hem. How wide should the material be cut? See Illustration 2.

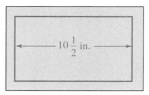

ILLUSTRATION 2

15. In Illustration 3, find the perimeter and the area of the triangle.

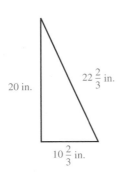

ILLUSTRATION 3

16. Evaluate:

$$\left(\frac{2}{3} \cdot \frac{5}{16}\right) - \left(-1\frac{3}{5} \div 4\frac{4}{5}\right)$$

17. Simplify the complex fraction.

$$\frac{-\dfrac{5}{6}}{\dfrac{7}{8}}$$

18. Simplify the complex fraction.

$$\frac{\dfrac{1}{2} + \dfrac{1}{3}}{-\dfrac{1}{6} - \dfrac{1}{3}}$$

19. Explain what mathematical concept is being shown.

a. $\dfrac{6}{8} = \dfrac{3 \cdot \cancel{2}^{1}}{4 \cdot \cancel{2}_{1}} = \dfrac{3}{4}$

b.

c. $\dfrac{3}{5} = \dfrac{3 \cdot 4}{5 \cdot 4} = \dfrac{12}{20}$

20. JOB APPLICANTS   Three-fourths of the applicants for a position had previous experience. The number who did not have prior experience was 36. How many people applied?

21. Show that finding $\frac{1}{2}$ of a number is the same as dividing the number by 2.

22. Show each step in the process that is used to find the LCD for two fractions whose denominators are 24 and 72.

Decimals

4

DECIMALS PROVIDE ANOTHER WAY TO REPRESENT FRACTIONS AND MIXED NUMBERS. THEY ARE USED IN MEASUREMENT BECAUSE IT IS EASY TO PUT THEM IN ORDER AND TO COMPARE THEM.

4.1   An Introduction to Decimals

In this section, you will learn about

- **Decimal fractions**
- **The place value system for decimal numbers**
- **Reading and writing decimals**
- **Comparing decimals and the number line**
- **Rounding**

INTRODUCTION. This section introduces the **decimal numeration system**—an extension of the place value system that we used when working with whole numbers. You may not realize it, but you have often worked with the decimal numeration system, because we use decimal notation with money.

Decimal fractions

Decimal fractions are fractions with a denominator of 10, 100, 1,000, 10,000, and so on. However, when writing a decimal, we don't use a fraction bar, nor is a denominator shown.

In Figure 4-1, a rectangle is divided into 10 equal parts. One-tenth of the figure is shaded. We can use either the fraction $\frac{1}{10}$ or the decimal 0.1 to describe the shaded region.

$$\frac{1}{10} = 0.1$$

We have expressed the amount of the figure that is shaded in two different ways. Both are read as "one-tenth."

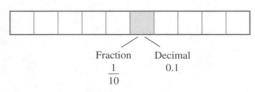

Fraction   Decimal
$\frac{1}{10}$ 0.1

FIGURE 4-1

In Figure 4-2, a square is divided into 100 equal parts. One of the 100 parts is shaded; it can be represented by the fraction $\frac{1}{100}$ or by the decimal 0.01. Both are read as "one one-hundredth."

$$\frac{1}{100} = 0.01$$

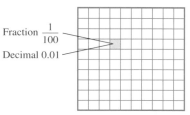

Fraction $\frac{1}{100}$

Decimal 0.01

FIGURE 4-2

The place value system for decimal numbers

Decimal numbers are written by placing digits (0, 1, 2, 3, 4, 5, 6, 7, 8, 9) into place value columns that are separated by a **decimal point.** See Figure 4-3. The place value names of all the columns to the right of the decimal point end in "th." The "th" tells us that the value of the column is a fraction whose denominator is a power of ten. Columns to the left of the decimal point have a value greater than or equal to 1; columns to the right of the decimal point have a value less than 1. We can show the value represented by each digit of a decimal by using **expanded notation.**

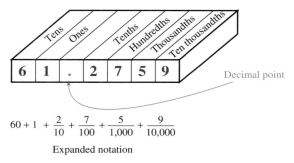

Decimal point

$$60 + 1 + \frac{2}{10} + \frac{7}{100} + \frac{5}{1,000} + \frac{9}{10,000}$$

Expanded notation

FIGURE 4-3

Decimal points are used to separate the whole-number part of a decimal from its fractional part.

12.37

Whole-number part. — Fractional part.

Decimal point.

When there is no whole-number part of a decimal, we show that by entering a zero to the left of the decimal point.

.85 = 0.85

No whole number part. Enter a zero here.

We can write a whole number in decimal notation by placing a decimal point to its right and then adding a zero, or zeros, to the right of the decimal point.

99 = 99.0 = 99.00

A whole number. Place a decimal point here and enter a zero, or zeros, to the right of it.

Writing additional zeros to the right of the decimal point *following the last digit* does not change the value of the decimal.

$$12.57 = 12.570 = 12.5700$$

These additional zeros do not change the value of the decimal.

Reading and writing decimals

The decimal 12.37 can be read as "twelve point three seven." Another way of reading a decimal states the whole-number part first and then the fractional part.

Reading a decimal	To read a decimal:
	1. Look to the left of the decimal point and say the name of the whole number.
	2. The decimal point is then read as "and."
	3. Say the fractional part of the decimal as a whole number followed by the name of the place value column of the digit that is farthest to the right.

Using this procedure, here is the other way to read 12.37.

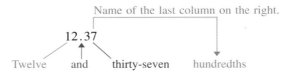

When we read a decimal in this way, it is easy to write it in words and as a mixed number.

12.37	Twelve and thirty-seven hundredths	$12\frac{37}{100}$
Decimal	Words	Mixed number

EXAMPLE 1 *Writing a decimal in other forms.* Write each decimal in words and then as a fraction or mixed number. **Do not simplify the fraction.**

a. The world speed record for a human-powered vehicle is 65.484 miles per hour, set in 1986.

b. The smallest fresh-water fish is the dwarf pygmy goby, found in the Philippines. Adult males weigh 0.00014 ounce.

Solution

a. In 65.48**4**, the red 4 is in the thousandths column. We can express this decimal as sixty-five *and* four hundred eighty-four thousandths, or $65\frac{484}{1,000}$.

b. In 0.0001**4**, the red 4 is in the hundred-thousandths column. We can express this decimal as fourteen hundred-thousandths, or $\frac{14}{100,000}$.

Self Check

Write each decimal in words and then as a mixed number.

a. Sputnik 1, the first artificial satellite, weighed 184.3 pounds.

b. The planet Mercury makes one revolution every 87.9687 days.

Answer:
a. One hundred eighty-four and three-tenths, or $184\frac{3}{10}$
b. Eighty-seven and nine thousand six hundred eighty-seven ten thousandths, or $87\frac{9,687}{10,000}$ ∎

Decimals can be negative. For example, a record low temperature of $-128.6°$ F was recorded in Vostok, Antarctica on July 21, 1983. This is read as "negative one hundred twenty-eight and six tenths." Written as a mixed number, it is $-128\frac{6}{10}$.

Comparing decimals and the number line

The relative sizes of a set of decimals can be determined by scanning their place value columns from left to right, column by column, looking for a difference in the digits. For example,

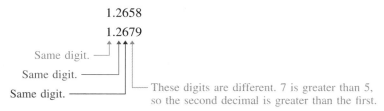

1.2658

1.2679

Same digit.

Same digit.

Same digit.

These digits are different. 7 is greater than 5, so the second decimal is greater than the first.

Thus, 1.2679 is greater than 1.2658. We write $1.2679 > 1.2658$.

Comparing positive decimals	To compare two positive decimals:
	1. Make sure both numbers have the same number of decimal places to the right of the decimal point. Write any additional zeros necessary to achieve this.
	2. Compare the digits of each decimal, column by column, working from left to right.
	3. When two digits differ, the decimal with the greater digit is the greater number.

EXAMPLE 2   *Comparing positive decimals.*   Which is greater, 54.9 or 54.929?

Solution

   54.900   Write two zeros after 9 so that the decimals have the same number of digits to
   54.929   the right of the decimal point.
     ↑

This is the first column in which the digits differ. Since 2 is greater than 0, we can conclude that $54.929 > 54.9$.

Self Check

Which is greater, 113.7 or 113.657?

Answer: 113.7

Comparing negative decimals	To compare two negative decimals:
	1. Make sure both numbers have the same number of decimal places to the right of the decimal point. Write any additional zeros necessary to achieve this.
	2. Compare the digits of each decimal, column by column, working from left to right.
	3. When two digits differ, the decimal with the smaller digit is the greater number.

EXAMPLE 3   *Comparing negative decimals.*   Which is greater, -10.45 or -10.419?

Solution

  -10.450 Write a 0 after 5 to help in the comparison.
  -10.419
     ↑

Working from left to right, this is the first column where the digits differ. Since 1 is less than 5, we conclude that $-10.419 > -10.45$.

Self Check

Which is greater, -703.8 or -703.78?

Answer: -703.78

A number line can be used to show the relative sizes of decimals.

EXAMPLE 4 **Graphing decimals.** Graph −1.8, −1.23, −0.3, and 1.89.

Solution

−1.8 −1.23 −0.3   1.89

−2   −1   0   1   2

Self Check

Graph −1.1, −0.6, 0.8, and 1.9.

−1.1 −0.6   0.8   1.9

−2   −1   0   1   2

Rounding

When working with decimals, we often round answers to a specific number of decimal places.

Rounding a decimal	1. To round a decimal to a specified decimal place, locate the digit in that place. Call it the *rounding digit*. 2. Look at the *test digit* to the right of the rounding digit. 3. If the test digit is 5 or greater, round up by adding 1 to the rounding digit and dropping all the digits to its right. If the test digit is less than 5, round down by keeping the rounding digit and dropping all the digits to its right.

EXAMPLE 5 **Chemistry.** In a chemistry class, a student uses a balance to weigh a compound. The digital readout on the scale shows 1.2387 g. Round this decimal to the nearest thousandth of a gram.

Solution We are asked to round to the nearest thousandth.

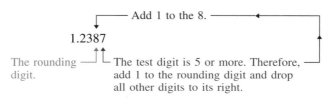

Add 1 to the 8.

1.2387

The rounding digit.   The test digit is 5 or more. Therefore, add 1 to the rounding digit and drop all other digits to its right.

The compound weighs approximately 1.239 g.

EXAMPLE 6 **Rounding decimals.** Round each decimal to the indicated place value.

a. −645.13   nearest tenth

b. 33.097   nearest hundredth

Solution

a.   −645.13

Rounding digit.   Since the test digit is less than 5, drop it and all the digits to its right.

The result is −645.1

b.   33.097

Rounding digit.   Since the test digit is 7, we add 1 to 9 and drop all the digits to the right.

1 0
33.09   Adding a 1 to the 9 requires that we carry a 1 to the next column.

When we are asked to round to the nearest hundredth, we must have a digit in the hundredths column, even if it is a zero. Therefore, the result is 33.10.

Self Check

Round each decimal to the indicated place value.

a. −708.522   nearest tenth

b. 9.1198   nearest thousandth

Answer: **a.** −708.5,
b. 9.120

VOCABULARY   *In Exercises 1–4, fill in the blanks to make a true statement.*

1. Give the name of each place value column.

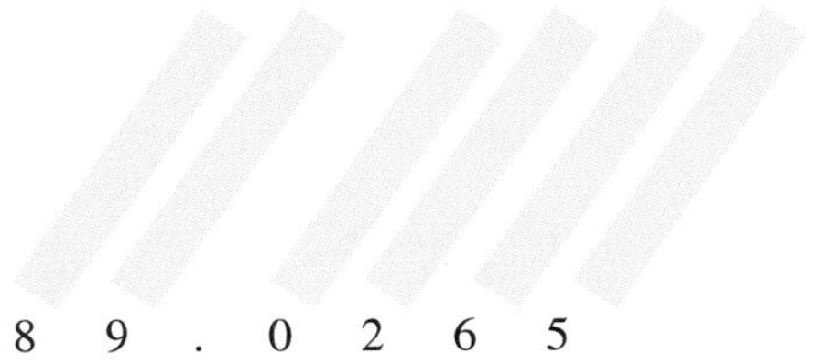

8   9   .   0   2   6   5

2. We can show the value represented by each digit of the decimal 98.6213 by using _____ notation.

$$98.6213 = 90 + 8 + \frac{6}{10} + \frac{2}{100} + \frac{1}{1,000} + \frac{3}{10,000}$$

3. We can approximate a decimal number using the process called _____.

4. When reading a decimal number, the decimal point can be read as "_____" or "_____."

CONCEPTS

5. Given: 32.415.
   a. Write this decimal in words.

   b. What is its whole-number part?
   c. What is its fractional part?
   d. Write this decimal in expanded notation.

6. Given: −0.0061.
   a. Write this decimal in words.

   b. What is its whole-number part?
   c. What is its fractional part?
   d. Write this decimal in expanded notation.

7. Graph $\frac{7}{10}$, −0.7, $-1\frac{1}{100}$, and 1.01.

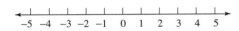

8. Graph −1.21, −3.29, and −4.25.

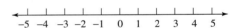

9. True or false?
   a.   0.9 = 0.90
   b.   1.260 = 1.206
   c. −1.2800 = −1.280
   d.   0.001 = .0010

10. Consider the decimal number 0.3.
   a. Enter a zero to the right of 0.3 and write this new decimal as a fraction.
   b. Simplify the fraction from part a.
   c. What can you conclude from the results of parts a and b?

11. Represent the shaded part of the square in Illustration 1 using a fraction and a decimal.

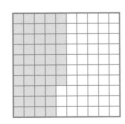

ILLUSTRATION 1

12. Represent the shaded part of the figure in Illustration 2 using a fraction and a decimal.

ILLUSTRATION 2

13. The line segment in Illustration 3 is one inch long. Show a length of 0.3 inch on it.

_____

ILLUSTRATION 3

14. Read the meter in Illustration 4. What decimal is indicated by the arrow?

ILLUSTRATION 4

NOTATION

15. Construct a decimal number by writing
0 in the tenths column,
4 in the thousandths column,
1 in the tens column,
9 in the thousands column,
8 in the hundreds column,
2 in the hundredths column,
5 in the ten thousandths column, and
6 in the ones column.

16. Represent each of these situations using a signed number.
a. A deficit of $15,600.55
b. A river 6.25 feet under flood stage
c. A state budget $6.4 million in the red
d. 3.9 degrees below zero
e. 17.5 seconds prior to liftoff
f. A checking account overdrawn by $33.45

PRACTICE *In Exercises 17–24, write each decimal in words and then as a fraction or mixed number.*

17. 50.1   **18.** 0.73   **19.** −0.0137   **20.** −76.09

21. 304.0003   **22.** 68.91   **23.** −72.493   **24.** −31.5013

In Exercises 25–28, write each decimal using numbers.

25. Negative thirty-nine hundredths

26. Negative twenty-seven and forty-four hundredths

27. Six and one hundred eighty-seven thousandths

28. Ten and fifty-six ten-thousandths

In Exercises 29–32, round to the nearest tenth.

29. 506.098   **30.** 0.441   **31.** 77.21   **32.** 3,987.8911

In Exercises 33–36, round to the nearest hundredth.

33. −0.137   **34.** −808.0897   **35.** 33.0032   **36.** 64.0059

In Exercises 37–40, round to the nearest thousandth.

37. 3.2325   **38.** 16.0995   **39.** 55.0391   **40.** 2,300.9998

In Exercises 41–44, round to the nearest whole number.

41. 38.901   **42.** 405.64   **43.** 2,988.399   **44.** 10,453.27

In Exercises 45–46, round each dollar amount to the indicated value.

45. $3,090.28
     a. dollar   **b.** ten cents

46. $289.73
     a. dollar   **b.** ten cents

In Exercises 47–50, fill in the blanks with the proper symbol (<, >, or =) to make a true statement.

47. −23.45 ___ −23.1   **48.** −301.98 ___ −302.45   **49.** −.065 ___ −.066   **50.** −3.99 ___ −3.9888

In Exercises 51–52, arrange the decimals in order, from least to greatest.

51. 132.64, 132.6499, 132.6401   **52.** 0.007, 0.00697, 0.00689

APPLICATIONS *In Exercises 53–62, solve each problem.*

53. GEOLOGY Geologists classify types of soil according to the grain size of the particles that make up the soil. The four major classifications are:

clay   0.00008 in. and under
silt   0.00008 in. to 0.002 in.
sand   0.002 in. to 0.08 in.
granule 0.08 in. to 0.15 in.

Complete the chart in Illustration 5 by classifying each sample.

Sample	Location	Size (in.)	Classification
A	NE corner	0.095	
B	dry lake	0.00003	
C	hilltop	0.0007	
D	riverbank	0.009	

ILLUSTRATION 5

54. MICROSCOPE A microscope used in a lab is capable of viewing structures that range in size from 0.1 to 0.0001 centimeter. Which of the structures listed in Illustration 6 would be visible through this microscope?

Structure	Size (in cm)
bacterium	0.00011
plant cell	0.015
virus	0.000017
animal cell	0.00093
asbestos fiber	0.0002

ILLUSTRATION 6

55. HORSE RACING During a workout of a racehorse, the trainer recorded the time it took the horse to run each successive quarter mile. These "splits" were 24.15, 24.78, 24.82, and 24.07 seconds. From what you can learn from these times, describe the horse's strengths and weaknesses.

56. EXCHANGE RATES When changing money from one country into money of another country, an exchange rate is used. See Illustration 7. For which currencies did the exchange rate rise over the three-month period?

Value of $1		
January	**March**	**Foreign currency**
7.6550	7.6650	Mexican pesos
0.6450	0.6530	British pounds
101.68	101.41	Japanese yen
1.4380	1.4355	German marks

ILLUSTRATION 7

57. METRIC SYSTEM The metric system is widely used in science to measure length (meters), weight (grams), and capacity (liters). Round each decimal to the nearest hundredth.
 a. 1 yd is 0.9144 m.
 b. 1 ft is 0.3048 m.
 c. 1 mi is 1,609.344 m.
 d. 1 lb is 453.59237 g.
 e. 1 oz is 28.349523 g.
 f. 1 gal is 3.785306 l.

58. DEWEY DECIMAL SYSTEM A widely used system for classifying books in a library is the Dewey Decimal System. Books on the same subject are grouped together by number. For example, books about the arts are assigned numbers between 700 and 799. When stacked on the shelves, the books are to be in numerical order, from left to right. How should the titles in Illustration 8 be rearranged to be in the proper order?

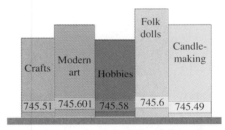

ILLUSTRATION 8

59. GASOLINE PRICES Use the data in Illustration 9 to construct a bar graph showing the national average price per gallon for self-serve regular gasoline.

Year	1992	1993	1994	1995	1996
Price	$1.08	$1.12	$1.09	$1.13	$1.24

ILLUSTRATION 9

60. TURKEY CONSUMPTION   Use the data in Illustration 10 to construct a line graph showing the average turkey consumption per person per year (in pounds).

Year	1988	1990	1992	1994
Pounds	15.7	17.6	18.0	18.0

ILLUSTRATION 10

61. MONEY   We use a decimal point when working with dollars, but the decimal point is not necessary when working with cents. For each dollar amount in Illustration 11, give the equivalent amount expressed as cents.

Dollars	Cents
$0.50	
$0.05	
$0.55	
$5.00	
$0.01	

ILLUSTRATION 11

62. OLYMPICS   The results of the women's all-around gymnastic competition in the 1988 Los Angeles Olympic Games are shown in Illustration 12. Which gymnasts won the gold, silver, and bronze medals?

Name	Country	Score
Simona Pauca	Romania	78.675
Ma Yanhong	China	77.85
Julianne McNamara	U.S.A.	78.4
Mary Lou Retton	U.S.A.	79.175
Ecaterina Szabo	Romania	79.125
Laura Cutina	Romania	78.3

ILLUSTRATION 12

WRITING   *Write a paragraph using your own words.*

63. Explain the difference between ten and a tenth.

64. "The more digits a number contains, the larger it is." Is this statement true? Explain your response.

65. How are fractions and decimals related?

66. Explain the benefits of a monetary system that is based on decimals instead of fractions.

REVIEW

67. Add: $75\frac{3}{4} + 88\frac{4}{5}$.

68. Multiply: $\frac{2}{15}\left(-\frac{5}{4}\right)$.

69. Evaluate: $\left(\frac{2}{3}\right)^5$.

70. Evaluate: $\frac{6 + (5-2)^2}{3(2-1)}$.

71. Find the area of a triangle with base 16 in. and height 9 in.

72. Express the fraction $\frac{2}{3}$ as an equivalent fraction with denominator 12.

73. Add: $-2 + (-3) + 4$.

74. Subtract: $-15 - (-6)$.

4.2   *Addition and Subtraction with Decimals*

In this section, you will learn about

- **Adding decimals**
- **Subtracting decimals**
- **Adding and subtracting signed decimals**

INTRODUCTION.   We will now study the methods used to add and subtract positive and negative decimals.

Adding decimals

When adding decimals, we line up the columns so that ones are added to ones, tenths are added to tenths, hundredths are added to hundredths, and so on. As an example, consider the following problem.

Line up the columns and the decimal points vertically. Then add the numbers.

$$
\begin{array}{r}
12.140 \\
3.026 \\
4.000 \\
+\ 0.700 \\
\hline
19.866
\end{array}
$$

Write the decimal point in the result directly under the decimal points in the problem.

Adding decimals	To add decimal numbers:
	1. Line up the decimal points, using the vertical column format.
	2. Add the numbers as you would add whole numbers.
	3. Write the decimal point in the result directly below the decimal points of the problem.

EXAMPLE 1 *Adding decimals.* Add: $1.903 + 0.6 + 8 + 0.78$.

Solution

$$
\begin{array}{r}
2 \\
1.903 \\
0.600 \\
8.000 \\
+\ 0.780 \\
\hline
11.283
\end{array}
$$

To make the addition by columns easier, write two zeros after 6, a decimal point and three zeros after 8, and one zero after 8 in 0.78.

Add column by column. Write the decimal point.

The result is 11.283.

Self Check

Add: $0.07 + 35 + 0.888 + 4.1$.

Answer: 40.058

Accent on Technology ***Preventing heart attacks***

The bar graph in Figure 4-4 shows the number of grams of fiber in a standard serving of each of several foods. It is believed that men can significantly cut their risk of heart attack by eating at least 28 grams of fiber a day. Does this diet meet or exceed the 28-gram requirement?

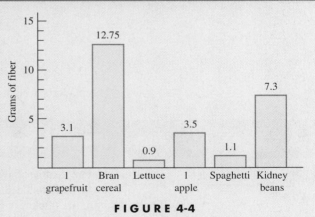

FIGURE 4-4

Solution: To find the total fiber intake, we will add the fiber content of each of the foods. We can use a scientific calculator to add the decimals.

Keystrokes: 3.1 $+$ 12.75 $+$ .9 $+$ 3.5 $+$ 1.1 $+$ 7.3 $=$ | 28.65 |

Since $28.65 > 28$, this diet exceeds the daily fiber requirement of 28 grams.

Subtracting decimals

To subtract decimals, we line up the decimal points and corresponding columns so that we subtract like objects—tenths from tenths, hundredths from hundredths, and so on.

Subtracting decimals	To subtract decimal numbers:
	1. Line up the decimal points using the vertical column format.
	2. Subtract the numbers as you would subtract whole numbers.
	3. Write the decimal point in the result directly below the decimal points of the problem.

EXAMPLE 2 *Subtracting decimals.* Subtract: **a.** 279.6 − 138.7 and **b.** 15.4 − 13.059.

Solution

a.
$$\begin{array}{r} 8\ 16 \\ 279.\cancel{6} \\ -138.7 \\ \hline 140.9 \end{array}$$
Borrow from the ones column to subtract in the tenths column.

b.
$$\begin{array}{r} 9 \\ 3\ 10\ 10 \\ 15.\cancel{4}\ \cancel{0}\ \cancel{0} \\ -13.0\ 5\ 9 \\ \hline 2.3\ 4\ 1 \end{array}$$
Add two zeros to the right of 15.4 to make borrowing easier.
First, borrow from the tenths column; then borrow from the hundredths column.

Self Check

Subtract:

a. 382.5 − 227.1

b. 30.1 − 27.122

Answer: **a.** 155.4, **b.** 2.978 ∎

EXAMPLE 3 *Conditioning program.* A 350-pound football player lost 15.7 pounds during the first week of football practice. During the second week, he gained 4.9 pounds. What is his weight after the first two weeks of practice?

Solution The word *lose* indicates subtraction. The phrase *gained back* indicates addition.

Beginning weight	−	first week weight loss	+	second week weight gain	=	weight after two weeks of practice.

$$350 - 15.7 + 4.9 = 334.3 + 4.9$$
Working from left to right, do the subtraction first: 350 − 15.7 = 334.3.

$$= 339.2$$
Do the addition.

The player's weight is 339.2 pounds after two weeks of practice. ∎

Accent on Technology **Weather balloons**

A giant weather balloon is made of neoprene, a flexible rubberized substance, that has an uninflated thickness of 0.011 inch. When the balloon is inflated with helium, the thickness becomes 0.0018 inch. Find the change in thickness of the neoprene after the balloon is inflated.

Solution: To find the change in thickness, we need to subtract. We can use a scientific calculator to subtract the decimals.

Keystrokes:  . 0 1 1 − . 0 0 1 8 = `0.0092`

After the balloon is inflated, the neoprene loses 0.0092 of an inch in thickness.

Adding and subtracting signed decimals

To add signed decimals, we use the same rules that we used for adding integers.

Adding two decimals	**With like signs:** Add their absolute values and attach their common sign to the sum.
	With unlike signs: Subtract their absolute values (the smaller from the larger) and attach the sign of the number with the largest absolute value to the sum.

EXAMPLE 4   *Addition of signed decimals.*   Add: $-6.1 + (-4.7)$.

Solution

Since the decimals are both negative, we add their absolute values and attach a negative sign to the sum.

$$-(6.1 + 4.7) = -10.8$$

Self Check
Add: $-5.04 + (-2.32)$.

Answer: -7.36

EXAMPLE 5   *Addition of signed decimals.*   Add: $5.3 + (-12.9)$.

Solution

In this example, the signs are unlike. Since -12.9 has the largest absolute value, we subtract 5.3 from 12.9 and attach a negative sign to the result.

$$-(12.9 - 5.3) = -7.6$$

Self Check
Add: $-21.4 + 16.7$.

Answer: -4.7

To subtract signed decimals, we add the opposite of the decimal that is being subtracted.

EXAMPLE 6   *Subtraction of signed decimals.*   Subtract: $-4.3 - 5.2$.

Solution

$-4.3 - 5.2 = -4.3 + (-5.2)$ Add the opposite of 5.2.

$\qquad\quad = -9.5$ Add the absolute values. The sum of two negative numbers is negative.

Self Check
Subtract: $-1.1 - 2.8$

Answer: -3.9

EXAMPLE 7   *Subtraction with signed decimals.*
Subtract: $-8.3 - (-6.2)$.

Solution

$-8.3 - (-6.2) = -8.3 + 6.2$ Add the opposite of -6.2.

$\qquad\qquad\quad = -2.1$ Subtract the smaller absolute value from the larger. Since -8.3 has the larger absolute value, the result is negative.

Self Check
Subtract: $-2.5 - (-4.4)$.

Answer: 1.9

If a numerical expression has grouping symbols, we do the operations inside them first.

EXAMPLE 8 *Grouping symbols.* Evaluate: $-12.2 - (-14.5 + 3.8)$.

Self Check

Evaluate: $-4.9 - (-1.2 + 5.6)$.

Solution

$$-12.2 - (-14.5 + 3.8) = -12.2 - (-10.7) \quad \text{Do the addition inside the grouping symbols: } -14.5 + 3.8 = -10.7.$$
$$= -12.2 + 10.7 \quad \text{Add the opposite of } -10.7.$$
$$= -1.5 \quad \text{Do the addition.}$$

Answer: -9.3

STUDY SET Section 4.2

VOCABULARY *In Exercises 1–4, fill in the blanks to make a true statement.*

1. The answer to an addition problem is called the _____.

2. The answer to a subtraction problem is called the _____.

3. Every whole number has an _____ decimal point to its right.

4. The symbols $|\quad|$ indicate that we are to find the _____ value of a number.

CONCEPTS

5. a. Add: $0.3 + 0.17$.
   b. Write 0.3 and 0.17 as fractions.
   c. Find a common denominator for the fractions of part b, and then add them.
   d. Express your answer to part c as a decimal.
   e. Compare your answers from part a and part d.

6. In this subtraction problem, we must borrow. How much is borrowed from the 3, and in what form is it borrowed?

$$\begin{array}{r} \overset{2\ 11}{29.3\cancel{1}} \\ -25.16 \\ \hline \end{array}$$

PRACTICE *In Exercises 7–18, do each addition.*

7.
$$\begin{array}{r} 32.5 \\ + \ 7.4 \\ \hline \end{array}$$

8.
$$\begin{array}{r} 6.3 \\ +13.5 \\ \hline \end{array}$$

9.
$$\begin{array}{r} 21.6 \\ +33.12 \\ \hline \end{array}$$

10.
$$\begin{array}{r} 19.4 \\ +31.95 \\ \hline \end{array}$$

11. $12 + 3.9$

12. $0.01 + 3.6$

13. $0.03034 + 0.2003$

14. $19.9 + 19.9$

15.
$$\begin{array}{r} 247.9 \\ 40 \\ + \ \ 0.56 \\ \hline \end{array}$$

16.
$$\begin{array}{r} 0.0053 \\ 1.78 \\ +6 \\ \hline \end{array}$$

17. $45 + 9.9 + 0.12 + 3.02$

18. $505.01 + 23 + 0.989 + 12.07$

In Exercises 19–26, do each subtraction.

19.
$$\begin{array}{r} 12.98 \\ - \ 3.45 \\ \hline \end{array}$$

20.
$$\begin{array}{r} 1.6 \\ -0.16 \\ \hline \end{array}$$

21.
$$\begin{array}{r} 78.1 \\ - \ 7.81 \\ \hline \end{array}$$

22.
$$\begin{array}{r} 202.234 \\ - \ 19.34 \\ \hline \end{array}$$

23. $5 - 0.023$

24. $30 - 11.98$

25. $24 - 23.81$

26. $7.001 - 5.9$

In Exercises 27–34, do each addition.

27. $-45.6 + 34.7$ **28.** $-19.04 + 2.4$ **29.** $46.09 + (-7.8)$ **30.** $34.7 + (-30.1)$

31. $-7.8 + (-6.5)$ **32.** $-5.78 + (-33.1)$ **33.** $-0.0045 + (-0.031)$ **34.** $-90.09 + (-0.087)$

In Exercises 35–42, do each subtraction.

35. $-9.5 - 7.1$ **36.** $-7.08 - 14.3$ **37.** $30.03 - (-17.88)$ **38.** $143.3 - (-64.01)$

39. $-2.002 - (-4.6)$ **40.** $-0.005 - (-8)$ **41.** $-7 - (-18.01)$ **42.** $-63.04 - (-8.911)$

In Exercises 43–50, evaluate each expression.

43. $3.4 - 6.6 + 7.3$ **44.** $3.4 - (6.6 + 7.3)$

45. $(-9.1 - 6.05) - (-51)$ **46.** $-9.1 - (-6.05) + 51$

47. $16 - (67.2 + 6.27)$ **48.** $-43 - (0.032 - 0.045)$

49. $(-7.2 + 6.3) - (-3.1 - 4)$ **50.** $2.3 + [2.4 - (2.5 - 2.6)]$

In Exercises 51–52, add or subtract as indicated.

51. Find the sum of *two and forty-three hundredths* and *five and six-tenths.*

52. Find the difference of *nineteen-hundredths* and *six-thousandths.*

APPLICATIONS   *In Exercises 53–66, solve each problem.*

53. SPORTS PAGE   In the sports pages of any newspaper, decimal numbers are used quite often.
  a. "German bobsledders set a world record today with a final run of 53.03, finishing ahead of the Italian team by only fourteen-thousandths of a second." What was the time for the Italian bobsled team?

  b. "The women's figure skating title was decided by only thirty-three hundredths of a point." If the winner's point total was 102.71, what was the second-place finisher's total?

54. NURSING   Illustration 1 shows a patient's health chart. A nurse failed to fill in certain portions. (98.6° Fahrenheit is considered normal.) Complete the chart.

Day of week	Patient's temperature	How much above normal
Monday	99.7°	
Tuesday		2.5°
Wednesday	98.6°	
Thursday	100.0°	
Friday		0.9°

ILLUSTRATION 1

55. VEHICLE SPECIFICATIONS   Certain dimensions of a compact car are shown in Illustration 2. What is the wheelbase of the car?

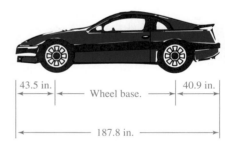

43.5 in.   Wheel base.   40.9 in.

187.8 in.

ILLUSTRATION 2

56. pH SCALE   The pH scale shown in Illustration 3 is used to measure the strength of acids and bases in chemistry. Find the difference in pH readings between
  a. Bleach and stomach acid
  b. Ammonia and coffee
  c. Blood and coffee

Strong acid   Neutral   Strong base

0 1 2 3 4 5 6 7 8 9 10 11 12 13 14

Stomach acid   Coffee   Blood   Ammonia   Bleach
1.75   5.01   7.38   12.03   12.7

ILLUSTRATION 3

57. BAROMETRIC PRESSURE Barometric pressure readings are recorded on the weather map in Illustration 4. In a low pressure area (L on the map), the weather is often stormy. The weather is usually fair in a high pressure area (H). What is the difference in readings between the areas of highest and lowest pressure? In what part of the country would you expect the weather to be fair?

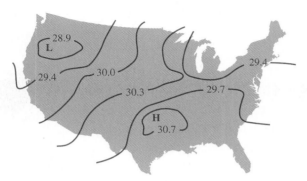

ILLUSTRATION 4

58. QUALITY CONTROL An electronics company has strict specifications for silicon chips used in a computer. The company will install only chips that are within 0.05 centimeters of the specified thickness. Illustration 5 gives that specification for two types of chip. Fill in the blanks to complete the chart.

Chip type	Thickness specification	Acceptable range	
		Low	High
A	0.78 cm		
B	0.643 cm		

ILLUSTRATION 5

59. OFFSHORE DRILLING A company needs to construct a pipeline from an offshore oil well to a refinery located on the coast. Company engineers have come up with two plans for consideration, as shown in Illustration 6. Use the information in the illustration to complete the following table.

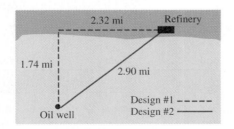

ILLUSTRATION 6

	Pipe underwater	Pipe underground	Total pipe
Design 1			
Design 2			

60. TV RATINGS The bar graph in Illustration 7 shows the prime-time ratings for the major television networks as of November 1995.

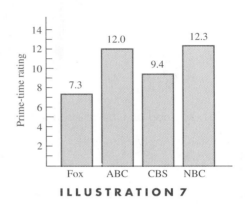

ILLUSTRATION 7

a. Which network had the highest rating? Which had the lowest rating?
b. What is the difference between the highest and the lowest rating?
c. How far was CBS behind the second-rated network?

61. AMERICAN RECORDHOLDERS Florence Griffith-Joyner holds the United States national and world record in the 100-meter sprint: 10.49 seconds. Jenny Thompson holds the national record in the 100-meter freestyle swim: 54.48 seconds. How much faster did Griffith-Joyner run the 100 meters than Thompson swam it?

62. FLIGHT PATH See Illustration 8. Find the added distance a plane must travel to avoid flying through the storm.

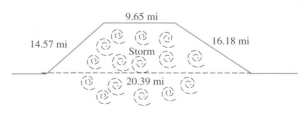

ILLUSTRATION 8

63. DEPOSIT SLIP A deposit slip for a savings account is shown in Illustration 9. Find the subtotal and then the total deposit.

Deposit		
Cash	242	50
Checks (properly endorsed)	116	10
	47	93
Total from reverse side	359	16
Subtotal		
Less cash	25	00
Total deposit		

ILLUSTRATION 9

64. MOTION Forces such as water current or wind can increase or decrease the speed of an object in motion. Find the speed of each object.

a. An airplane's speed in still air is 450 mph, and it has a tail wind of 35.5 mph helping it along.

b. A man can paddle a canoe at 5 mph in still water, but he is going upstream. The speed of the current against him is 1.5 mph.

65. PROFIT Illustration 10 shows a company's revenue and costs for 1994, 1995, and 1996. Use the graph to determine the yearly profit (or loss) of this company for each year.

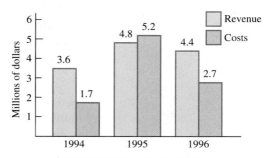

ILLUSTRATION 10

66. RETAILING Complete Illustration 11 by filling in the retail price of each appliance, given its cost to the dealer and the store markup.

Item	Cost	Markup	Retail price
Refrigerator	$510.80	$105.00	
Washing machine	$189.50	$55.50	
Dryer	$163	$23.12	

ILLUSTRATION 11

In Exercises 67–72, use a calculator to evaluate each expression.

67. 2,367.909 + 5,789.0253

68. 0.00786 + 0.3423

69. 9,000.09 − 7,067.445

70. 1 − 0.004999

71. 3,434.768 − (908 − 2.3 + .0098)

72. 12 − (0.723 + 3.05611)

WRITING *Write a paragraph using your own words.*

73. Explain why we line up the decimal points and corresponding columns when adding decimals.

74. Explain why we can write additional zeros to the right of a decimal such as 7.89 without affecting its value.

REVIEW

75. Simplify: $(-5)^3$.

76. Simplify: $\dfrac{-\dfrac{3}{4}}{\dfrac{5}{16}}$.

77. Multiply: $-\dfrac{15}{26} \cdot 1\dfrac{4}{9}$.

78. Simplify: $2 + 5[-2 - (6 + 1)]$.

Multiplication with Decimals

In this section, you will learn about

- **Multiplying decimals**
- **Multiplying decimals by powers of 10**
- **Multiplying signed decimals**
- **Order of operations**

INTRODUCTION. In our study of decimals, we now focus on the operation of multiplication. First, we develop the method used to multiply decimals. Then we use that procedure to evaluate numerical expressions and to solve problems involving decimals.

Multiplying decimals

Let's examine the multiplication $0.3 \cdot 0.17$. If we write each decimal as a fraction and multiply them, we can make some observations about multiplying decimals.

$$0.3 \cdot 0.17 = \frac{3}{10} \cdot \frac{17}{100} \qquad \text{Express 0.3 and 0.17 as fractions.}$$

$$= \frac{3 \cdot 17}{10 \cdot 100} \qquad \text{Multiply the numerators and multiply the denominators.}$$

$$= \frac{51}{1,000} \qquad \text{Multiply in the numerator and denominator.}$$

$$= 0.051 \qquad \text{Write } \tfrac{51}{1,000} \text{ as a decimal.}$$

Observations   **1.** The digits in the product are found by multiplying 3 and 17.

$$0.3 \quad \cdot \quad 0.17 \quad = \quad 0.051$$
$$3 \cdot 17 = 51$$

2. The product has 3 decimal places. The *sum* of the decimal places of the factors 0.3 and 0.07 is also 3.

$$0.3 \quad \cdot \quad 0.17 \quad = \quad 0.051$$

| 1 decimal place. | 2 decimal places. | 3 decimal places. |

These observations suggest the following rule for multiplying decimals.

Multiplying decimals

To multiply decimals:

1. Multiply them as if they were whole numbers.

2. Place the decimal point in the answer so that the result has the same number of decimal places as the sum of the decimal places of the factors.

EXAMPLE 1   *Multiplying decimals.*   Multiply: 5.9 · 3.4.

Solution

We ignore the decimal points and multiply the decimals as if they were whole numbers. Initially, we think of this problem as 59 times 34.

$$
\begin{array}{r}
59 \\
\times\ 34 \\
\hline
236 \\
177 \\
\hline
2006
\end{array}
$$

To place the decimal point in the product, we find the total number of digits to the right of the decimal points of the factors.

$$
\begin{array}{r}
5.9 \\
\times\ 3.4 \\
\hline
236 \\
177 \\
\hline
20.06
\end{array}
$$

5.9 ← 1 decimal place
× 3.4 ← 1 decimal place } The answer will have 1 + 1 = 2 decimal places.

Finally, locate the decimal point so that the answer has 2 decimal places.

When multiplying decimals, it is not necessary to line up the decimal points, as the next example illustrates.

Self Check

Multiply: 2.74 · 4.3.

Answer: 11.782

EXAMPLE 2   *Inserting placeholder zeros.*   Multiply: 1.3(0.005).

Solution

We multiply 13 by 5.

1.3 ← 1 decimal place
×0.005 ← 3 decimal places } The answer will have 1 + 3 = 4 decimal places.

$$
\begin{array}{r}
1.3 \\
\times 0.005 \\
\hline
65
\end{array}
$$

We then place the decimal point.

$$
\begin{array}{r}
1.3 \\
\times\ 0.005 \\
\hline
0.0065
\end{array}
$$

Add 2 placeholder zeros and position the decimal point so that the product has 4 decimal places.

Self Check

Multiply: (0.0002)7.2.

Answer: 0.00144

Accent on Technology   **Heating costs**

When billing a household, a gas company converts the amount of natural gas used into units of heat energy called *therms*. The number of therms used by a household in one month and the cost per therm are shown below.

Customer charge . . . . .39 therms @ $0.72264

To find the total charges for the month, we multiply the number of therms by the cost per therm.

Evaluate: 39 · 0.72264.

Keystrokes: ⟨3⟩⟨9⟩⟨×⟩⟨.⟩⟨7⟩⟨2⟩⟨2⟩⟨6⟩⟨4⟩⟨=⟩ 28.18296

Rounding to the nearest cent, we see that the total charge is $28.18.

EXAMPLE 3

Multiplying a decimal and a whole number.
Multiply: 234(3.1).

Solution

$$
\begin{array}{r}
234 \leftarrow \text{No decimal places} \\
\times\ 3.1 \leftarrow \text{1 decimal place} \\
\hline
23\ 4 \\
702 \\
\hline
725.4
\end{array}
$$

234 ← No decimal places ⎫
× 3.1 ← 1 decimal place ⎬ The answer will have 0 + 1 = 1 decimal place.

725.4   Locate the decimal point so that the answer has 1 decimal place.

Self Check

Multiply: 178(2.7).

Answer: 480.6

Multiplying decimals by powers of 10

The numbers 10, 100, and 1,000 are called *powers of 10,* because they result when we evaluate 10^1, 10^2, and 10^3, respectively. To develop a rule to determine the product when multiplying a decimal and a power of 10, we will do the following multiplications and make some observations about the results.

Multiply: $8.675 \cdot 10$

$$
\begin{array}{r}
8.675 \\
\times\ \ \ \ 10 \\
\hline
0000 \\
8675 \\
\hline
86.750
\end{array}
$$

The product is 86.75.

Multiply: $8.675 \cdot 100$

$$
\begin{array}{r}
8.675 \\
\times\ \ \ 100 \\
\hline
0000 \\
0000 \\
8675 \\
\hline
867.500
\end{array}
$$

The product is 867.5.

Multiply: $8.675 \cdot 1,000$

$$
\begin{array}{r}
8.675 \\
\times\ \ 1000 \\
\hline
0000 \\
0000 \\
0000 \\
8675 \\
\hline
8675.000
\end{array}
$$

The product is 8,675.

Observations

1. In each case, the product contains the same digits as the factor 8.675.

2. When inspecting the answers, the decimal point in the first factor 8.675 appears to be moved to the right by the multiplication process. The number of decimal places it moves depends on the power of 10 by which 8.675 is multiplied.

One zero in 10

$8.675 \cdot 10 = 86.75$

It moves one place
to the right.

Two zeros in 100

$8.675 \cdot 100 = 867.5$

It moves two places
to the right.

Three zeros in 1,000

$8.675 \cdot 1,000 = 8675$

It moves three places
to the right.

These observations suggest the following rule.

Multiplying a decimal by a power of 10

To multiply a decimal by a power of 10, move the decimal point in the decimal the same number of places to the right as the number of zeros in the power of 10.

EXAMPLE 4

Multiplying decimals by powers of 10.
Find the product: **a.** $2.81 \cdot 10$ and **b.** $0.076 \cdot 10,000$.

Solution

a. $2.81 \cdot 10 = 28.1$ Since 10 has 1 zero, move the decimal point 1 place to the right.

b. $0.076 \cdot 10,000 = 0760.$ Since 10,000 has 4 zeros, move the decimal point 4 places to the right. Write a placeholder zero (shown in blue).

$= 760$

Self Check

Find the product:

a. $0.721 \cdot 100$

b. $6.08\ (1,000)$

Answer: **a.** 72.1, **b.** 6,080

Multiplying signed decimals

Recall that the product of two numbers with like signs is positive, and the product of two numbers with unlike signs is negative.

EXAMPLE 5 *Multiplying signed decimals.* Multiply: **a.** $-1.8(4.5)$ and **b.** $(-1,000)(-59.08)$.

Solution

a. $-1.8(4.5) = -(1.8 \cdot 4.5)$ The product of two numbers with unlike signs is negative.

       $= -8.1$ $1.8 \cdot 4.5 = 8.10 = 8.1$.

b. The product of two decimals with like signs is positive.

     $(-1,000)(-59.08) = 59,080$ Since 1,000 has 3 zeros, move the decimal point 3 places to the right. Write a placeholder zero.

Self Check
Multiply:

a. $6.6(-5.5)$

b. $(-44.968)(-100)$

Answer: **a.** -36.3, **b.** $4,496.8$ ∎

EXAMPLE 6 *Evaluating powers of decimals.* Evaluate: **a.** $(2.4)^2$ and **b.** $(-0.05)^2$.

Solution

a. $(2.4)^2 = 2.4 \cdot 2.4$ Write 2.4 as a factor 2 times.

       $= 5.76$ Do the multiplication.

b. $(-0.05)^2 = (-0.05)(-0.05)$ Write -0.05 as a factor 2 times.

       $= 0.0025$ Do the multiplication. The product of two numbers with like signs is positive.

Self Check
Evaluate:

a. $(-1.3)^2$

b. $(0.09)^2$

Answer: **a.** 1.69, **b.** 0.0081 ∎

Order of operations

In the remaining examples, we apply the rules for order of operations to evaluate expressions involving decimals.

EXAMPLE 7 *Order of operations.* Evaluate: $(0.6)^2 + 5(-3.6 + 1.9)$.

Solution

$(0.6)^2 + 5(-3.6 + 1.9) = (0.6)^2 + 5(-1.7)$ Do the addition inside the parentheses.

            $= 0.36 + 5(-1.7)$ Find the power: $(0.6)^2 = 0.36$.

            $= 0.36 + (-8.5)$ Do the multiplication: $5(-1.7) = -8.5$.

            $= -8.14$ Do the addition.

Self Check
Evaluate:
$-2(-4.4 + 5.6) + (-0.8)^2$.

Answer: -1.76 ∎

EXAMPLE 8 *Evaluating a formula.* Evaluate the formula $A = p + prt$ if $p = 23.75$, $r = 0.02$, and $t = 3.5$.

Solution

$A = p + prt$

$A = 23.75 + 23.75(0.02)(3.5)$ Substitute 23.75 for p, 0.02 for r, and 3.5 for t.

   $= 23.75 + 1.6625$ $23.75(0.02)(3.5) = 1.6625$.

   $= 25.4125$ Do the addition.

Self Check
Evaluate the formula $C = \pi r^2$ if $\pi = 3.14$ and $r = 2.1$.

Answer: 13.8474 ∎

EXAMPLE 9   *Weekly earnings.* A cashier's work week is 40 hours. After his daily shift is over, he can work overtime at a rate 1.5 times his regular rate of $7.50 per hour. How much money will he earn in a week if he works 6 hours of overtime?

Solution   First, we need to find his overtime rate, which is 1.5 times his regular rate of $7.50 per hour.

$$1.5(7.50) = 11.25$$

His overtime rate is $11.25 per hour.

To find his total weekly earnings, we use the following fact.

The regular rate	·	40 hours	+	the overtime rate	·	overtime hours worked	=	his total earnings.

$$7.50(40) + 11.25(6) = 300 + 67.50 \quad \text{Do the multiplications.}$$
$$= 367.50 \quad \text{Do the addition.}$$

The cashier's earnings for the week are $367.50.

STUDY SET   Section 4.3

VOCABULARY   *In Exercises 1–2, fill in each blank to make a true statement.*

1. In the multiplication problem 2.89 · 15.7, 2.89 and 15.7 are called _____. The answer, 45.373, is called the _____.

2. Numbers such as 10, 100, and 1,000 are called _____ of 10.

CONCEPTS   *In Exercises 3–4, fill in each blank to make a true statement.*

3. To multiply decimals, multiply them as if they were _____ numbers. The number of decimal places in the product is the same as the _____ of the decimal places of the factors.

4. To multiply a decimal by a power of 10, move the decimal point to the _____ the same number of decimal places as the number of _____ in the power of 10.

5. When we move the decimal point to the right, does the decimal number get larger or smaller?

6. Suppose that the result of multiplying two decimals is 2.300. Write this result using two digits.

7. a. Multiply $\frac{3}{10}$ and $\frac{7}{100}$.
   b. Now write both fractions from part a as decimals. Multiply them in that form. Compare your results from parts a and b.

8. a. Multiply 0.11 and 0.3
   b. Now write both decimals in part a as fractions. Multiply them in that form. Compare your results from parts a and b.

PRACTICE   *In Exercises 9–48, do each multiplication.*

9. (0.4)(0.2)

10. (0.2)(0.3)

11. (−0.5)(0.3)

12. (0.6)(−0.7)

13. (1.4)(0.7)

14. (2.1)(0.4)

15. (0.08)(0.9)

16. (0.003)(0.9)

17. (−5.6)(−2.2)

18. (−7.1)(−4.1)

19. (−4.9)(0.001)

20. (0.001)(−7.09)

21. (−0.35)(0.24)

22. (−0.85)(0.42)

23. (−2.13)(4.05)

24. (3.06)(−1.82)

25. 16 · 0.6

26. 24 · 0.8

27. −7(8.1)

28. −5(4.7)

29. 0.04(306)

30. 0.02(417)

31. 60.61(−0.3)

32. −70.07 · 0.6

33. −0.2(0.3)(−0.4)

34. −0.1(−2.2)(0.5)

35. 5.5(10)(−0.3)

36. 6.2(100)(−0.8)

37. 4.2 · 10

38. 10 · 7.1

39. 67.164 · 100

40. 708.199 · 100

41. $-0.056(10)$ **42.** $-100(0.0897)$ **43.** $1,000(8.05)$ **44.** $23.7(1,000)$

45. $0.098(10,000)$ **46.** $3.63(10,000)$ **47.** $-0.2 \cdot 1,000$ **48.** $-1,000 \cdot 1.9$

In Exercises 49–50, complete each table.

49.

Decimal	Its square
0.1	
0.2	
0.3	
0.4	
0.5	
0.6	
0.7	
0.8	
0.9	

50.

Decimal	Its cube
0.1	
0.2	
0.3	
0.4	
0.5	
0.6	
0.7	
0.8	
0.9	

In Exercises 51–54, find each power.

51. $(1.2)^2$ **52.** $(2.3)^2$ **53.** $(-1.3)^2$ **54.** $(-2.5)^2$

In Exercises 55–62, evaluate each expression.

55. $-4.6(23.4 - 19.6)$ **56.** $6.9(9.8 - 8.9)$ **57.** $(-0.2)^2 + 2(7.1)$ **58.** $(-6.3)(3) - (1.2)^2$

59. $(-0.7 - 0.5)(2.4 - 3.1)$ **60.** $(-8.1 - 7.8)(0.3 + 0.7)$ **61.** $(0.5 + 0.6)^2(-3.2)$ **62.** $(-5.1)(4.9 - 3.4)^2$

In Exercises 63–66, evaluate each formula.

63. $A = lw$ for $l = 5.3$ and $w = 7.2$

64. $P = 2l + 2w$ for $l = 3.7$ and $w = 3.6$

65. $A = \frac{1}{2}bh$ for $b = 7.5$ and $h = 6.8$

66. $C = 2\pi r$ for $\pi = 3.14$ and $r = 2.5$

APPLICATIONS *In Exercises 67–76, solve each problem.*

67. CONCERT SEATING   Two types of tickets were sold for a concert. Floor seating cost $12.50 a ticket, and balcony seats were $15.75.

    a. Complete the table in Illustration 1 and find the receipts from each type of ticket.

    b. Find the total receipts from the sale of both types of tickets.

Ticket type	Price	Number sold	Income
Floor		1,000	
Balcony		100	

ILLUSTRATION 1

68. CITY PLANNING   In the city map in Illustration 2, the streets form a grid. They are 0.35 mile apart. Find the distance of each trip.

    a. The airport to the Convention Center

    b. City Hall to the Convention Center

    c. The airport to City Hall

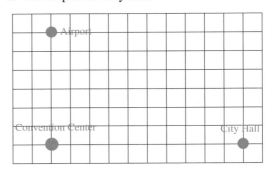

ILLUSTRATION 2

69. STORM DAMAGE   After a rainstorm, the saturated ground under a hilltop house began to give way. A survey team noted that the house dropped 0.57 inch initially. In the next two weeks, the house only fell 0.09 inch per week. What is the change in the elevation of the house during this three-week period?

70. WATER USAGE   In May, the water level of a reservoir reached its high mark for the year. During the summer months, as water usage increased, the level dropped. In the months of May and June, it fell 4.3 feet each month. In August, because of high temperatures, it fell another 8.7 feet. By September, what is the change from the year's high water mark?

71. WEIGHTLIFTING   The barbell in Illustration 3 is evenly loaded with iron plates. How much plate weight is loaded on the barbell?

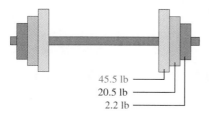

45.5 lb
20.5 lb
2.2 lb

ILLUSTRATION 3

72. PLUMBING BILL   In Illustration 4, an invoice for plumbing work is torn. What is the charge for the 4 hours of work? What is the total charge?

Carter Plumbing 100 W. Dalton Ave.	Invoice #210
Standard sevice charge 4 hr @ $40.55/hr	$25.75
Total	

ILLUSTRATION 4

73. BAKERY SUPPLIES   A bakery buys various types of nuts as ingredients for cookies. Complete Illustration 5 by filling in the cost of each purchase.

Type of nut	Price per pound	Pounds	Cost
Almonds	$3.25	16	
Walnuts	$2.10	25	
Peanuts	$1.85	3.6	

ILLUSTRATION 5

74. RETROFIT   Illustration 6 shows the width of the three columns of an existing freeway overpass. A computer analysis indicates that each column needs to be increased in width by a factor of 1.4 to ensure stability during an earthquake. According to the analysis, how wide should each of the columns be?

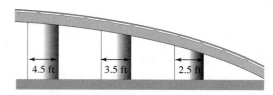

4.5 ft   3.5 ft   2.5 ft

ILLUSTRATION 6

75. SWIMMING POOL CONSTRUCTION   Long bricks, called *coping,* can be used to outline the edge of a swimming pool. How much coping will be needed in the construction of the swimming pool shown in Illustration 7?

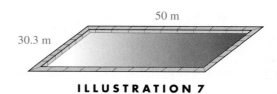

50 m

30.3 m

ILLUSTRATION 7

76. SOCCER   A soccer goal measures 24 feet wide by 8 feet high. Major League Soccer officials are proposing to increase its width by 1.5 feet and increase its height by 0.75 foot.

a. What is the area of the goal opening now?

b. What would it be if their proposal is adopted?

c. How much area would be added?

In Exercises 77–82, use a calculator to answer each problem.

77. $(-9.0089 + 10.0087)(15.3)$

78. $(-4.32)^3 - 78.969$

79. $(18.18 + 6.61)^2 + (5 - 9.09)^2$

80. $304 - 3.780876(100)$

81. ELECTRIC BILL   When billing a household, a utility company charges for the number of kilowatt-hours used. A kilowatt-hour (kwh) is a standard measure of electricity. If the cost of 1 kwh is $0.14277, what is the electric bill for a household using 719 kwh in a month? Round the answer to the nearest cent.

82. UTILITY TAX   Some gas companies are required to tax the number of therms used each month by the customer. What are the taxes collected on a monthly usage of 31 therms if the tax rate is $0.00566 per therm? Round the answer to the nearest cent.

WRITING   *Write a paragraph using your own words.*

83. Explain how to determine where to place the decimal point in the answer when multiplying two decimals.

84. List the similarities and differences between whole-number multiplication and decimal multiplication.

85. What is a decimal place?

86. What is the purpose of a rule for order of operations?

REVIEW

87. Round 7,346 to the nearest hundred.

88. Use $C = \dfrac{5(F - 32)}{9}$ to convert 86° F to Celsius.

89. Multiply: $3\dfrac{1}{3}\left(-1\dfrac{4}{5}\right)$.

90. Simplify: $(-3)^4$.

91. Write this notation in words: $|-3|$

92. What is the LCD of fractions with denominators of 4, 5, and 6?

93. Simplify: $-\dfrac{8}{8}$

94. Find one-half of 7 and square the result.

4.4   *Division with Decimals*

In this section, you will learn about

- **Dividing a decimal by a whole number**
- **Divisors that are decimals**
- **Rounding when dividing**
- **Dividing decimals by powers of 10**
- **Order of operations**

INTRODUCTION.   Every division is composed of three parts: the divisor, the dividend, and the quotient.

Long division form

$$\text{Divisor} \longrightarrow 5\overline{)10} \begin{array}{l} \underset{\longleftarrow \text{ Quotient}}{\overset{2}{}} \\ \longleftarrow \text{ Dividend} \end{array}$$

Fraction form

$$\text{Dividend} \longrightarrow \dfrac{10}{5} = 2 \longleftarrow \text{Quotient}$$
$$\text{Divisor} \longrightarrow$$

In this section, we examine division problems in which the divisor and/or the dividend are decimals.

Dividing a decimal by a whole number

To use long division to divide 47 by 10, we proceed as follows.

$$10\overline{)47} \quad \overset{4\frac{7}{10}}{}$$

Here the result is written in quotient $+ \frac{\text{remainder}}{\text{divisor}}$ form.

$$\underline{40}$$
$$7$$

To do this same division using decimals, we write 47 as 47.0 and divide as we would divide whole numbers.

$$
\begin{array}{r}
4.7 \\
10\overline{)47.0} \\
\underline{40} \\
7\,0 \\
\underline{7\,0} \\
0
\end{array}
$$

Note that the decimal point in the result is placed directly above the decimal point in the dividend.

Since $4\frac{7}{10} = 4.7$, either method gives the same result. The second part of this discussion suggests the following method for dividing a decimal by a whole number.

Dividing a decimal by a whole number	1. Write the problem in long division form. 2. Divide as if working with whole numbers. 3. Write the decimal point in the result directly above the decimal point in the dividend. If necessary, additional zeros can be written to the right of the dividend to allow the division to proceed.

EXAMPLE 1 *Dividing a decimal by a whole number.*
Divide: $71.68 \div 28$.

Solution

$$
28\overline{)71.68}
$$

Write the decimal point in the answer directly above the decimal point in the dividend.

$$
\begin{array}{r}
2.56 \\
28\overline{)71.68} \\
\underline{56} \\
15\,6 \\
\underline{14\,0} \\
1\,68 \\
\underline{1\,68} \\
0
\end{array}
$$

Divide as if working with whole numbers.

The remainder is 0.

The answer is 2.56. We can check this result by multiplying the divisor and the quotient; their product should equal the dividend. Since $28 \cdot 2.56 = 71.68$, the result is correct.

EXAMPLE 2 *Writing extras zeros.* Divide: $19.2 \div 5$.

Solution

$$
\begin{array}{r}
3.8 \\
5\overline{)19.2} \\
\underline{15}\downarrow \\
4\,2 \\
\underline{4\,0} \\
2
\end{array}
$$

All the digits in the dividend have been used, but the remainder is not 0.

We can write a zero to the right of 2 in the dividend and continue the division process. Recall that writing additional zeros to the right of the decimal point does not change the value of the decimal.

Self Check
Divide: $101.44 \div 32$.

Answer: 3.17

Self Check
Divide: $3.4 \div 4$.

$$\begin{array}{r} 3.84 \\ 5\overline{)19.20} \\ \underline{15} \\ 4\,2 \\ \underline{4\,0}\!\downarrow \\ 20 \\ \underline{20} \\ 0 \end{array}$$

Write a zero to the right of the 2 and bring it down.

Continue to divide.

The remainder is 0.

The answer is 3.84.

Answer: 0.85

Divisors that are decimals

When the divisor is a decimal, we change it to a whole number and proceed as in division of whole numbers. To justify this procedure, we consider the problem $0.36\overline{)0.2592}$, where the divisor is a decimal. First, we express the division in another form.

$0.36\overline{)0.2592}$ can be represented by $\dfrac{0.2592}{0.36}$

To write the divisor, 0.36, as a whole number, its decimal point needs to be moved two places to the right. This can be accomplished by multiplying it by 100. However, if the denominator of the fraction is multiplied by 100, the numerator must also be multiplied by 100 to maintain the same value.

$\dfrac{0.2592}{0.36} = \dfrac{0.2592 \cdot \mathbf{100}}{0.36 \cdot \mathbf{100}}$ Multiply numerator and denominator by 100,

$= \dfrac{25.92}{36}$ Multiplying by 100 moves both decimal points two places to the right.

This fraction represents the division problem $36\overline{)25.92}$.

Observations

1. The division problem $0.36\overline{)0.2592}$ is equivalent to $36\overline{)25.92}$. That is, they have the same answer.
2. The decimal point in *both* the divisor and the dividend of the first division problem have been moved two decimal places to the right to create the second division problem.

$0.36\overline{)0.2592}$ becomes $36\overline{)25.92}$

These observations suggest the following rule for division with decimals.

Division with a decimal divisor

To divide with a decimal divisor:

1. Move the decimal point of the divisor so that it becomes a whole number.
2. Move the decimal point of the dividend the same number of places to the right.
3. Divide as if working with whole numbers. Write the decimal point of the answer directly above the decimal point in the dividend.

EXAMPLE 3   *Dividing decimals.*   Divide: $\dfrac{0.2592}{0.36}$.

Self Check

Divide: $\dfrac{0.6045}{0.65}$.

Solution

$$0.36\overline{)0.25.92}$$ Move the decimal point 2 places to the right in the divisor and dividend.

$$
\begin{array}{r}
0.72 \\
36\overline{)25.92} \\
\underline{25\ 2} \\
72 \\
\underline{72} \\
0
\end{array}
$$

Now divide as with whole numbers. Write the decimal point of the answer directly above that of the dividend.

The result is 0.72.

Answer: 0.93

Rounding when dividing

In Example 3, the division process ended after we obtained a zero from the second subtraction. We say that the division process *terminated*. Sometimes when dividing, the subtractions never give a zero remainder, and the division process continues forever. In this case, we can round the result.

EXAMPLE 4   ***Rounding when dividing.***   Divide: $\dfrac{2.35}{0.7}$. Round to the nearest hundredth.

Solution   Using long division form, we have $0.7\overline{)2.35}$.

$$0.7\overline{)2.3.5}$$

To write the divisor as a whole number, move the decimal point one place to the right. Do the same for the dividend. Place the decimal point in the answer directly above that of the dividend.

$$7\overline{)23.500}$$

To round off to the hundredths column, we must divide to the thousandths column. We write two zeros on the right of the dividend.

$$
\begin{array}{r}
3.357 \\
7\overline{)23.500} \\
\underline{21} \\
2\ 5 \\
\underline{2\ 1} \\
40 \\
\underline{35} \\
50 \\
\underline{49} \\
1
\end{array}
$$

After dividing to the thousandths column, round to the hundredths column.

To the nearest hundredth, the answer is 3.36.

Accent on Technology   *The nucleus of a cell*

The nucleus of a cell contains vital information about the cell in the form of DNA. The nucleus is very small in size: A typical animal cell has a nucleus that is only 0.00023622 inch across. How many nuclei would have to be laid end-to-end to extend to a length of 1 inch?

To find how many 0.00023622-inch lengths there are in 1 inch, we must use division.

Evaluate: $1 \div 0.00023622$.

Keystrokes: $\boxed{1}\ \boxed{\div}\ \boxed{.}\ \boxed{0}\ \boxed{0}\ \boxed{0}\ \boxed{2}\ \boxed{3}\ \boxed{6}\ \boxed{2}\ \boxed{2}\ \boxed{=}$ $\boxed{4233.3418}$

It would take approximately 4,233 nuclei laid end-to-end to extend to a length of 1 inch.

Dividing decimals by powers of 10

To develop a set of rules for division by a power of 10, we consider the problem $8.13 \div 10$.

$$
\begin{array}{r}
0.813 \\
10)\overline{8.130} \\
\end{array}
$$

Write one 0 to the right of 3 in the dividend.

$$
\begin{array}{r}
0 \\
\overline{8\ 1} \\
8\ 0 \\
\overline{13} \\
10 \\
\overline{30} \\
30 \\
\overline{0}
\end{array}
$$

Observation The quotient, 0.813, and the dividend, 8.13, are the same except for the location of the decimal points. The quotient can be easily obtained by moving the decimal point of the dividend 1 place to the *left*.

This observation suggests the following rule for dividing a decimal by a power of 10.

Dividing a decimal by a power of 10	To divide a decimal by a power of 10, move the decimal point of the dividend to the left the same number of decimal places as the number of zeros in the power of 10.

EXAMPLE 5 *Dividing decimals by powers of 10.* Find the quotient:
a. $16.74 \div 10$ and **b.** $8.6 \div 10{,}000$.

Solution

a. $16.74 \div 10 = 1.674$ Since 10 has 1 zero, move the decimal point 1 place to the left.

b. $8.6 \div 10{,}000 = .00086$ Since 10,000 has 4 zeros, move the decimal point 4 places to the left. Write 3 placeholder zeros.

$\qquad\qquad = 0.00086$

Self Check

Find the quotient:

a. $721.3 \div 100$

b. $\dfrac{1.07}{1{,}000}$

Answer: **a.** 7.213,
b. 0.00107

Order of operations

In the next example, we use the rule for order of operations to evaluate an expression involving division by a decimal.

EXAMPLE 6 *Order of operations:* Evaluate: $\dfrac{2(0.351) + 0.5592}{-0.4}$.

Solution

$$
\frac{2(0.351) + 0.5592}{-0.4} = \frac{0.702 + 0.5592}{-0.4}
$$
Do the multiplication first: $2(0.351) = 0.702$.

$$
= \frac{1.2612}{-0.4}
$$
Do the addition: $0.702 + 0.5592 = 1.2612$.

$$
= -3.153
$$
Do the division. The quotient of two numbers with unlike signs is negative.

Self Check

Evaluate: $\dfrac{2.7756 + 3(-0.63)}{-0.8}$.

Answer: -1.107

ESTIMATION

Decimals

In this section, we will use estimation procedures to approximate the answers to addition, subtraction, multiplication, and division problems involving decimals. You will recall that we use rounding when estimating to help simplify the computations so that they can be performed quickly and easily.

EXAMPLE 1 — *Estimating sums and differences.*

a. Estimate to the nearest ten the sum of 261.76 and 432.94.

Solution We round each number to the nearest ten.

$$261.76 + 432.94$$
$$\downarrow \text{ Round } \downarrow$$
$$260 \;\;+\;\; 430 \;\; = 690$$

The estimate is 690.

b. Estimate using front-end rounding: 381.77 − 57.01.

Solution

$$381.77 - 57.01$$
$$\downarrow \qquad\quad \downarrow$$
$$400 \;\;-\;\; 60 \;\;= 340$$

Each number is rounded to its largest place value: 381.77 to the nearest hundred and 57.01 to the nearest ten.

The estimate is 340.

Self Check

a. Estimate to the nearest ten the sum of 526.93 and 284.03.

b. Estimate using front-end rounding: 512.33 − 36.47.

Answer: **a.** 810, **b.** 460 ■

EXAMPLE 2 — *Estimating products.*

a. Estimate the product: 6.41 · 27.

Solution We use front-end rounding.

$$6.41 \cdot 27 \approx 6 \cdot 30 \qquad \text{The symbol} \approx \text{means "is approximately equal to."}$$

The estimate is 180.

b. Estimate the product: 5.2 · 13.91.

Solution Use front-end rounding.

$$5.2 \cdot 13.91 \approx 5 \cdot 10$$

The estimate is 50.

c. Estimate the product: 0.124 · 98.6.

Solution Notice that 98.6 ≈ 100

$$0.124 \cdot 98.6 \approx 0.124 \cdot 100 \qquad \text{To multiply a decimal by 100, move the decimal point 2 places to the right.}$$

The estimate is 12.4.

Self Check

a. Estimate the product:

$$42 \cdot 17.65$$

b. Estimate the product:

$$182 \cdot 24.04$$

c. Estimate the product:

$$979.3 \cdot 2.3485$$

Answer: **a.** 800, **b.** 4,000, **c.** 2,348.5 ■

When estimating a quotient, we round the divisor and the dividend so that they will divide evenly. Try to round both numbers up or both numbers down.

EXAMPLE 3 *Estimating quotients.* Estimate the quotient: 246.03 ÷ 4.31.

Solution

4.31 is close to 4. A multiple of 4 close to 246.03 is 240. (Note that both the divisor and dividend were rounded down.)

246.03 ÷ 4.31 ≈ 240 ÷ 4   Do the division mentally.

The estimate is 60.

Self Check

Estimate: 6,429.6 ÷ 7.19.

Answer: 900

STUDY SET *In Exercises 1–10, use the following information about refrigerators to estimate the answers to each question. Remember that answers may vary depending on the rounding method used.*

Deluxe model	**Standard model**	**Economy model**
Price: $978.88	Price: $739.99	Price: $599.95
Capacity: 25.2 cubic feet	Capacity: 20.6 cubic feet	Capacity: 18.8 cubic feet
Energy cost: $6.79 a month	Energy cost: $5.61 a month	Energy cost: $4.39 a month

1. How much more expensive is the deluxe model than the standard model?

2. A couple wants to buy two standard models, one for themselves and one for their newly married son and daughter-in-law. What is the total cost?

3. How much less storage capacity does the economy model have than the standard model?

4. The owner of a duplex apartment wants to purchase a standard model for one unit and an economy model for the other. What will be the total cost?

5. A stadium manager has a budget of $20,000 to furnish the luxury boxes at a football stadium with refrigerators. How many standard models can she purchase for this amount?

6. How many more cubic feet of storage do you get with the deluxe model as compared to the economy model?

7. Three roommates are planning on purchasing the deluxe model and splitting the cost evenly. How much will each have to pay?

8. What is the energy cost per year to run the deluxe model?

9. If you make a $220 down payment on the standard model, how much of the cost is left to finance?

10. The economy model can be expected to last for 10 years. What would be the total energy cost over that period?

In Exercises 11–12, estimate the answer to each problem. Does the calculator result seem reasonable? That is, does it appear that the problem was entered into the calculator correctly?

11. 53 · 5.61 ⎡ 241.23 ⎤

12. 89.11 ÷ 22.707 ⎡ 39.24340 ⎤

STUDY SET   Section 4.4

VOCABULARY *In Exercises 1–2, fill in the blanks to make a true statement.*

1. In the division $2.5\overline{)4.075} = 1.63$, 4.075 is called the _____, 2.5 is the _____, and 1.63 is the _____.

2. In $\dfrac{33.6}{0.3}$, the _____ indicates division.

In Exercises 3–4, fill in the blanks to make a true statement.

3. To divide by a decimal, move the decimal point of the divisor so that it becomes a _____ number. The decimal point of the _____ is then moved the same number of places to the _____. The decimal point of the _____ is written directly above that of the dividend.

4. To divide a decimal by a power of 10, move the decimal point of the dividend to the _____ the same number of decimal places as the number of _____ in the power of 10.

5. Is this statement true or false?

$45 = 45.0 = 45.000$

6. When a decimal is divided by 10, is the answer smaller or larger than the original number?

7. To complete the division $7.8\overline{)14.562}$, the decimal points of the divisor and dividend are moved 1 place to the right. This is equivalent to multiplying the numerator and the denominator of $\frac{14.562}{7.8}$ by what number?

8. a. When dividing decimals with like signs, what is the sign of the quotient?
b. When dividing decimals with unlike signs, what is the sign of the quotient?

9. How can we check the result of this division?

$\frac{1.917}{0.9} = 2.13$

10. When rounding a decimal to the hundredths column, to what other column must we refer?

NOTATION

11. Explain what the arrows are illustrating.

$4.67\overline{)32.08.7}$

12. What is this arrow illustrating?

$$
\begin{array}{r}
0.7 \\
4\overline{)3.100} \\
-28\downarrow \\
\hline
40
\end{array}
$$

PRACTICE In Exercises 13–36, do each division.

13. $8\overline{)36}$

14. $4\overline{)10}$

15. $-39 \div 4$

16. $-26 \div 8$

17. $49.6 \div 8$

18. $23.5 \div 5$

19. $9\overline{)288.9}$

20. $6\overline{)337.8}$

21. $(-14.76) \div (-6)$

22. $(-13.41) \div (-9)$

23. $\frac{-55.02}{7}$

24. $\frac{-24.24}{8}$

25. $45\overline{)119.7}$

26. $41\overline{)146.37}$

27. $250.95 \div 35$

28. $241.86 \div 29$

29. $41.6 \div 0.32$

30. $31.8 \div 0.15$

31. $(-199.5) \div (-0.19)$

32. $(-2,381.6) \div (-0.26)$

33. $\frac{0.0102}{0.017}$

34. $\frac{0.0092}{0.023}$

35. $\frac{0.0186}{0.031}$

36. $\frac{0.416}{0.52}$

In Exercises 37–40, divide and round each result to the nearest tenth.

37. $3\overline{)16}$

38. $7\overline{)20}$

39. $-5.714 \div 2.4$

40. $-21.21 \div 3.8$

In Exercises 41–44, divide and round each result to the nearest hundredth.

41. $12.243 \div 0.9$

42. $13.441 \div 0.6$

43. $0.04\overline{)0.03164}$

44. $0.08\overline{)0.02201}$

In Exercises 45–52, do the division mentally.

45. $7.895 \div 100$

46. $23.05 \div 10$

47. $0.064 \div (-100)$

48. $0.0043 \div (-10)$

49. $1000\overline{)34.8}$

50. $100\overline{)678.9}$

51. $\frac{45.04}{10}$

52. $\frac{22.32}{100}$

In Exercises 53–56, evaluate each expression. Round each result to the nearest hundredth.

53. $\frac{-1.2 - 3.4}{3(1.6)}$

54. $\frac{(-1.3)^2 + 6.7}{-0.9}$

55. $\frac{40.7(-5.3)}{0.4 - 0.61}$

56. $\frac{(0.5)^2 - (0.3)^2}{0.005 + 0.1}$

In Exercises 57–60, evaluate each formula.

57. $t = \dfrac{d}{r}$ for $d = 211.75$ and $r = 60.5$

58. $r = \dfrac{d}{t}$ for $d = 219.375$ and $t = 3.75$

59. $h = \dfrac{2A}{b}$ for $A = 9.62$ and $b = 3.7$

60. $\pi = \dfrac{C}{d}$ for $C = 14.4513$ and $d = 4.6$ (Round to the nearest hundredth.)

APPLICATIONS *In Exercises 61–70, solve each problem.*

61. OIL WELL Geologists have mapped out the substances through which engineers must drill to reach an oil deposit. (See Illustration 1.) What is the average depth that must be drilled *each week* if this is to be a *four-week* project?

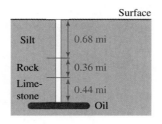

ILLUSTRATION 1

62. ICE SKATING In an ice skating competition, a performer received the following marks from the 6 international judges:

5.6, 5.9, 5.6, 5.5, 5.7, 5.7

To compute the skater's overall score, the highest and lowest marks are thrown out. The remaining scores are then added. Finally, that sum is divided by 4. Find the skater's overall score.

63. BUTCHER SHOP A meat slicer is designed to trim 0.05-inch-thick pieces from a sausage. If the sausage is 14 inches long, how many slices will result?

64. COMPUTERS A computer can do an arithmetic computation in 0.00003 second. How many of these computations could it do in 60 seconds?

65. HIKING Use the information in Illustration 2 to find the time of arrival for the hiker.

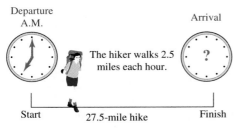

ILLUSTRATION 2

66. VOLUME CONTROL A volume control is shown in Illustration 3. If the distance between the Low and High settings is 21 cm, how far apart are the equally spaced volume settings?

ILLUSTRATION 3

67. SPRAY BOTTLE Production planners have found that each squeeze of the trigger of a spray bottle emits 0.015 ounce of liquid. How many squeezes would there be in a 8.5-ounce bottle?

68. CAR LOAN See the loan statement in Illustration 4. How many more monthly payments must be made to pay off the loan?

American Finance Company		June
Monthly payment: $42.10	Paid to date: $547.30	
	Loan balance: $631.50	

ILLUSTRATION 4

69. HOURLY WAGE Illustration 5 shows how much three restaurant workers earn for a 40-hour work week. Find the hourly wage of each worker.

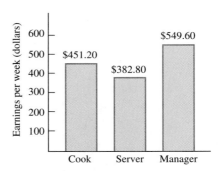

ILLUSTRATION 5

70. AVERAGE PROFIT The line graph in Illustration 6 shows a company's annual profits and losses for five years. Find the average yearly profit or loss.

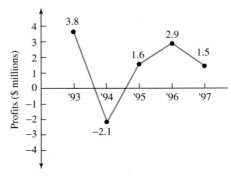

ILLUSTRATION 6

71. $\dfrac{8.6 + 7.99 + (4.05)^2}{4.56}$ **72.** $\dfrac{0.33 + (-0.67)(1.3)^3}{0.0019}$ **73.** $\left(\dfrac{45.9098}{-234.12}\right)^2 - 4$ **74.** $\left(\dfrac{6.0007}{3.002}\right) - \left(\dfrac{78.8}{12.45}\right)$

WRITING   *Write a paragraph using your own words.*

75. Explain the process used to divide two numbers when both the divisor and the dividend are decimals.

76. Explain why we must sometimes use rounding when writing the answer to a division problem.

77. $.5)\overline{2.005}$ is equivalent to $5)\overline{20.05}$. Explain what *equivalent* means in this case.

78. In $3)\overline{0.7}$, why can additional zeros be placed to the right of 0.7 without affecting the outcome?

REVIEW

79. Simplify the complex fraction: $\dfrac{\frac{7}{8}}{\frac{3}{4}}$

80. Express the fraction $\frac{3}{4}$ as an equivalent fraction with denominator 36.

81. List the integers.

82. Add: $-\dfrac{3}{4} + \dfrac{2}{3}$.

83. Multiply: $7\frac{1}{3} \cdot 5\frac{1}{4}$.

84. Evaluate: $\left(\dfrac{1}{2}\right)^3 - \left(\dfrac{1}{2}\right)^2$.

85. Multiply: $(-5.8)(-7.25)$.

86. What is the opposite of 9?

4.5   *Fractions and Decimals*

In this section, you will learn about

- **Writing fractions as equivalent decimals**
- **Repeating decimals**
- **Rounding repeating decimals**
- **Graphing fractions and decimals**
- **Problems involving fractions and decimals**

INTRODUCTION.   In this section, we will further investigate the relationship between fractions and decimals.

Writing fractions as equivalent decimals

To write $\frac{5}{8}$ as a decimal, we use the fact that $\frac{5}{8}$ indicates the division $5 \div 8$. We can convert $\frac{5}{8}$ to decimal form by doing the division.

$$
\begin{array}{r}
.625 \\
8)\overline{5.000} \\
\underline{4\ 8} \\
20 \\
\underline{16} \\
40 \\
\underline{40} \\
0
\end{array}
$$

Write a decimal point and additional zeros to the right of 5.

←— The remainder is zero.

Thus, $\frac{5}{8} = 0.625$.

Writing a fraction as a decimal	To write a fraction as a decimal, divide the numerator of the fraction by its denominator.

EXAMPLE 1   *Writing a fraction as a decimal.*   Write $\frac{3}{4}$ as a decimal.

Solution

We divide the numerator by the denominator.

$$
\begin{array}{r}
.75 \\
4\overline{)3.00} \\
\underline{2\ 8} \\
20 \\
\underline{20} \\
0
\end{array}
$$

Write a decimal point and two zeros to the right of 3.

$\longleftarrow$ The remainder is zero.

Thus, $\frac{3}{4} = 0.75$.

Self Check

Write $\frac{3}{16}$ as a decimal.

Answer: 0.1875

In Example 1, the division process ended because a remainder of 0 was obtained. In this case, we call the quotient, 0.75, a **terminating decimal.**

Repeating decimals

Sometimes, when finding a decimal equivalent of a fraction, the division process never gives a zero remainder. In this case, the result is a **repeating decimal.** Examples of repeating decimals are 0.4444. . . and 1.373737. . . . The three dots tell us that a block of digits continues to repeat in the pattern shown. Repeating decimals can be written using a bar over the repeating block of digits. For example, 0.4444. . . can be written as $0.\overline{4}$, and 1.373737. . . can be written as $1.\overline{37}$.

EXAMPLE 2   *Repeating decimals.*   Write $\frac{5}{12}$ as a decimal.

Solution

We use division to find the decimal equivalent.

$$
\begin{array}{r}
.4166 \\
12\overline{)5.0000} \\
\underline{4\ 8} \\
20 \\
\underline{12} \\
80 \\
\underline{72} \\
80 \\
\underline{72} \\
8
\end{array}
$$

Write a decimal point and four zeros to the right of 5.

It is apparent that 8 will continue to reappear as the remainder. Therefore, 6 will continue to reappear in the quotient. Since the repeating pattern is now clear, we may stop the division.

Thus, $\frac{5}{12} = 0.41\overline{6}$.

Self Check

Write $\frac{3}{11}$ as a decimal.

Answer: $0.\overline{27}$

Every fraction can be written as either a terminating decimal or a repeating decimal. For this reason, the set of fractions (**rational numbers**) form a subset of the set of decimals (called the set of **real numbers**).

Not all decimals are terminating or repeating decimals. For example,

0.2 02 002 0002 · · ·

does not terminate, and it has no repeating block of digits. This decimal cannot be written as a fraction with an integer as its numerator and a nonzero integer as its denominator. Thus, it is not a rational number. It is an example from the set of **irrational numbers.**

Rounding repeating decimals

When a fraction is written in decimal form, the result is either a terminating or a repeating decimal. Repeating decimals are often rounded to a specified place value.

EXAMPLE 3   *Rounding the decimal equivalent.*   Write $-\frac{1}{3}$ as a decimal and round to the nearest hundredth.

Solution   First, we divide the numerator by the denominator to find the decimal equivalent of $\frac{1}{3}$.

$$
\begin{array}{r}
0.333 \\
3\overline{)1.000} \\
\underline{9} \\
10 \\
\underline{9} \\
10 \\
\underline{9} \\
1
\end{array}
$$

Write a decimal point and additional zeros to the right of 1.

To round to the hundredths columns, we must divide to the thousandths column.

To find the decimal equivalent of $-\frac{1}{3}$, we attach a negative sign to the result

Round 0.333 to the nearest hundredth by examining the test digit in the thousandths column.

$-0.33\overset{\downarrow}{3}$

Since 3 is less than 5, we round down, and $-\frac{1}{3} \approx -0.33$. (The symbol $\approx$ means "is approximately equal to.")

EXAMPLE 4   *Rounding a decimal equivalent.*   Write $\frac{2}{7}$ as a decimal and round to the nearest thousandth.

Solution

$$
\begin{array}{r}
.2857 \\
7\overline{)2.0000} \\
\underline{1\,4} \\
60 \\
\underline{56} \\
40 \\
\underline{35} \\
50 \\
\underline{49} \\
1
\end{array}
$$

Write a decimal point and additional zeros to the right of 2.

To round to the thousandths column, we must divide to the ten thousandths column.

Round 0.2857 to the nearest thousandth by examining the test digit in the ten thousandths column.

$0.285\overset{\downarrow}{7}$

Since 7 is greater than 5, we round up, and $\frac{2}{7} \approx 0.286$.

Self Check
Write $\frac{7}{15}$ as a decimal and round to the nearest thousandth.

Answer: 0.467

The problem in Example 4 can also be solved using a scientific calculator.

After performing a calculation, a scientific calculator can round the result to a given decimal place. This is done using the *fixed-point key*. As an example, let's find the decimal equivalent of $\frac{2}{7}$ and round to the nearest thousandth.

Keystrokes: First, set your calculator to round to the third decimal place (thousandths) by pressing $\boxed{\text{FIX}}$ $\boxed{3}$. Then press $\boxed{2}$ $\boxed{\div}$ $\boxed{7}$ $\boxed{=}$.

$$\boxed{0.286}$$

Thus, $\frac{2}{7} \approx 0.286$. To round to the nearest tenth, we would fix 1; to round to the hundredths, we would fix 2, and so on.

To write a mixed number in decimal form, recall that a mixed number is made up of a whole-number part and a fractional part.

EXAMPLE 5   *Writing a mixed number as a decimal.*   Write $5\frac{3}{8}$ in decimal form.

Solution

Recall that $5\frac{3}{8} = 5 + \frac{3}{8}$. Therefore, we need only consider how to write $\frac{3}{8}$ as a decimal.

$$
\begin{array}{r}
.375 \\
8\overline{)3.000} \\
\end{array}
$$
Write a decimal point and three zeros to the right of 3.

$$
\begin{array}{r}
\underline{2\,4} \\
60 \\
\underline{56} \\
40 \\
\underline{40} \\
0
\end{array}
$$

Thus, $5\frac{3}{8} = 5 + \frac{3}{8} = 5 + 0.375 = 5.375$.

Self Check

Write $8\frac{19}{20}$ in decimal form.

Answer: 8.95

Graphing fractions and decimals

A number line can be used to show the relationship between fractions and their respective decimal equivalents. Figure 4-5 shows some commonly used fractions that have terminating decimal equivalents.

FIGURE 4-5

The number line in Figure 4-6 shows some commonly used fractions that have repeating decimal equivalents.

FIGURE 4-6

Problems involving fractions and decimals

Numerical expressions can contain both fractions and decimals. In the following examples, we show how different methods can be used to solve problems of this type.

EXAMPLE 6   *Expressions containing fractions and decimals.*
Evaluate: $\frac{1}{3} + 0.27$.

Solution

We will work in terms of fractions. Write 0.27 as a fraction and add it to $\frac{1}{3}$.

$$\frac{1}{3} + 0.27 = \frac{1}{3} + \frac{27}{100}$$ Replace 0.27 with $\frac{27}{100}$.

$$= \frac{1 \cdot 100}{3 \cdot 100} + \frac{27 \cdot 3}{100 \cdot 3}$$ Express each fraction in terms of 300ths.

$$= \frac{100}{300} + \frac{81}{300}$$ Multiply in the numerators and in the denominators.

$$= \frac{181}{300}$$ Add the numerators and write the sum over the common denominator, 300.

In the next example, we solve the problem in Example 6 using a different method. This time we work with decimals.

EXAMPLE 7   *Expressions containing fractions and decimals.*
Evaluate: $\frac{1}{3} + 0.27$.

Solution

We will work in terms of decimals. We have seen that the decimal equivalent of $\frac{1}{3}$ is the repeating decimal $0.\overline{3}$. To add it to 0.27, we round it to the nearest hundredth.

First, we have $\frac{1}{3} = 0.333\ldots \approx \mathbf{0.33}$

$$\frac{1}{3} + 0.27 \approx \mathbf{0.33} + 0.27$$ Approximate $\frac{1}{3}$ with the decimal 0.33.

$$\approx 0.60$$ Do the addition.

Note that this answer is an approximation, whereas the result in Example 6 was exact.

⊙⊙

EXAMPLE 8   *Expressions containing fractions and decimals.*
Evaluate: $\left(\frac{4}{5}\right)(1.35) + (0.5)^2$.

Solution

Since $\frac{4}{5}$ can be expressed as a terminating decimal, it appears simplest to work in terms of decimals. We use division to find the decimal equivalent of $\frac{4}{5}$.

$$\begin{array}{r} .8 \\ 5\overline{)4.0} \\ \underline{4\,0} \\ 0 \end{array}$$ Write a decimal point and one zero to the right of the 4.

$$\left(\frac{4}{5}\right)(1.35) + (0.5)^2 = \mathbf{(0.8)}(1.35) + (0.5)^2$$ Replace $\frac{4}{5}$ with its decimal equivalent, 0.8.

$$= (0.8)(1.35) + 0.25$$ Find the power: $(0.5)^2 = 0.25$.

$$= 1.08 + 0.25$$ Do the multiplication: $(0.8)(1.35) = 1.08$.

$$= 1.33$$ Do the addition.

EXAMPLE 9 **Shopping.** During a trip to the grocery store, a shopper purchases $\frac{3}{4}$ pound of fruit, priced at $0.88 a pound, and $\frac{1}{3}$ pound of fresh-ground coffee, selling for $6.60 a pound. Find the total cost of these items.

Solution To find the cost of each item, we multiply the amount purchased by its unit price. Then we add the two individual costs to obtain the total cost.

Cost of fruit	+	cost of coffee	=	total cost.

$$\left(\frac{3}{4}\right)(0.88) \quad + \quad \left(\frac{1}{3}\right)(6.60) \quad = \quad \text{total cost}$$

We can work with the decimals and fractions in this form; no converting is necessary.

$$\left(\frac{3}{4}\right)(0.88) + \left(\frac{1}{3}\right)(6.60) = \left(\frac{3}{4}\right)\left(\frac{0.88}{1}\right) + \left(\frac{1}{3}\right)\left(\frac{6.60}{1}\right)$$ Express 0.88 as $\frac{0.88}{1}$ and 6.60 as $\frac{6.60}{1}$.

$$= \frac{2.64}{4} + \frac{6.60}{3}$$ Multiply the numerators and the denominators.

$$= 0.66 + 2.20$$ Do each division.

$$= 2.86$$ Do the addition.

The total cost of the items is $2.86.

STUDY SET Section 4.5

VOCABULARY *In Exercises 1–4, fill in the blanks to make a true statement.*

1. The decimal form of the fraction $\frac{1}{3}$ is a _____ decimal, which is written $0.\overline{3}$ or $0.3333. \ldots$

2. The decimal form of the fraction $\frac{3}{4}$ is a _____ decimal, which is written 0.75.

3. The set of decimals is called the set of _____ numbers.

4. A decimal that cannot be written as a fraction with an integer numerator and a nonzero integer denominator is called an _____ number.

CONCEPTS

5. What are two ways in which $\frac{7}{8}$ can be interpreted?

6. Insert the proper symbol $<$ or $>$ in the blank to make the statement true.
  a. $0.\overline{6}$ ___ 0.7   **b.** $0.\overline{6}$ ___ 0.6

7. When rounding 0.272727. . . to the nearest hundredth, is the result larger or smaller than the original number?

8. Write each decimal in fraction form.
  a. 0.7   **b.** 0.77

9. Graph $1\frac{1}{4}$ and -0.75 on the number line.

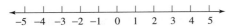

11. Graph 3.3 and -2.5 on the number line.

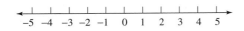

10. Graph $2\frac{7}{8}$ and -2.375 on the number line.

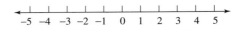

12. Graph -0.8 and 3.7 on the number line.

NOTATION

13. Examine the color portion of this long division. What can be deduced about the decimal equivalent of $\frac{5}{6}$?

$$\begin{array}{r} .833 \\ 6\overline{)5.000} \\ \underline{4\ 8} \\ 20 \\ \underline{18} \\ 20 \end{array}$$

14. a. Using fraction notation, evaluate the numerical expression

$$\frac{1}{3} + 0.11$$

b. Using decimal notation, evaluate the expression in part a. Round to the nearest hundredth.

PRACTICE *In Exercises 15–30, write each fraction in decimal form.*

15. $\dfrac{1}{2}$ **16.** $\dfrac{1}{4}$ **17.** $\dfrac{-5}{8}$ **18.** $-\dfrac{3}{5}$

19. $\dfrac{9}{16}$ **20.** $\dfrac{3}{32}$ **21.** $-\dfrac{17}{32}$ **22.** $-\dfrac{15}{16}$

23. $\dfrac{11}{20}$ **24.** $\dfrac{19}{25}$ **25.** $\dfrac{31}{40}$ **26.** $\dfrac{17}{20}$

27. $-\dfrac{3}{200}$ **28.** $-\dfrac{21}{50}$ **29.** $\dfrac{1}{500}$ **30.** $\dfrac{1}{250}$

In Exercises 31–38, write each fraction in decimal form.

31. $\dfrac{2}{3}$ **32.** $\dfrac{7}{9}$ **33.** $\dfrac{5}{11}$ **34.** $\dfrac{4}{15}$

35. $-\dfrac{7}{12}$ **36.** $-\dfrac{17}{22}$ **37.** $\dfrac{1}{30}$ **38.** $\dfrac{1}{60}$

In Exercises 39–42, write each fraction in decimal form. Round to the nearest hundredth.

39. $\dfrac{7}{30}$ **40.** $\dfrac{14}{15}$ **41.** $\dfrac{17}{45}$ **42.** $\dfrac{8}{9}$

In Exercises 43–46, write each fraction in decimal form. Round to the nearest thousandth.

43. $\dfrac{5}{33}$ **44.** $\dfrac{5}{12}$ **45.** $\dfrac{10}{27}$ **46.** $\dfrac{17}{21}$

In Exercises 47–50, write each fraction in decimal form. Round to the nearest hundredth.

47. $\dfrac{4}{3}$ **48.** $\dfrac{10}{9}$ **49.** $-\dfrac{34}{11}$ **50.** $-\dfrac{25}{12}$

In Exercises 51–58, write each mixed number in decimal form. Round to the nearest hundredth when the result is a repeating decimal.

51. $3\dfrac{3}{4}$ **52.** $5\dfrac{4}{5}$ **53.** $-8\dfrac{2}{3}$ **54.** $-1\dfrac{7}{9}$

55. $12\dfrac{11}{16}$ **56.** $32\dfrac{1}{8}$ **57.** $203\dfrac{11}{15}$ **58.** $568\dfrac{23}{30}$

In Exercises 59–62, fill in the correct symbol ($<$ or $>$) to make a true statement. (Hint: Express each number as a decimal.)

59. $\dfrac{7}{8}$ ___ 0.895 **60.** 4.56 ___ $4\dfrac{2}{5}$ **61.** $-\dfrac{11}{20}$ ___ $-0.\overline{4}$ **62.** $-9.0\overline{9}$ ___ $-9\dfrac{1}{11}$

In Exercises 63–70, evaluate each expression. Express each result as a fraction in lowest terms.

63. $-\dfrac{3}{4} \cdot 5.1$ **64.** $-1.8\left(\dfrac{7}{25}\right)$ **65.** $\dfrac{1}{9} + 0.3$ **66.** $\dfrac{2}{3} + 0.1$

67. $\dfrac{5}{11}(0.3)$ **68.** $(0.9)\left(\dfrac{1}{27}\right)$ **69.** $\dfrac{1}{3}\left(-\dfrac{1}{15}\right)(0.5)$ **70.** $(-0.4)\left(\dfrac{5}{18}\right)\left(-\dfrac{1}{3}\right)$

In Exercises 71–78, evaluate each expression. Express each result as a decimal.

71. $(3.5 + 6.7)\left(-\dfrac{1}{4}\right)$ **72.** $\left(-\dfrac{5}{8}\right)(5.3 - 3.9)$ **73.** $\left(\dfrac{1}{5}\right)^2(1.7)$ **74.** $(2.35)\left(\dfrac{2}{5}\right)^2$

75. $7.5 - (0.78)\left(\dfrac{1}{2}\right)$ **76.** $8.1 - \left(\dfrac{3}{4}\right)(0.12)$

77. $\dfrac{3}{8}(-3.2) + (4.5)\left(-\dfrac{1}{9}\right)$ **78.** $(-0.8)\left(\dfrac{1}{4}\right) + \left(\dfrac{1}{3}\right)(0.39)$

In Exercises 79–80, evaluate each formula. Round to the nearest tenth.

79. $C = \dfrac{5(F - 32)}{9}$ for $F = 64.5$ **80.** $F = \dfrac{9}{5}C + 32$ for $C = 0.58$

 In Exercises 81–84, use a calculator to write each fraction in decimal form.

81. $\dfrac{23}{101}$ **82.** $\dfrac{1}{99}$ **83.** $\dfrac{1,736}{50}$ **84.** $-\dfrac{11}{128}$

APPLICATIONS   In Exercises 85–92, solve each problem.

85. DRAFTING   The architect's scale has several measuring edges. The edge marked 16 divides each inch into 16 equal parts. (See Illustration 1.) Find the decimal form for each fractional part of one inch that is highlighted on the scale.

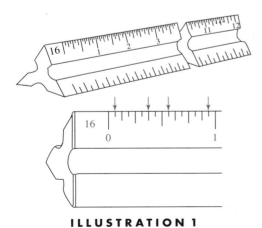

ILLUSTRATION 1

86. FREEWAY SIGNS   The freeway sign in Illustration 2 gives the number of miles to the next three exits. Convert the mileages to decimal notation.

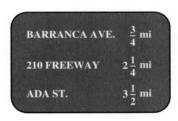

BARRANCA AVE.	$\dfrac{3}{4}$ mi
210 FREEWAY	$2\dfrac{1}{4}$ mi
ADA ST.	$3\dfrac{1}{2}$ mi

ILLUSTRATION 2

87. GARDENING   Two brands of replacement line for a lawn trimmer are labeled in different ways. (See Illustration 3.) On one package, the line's thickness is expressed as a decimal; on the other, as a fraction. Which line is thicker?

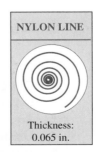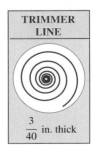

NYLON LINE — Thickness: 0.065 in.

TRIMMER LINE — $\dfrac{3}{40}$ in. thick

ILLUSTRATION 3

88. AUTO MECHANICS While doing a tuneup, a mechanic checks the gap on one of the spark plugs of a car to be sure it is firing correctly. The owner's manual states that the gap should be $\frac{2}{125}$ inch. The gauge the mechanic uses to check the gap is in decimal notation; it registers 0.025 inch. Is the spark plug gap too large or too small?

89. HORSE RACING In thoroughbred racing, the time a horse takes to run a given distance is measured using fifths of a second. For example, 55^2 (read "fifty-five and two") means $55\frac{2}{5}$ seconds. Illustration 4 lists four split times for a horse. Express the times in decimal form.

Speedy Flight Turfway Park, Ky 3-year–old
17 May 97 $1\frac{1}{16}$ mile $:23^2$ $:23^4$ $:24^1$ $:32^3$

ILLUSTRATION 4

90. GEOLOGY A geologist weighed a rock sample at the site where it was discovered and found it to weigh $17\frac{7}{8}$ lb. Later, a more accurate digital scale in the laboratory gave the weight as 17.671 lb. What is the difference in the two measurements?

91. WINDOW REPLACEMENT The amount of sunlight that comes into a room depends on the area of the windows in the room. What is the area of the window in Illustration 5?

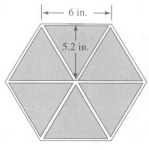

ILLUSTRATION 5

92. FOREST FIRE CONTAINMENT A command post asked each of three fire crews to estimate the length of the fire line they were fighting. Their reports came back in different forms, as indicated in Illustration 6. Find the perimeter of the fire.

ILLUSTRATION 6

WRITING

93. Explain the procedure used to write a fraction in decimal form.

94. Compare and contrast the two numbers 0.5 and $0.\overline{5}$.

REVIEW

95. Add: $-2 + (-3) + 10 + (-6)$.

96. Evaluate: $-3 + 2[-3 + (2 - 7)]$.

97. List the first eight whole numbers.

98. Simplify: $\frac{20}{55}$.

99. Multiply: $3\frac{1}{3} \cdot 4\frac{1}{2}$.

100. Divide: $3\frac{1}{3} \div 4\frac{1}{2}$.

4.6 *Square Roots*

In this section, you will learn about

- **Square roots**
- **Evaluating numerical expressions containing radicals**
- **Square roots of fractions and decimals**
- **Using a calculator to find square roots**
- **Approximating square roots**

INTRODUCTION. There are six basic operations of arithmetic. We have seen the relationships between addition and subtraction and between multiplication and division. In this section, we will explore the relationship between raising a number to a power and finding a root. Decimals will play an important role in this discussion.

Square roots

When we raise a number to the second power, we are squaring it, or finding its **square.**

The square of 6 is 36, because $6^2 = 36$.
The square of -6 is 36, because $(-6)^2 = 36$.

The **square root** of a given number is a number whose square is the given number. For example, the square roots of 36 are 6 or -6, because either number, when squared, yields 36.

Square root	A number is a **square root** of a second number if the square of the first number equals the second number.

EXAMPLE 1 *Finding square roots.* Find the square roots of 49.

Solution

Ask yourself, "What number was squared to obtain 49?"

$$7^2 = 49 \quad \text{and} \quad (-7)^2 = 49$$

Thus, 7 and -7 are the square roots of 49.

Self Check
Find the square roots of 64.

Answer: 8 and -8 ■

In Example 1, we saw that 49 has two square roots—one positive and one negative. The symbol $\sqrt{}$ is called a **radical sign** and is used to indicate a positive square root.

When a number, called the **radicand,** is inserted under a radical sign, we are to find the positive square root of that number. For example, if we evaluate (or simplify) $\sqrt{36}$ (read as "the square root of 36"), the result is

$$\sqrt{36} = 6$$

The negative square root of 36 is denoted as $-\sqrt{36}$; therefore,

$$-\sqrt{36} = -6$$

EXAMPLE 2 *Evaluating radical expressions.* Simplify each radical expression: **a.** $\sqrt{81}$ and **b.** $-\sqrt{100}$.

Solution
a. $\sqrt{81}$ means the positive square root of 81. Since $9^2 = 81$, we get $\sqrt{81} = 9$.

b. $-\sqrt{100}$ means the negative square root of 100. Since $10^2 = 100$, we get $-\sqrt{100} = -10$.

Self Check
Simplify each radical expression:

a. $\sqrt{144}$ and **b.** $-\sqrt{81}$.

Answer: **a.** 12, **b.** -9 ■

Evaluating numerical expressions containing radicals

Numerical expressions can contain radical expressions. When applying the rules for order of operations, we treat a radical expression as we would a power.

EXAMPLE 3 *Evaluating expressions containing radicals.*
Evaluate: **a.** $\sqrt{64} + \sqrt{9}$ and **b.** $-\sqrt{25} - \sqrt{4}$.

Solution

a. $\sqrt{64} + \sqrt{9} = 8 + 3$ Evaluate each radical expression first.

$\qquad\qquad\qquad = 11$ Do the addition.

b. $-\sqrt{25} - \sqrt{4} = -5 - 2$ Evaluate each radical expression first.

$\qquad\qquad\qquad = -7$ Do the subtraction.

Self Check

Evaluate each expression:

a. $\sqrt{121} + \sqrt{1}$ and
b. $-\sqrt{9} - \sqrt{16}$.

Answer: **a.** 12, **b.** -7 ■

EXAMPLE 4 *Evaluating expressions containing radicals.*
Evaluate: **a.** $6\sqrt{100}$ and **b.** $-5\sqrt{16} + 3\sqrt{9}$.

Solution

We apply the rules for order of operations.

a. $6\sqrt{100}$ means $6 \cdot \sqrt{100}$.

$\qquad 6\sqrt{100} = 6(10)$ Simplify the radical first.

$\qquad\qquad\quad = 60$ Do the multiplication.

b. $-5\sqrt{16} + 3\sqrt{9} = -5(4) + 3(3)$ Simplify each radical.

$\qquad\qquad\qquad\qquad = -20 + 9$ Do the multiplications.

$\qquad\qquad\qquad\qquad = -11$ Do the addition.

Self Check

Evaluate each expression:

a. $8\sqrt{121}$ and

b. $-6\sqrt{25} + 2\sqrt{36}$.

Answer: **a.** 88, **b.** -18 ■

Square roots of fractions and decimals

So far, we have found square roots of whole numbers. We can also find square roots of fractions and decimals.

EXAMPLE 5 *Square roots of fractions and decimals.*
Simplify: **a.** $\sqrt{\dfrac{25}{64}}$ and **b.** $\sqrt{0.81}$.

Solution

a. $\sqrt{\dfrac{25}{64}} = \dfrac{5}{8}$, because $\left(\dfrac{5}{8}\right)^2 = \dfrac{25}{64}$.

b. $\sqrt{0.81} = 0.9$, because $(0.9)^2 = 0.81$.

Self Check

Simplify:

a. $\sqrt{\dfrac{16}{49}}$ and **b.** $\sqrt{0.04}$.

Answer: **a.** $\dfrac{4}{7}$, **b.** 0.2 ■

Using a calculator to find square roots

We can also use a calculator to find square roots.

We use the $\boxed{\sqrt{}}$ key (square root key) on a scientific calculator to find square roots.

Evaluate: $\sqrt{729}$

Keystrokes: $\boxed{7}$ $\boxed{2}$ $\boxed{9}$ $\boxed{\sqrt{}}$ $\boxed{\qquad\qquad 27}$

We have found that $\sqrt{729} = 27$. To check this result, we need to square 27. This can be done by pressing the keys $\boxed{2}$, $\boxed{7}$, and $\boxed{x^2}$. We obtain 729. Thus, 27 is the square root of 729.

Approximating square roots

Numbers whose square roots are whole numbers are called **perfect squares.** The perfect squares that are less than or equal to 100 are

0, 1, 4, 9, 16, 25, 36, 49, 64, 81, 100

To find the square root of a number that is not a perfect square, we can use a calculator. For example, to find $\sqrt{17}$, we press these keys:

$\boxed{1}$ $\boxed{7}$ $\boxed{\sqrt{}}$

The display reads 4.123105626. This result is not exact, because $\sqrt{17}$ is a **nonterminating decimal** that never ends and never repeats. $\sqrt{17}$ is an **irrational number.** Together, the rational and the irrational numbers form the set of **real numbers.** Rounded to the nearest thousandth, we have

$\sqrt{17} = 4.123$

EXAMPLE 6   *Approximating square roots.*   Use a scientific calculator to find each square root. Round to the nearest hundredth.

    a. 373   **b.** 56.2   **c.** 0.0045

Solution

a. From the calculator, we get $\sqrt{373} \approx 19.31320792$. Rounding to the nearest hundredth, $\sqrt{373} = 19.31$.

b. From the calculator, we get $\sqrt{56.2} \approx 7.496665926$. Rounding to the nearest hundredth, $\sqrt{56.2} = 7.50$.

c. From the calculator, we get $\sqrt{0.0045} \approx 0.067082039$. Rounding to the nearest hundredth, $\sqrt{0.0045} = 0.07$.

Self Check

Use a calculator to find the square root. Round to the nearest hundredth.

a. 607.8

b. 0.076

Answer: **a.** 24.65, **b.** 0.28

STUDY SET   Section 4.6

VOCABULARY   *In Exercises 1–6, fill in the blanks to make a true statement.*

1. When we find what number is squared to obtain a given number, we are finding the _____ of the given number.

2. Whole numbers such as 25, 36, and 49 are called perfect _____ because their square roots are whole numbers.

3. The symbol $\sqrt{}$ is called a _____. It indicates that we are to find a _____ square root.

4. The decimal number that represents $\sqrt{17}$ is a _____ decimal—it never ends.

5. In $\sqrt{26}$, 26 is called the _____.

6. The symbol $\approx$ means "is _____ equal to."

7. The square of 5 is ___ , because $(5)^2 =$ ___ .

8. The square of $\frac{1}{4}$ is ___ , because $\left(\frac{1}{4}\right)^2 =$ ___ .

9. The two square roots of 49 are 7 and -7, because ___ $= 49$ and ___ $= 49$.

10. The two square roots of 4 are 2 and -2, because ___ $= 4$ and ___ $= 4$.

11. Since $\left(\frac{3}{4}\right)^2 = \frac{9}{16}$, $\sqrt{\frac{9}{16}} =$ ___ .

12. Since $(0.4)^2 = 0.16$, $\sqrt{0.16} =$ ___ .

13. Without evaluating the following square roots, write them in order, from smallest to largest: $\sqrt{23}$, $\sqrt{11}$, $\sqrt{27}$, $\sqrt{6}$.

14. Without evaluating the square roots, write them in order from smallest to largest: $-\sqrt{13}$, $-\sqrt{5}$, $-\sqrt{17}$, $-\sqrt{37}$.

15. Simplify.
   a. $\sqrt{1}$
   b. $\sqrt{0}$

16. Multiplication can be thought of as the opposite of division. What is the opposite of finding the square root of a number?

17. **a.** Use a calculator to approximate $\sqrt{6}$ to the nearest tenth.
   b. Square the result from part a.
   c. Find the difference between 6 and the answer to part b.

18. **a.** Use a calculator to approximate $\sqrt{6}$ to the nearest hundredth.
   b. Square the result from part a.
   c. Find the difference between the answer to part b and 6.

19. Graph $\sqrt{9}$ and $-\sqrt{5}$ on the number line.

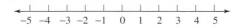

20. Graph $-\sqrt{3}$ and $\sqrt{7}$ on the number line.

21. Between what two whole numbers would each square root be located when graphed on a number line?
   a. $\sqrt{19}$
   b. $\sqrt{87}$

22. Between what two whole numbers would each square root be located when graphed on a number line?
   a. $\sqrt{50}$
   b. $\sqrt{33}$

23. Simplify: $-\sqrt{49} + \sqrt{64}$.
$$-\sqrt{49} + \sqrt{64} = \underline{\ \ } + \underline{\ \ } \quad \text{Simplify each radical.}$$
$$= 1 \qquad\qquad\quad \text{Do the addition.}$$

24. Simplify: $2\sqrt{100} - 5\sqrt{25}$.
$$2\sqrt{100} - 5\sqrt{25} = 2(\underline{\ \ }) - 5(\underline{\ \ }) \quad \text{Simplify each radical.}$$
$$= \underline{\ \ } - 25 \qquad\qquad \text{Do the multiplication.}$$
$$= -5 \qquad\qquad\qquad \text{Do the subtraction.}$$

25. $\sqrt{16}$

26. $\sqrt{64}$

27. $-\sqrt{121}$

28. $-\sqrt{144}$

29. $-\sqrt{0.49}$

30. $-\sqrt{0.64}$

31. $\sqrt{0.25}$

32. $\sqrt{0.36}$

33. $\sqrt{0.09}$

34. $\sqrt{0.01}$

35. $-\sqrt{\frac{1}{81}}$

36. $-\sqrt{\frac{1}{4}}$

37. $-\sqrt{\frac{16}{9}}$

38. $-\sqrt{\frac{64}{25}}$

39. $\sqrt{\frac{4}{25}}$

40. $\sqrt{\frac{36}{121}}$

41. $5\sqrt{36} + 1$

42. $2 + 6\sqrt{16}$

43. $-4\sqrt{36} + 2\sqrt{4}$

44. $-6\sqrt{81} + 5\sqrt{1}$

45. $\sqrt{\frac{1}{16}} - \sqrt{\frac{9}{25}}$

46. $\sqrt{\frac{25}{9}} - \sqrt{\frac{64}{81}}$

47. $5(\sqrt{49})(-2)$

48. $(-\sqrt{64})(-2)(3)$

49. $\sqrt{0.04} + 2.36$

50. $\sqrt{0.25} + 4.7$

51. $-3\sqrt{1.44}$

52. $-2\sqrt{1.21}$

In Exercises 53–54, use a calculator to complete each square root table. Round to the nearest thousandth when an answer is not exact.

53.

Number	Square root
1	
2	
3	
4	
5	
6	
7	
8	
9	
10	

54.

Number	Square root
10	
20	
30	
40	
50	
60	
70	
80	
90	
100	

 In Exercises 55–58, use a calculator to simplify each of the following.

55. $\sqrt{1,369}$ **56.** $\sqrt{841}$ **57.** $\sqrt{3,721}$ **58.** $\sqrt{5,625}$

 In Exercises 59–62, use a calculator to approximate each of the following to the nearest hundredth.

59. $\sqrt{15}$ **60.** $\sqrt{51}$ **61.** $\sqrt{66}$ **62.** $\sqrt{204}$

 In Exercises 63–66, use a calculator to approximate each of the following to the nearest thousandth.

63. $\sqrt{24.05}$ **64.** $\sqrt{70.69}$ **65.** $-\sqrt{11.1}$ **66.** $\sqrt{0.145}$

APPLICATIONS In Exercises 67–72, square roots have been used to express various lengths. Solve each problem by simplifying any square roots. You may need to use a calculator. If this is the case, round to the nearest tenth.

67. CARPENTRY   Find the length of the slanted part of each roof truss.

a.

b.

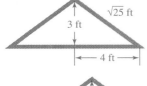

68. RADIO ANTENNA   See Illustration 1. How far from the base of the antenna is each guy wire anchored to the ground? (The measurements are in feet.)

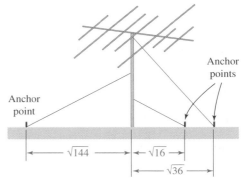

ILLUSTRATION 1

69. BASEBALL DIAMOND Illustration 2 shows some dimensions of a major league baseball field. How far is it from home plate to second base?

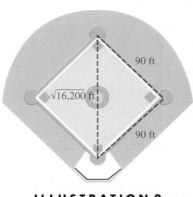

90 ft

$\sqrt{16,200}$ ft

90 ft

ILLUSTRATION 2

70. SURVEYING Use the imaginary triangles set up by a surveyor to find the length of each lake. (The measurements are in meters.)

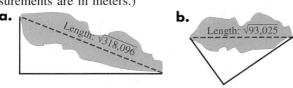

a. Length: $\sqrt{318,096}$

b. Length: $\sqrt{93,025}$

71. BIG-SCREEN TELEVISION The picture screen on a television set is measured diagonally. What size screen is shown in Illustration 3?

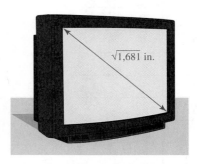

$\sqrt{1,681}$ in.

ILLUSTRATION 3

72. LADDER A painter's ladder is shown in Illustration 4. How long are the legs of the ladder?

$\sqrt{225}$ ft $\sqrt{169}$ ft

ILLUSTRATION 4

In Exercises 73–76, use a calculator to evaluate each radical expression. If an answer is not exact, round to the nearest ten thousandth.

73. $\sqrt{24,000,201}$

74. $-\sqrt{4.012009}$

75. $-\sqrt{0.00111}$

76. $\sqrt{\dfrac{27}{44}}$

WRITING Write a paragraph using your own words.

77. When asked to find $\sqrt{16}$, one student's answer was 8. Explain his misunderstanding of the concept of square root.

78. Explain the difference between the square and the square root of a number.

79. What is a nonterminating decimal? Use an example in your explanation.

80. What do you think might be meant by the term *cube root?*

REVIEW

81. Multiply: $6.75 \cdot 12.2$.

82. Simplify: $\dfrac{\frac{-2}{3}}{8}$.

83. Evaluate: $5(-2)^2 - \dfrac{16}{4}$.

84. Divide: $5.7\overline{)18.525}$.

85. List the natural numbers.

86. Evaluate: $(3.4)^3$.

87. Simplify: $\left(\dfrac{2}{3}\right)^2 - \left(-\dfrac{3}{4}\right)^2$.

88. Insert the proper symbol, $<$ or $>$, in the blank to make a true statement: -15 ___ -14.

The Real Numbers

A **real number** is any number that can be expressed as a decimal. All of the types of numbers that we have discussed in this book are real numbers. As we have seen, the set of real numbers is made up of several subsets of numbers.

In Exercises 1–5, if possible, list the numbers belonging to each set. If it is not possible to list them, define the set in words.

1. Natural numbers

2. Whole numbers

3. Integers

4. Rational numbers

5. Irrational numbers

This diagram shows how the set of real numbers is made up of two distinct sets: the rational and the irrational numbers. Since every natural number is a whole number, we show the set of natural numbers included in the whole numbers. Because every whole number is an integer, the whole numbers are shown contained in the integers. Since every integer is a rational number, we show the integers included in the rational numbers.

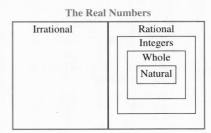

In Exercises 6–15, tell whether each statement is true or false.

6. Every integer is a real number.

7. Every fraction can be written as a terminating decimal.

8. Every real number is a whole number.

9. Some irrational numbers are integers.

10. Some rational numbers are natural numbers.

11. No numbers are both rational and irrational numbers.

12. All real numbers can be graphed on a number line.

13. The set of whole numbers is a subset of the irrational numbers.

14. All decimals either terminate or repeat.

15. Every natural number is an integer.

16. List the numbers in the set -2, -1.2, $-\frac{7}{8}$, 0, $1\frac{2}{3}$, 2.75, $\sqrt{23}$, 10, $1.161661666\ldots$ that are

   a. Natural numbers

   b. Whole numbers

   c. Integers

   d. Rational numbers

   e. Irrational numbers

   f. Real numbers

An Introduction to Decimals

CONCEPTS

Decimal fractions are fractions with a denominator of 10, 100, 1,000, 10,000, and so on.

Expanded notation is used to show the value represented by each digit in the *decimal numeration system.*

$$5.6791 =$$

$$5 + \frac{6}{10} + \frac{7}{100} + \frac{9}{1,000} + \frac{1}{10,000}$$

To express a decimal in words, say:
1. the whole number to the left of the decimal point;
2. "and" for the decimal point;
3. the whole number to the right of the decimal point, followed by the name of the last place value column on the right.

To compare the size of two decimals, compare the digits of each decimal, column by column, working from left to right.

A decimal point and additional zeros may be written to the right of a whole number.

REVIEW EXERCISES

1. Represent the amount of the square that is shaded in Illustration 1, using a decimal and a fraction.

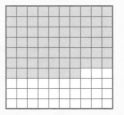

ILLUSTRATION 1

2. In Illustration 2, shade 0.8 of the rectangle.

ILLUSTRATION 2

3. Write 16.4523 in expanded notation.

4. Write each decimal in words and then as a fraction or mixed number.
a. 2.3
b. −15.59
c. 0.0601
d. 0.00001

5. Graph 1.55 and −2.7 on a number line.

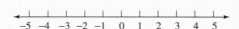

6. Graph 1.4 and −0.8 on a number line.

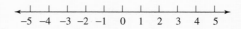

7. VALEDICTORIAN   At the end of the school year, the five students listed in Illustration 3 were in the running to be class valedictorian. Rank the students in order by GPA, beginning with the valedictorian.

Name	GPA
Diaz, Cielo	3.9809
Chou, Wendy	3.9808
Washington, Shelly	3.9865
Gerbac, Lance	3.899
Singh, Amani	3.9713

ILLUSTRATION 3

8. True or false: 78 = 78.0.

9. Place the proper symbol (<, >, or =) in the blank to make a true statement.
a. 4.5 ___ 4.6
b. −2.35 ___ −2.53
c. 10.90 ___ 10.9
d. 0.027894 ___ 0.034

To round a decimal, first locate the rounding digit and the test digit.
1. If the test digit is a number less than 5, drop it and all digits to the right of the rounding digit.
2. If it is 5 or larger, add 1 to the rounding digit and drop all digits to its right.

10. Round each decimal to the specified place value column.
 a. 4.578: hundredths
 b. 3,706.0895: thousandths
 c. −0.0614: tenths
 d. 88.12: tenths

SECTION 4.2

Addition and Subtraction with Decimals

To add (or subtract) decimal numbers:
1. Line up their decimal points.
2. Add (or subtract) as you would with whole numbers.
3. Write the decimal point in the result directly below the decimal points of the problem.

11. Do each addition.
 a. $19.5 + 34.4 + 12.8$
 b. $3.4 + 6.78 + 35 + 0.008$

12. Do each subtraction.
 a. $68.47 − 53.3$
 b. $45.08 − 17.37$

13. Evaluate each expression.
 a. $−16.1 + 8.4$
 b. $−4.8 − (−7.9)$
 c. $−3.55 + (−1.25)$
 d. $−15.1 − 13.99$

14. Evaluate each expression.
 a. $−8.8 + (−7.3 − 9.5)$
 b. $(5 − 0.096) − (−0.035)$

15. SALE PRICE   A calculator normally sells for $52.20. If it is being discounted $3.99, what is the sale price?

16. MICROWAVE OVEN A microwave oven is shown in Illustration 4. How tall is the window?

ILLUSTRATION 4

SECTION 4.3

Multiplication with Decimals

To multiply decimals,
1. Multiply as if working with whole numbers.
2. Place the decimal point in the answer so that the result has the same number of decimal places as the sum of the decimal places of the factors.

17. Do each multiplication.
 a. $(−0.6)(0.4)$
 b. $2.3 \cdot 0.9$
 c. $5.5(−3.1)$
 d. $32.45(6.1)$
 e. $(−0.003)(−0.02)$
 f. $7 \cdot 0.6$

To multiply a decimal by 10, 100, 1,000, and so on, move the decimal point the same number of places to the right as the number of zeros in the power of 10.

Exponents are used to represent repeated multiplication.

Use the rules for order of operations to evaluate numerical expressions containing decimals.

To evaluate an algebraic expression, substitute specific numbers for the variables in the expression and apply the order of operations rules.

18. Do each multiplication mentally.
  a. $1,000(90.1452)$
  b. $(-10)(-2.897)(100)$

19. Find each power.
  a. $(0.2)^2$
  b. $(-0.15)^2$
  c. $(3.3)^2$
  d. $(0.1)^3$

20. Evaluate each numerical expression.
  a. $(0.6 + 0.7)^2 - 12.3$
  b. $3(7.8) + 2(1.1)^2$

21. Evaluate the formula $A = lw$ for $l = 32.5$ and $w = 21.3$.

22. WORD PROCESSOR   The Page Setup screen for a word processor is shown in Illustration 5. Find the area on a 8.5-inch-by-11-inch piece of paper that can be filled with text if the margins are set as shown.

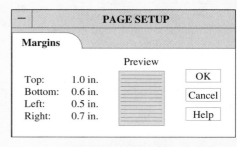

ILLUSTRATION 5

23. AUTO PAINTING   A manufacturer uses a three-part process to finish the exterior of the cars it produces.

  Step 1: A 0.03-inch-thick rust-prevention undercoat is applied.

  Step 2: Three layers of color coat, each 0.015 of an inch thick, are sprayed on.

  Step 3: The finish is then buffed down, losing 0.005 of an inch of its thickness.

What is the resulting thickness of the automobile's finish?

SECTION 4.4 — *Division with Decimals*

To divide a decimal by a whole number:
1. Divide as if working with whole numbers.
2. Write the decimal point in the result directly above the decimal point in the dividend.

24. Do each division.
  a. $12\overline{)15}$
  b. $-41.8 \div 4$
  c. $\dfrac{-29.67}{-23}$
  d. $24.618 \div 6$

25. Do each division.
  a. $12.47 \div (-4.3)$
  b. $\dfrac{0.0742}{1.4}$
  c. $\dfrac{15.75}{0.25}$
  d. $\dfrac{-0.03726}{-0.046}$

To divide with a decimal divisor:
1. Move the decimal point in the divisor so it becomes a whole number.
2. Move the decimal point in the dividend the same number of places to the right.
3. Use the process for dividing a decimal by a whole number.

To divide a decimal by a power of 10, such as 10, 100, or 1,000, move the decimal point of the dividend to the left the same number of places as the number of zeros in the power of 10.

26. Divide and round each result to the nearest tenth.

a. $78.98 \div 6.1$

b. $\dfrac{-5.338}{0.008}$

27. Evaluate the formula $C = \dfrac{5(F - 32)}{9}$ for $F = 68.4$ and round to the nearest hundredth.

28. THANKSGIVING DINNER   The cost of purchasing the ingredients for a Thanksgiving turkey dinner for a family of 5 was \$41.70. What was the cost of the dinner per person?

29. Do each division mentally.

a. $89.76 \div 100$

b. $\dfrac{0.0112}{-10}$

30. Evaluate the numerical expression $\dfrac{(1.4)^2 + 2(4.6)}{0.5 + 0.3}$.

31. SERVING SIZE   Illustration 6 shows the package labeling on a box of children's cereal. Use the information given to find the number of servings.

Nutrition Facts	
Serving size	1.1 ounce
Servings per container	?
Package weight	15.5 ounces

ILLUSTRATION 6

32. TELESCOPE   To change the position of a focusing mirror on a telescope, an adjustment knob is used. The mirror moves 0.025 inch with each revolution of the knob. The mirror needs to be moved 0.2375 inch to improve the sharpness of the image. How many revolutions of the adjustment knob does this require?

SECTION 4.5   *Fractions and Decimals*

To write a fraction as a decimal, divide the numerator by the denominator.

We obtain either a *terminating* or a *repeating* decimal when using division to write a fraction as a decimal.

An overbar can be used instead of the three dots . . . to represent the repeating pattern in a repeating decimal.

The set of *real numbers* is the set of decimals.

33. Write each fraction in decimal form.

a. $\dfrac{7}{8}$ **b.** $-\dfrac{2}{5}$ **c.** $\dfrac{9}{16}$ **d.** $\dfrac{3}{50}$

34. Write each fraction in decimal form.

a. $\dfrac{6}{11}$ **b.** $-\dfrac{2}{3}$

35. Write each fraction in decimal form. Round to the nearest hundredth.

a. $\dfrac{19}{33}$ **b.** $\dfrac{31}{30}$

36. Place the proper symbol ($<$ or $>$) in the blank to make a true statement.

a. $\dfrac{13}{25}$ ___ 0.499 **b.** $-0.\overline{26}$ ___ $-\dfrac{4}{15}$

37. Graph $1\frac{1}{8}$, $-\frac{1}{3}$, $2\frac{3}{4}$, and $-\frac{9}{10}$ on the number line. Label each using its decimal equivalent.

38. Evaluate each numerical expression. Find the exact answer.

a. $\frac{1}{3} + 0.4$ **b.** $\frac{4}{5}(-7.8)$

c. $\frac{1}{2}(9.7 + 8.9)(10)$ **d.** $\frac{1}{3}(3.14)(3)^2(4.2)$

39. Evaluate $\frac{4}{3}(3.14)(2)^3$. Round the result to the nearest hundredth.

40. ROADSIDE EMERGENCY   In case of trouble, truckers carry reflectors to be placed on the highway shoulder to warn approaching cars of a stalled vehicle. (See Illustration 7.) What is the area of one of these triangular-shaped reflectors?

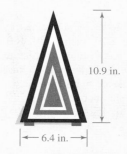

10.9 in.

6.4 in.

ILLUSTRATION 7

A number line can be used to show the relationship between fractions and decimals.

SECTION 4.6

Square Roots

A number is a *square root* of a second number if the square of the first is equal to the second.

A *radical sign* $\sqrt{}$ is used to indicate a positive square root. The square root of a *perfect square* is a whole number.

A square root can be approximated using a calculator.

When evaluating an expression containing square roots, treat a radical as you would a power when applying the rules for order of operations.

41. Fill in the blanks to make a true statement. Two square roots of 64 are 8 and -8, because ___ = 64 and ___ = 64.

42. Simplify each expression without using a calculator.

a. $\sqrt{49}$ **b.** $-\sqrt{16}$ **c.** $\sqrt{100}$ **d.** $\sqrt{0.09}$

e. $\sqrt{\frac{64}{25}}$ **f.** $\sqrt{0.81}$ **g.** $-\sqrt{\frac{1}{36}}$ **h.** $\sqrt{0}$

43. Between what two whole numbers would $\sqrt{83}$ be located when graphed on a number line?

44. Use a calculator to find $\sqrt{11}$ and round to the nearest tenth. Now square the approximation. How close is it to 11?

45. Graph each square root on the number line: $\sqrt{3}$, $-\sqrt{2}$, and $\sqrt{0}$.

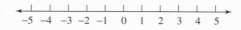

46. Simplify each expression without using a calculator.

a. $-3\sqrt{100}$ **b.** $5\sqrt{0.25}$

c. $-3\sqrt{49} - \sqrt{36}$ **d.** $\sqrt{\frac{9}{100}} + \sqrt{1.44}$

47. Use a calculator to find each square root to the nearest hundredth.

a. $\sqrt{19}$ **b.** $\sqrt{59}$

1. Express the amount of the square in Illustration 1 that is shaded, using a fraction and a decimal.

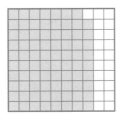

ILLUSTRATION 1

2. WATER PURITY   A county health department sampled the pollution content of tap water in 6 cities, with the results shown in Illustration 2. Rank the cities in order, from dirtiest tap water to cleanest.

City	Pollution parts per million
Monroe	0.0909
Covington	0.0899
Paston	0.0901
Cadia	0.0890
Selway	0.1001

ILLUSTRATION 2

3. Write 0.271 as a fraction.

4. Round to the nearest thousandth: 33.0495.

5. SKATING RECEIPTS   At an ice-skating complex, receipts on Friday were $30.25 for indoor skating and $62.25 for outdoor skating. On Saturday, the corresponding amounts were $40.50 and $75.75. Find the total income for the two days.

6. Do each operation mentally.
  a. $567.909 \div 1,000$ **b.** $0.00458 \cdot 100$

7. EARTHQUAKE FAULT LINE   After an earthquake, geologists found that the ground on the west side of the fault line had dropped 0.83 inch. The next week, a strong aftershock caused the same area to sink 0.19 inch deeper. How far did the ground on the west side of the fault drop because of the seismic activity?

8. Do each operation.
  a. $2 + 4.56 + 0.89 + 3.3$
  b. $45.2 - 39.079$
  c. $(0.32)^2$
  d. $-6.7(-2.1)$

9. Find the area of the lot in Illustration 3.

1.8 mi

3.6 mi

ILLUSTRATION 3

10. TELEPHONE BOOK   To print a telephone book, 565 sheets of paper were used. If the book is 2.3 inches thick, what is the thickness of each sheet of paper, to the nearest thousandth of an inch?

11. Evaluate: $4.1 - (3.2)(0.4)^2$.

12. Write each fraction as a decimal.

    a. $\dfrac{17}{50}$ **b.** $\dfrac{5}{12}$

13. Do the division and round to the nearest hundredth: $\dfrac{12.146}{-5.3}$.

14. Find $11\overline{)13}$.

15. Graph $\frac{3}{5}$ and $-\frac{4}{5}$ on the number line. Label each using its decimal equivalent.

16. Find the exact answer: $\dfrac{2}{3} + 0.7$.

17. CHEMISTRY   In a lab experiment, a chemist mixed three compounds together to form a mixture weighing 4.37 g. Later, she discovered that she had forgotten to record the weight of compound C in her notes. Find the weight of compound C used in the experiment.

	Weight
Compound A	1.86 g
Compound B	2.09 g
Compound C	?
Mixture total	4.37 g

18. Fill in the blanks to make a true statement. Two square roots of 100 are 10 and -10, because ___ = 100 and _____ = 100.

19. Graph $\sqrt{2}$ and $-\sqrt{5}$ on the number line.

20. Simplify.

    a. $-2\sqrt{25} + 3\sqrt{49}$ **b.** $\sqrt{\dfrac{1}{36}} - \sqrt{\dfrac{1}{25}}$

21. Insert the proper symbol ($<$ or $>$) to make a true statement.

    a. -6.78 ___ -6.79 **b.** $\dfrac{3}{8}$ ___ 0.3

    c. $\sqrt{\dfrac{16}{81}}$ ___ $\dfrac{16}{81}$ **d.** $0.\overline{45}$ ___ 0.45

22. Simplify each square root.

    a. $-\sqrt{0.04}$ **b.** $\sqrt{1.69}$

23. Although the decimal 3.2999 contains more digits than 3.3, it is smaller than 3.3. Explain why this is so.

24. What is a repeating decimal? Give an example.

Chapters 1–4   *Cumulative Review Exercises*

In Exercises 1–2, consider the number 5,434,679.

1. Round to the nearest hundred.

2. Round to the nearest ten thousand.

In Exercises 3–6, do each operation.

3. 4,679
    +3,457

4. 7,897
    −4,378

5. 5,345
    ×   56

6. $35\overline{)34,685}$

In Exercises 7–8, refer to the rectangular swimming pool shown in Illustration 1.

7. Find the perimeter of the pool.

8. Find the area of the pool.

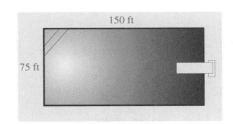

ILLUSTRATION 3

In Exercises 9–12, find the prime factorization of each number.

9. 84   **10.** 450   **11.** 360   **12.** 3,600

In Exercises 13–16, evaluate each expression.

13. $6 + (-2)(-5)$

14. $(-2)^3 - 3^3$

15. $\dfrac{2(-7) + 3(2)}{2(-2)}$

16. $\dfrac{2(3^2 - 4^2)}{-2(3) - 1}$

In Exercises 17–20, simplify each fraction.

17. $\dfrac{21}{28}$

18. $\dfrac{40}{16}$

19. $\dfrac{108}{144}$

20. $-\dfrac{75}{125}$

In Exercises 21–24, do the operations.

21. $\dfrac{6}{5} \cdot \dfrac{2}{3}$

22. $\dfrac{14}{8} \div \dfrac{7}{2}$

23. $\dfrac{2}{3} + \dfrac{3}{4}$

24. $\dfrac{4}{3} - \dfrac{3}{5}$

In Exercises 25–28, write each mixed number as an improper fraction.

25. $3\dfrac{5}{6}$

26. $-6\dfrac{5}{8}$

27. $-10\dfrac{5}{6}$

28. $5\dfrac{1}{20}$

In Exercises 29–32, write each improper fraction as a mixed number.

29. $\dfrac{5}{4}$ **30.** $\dfrac{21}{4}$ **31.** $-\dfrac{45}{7}$ **32.** $-\dfrac{111}{25}$

In Exercises 33–34, do each operation.

33. $4\dfrac{2}{3} + 5\dfrac{1}{4}$ **34.** $14\dfrac{2}{5} - 8\dfrac{2}{3}$

In Exercises 35–36, simplify each expression.

35. $\left(\dfrac{1}{4} - \dfrac{7}{8}\right) \div \left(-2\dfrac{3}{16}\right)$ **36.** $\dfrac{\dfrac{2}{3} - 7}{4\dfrac{5}{6}}$

In Exercises 37–38, round each decimal to the nearest tenth.

37. 37.356   **38.** 95.046

In Exercises 39–42, do the operations.

39. Add: 12.375
      4.89
      5.2

40. Subtract: 25.456
           13.227

41. Multiply: 3.45
          3.4

42. Divide: $5.4\overline{\smash{)}336.42}$

In Exercises 43–44, change each fraction into a decimal fraction.

43. $\dfrac{12}{15}$ **44.** $\dfrac{22}{30}$

In Exercises 45–46, find each square root.

45. $\sqrt{49}$ **46.** $\sqrt{\dfrac{625}{16}}$

In Exercises 47–48, use a calculator to find each square root to the nearest hundredth.

47. $\sqrt{17.3}$ **48.** $\sqrt{0.001}$

Percent

5

243

PERCENTS ARE BASED ON THE NUMBER 100. THEY OFFER US A STANDARDIZED WAY TO MEASURE AND DESCRIBE MANY SITUATIONS IN OUR DAILY LIVES.

5.1 Percents, Decimals, and Fractions

In this section, you will learn about

- **The meaning of percent**
- **Changing a percent to a fraction**
- **Changing a percent to a decimal**
- **Changing a decimal to a percent**
- **Changing a fraction to a percent**

INTRODUCTION. Percents are a popular way to present numeric information. Stores use them to advertise discounts, manufacturers use them to describe the content of their products, and banks use them to list interest rates for loans and savings accounts. Newspapers are full of statistics presented in percent form. In this section, we introduce percent and show how fractions, decimals, and percents are interrelated.

The meaning of percent

A percent tells us the number of parts per 100. You can think of a percent as the *numerator* of a fraction that has a denominator of 100.

Percent	**Percent** means per one hundred.

In Figure 5-1, 93 out of 100 equal-sized squares are shaded. Thus, $\frac{93}{100}$ or 93 percent of the figure is shaded. The word *percent* can be written using the symbol %, so 93% of Figure 5-1 is shaded.

$$\overset{\text{Numerator}}{\underset{\text{Per 100}}{\frac{93}{100} = 93\%}}$$

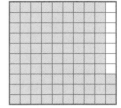

FIGURE 5-1

If the entire grid in Figure 5-1 had been shaded, we would say that 100 out of the 100 squares, or 100%, was shaded.

Changing a percent to a fraction

To change a percent into an equivalent fraction, we use the definition of percent.

Changing a percent to a fraction	To change a percent to a fraction, drop the % sign and write the given number over 100.

EXAMPLE 1 ***Changing a percent to a fraction.*** The chemical makeup of the earth's atmosphere is 78% nitrogen, 21% oxygen, and 1% other gases. Change each percent to a fraction.

Self Check

Change 48% to a fraction.

Solution
We begin with nitrogen.

$78\% = \dfrac{78}{100}$ Use the definition of percent. 78% means 78 per one hundred. This fraction can be simplified.

$= \dfrac{39 \cdot \overset{1}{\cancel{2}}}{50 \cdot \underset{1}{\cancel{2}}}$ Factor 78 as $39 \cdot 2$ and 100 as $50 \cdot 2$. Divide out the common factor of 2.

$= \dfrac{39}{50}$

Nitrogen makes up $\frac{78}{100}$, or $\frac{39}{50}$, of the earth's atmosphere.

Oxygen makes up 21% or $\frac{21}{100}$ of the earth's atmosphere. Other gases make up 1% or $\frac{1}{100}$ of the atmosphere.

Answer: $\dfrac{12}{25}$

EXAMPLE 2 *Changing a percent to a fraction.* In the United States, 61.7% of all married women with children under 6 years old are part of the labor force. Change this percent to a fraction.

Self Check

Change 41.5% to a fraction.

Solution

$$61.7\% = \frac{61.7}{100}$$ Drop the % sign and write 61.7 over 100.

$$= \frac{61.7 \cdot 10}{100 \cdot 10}$$ To obtain a whole number in the numerator, multiply by 10. This will move the decimal point 1 place to the right. Multiply the denominator by 10 as well.

$$= \frac{617}{1,000}$$ Do the multiplication in the numerator and in the denominator.

On average, 617 out of every 1,000 married women with children under 6 years old are part of the labor force.

Answer: $\frac{83}{200}$

EXAMPLE 3 *Changing a percent to a fraction.* Change $66\frac{2}{3}\%$ to a fraction.

Self Check

Change $83\frac{1}{3}\%$ to a fraction.

Solution

$$66\frac{2}{3}\% = \frac{66\frac{2}{3}}{100}$$ Drop the % sign and write $66\frac{2}{3}$ over 100.

$$= 66\frac{2}{3} \div 100$$ The fraction bar indicates division.

$$= \frac{200}{3} \cdot \frac{1}{100}$$ Change $66\frac{2}{3}$ to a mixed number and then multiply by the reciprocal of 100.

$$= \frac{2 \cdot 100 \cdot 1}{3 \cdot 100}$$ Multiply the numerators and the denominators. Factor 200 as $2 \cdot 100$.

$$= \frac{2 \cdot \overset{1}{\cancel{100}} \cdot 1}{3 \cdot \underset{1}{\cancel{100}}}$$ Divide out the common factor of 100.

$$= \frac{2}{3}$$ Multiply in the numerator. Multiply in the denominator.

Answer: $\frac{5}{6}$

Changing a percent to a decimal

To write a percent as a decimal, recall that a percent can be written as a fraction with denominator 100, and that a denominator of 100 indicates division by 100.

Consider 14.25%, which means 14.25 per 100.

$$14.25\% = \frac{14.25}{100}$$ Use the definition of percent: write 14.25 over 100.

$$= 14.25 \div 100$$ The fraction bar indicates division.

$$= 0.14.25$$ To divide a decimal by 100, move the decimal point 2 places to the left.

$$14.25\% = 0.1425$$

This example suggests the following procedure.

Changing a percent to a decimal	To change a percent to a decimal, drop the % sign and move the decimal point 2 places to the left.

EXAMPLE 4 *Changing a percent to a decimal.* The most common blood type in the United States, O+, occurs in about 37% of the population. Change this percent to a decimal.

Solution

The number 37 has an understood decimal point to the right of the 7.

$37\% = 37.0\%$ Write a decimal point and a 0 on the right of 37.
  $= .37.0$ Drop the % sign and move the decimal point 2 places to the left.
  $= 0.37$ Write a 0 to the left of the decimal point.

Self Check

Change 78% to a decimal.

Answer: 0.78

EXAMPLE 5 *Changing a percent to a decimal.* Change 310% to a decimal.

Solution

The whole number 310 has an understood decimal point to the right of 0.

$310\% = 310.0\%$ Write a decimal point and a 0 on the right of 310.
  $= 3.10.0$ Drop the % sign and move the decimal point 2 places to the left.
  $= 3.100$
  $= 3.1$ Drop the unnecessary 0's to the right of the 1.

Self Check

Change 600% to a decimal.

Answer: 6

EXAMPLE 6 *Changing a percent to a decimal.* The population of the state of Oklahoma is approximately $1\frac{1}{4}\%$ of the population of the United States. Change this percent to a decimal.

Solution

$1\frac{1}{4}\% = 1.25\%$ Write $1\frac{1}{4}$ as 1.25.

  $= 0.01.25$ Drop the % sign and move the decimal point 2 places to the left.
  $= 0.0125$

Self Check

Change $15\frac{3}{4}\%$ to a decimal.

Answer: 0.1575

Changing a decimal to a percent

To change a percent to a decimal, we drop the % sign and move the decimal point 2 places to the left. To write a decimal as a percent, we do the opposite: we move the decimal point two places to the right and insert a % sign.

Changing a decimal to a percent	To change a decimal to a percent, move the decimal point 2 places to the right and insert a % sign.

EXAMPLE 7 *Changing a decimal to a percent.* Land areas make up 0.291 of the world's surface. Change this decimal to a percent.

Solution

$0.291 = 0.29.1\%$ Move the decimal point 2 places to the right and insert the % sign.
  $= 29.1\%$

Self Check

Change 0.5343 to a percent.

Answer: 53.43%

Changing a fraction to a percent

We will use a two-step process to change a fraction to a percent. First, we write the fraction as a decimal. Then we change that decimal to a percent.

$$\text{Fraction} \longrightarrow \text{Decimal} \longrightarrow \text{Percent}$$

Changing a fraction to a percent	To change a fraction to a percent, 1. write the fraction as a decimal by dividing its numerator by its denominator; 2. move the decimal point 2 places to the right and insert a % sign.

EXAMPLE 8   ***Changing a fraction to a percent.***   The highest-rated television show of all time was a special episode of M*A*S*H that aired February 28, 1983. Surveys found that three out of every five American households watched this show. Express the rating as a percent.

Solution

3 out of 5 can be expressed as $\frac{3}{5}$. We need to change this fraction to a decimal.

$$\begin{array}{r} 0.6 \\ 5\overline{)3.0} \\ \underline{3\,0} \\ 0 \end{array}$$

Write 3 as 3.0 and then divide the numerator by the denominator.

$\dfrac{3}{5} = 0.6$ The result is a terminating decimal.

$0.6 = 0.60.\%$ Write a placeholder 0 to the right of the 6. Move the decimal point 2 places to the right and insert a % sign.

$ = 60\%$

Self Check

Change $\dfrac{7}{8}$ to a percent.

Answer: 87.5%

In Example 8, the result of the division was a terminating decimal. Sometimes, when changing a fraction to a decimal, the result of the division is a repeating decimal.

EXAMPLE 9   ***Changing a fraction to a percent.***   Change $\dfrac{5}{6}$ to a percent.

Solution   The first step is to change $\dfrac{5}{6}$ to a decimal.

$$\begin{array}{r} 0.8333 \\ 6\overline{)5.0000} \\ \underline{4\,8} \\ 20 \\ \underline{18} \\ 20 \\ \underline{18} \\ 20 \end{array}$$

Write 5 as 5.0000. Divide the numerator by the denominator.

$\dfrac{5}{6} = 0.8333\ldots$ The result is a repeating decimal.

$\phantom{\dfrac{5}{6}} = 0.83.33\ldots\%$ Change this decimal to a percent by moving the decimal point 2 places to the right and inserting a % sign.

$\phantom{\dfrac{5}{6}} = 83.33\ldots\%$

At this stage, we must decide whether we want an approximation or an exact answer. For an approximation, we round the decimal to a specific place value. For an exact answer, we represent the repeating part of the decimal with a fraction.

Approximation	*Exact answer*
$\dfrac{5}{6} = 83.33...\%$	$\dfrac{5}{6} = 83.3333...\%$
$\approx 83.3\%$ Round off to the nearest tenth.	$= 83\dfrac{1}{3}\%$ Use a fraction.
$\dfrac{5}{6} \approx 83.3\%$	$\dfrac{5}{6} = 83\dfrac{1}{3}\%$

Some percents occur so frequently that it is useful to memorize their fractional and decimal equivalents. Study the information in this table and memorize it for future use.

Percent	Decimal	Fraction
1%	0.01	$\dfrac{1}{100}$
10%	0.1	$\dfrac{1}{10}$
20%	0.2	$\dfrac{1}{5}$
25%	0.25	$\dfrac{1}{4}$
$33\frac{1}{3}\%$	0.3333...	$\dfrac{1}{3}$
50%	0.5	$\dfrac{1}{2}$
$66\frac{2}{3}\%$	0.6666...	$\dfrac{2}{3}$
75%	0.75	$\dfrac{3}{4}$

STUDY SET   Section 5.1

VOCABULARY   *In Exercises 1–2, fill in the blanks to make a true statement.*

1. _____ means per one hundred.

2. When changing a fraction to a decimal, the result is either a _____ or a repeating decimal.

CONCEPTS   *In Exercises 3–6, fill in the blanks to make a true statement.*

3. To change a percent to a fraction, drop the percent sign and write the given number over ___ .

4. To change a percent to a decimal, drop the % sign and move the decimal point two places to the _____ .

5. To change a decimal to a percent, move the decimal point two places to the _____ and insert a % sign.

6. To change a fraction to a percent, first write the fraction as a _____ . Then move the decimal point two places to the _____ and insert a % sign.

7. a. See Illustration 1. Express the amount of the figure that is shaded as a decimal, a percent, and a fraction.

b. What percent of the figure is not shaded?

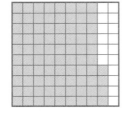

ILLUSTRATION 1

8. In Illustration 2, each set of 100 squares represents 100%. What percent is shaded?

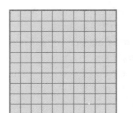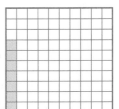

ILLUSTRATION 2

PRACTICE In Exercises 9–24, change each percent to a fraction. Simplify when necessary.

9. 17%
10. 31%
11. 5%
12. 4%
13. 60%
14. 40%
15. 125%
16. 210%
17. $\frac{2}{3}$%
18. $\frac{1}{5}$%
19. $5\frac{1}{4}$%
20. $6\frac{3}{4}$%
21. 0.6%
22. 0.5%
23. 1.9%
24. 2.3%

In Exercises 25–36, change each percent to a decimal.

25. 19%
26. 83%
27. 6%
28. 2%
29. 40.8%
30. 34.2%
31. 250%
32. 600%
33. 0.79%
34. 0.01%
35. $\frac{1}{4}$%
36. $8\frac{1}{5}$%

In Exercises 37–48, change each decimal to a percent.

37. 0.93
38. 0.44
39. 0.612
40. 0.727
41. 0.0314
42. 0.0021
43. 8.43
44. 7.03
45. 50
46. 3
47. 9.1
48. 8.7

In Exercises 49–60, change each fraction to a percent.

49. $\frac{17}{100}$
50. $\frac{29}{100}$
51. $\frac{4}{25}$
52. $\frac{47}{50}$
53. $\frac{2}{5}$
54. $\frac{21}{50}$
55. $\frac{21}{20}$
56. $\frac{33}{20}$
57. $\frac{5}{8}$
58. $\frac{3}{8}$
59. $\frac{3}{16}$
60. $\frac{1}{32}$

In Exercises 61–64, find the exact equivalent percent for each fraction.

61. $\frac{2}{3}$
62. $\frac{1}{6}$
63. $\frac{5}{6}$
64. $\frac{4}{3}$

In Exercises 65–68, express each of the given fractions as a percent. Round to the nearest hundredth.

65. $\frac{1}{9}$
66. $\frac{2}{3}$
67. $\frac{5}{9}$
68. $\frac{7}{3}$

APPLICATIONS In Exercises 69–80, solve each problem.

69. U.N. SECURITY COUNCIL   The United Nations has 184 members. The United States, the Russian Federation, Britain, France, and China, along with 10 other nations, make up the Security Council.
   a. What fraction of the members of the United Nations belong to the Security Council?
   b. Write your answer to part a in percent form. (Round to the nearest one percent.)

70. ECONOMIC FORECAST   One economic indicator of the national economy is the number of orders placed by manufacturers. One month, the number of orders rose one-fourth of one percent.
   a. Write this using a % symbol.
   b. Express it as a fraction.
   c. Express it as a decimal.

71. PIANO KEYS   Of the 88 keys on a piano, 36 are black.
   a. What fraction of the keys are black?
   b. What percent of the keys are black? (Round to the nearest one percent.)

72. INTEREST RATES Write the interest rate associated with each of these accounts as a decimal.
  a. Home loan: 7.75%
  b. Savings account: 5%
  c. Credit card: 14.25%

73. THE SPINE The human spine consists of a group of bones (vertebrae). (See Illustration 3.)
  a. What fraction of the vertebrae are lumbar?
  b. What percent of the vertebrae are lumbar? (Round to the nearest one percent.)
  c. What percent of vertebrae are cervical? (Round to the nearest one percent.)

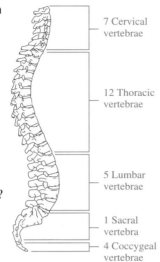

7 Cervical vertebrae

12 Thoracic vertebrae

5 Lumbar vertebrae

1 Sacral vertebra

4 Coccygeal vertebrae

ILLUSTRATION 3

74. REGIONS OF THE COUNTRY The continental United States is divided into seven regions. (See Illustration 4.)
  a. What percent of the 50 states are in the Rocky Mountain region?
  b. What percent of the 50 states are in the Midwestern region? (*Hint:* Don't forget Iowa.)
  c. What percent of the 50 states are not located in any of these regions?

New England States

Rocky Mountain States

Midwestern States

Middle Atlantic States

Southwestern States

Southern States

Pacific Coast States

ILLUSTRATION 4

75. STEEP GRADE Sometimes, signs are used to warn truckers when they are approaching a steep grade on the highway. (See Illustration 5.) For a 5% grade, how many feet does the road rise over a 100-foot run?

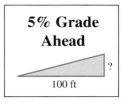

5% Grade Ahead

100 ft

ILLUSTRATION 5

76. COMPANY LOGO In Illustration 6, what part of the company's logo is shaded red? Express your answer as a percent, a fraction, and a decimal. Do not round.

Recycling Industries Inc.

ILLUSTRATION 6

77. IVORY SOAP A popular soap claims to be $99\frac{44}{100}\%$ pure. Write this percent as a decimal.

78. DRUNK DRIVING In most states, it is illegal to drive with a blood alcohol concentration of 0.08% or more. Change this percent to a fraction. Do not simplify. Explain what the numerator and the denominator of the fraction represent.

79. BASKETBALL STANDINGS In the standings, we see that Chicago has won 60 of 67, or $\frac{60}{67}$ of its games. In what form is the team's winning percentage presented in the newspaper? Express it as a percent.

Eastern Conference			
Team	**W**	**L**	**Pct.**
Chicago	60	7	.896

80. WON–LOST RECORD In sports, when a team wins as many as it loses, it is said to be playing "500 ball." Examine the standings and explain the significance of the number 500, using concepts studied in this section.

Eastern Conference			
Team	**W**	**L**	**Pct.**
Orlando	33	33	.500

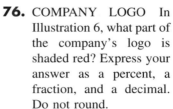

In Exercises 81–82, use a calculator to solve each problem.

81. BIRTHDAY If the day of your birthday represents $\frac{1}{365}$ of a year, what percent of the year is it? Round to the nearest hundredth of a percent.

82. POPULATION As a fraction, each resident of the United States represents approximately $\frac{1}{255,000,000}$ of the population. Express this as a percent. Round to one nonzero digit.

Write a paragraph using your own words.

83. If you were writing advertising, which form do you think would attract more customers: "25% off" or "$\frac{1}{4}$ off"? Explain your reasoning.

85. Explain how to change a fraction to a percent.

84. Many coaches ask their players to give a 110% effort during practices and games. What do you think this means? Is it possible?

86. Explain how an amusement park could have an attendance that is 103% of capacity.

REVIEW

87. Subtract: $\frac{2}{3} - \frac{3}{4}$.

89. Evaluate: $3 - |4 - 8|$.

91. Find the area of the rectangle discussed in Exercise 90.

88. Add: $\frac{1}{3} + \frac{1}{4} + \frac{1}{2}$.

90. Find the perimeter of a rectangle with a length of 10.5 inches and a width of 6.5 inches.

92. Subtract: $41 - 10.287$.

5.2 *Solving Percent Problems*

In this section, you will learn about

- **Equations**
- **Three types of percent problems**
- **Finding the amount**
- **Finding the percent**
- **Finding the base**
- **Circle graphs**

INTRODUCTION. Percent problems occur in three forms. In this section, we will study a method that can be used to solve all three types. This method involves solving simple equations.

Equations

An **equation** is statement that two quantities are equal. Here are some examples of equations:

$$4 + 4 = 8, \quad \frac{1}{2} \cdot 12 = 6, \quad \text{and} \quad A = l \cdot w$$

We note that every equation contains an $=$ sign. The following statements are not equations, because they do not contain an $=$ sign.

$$3 \times 5, \quad \frac{1}{3} + \frac{3}{4}, \quad \text{and} \quad 2p + 2l$$

Three types of percent problems

The front page of the newspaper in Figure 5-2 suggests three different types of percent problems.

Type 1: What number is 84% of 500?   (the labor article)

Type 2: 38 is what percent of 40?   (the drinking water article)

Type 3: 6 is 75% of what number?   (the new appointees article)

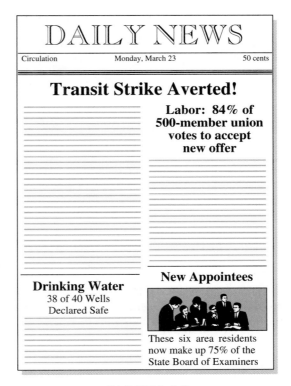

FIGURE 5-2

These percent problems have several things in common.

<table>
<tr><td>Observations</td><td>1. Each problem contains the word is. In mathematics, is can be translated to an = sign.</td></tr>
<tr><td></td><td>2. Each of the problems contains a phrase such as what number or what percent. In other words, there is an unknown quantity to be found.</td></tr>
<tr><td></td><td>3. Each problem contains the word of. In mathematics, of usually means multiply.</td></tr>
</table>

These observations suggest that each of the percent problems can be translated into an equation. The equation will have one unknown, and the operation of multiplication will be involved.

Finding the amount

To solve the first type of problem (*What number is 84% of 500?*), we multiply.

84% of 500 = 84% · 500   The word *of* means multiply.

= (0.84)(500)   To do the multiplication, change 84% to the decimal 0.84.

= 420

We have found that 420 is 84% of 500.

In the statement "420 is 84% of 500," the number 420 is called the **amount,** 84% is the percent, and 500 is called the **base.** In words, the relationship between the amount, the percent, and the base is *Amount is percent of base.* This relationship is shown in the **percent formula.**

The percent formula	Amount = percent · base

EXAMPLE 1   *Finding the amount.*   What number is 160% of 15.8?

Solution

In this example, the percent is 160 and the base is 15.8, the number following the word *of.* We can let A stand for the amount and use the percent formula.

Amount	=	percent	·	base
A	=	160%	·	15.8

Substitute 160 for the percent and 15.8 for the base.

The statement $A = 160\% \cdot 15.8$ is an equation, with the amount being the unknown. To solve this equation, we must find the value of the unknown quantity. In this equation, we can find the unknown amount by multiplication.

$A = 1.6 \cdot 15.8$ Change 160% to a decimal.

$= 25.28$ Do the multiplication.

Thus, 25.28 is 160% of 15.8.

Self Check

What number is 240% of 80?

Answer: 192

Finding the percent

In the second type of problem (*38 is what percent of 40?*), we must find the percent. In this problem, 38 is the amount and 40 is the base. Once again, we form an equation by using the percent formula.

Amount	=	percent	·	base
38	=	p	·	40

38 is the amount, p is the percent, and 40 is the base.

$38 = p \cdot 40$ The equation to solve.

$\dfrac{38}{40} = \dfrac{p \cdot 40}{40}$ To undo the multiplication of p by 40, divide both sides by 40.

$0.95 = p$ $\frac{38}{40} = 0.95$ and $\frac{40}{40} = 1$.

$p = 95\%$ To change a decimal to a percent, move the decimal point 2 places to the right and insert a % symbol.

Thus, 38 is 95% of 40.

EXAMPLE 2 *Finding the percent.* 14 is what percent of 32?

Solution

In this example, 14 is the amount and 32 is the base. Once again, we use the percent formula and let p stand for the percent.

Amount	=	percent	·	base	
14	=	p	·	32	Substitute 14 for the amount and 32 for the base.

The statement $14 = p \cdot 32$ is an equation, with the percent being the unknown. In this equation, we can find the unknown amount by division.

$14 = p \cdot 32$ The equation to solve.

$\dfrac{14}{32} = \dfrac{p \cdot 32}{32}$ To undo the multiplication of p by 32, divide both sides by 32.

$0.4375 = p$ $\frac{14}{32} = 0.4375$ and $\frac{32}{32} = 1$.

$p = 43.75\%$ To change the decimal to a percent, move the decimal point 2 places to the right and insert a % symbol.

Thus, 14 is 43.75% of 32.

Self Check

9 is what percent of 16?

Answer: 56.25%

Accent on Technology   *Cost of an air bag*

An air bag is estimated to add an additional $500 to the cost of a car. The cost of an air bag is what percent of a car's $16,295 sticker price?

In this problem, 500 is the amount and 16,295 is the base. We can form an equation by using the percent formula.

Amount	=	percent	·	base
500	=	p	·	16,295

Substitute 500 for the amount and 16,295 for the base. Let p represent the percent.

Then we solve the equation.

$500 = p \cdot 16,295$ The equation to solve.

$\dfrac{500}{16,295} = \dfrac{p \cdot 16,295}{16,295}$ To undo the multiplication of p by 16,295, divide both sides by 16,295.

$\dfrac{500}{16,295} = p$ $\frac{16,295}{16,295} = 1$.

We can use a calculator to divide on the left-hand side.

Evaluate: $\dfrac{500}{16,295}$

Keystrokes: $\boxed{5}\,\boxed{0}\,\boxed{0}\,\boxed{÷}\,\boxed{1}\,\boxed{6}\,\boxed{2}\,\boxed{9}\,\boxed{5}\,\boxed{=}$ $\boxed{0.03068425\text{B}}$

This display gives the answer in decimal form. To change it to a percent, we multiply the result by 100 and insert a percent symbol. This moves the decimal point 2 places to the right. If we round to the nearest tenth of a percent, the cost of an air bag is about 3.1% of the sticker price.

Finding the base

In the third type of problem (*6 is 75% of what number?*), we must find the base. Again, we substitute into the percent formula.

Amount	=	percent	·	base
6	=	75%	·	b

6 is the amount and 75% is the percent. Let b stand for the base.

Then we can solve the equation.

$$6 = 0.75 \cdot b \quad \text{Change 75\% to 0.75.}$$

$$\frac{6}{0.75} = \frac{0.75 \cdot b}{0.75} \quad \text{To undo the multiplication of } b \text{ by 0.75, divide both sides by 0.75.}$$

$$8 = b \quad \tfrac{6}{0.75} = 8 \text{ and } \tfrac{0.75}{0.75} = 1.$$

Thus, 6 is 75% of 8.

EXAMPLE 3 ***Finding the base.*** In an apartment complex, 110 of the units are currently rented. If this represents an 88% occupancy rate, how many units are there in the complex?

Solution An occupancy rate of 88% means that 88% of the units are occupied. In this problem, the percent is 88% and the amount is 110. We are to find the base. We form an equation by using the percent formula.

Amount	=	percent	·	base
110	=	88%	·	b

Let b represent the base.

We can then solve the equation.

$$110 = 0.88 \cdot b \quad \text{Change 88\% to 0.88.}$$

$$\frac{110}{0.88} = \frac{0.88 \cdot b}{0.88} \quad \text{To undo the multiplication of } b \text{ by 0.88, divide both sides by 0.88.}$$

$$125 = b \quad \tfrac{110}{0.88} = 125 \text{ and } \tfrac{0.88}{0.88} = 1.$$

The complex has 125 units.

In the next example, the computations are easier if we change the percent to a fraction instead of a decimal.

EXAMPLE 4 ***Finding the base.*** 31.5 is $33\frac{1}{3}$% of what number?

Solution
In this example, 31.5 is the amount and $33\frac{1}{3}$ is the percent. To find the base (which we will call b), we form an equation using the percent formula.

Amount	=	percent	·	base
31.5	=	$33\frac{1}{3}$%	·	b

Substitute 31.5 for the amount and $33\frac{1}{3}$% for the percent.

Self Check
150 is $66\frac{2}{3}$% of what number?

The statement $31.5 = 33\frac{1}{3}\% \cdot b$ is an equation, with the base being the unknown. In this equation, we can find the unknown amount by division.

$$31.5 = 33\frac{1}{3}\% \cdot b \qquad \text{The equation to solve.}$$

$$31.5 = \frac{1}{3} \cdot b \qquad 33\frac{1}{3}\% = \frac{33\frac{1}{3}}{100} = \frac{1}{3}.$$

$$31.5 \div \frac{1}{3} = \frac{1}{3} \cdot b \div \frac{1}{3} \qquad \text{To undo the multiplication of } b \text{ by } \frac{1}{3}, \text{ divide both sides by } \frac{1}{3}.$$

$$31.5 \cdot \frac{3}{1} = \frac{1}{3} \cdot b \cdot \frac{3}{1} \qquad \text{To divide by } \frac{1}{3}, \text{ multiply by } \frac{3}{1}.$$

$$94.5 = b \qquad 31.5 \cdot 3 = 94.5 \text{ and } \frac{1}{3} \cdot \frac{3}{1} = 1.$$

Thus, 31.5 is $33\frac{1}{3}\%$ of 94.5.

Answer: 225

Circle graphs

Percents are used with **circle graphs** as a way of presenting data for comparison. In Figure 5-3, the entire circle represents the floor area of a children's clothing store. The sizes of the segments of the graph indicate the percents of the store's total square footage that each department occupies.

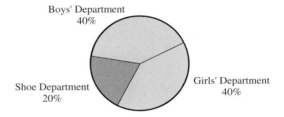

Boys' Department
40%

Shoe Department
20%

Girls' Department
40%

Square Footage Use, by Department

FIGURE 5-3

In the next example, we use information from a circle graph to solve a problem.

EXAMPLE 5

Presidential election results. Results from the 1992 presidential election are shown in Figure 5-4. Use the information to find the number of states won by Clinton.

Solution The circle graph shows that Clinton was victorious in 64% of the 50 states. Here, the percent is 64% and the base is 50. We use the percent formula and solve for the amount.

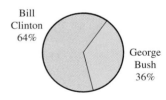

Bill Clinton 64%

George Bush 36%

1992 Presidential Election:
States won by each candidate

FIGURE 5-4

Amount	=	percent	·	base
A	=	64%	·	50

Substitute 64% for the percent and 50 for the base.

$A = 0.64 \cdot 50$ Change 64% to a decimal.

$A = 32$ Do the multiplication.

President Clinton won 32 states.

VOCABULARY   *In Exercises 1–4, use the percent formula to write each sentence as an equation.*

1. What number is 10% of 50?

_____

2. 16 is 55% of what number?

_____

3. 48 is what percent of 47?

_____

4. 12 is what percent of 20?

_____

In Exercises 5–6, fill in the blanks to make a true statement.

5. A _____ graph can be used to show the division of a whole quantity into its component parts.

6. In the statement "45 is 90% of 50," 45 is the _____ , 90% is called the _____ , and 50 is the _____ .

CONCEPTS

7. When computing with percents, the percent must be changed to a decimal or a fraction. Change each percent to a decimal.
  a. 12%
  b. 5.6%
  c. 125%
  d. $\frac{1}{4}\%$

8. When computing with percents, the percent must be changed to a decimal or a fraction. Change each percent to a fraction.
  a. $33\frac{1}{3}\%$
  b. $66\frac{2}{3}\%$
  c. $16\frac{2}{3}\%$
  d. $83\frac{1}{3}\%$

9. Without doing the calculation, tell whether 120% of 55 is more than 55 or less than 55.

10. Without doing the calculation, tell whether 12% of 55 is more than 55 or less than 55.

11. According to Illustration 1, what percent of the company's employees work on a part-time basis?

12. Illustration 2 shows a breakdown of the ticket sales for a movie theater. What percent of the theater's customers are seniors?

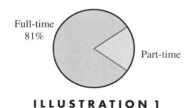

Full-time 81%

Part-time

ILLUSTRATION 1

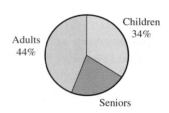

Children 34%

Adults 44%

Seniors

ILLUSTRATION 2

13. What is 100% of 25?

14. What percent of 32 is 32?

15. What is 200% of 25?

16. What is 300% of 25?

PRACTICE   *In Exercises 17–28, solve each problem.*

17. What number is 36% of 250?

18. What number is 82% of 300?

19. 16 is what percent of 20?

20. 13 is what percent of 25?

21. 7.8 is 12% of what number?

22. 39.6 is 44% of what number?

23. What number is 0.8% of 12?

24. What number is 5.6% of 40?

25. 0.5 is what percent of 40?

26. 0.3 is what percent of 15?

27. 3.3 is 7.5% of what number?

28. 8.4 is 20% of what number?

In Exercises 29–38, write each problem in the form of a Type 1, Type 2, or Type 3 problem and solve it. For example, "Find $7\frac{1}{4}\%$ of 600" can be written as "What number is $7\frac{1}{4}\%$ of 600?"

29. Find $7\frac{1}{4}\%$ of 600.

30. Find $1\frac{3}{4}\%$ of 800.

31. 102% of 105 is what number?

32. 210% of 66 is what number?

33. $33\frac{1}{3}\%$ of what number is 33?

34. $66\frac{2}{3}\%$ of what number is 28?

35. $9\frac{1}{2}\%$ of what number is 5.7?

36. $\frac{1}{2}\%$ of what number is 5?

37. What percent of 8,000 is 2,500?

38. What percent of 3,200 is 1,400?

In Exercises 39–40, use a circle graph to illustrate the given data. A circle divided into 100 sections is provided to aid in the graphing process.

39. Complete Illustration 3 to show the sources of energy in the United States in 1992.

40. Complete Illustration 4 to show the sources of energy in the United States in 1950.

Hydroelectric	4%
Nuclear	8%
Coal	23%
Natural gas	25%
Petroleum	40%

Hydroelectric	4%
Natural gas	18%
Coal	37%
Petroleum	41%

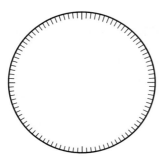

ILLUSTRATION 3

ILLUSTRATION 4

APPLICATIONS

41. CHILD CARE   After the first day of registration, 84 children had been enrolled in a new day care center. That represented 70% of the available slots. What was the maximum number of children the center could enroll?

42. RACING PROGRAM   One month before a stock car race, the sale of ads for the official race program was slow. Only 12 pages, or just 60% of the available pages, had been sold. What was the total number of pages devoted to advertising in the program?

43. FEDERAL OUTLAYS   Illustration 5 shows the breakdown of federal outlays for fiscal year 1994. If the total spending was approximately $1,500 billion, how much was spent on Social Security, Medicare, and other retirement programs?

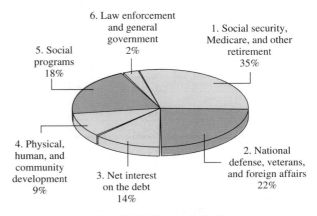

ILLUSTRATION 5

44. REBATE   A long-distance telephone company offered its customers a rebate of 20% of the cost of all long-distance calls made in the month of July. One customer's calls are listed in Illustration 6. What amount will this customer receive in the form of a rebate?

Date	Time	Place called	Min.	Amount
Jul 4	3:48 P.M.	Denver	11	$3.80
Jul 9	12:00 P.M.	Detroit	10	$7.50
Jul 20	8.59 A.M.	San Diego	26	$9.45

ILLUSTRATION 6

45. PRODUCT PROMOTION   To promote sales, a free 6-ounce bottle of shampoo is packaged with every large bottle. (See Illustration 7.) Use the information on the package to find how many ounces of shampoo the large bottle contains.

ILLUSTRATION 7

46. NUTRITION FACTS   The nutrition label on a package of corn chips is shown in Illustration 8.

Nutrition Facts
Serving Size: 1 oz. (28g/About 29 chips)
Servings Per Container: About 11

Amount Per Serving
Calories 160   Calories from Fat 90

% Daily Value

Total fat 10g	**15%**
Saturated fat 1.5 g	**7%**
Cholesterol 0mg	**0%**
Sodium 240mg	**12%**
Total carbohydrate 15g	**5%**
Dietary fiber 1g	**4%**
Sugars less than 1g	
Protein 2g	

ILLUSTRATION 8

a. How many milligrams of sodium are in one serving of chips?

b. According to the label, what percent of the daily value is this?

c. What daily value of sodium intake is deemed healthy?

47. DRIVER'S LICENSE   On the written part of his driving test, a man answered 28 out of 40 questions correctly. If 70% correct is passing, did he pass the test?

48. ALPHABET   What percent of the English alphabet do the vowels a, e, i, o, and u make up? (Round to the nearest one percent.)

49. MIXTURES   Complete the chart in Illustration 9 to find the number of milliliters of sulfuric acid in each of two beakers.

Milliliters of solution in beaker	% sulfuric acid	Milliliters of sulfuric acid in beaker
60	50%	
40	30%	

ILLUSTRATION 9

50. CUSTOMER GUARANTEE   To assure its customers of low prices, the Home Club offers a "10% Plus" guarantee. If the customer finds the same item selling for less somewhere else, he or she receives the difference in price, plus 10% of the difference. A woman bought miniblinds at the Home Club for $120 but later saw the same blinds on sale for $98 at another store. How much can she expect to be reimbursed?

51. MAKING COPIES   The zoom key on the control panel of a copier programs it to print a magnified or reduced copy of the original document. If the zoom is set at 180% and the original document contains type that is 1.5 inches tall, what will be the height of the type on the copy?

52. MAKING COPIES   The zoom setting for a copier is entered as a decimal: 0.98. Express it as a percent and find the resulting type size on the copy if the original has type 2 inches in height.

53. INSURANCE   The cost to repair a car after a collision was $4,000. The automobile insurance policy paid the entire bill except for a $200 deductible, which the driver paid. What percent of the cost did he pay?

54. FLOOR SPACE   A house has 1,200 square feet on the first floor and 800 square feet on the second floor. What percent of the square footage of the house is on the first floor?

55. REPEAT BUSINESS   A car dealer found that 39 of the 317 cars he sold one year were to customers who had previously purchased a car from him. What percent of his sales that year was repeat business? (Round to the nearest one percent.)

56. VOTER TURNOUT   In the 1992 presidential election, Bill Clinton received 44,909,899 votes, which was 43.0061628% of the popular vote. How many votes were cast for president in the 1992 election?

WRITING   *Write a paragraph using your own words.*

57. Explain the relationship in a percent problem between the amount, the percent, and the base.

58. Write a real-life situation that could be described by "9 is what percent of 20?"

59. Explain why 150% of a number is more than the number.

60. Explain why "Find 9% of 100" is an easy problem to solve.

REVIEW

61. Add: $2.78 + 6 + 9.09 + 0.3$.

62. Evaluate: $\sqrt{64} + 3\sqrt{9}$.

63. On a number line, which number is closer to 5, 4.9 or 5.001?

64. Subtract: $\dfrac{4}{5} - \dfrac{2}{3}$.

65. Multiply: $34.5464 \cdot 1,000$.

66. Find: $(0.2)^3$.

5.3   *Applications of Percent*

In this section, you will learn about

- **Taxes**
- **Commissions**
- **Percent of increase or decrease**
- **Discounts**

INTRODUCTION.   In this section, we discuss four applications of percent. Three of the four (taxes, commissions, and discounts) are directly related to purchasing. A solid understanding of these concepts will make you a better consumer. The fourth application uses percent to describe increases or decreases of such things as unemployment and grocery store sales.

Taxes

The sales receipt in Figure 5-5 gives a detailed account of what items were purchased, how many of each were purchased, and the price of each item.

The receipt shows that the $56.35 purchase price (labeled *subtotal*) was taxed at a **rate** of 5%. Sales tax of $2.82 was charged. The sales tax was then added to the subtotal to get the total price of $59.17.

Finding the total price	Total price = purchase price + sales tax

In Example 1, we verify that the amount of sales tax shown on the receipt in Figure 5-5 is correct.

```
                        BRADSHAW'S
                      Department Store

              4  @   1.05    GIFTS       $ 4.20
              1  @   1.39    BATTERIES   $ 1.39
              1  @  24.85    TOASTER     $24.85
              3  @   2.25    SOCKS       $ 6.75
              2  @   9.58    PILLOWS     $19.16

              SUBTOTAL                   $56.35
              SALES TAX @ 5.00%          $ 2.82
              TOTAL                      $59.17
```

The purchase price of the items bought

The sales tax on the items purchased

The total price

The sales tax rate

FIGURE 5-5

EXAMPLE 1   *Finding the sales tax.*   Find the sales tax on a purchase of $56.35 if the sales tax rate is 5%.

Solution

First we use the percent formula to form an equation. The percent is 5% and the base is 56.35. We are to find the amount of tax.

Amount	=	percent	·	base
A	=	5%	·	56.35

Substitute 5% for the percent and 56.35 for the base.

$A = 0.05 \cdot 56.35$ Change 5% to a decimal.

$A = 2.8175$ Do the multiplication.

Rounding to the nearest cent (hundredths), we find that the sales tax would be $2.82. The sales receipt in Figure 5-5 is correct.

Self Check

Find the sales tax on a purchase of $15.14 if the sales tax rate is 8%.

Answer: $1.21

In addition to sales tax, we pay many other types of taxes in our daily lives. Income tax, gasoline tax, and Social Security tax are just a few.

EXAMPLE 2   *Finding the tax rate.*   A waitress found that $11.04 was deducted from her weekly gross earnings of $240 for federal income tax. What withholding tax rate was used?

Solution

First, we use the percent formula to form an equation. The amount is 11.04 and the base is 240. We need to find the tax rate.

Amount	=	percent	·	base
11.04	=	p	·	240

$11.04 = p \cdot 240$

$\dfrac{11.04}{240} = \dfrac{p \cdot 240}{240}$ Divide both sides by 240.

$0.046 = p$ Do the divisions.

$4.6\% = p$ Change the decimal to a percent.

The withholding tax rate was 4.6%.

Self Check

$5,250 had to be paid on an inheritance of $15,000. What is the inheritance tax rate?

Answer: 35%

Commissions

Instead of working for a salary or getting paid at an hourly rate, many salespeople are paid on **commission.** They earn an amount based on the goods or services they sell.

EXAMPLE 3   *Finding a commission.*   The commission rate for a salesperson at an appliance store is 16.5%. What is his commission from the sale of a refrigerator costing $499.95?

Self Check
An insurance salesperson receives a 4.1% commission on each $120 premium paid by a client. What is the amount of the commission on this premium?

Solution

We use the percent formula to form an equation. The percent is 16.5% and the base is 499.95. We are to find the amount of the commission.

Amount	=	percent	·	base
A	=	16.5%	·	499.95

$A = 0.165 \cdot 499.95$ Change 16.5% to a decimal.

$A = 82.49175$ Use a calculator to do the multiplication.

To the nearest cent (hundredth), the commission is $82.49.

Answer: $4.92

Percent of increase or decrease

Percents can be used to describe how a quantity has changed. For example, consider Figure 5-6, which compares the number of hours of work the average worker took to earn enough to buy a washing machine in 1940 and 1990.

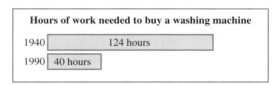

FIGURE 5-6

From the figure, we see that the number of hours needed has decreased. To describe this decrease using percent, we first subtract to find the amount of the decrease.

$$124 - 40 = 84$$

Next, we find what percent this difference is of the original amount. Here, the amount is 84 and the base is 124.

Amount	=	percent	·	base
84	=	p	·	124

$84 = p \cdot 124$

$\dfrac{84}{124} = \dfrac{p \cdot 124}{124}$ Divide both sides by 124.

$0.677419 \approx p$ Do the divisions.

$p \approx 67.7419\%$ Change the decimal to a percent.

$p \approx 68\%$ Round to the nearest one percent.

From 1940 to 1990, there was a 68% decrease in the number of hours it took the average worker to earn enough to buy a washing machine.

Finding the percent of increase or decrease	To find the percent of increase or decrease:
	1. Subtract the smaller number from the larger to find the amount of increase or decrease.
	2. Find what percent the difference is of the original amount.

EXAMPLE 4 *Finding the percent of increase.* The sales of bottled water at a grocery store increased from 4 cases to 15 cases a month over a two-year period. Find the percent of increase in bottled water sales for this store.

Solution

First, we find the amount of increase.

$15 - 4 = 11$ Find the difference in the number of cases sold each month.

Next, we use the percent formula to form an equation. Here, 11 is the amount and 4 is the base. We are to find the percent of increase.

Amount   =   percent   ·   base

11   = p ·   4

$11 = p \cdot 4$

$\dfrac{11}{4} = \dfrac{p \cdot 4}{4}$ Divide both sides by 4.

$2.75 = p$ Do the divisions.

$p = 275\%$ Change the decimal to a percent.

Sales of bottled water increased 275% over the two-year period.

Self Check

In one school district, the number of home-schooled children increased from 15 to 150 in 4 years. Find the percent of increase.

Answer: 900%

EXAMPLE 5 *Unemployment figures.* The Labor Department reported that the number of unemployed persons in a state had fallen 1.2% for the month of July. See Figure 5-7 and find the new unemployment figure.

Solution From the figure, we see that the unemployment figure for June was 204,000. We are told that this number fell in July, and we need to find out by what amount. To do so, we proceed as follows.

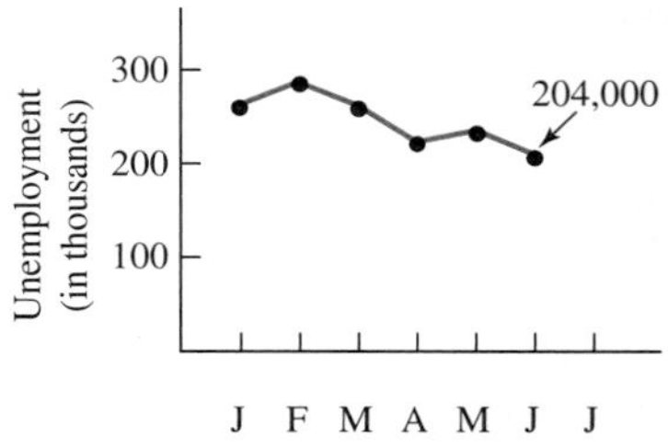

FIGURE 5-7

Amount   =   percent   ·   base

A =   1.2%   ·   204,000

$A = 0.012 \cdot 204{,}000$ Change 1.2% to a decimal.

$A = 2{,}448$ Do the multiplication.

Thus, 204,000 decreased by 2,448. To find the unemployment figure for July, we subtract.

$204{,}000 - 2{,}448 = 201{,}552$

The unemployment figure for July was 201,552.

Discounts

The difference between the original price and the sale price of an item is called the **discount.** If the discount is expressed as a percent of the selling price, it is called the

rate of discount. We will use the information in the advertisement shown in Figure 5-8 to discuss how to find a discount and how to find a discount rate.

Sidewalk Sale

Regularly $59^{80}

Men's Air light
Mid-top
Basketball Shoe

25% Off

Ladies' Shoe Sale
30-50% Off

AEROBIC
$21^{99}
reg. $39.99

Versatile fitness shoe for
every training need

FIGURE 5-8

EXAMPLE 6 *Finding the discount.* Find the amount of the discount on the pair of men's basketball shoes shown in Figure 5-8. Then find the sale price.

Solution

First, we will find 25% of the regular price, $59.80. Here, the percent is 25% and the base is 59.80.

Amount	=	percent	·	base
A	=	25%	·	59.80

$A = 0.25 \cdot 59.80$ Change 25% to a decimal.
$A = 14.95$ Do the multiplication.

The discount is $14.95. To find the sale price, we subtract the amount of the discount from the regular price.

$59.80 - 14.95 = 44.85$

The sale price of the men's basketball shoes is $44.85.

Self Check

Sunglasses, regularly selling for $15.40, are discounted 15%. Find the sale price.

Answer: $13.09

In Example 6, we used the following formula to find the sale price.

Finding the sale price	Sale price = original price − discount

EXAMPLE 7 *Finding the discount rate.* What is the rate of discount on the ladies' aerobic shoes advertised in Figure 5-8?

Solution

We can think of this as a percent-of-decrease problem. We first compute the amount of the discount. This decrease in price is found using subtraction.

$39.99 - 21.99 = 18$

The shoes are discounted $18. Now we find the percent of decrease in price. Here, the amount is 18 and the base is 39.99.

Amount	=	percent	·	base
18	=	p	·	39.99

Self Check

An early-bird special at a restaurant offers a $10.99 prime rib dinner for only $7.95 if it is ordered before 6 P.M. Find the rate of discount.

$$18 = p \cdot 39.99$$

$$\frac{18}{39.99} = \frac{p \cdot 39.99}{39.99} \qquad \text{Divide both sides by 39.99.}$$

$$0.450113 \approx p \qquad \text{Do the division.}$$

$$p \approx 45.0113\% \qquad \text{Change the decimal to a percent.}$$

Rounded to the nearest one percent, the discount rate is 45%.

Answer: 28%

STUDY SET Section 5.3

VOCABULARY *In Exercises 1–4, fill in the blanks to make a true statement.*

1. Some salespeople are paid on _____. It is based on a percent of the total dollar amount of the goods or services they sell.

2. When we use percent to describe how a quantity has increased when compared to its original value, we are finding the _____.

3. The difference between the original price and the sale price of an item is called the _____.

4. The _____ of a sales tax is expressed as a percent.

CONCEPTS

5. An organization experiences a 100% increase in membership. Represent the increase in another way.

6. The number of people watching a television show decreased by 50% over a ten-week period. Represent the decrease in another way.

APPLICATIONS *In Exercises 7–40, use a calculator to solve each problem. If a percent answer is not exact, round to the nearest one percent.*

7. SALES TAX   Find the sales tax on the purchase of a pair of pants costing $25 if the tax rate is 3%.

8. SALES TAX   Find the sales tax on the purchase of two lamps costing $45 each if the tax rate is 4.4%.

9. ROOM TAX   After checking out of a hotel, a man noticed that the hotel bill included an additional charge labeled *room tax*. If the price of the room was $129 plus a room tax of $10.32, find the room tax rate.

10. EXCISE TAX   While examining her monthly telephone bill, a woman noticed an additional charge of $1.24 labeled *federal excise tax*. If the basic service charges for that billing period were $42, what is the federal excise tax rate?

11. SALES RECEIPT   Complete the sales receipt in Illustration 1 by finding the subtotal, the sales tax, and the total.

12. SALES RECEIPT   Complete the sales receipt in Illustration 2 by finding the prices, the subtotal, the sales tax, and the total.

NURSERY CENTER
Your one-stop garden supply

3 @ 2.99	PLANTING MIX	$ 8.97
1 @ 9.87	GROUND COVER	$ 9.87
2 @ 14.25	SHRUBS	$28.50
SUBTOTAL		$
SALES TAX @ 6.00%		$
TOTAL		$

ILLUSTRATION 1

McCOY'S FURNITURE

1 @ 450.00	SOFA	$
2 @ 90.00	END TABLES	$
1 @ 350.00	LOVE SEAT	$
SUBTOTAL		$
SALES TAX @ 4.20%		$
TOTAL		$

ILLUSTRATION 2

13. SALES TAX HIKE In order to raise more revenue, some states raise the sales tax rate. How much additional money will be collected on the sale of a $15,000 car if the sales tax rate is raised 1%?

14. VALUE ADDED TAX In Canada, the value added tax, or VAT, is a 15% tax imposed by the government on every monetary transaction. What is the VAT for a dinner costing $48?

15. PAYCHECK Use the information on the paycheck stub in Illustration 3 to find the tax rate for the federal withholding and worker's compensation taxes that were deducted from the gross pay.

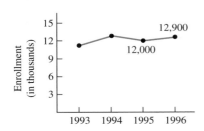

6286244
Issue date: 03-27-96
GROSS PAY $360.00
TAXES
FED. TAX $ 28.80
WORK. COMP. $ 4.32
NET PAY $326.88

ILLUSTRATION 3

16. GASOLINE TAX In one state, a gallon of unleaded gasoline sells for $1.29. This price includes federal and state taxes that total approximately $0.44. Therefore, the price of a gallon of gasoline, before taxes, is about $0.85. What is the tax rate on gasoline?

17. COLLEGE ENROLLMENT Use the information shown in Illustration 4 to find the percent of increase in enrollment for 1996 over 1995.

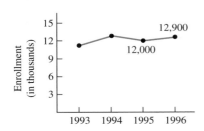

ILLUSTRATION 4

18. CROP DAMAGE After flooding damaged much of the crop, the cost of a head of lettuce jumped from $0.49 to $1.09. What percent of increase is this?

19. REDUCED CALORIES A company advertised its new, improved chips as having 36% less calories per serving than the original style. How many calories are in a serving of the new chips if a serving of the original style contained 150 calories?

20. OVERTIME Factory management wants to reduce the number of overtime hours by 25%. If the total number of overtime hours is 480 this month, what is the target number for next month?

21. COST-OF-LIVING INCREASE If a woman making $32,000 a year receives a cost-of-living increase of 2.4%, how much is her raise? What is her new salary?

22. POLICE FORCE A police department plans to increase its 80-person force by 5%. How many additional officers will be hired? What will be the new size of the department?

23. CAR INSURANCE A student paid a car insurance premium of $400 every six months. Then the premium dropped to $360, because she qualified for a good-student discount. What was the percent of decrease in the premium?

24. BUS PASS To increase the number of riders, a bus company reduced the price of a monthly pass from $112 to $98. What was the percent of decrease?

25. LAKE SHORELINE Because of a heavy spring run-off, the shoreline of a lake increased from 5.8 miles to 7.6 miles. What was the percent of increase in the shoreline?

26. PARKING The management of a mall has decided to increase the parking area. The plans are shown in Illustration 5. What will be the percent of increase in the parking area once the project is completed?

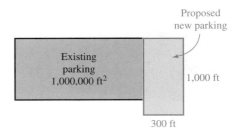

ILLUSTRATION 5

27. REAL ESTATE After selling a house for $98,500, a real estate agent split the 6% commission with another agent. How much did each person receive?

28. MEDICAL SUPPLIES A salesperson for a medical supplies company is paid a commission of 9% for orders under $8,000. For orders exceeding $8,000, she receives an additional 2% in commission on the total amount. What is her commission on a sale of $14,600?

29. SPORTS AGENT A sports agent charges his clients a fee to represent them during contract negotiations. The fee is based on a percent of the contract amount. If the agent earned $37,500 when his client signed a $2,500,000 professional football contract, what rate did he charge for his services?

30. ART GALLERY An art gallery displays paintings for artists and receives a commission from the artist when a painting is sold. What is the commission rate if a gallery received $135.30 when a painting was sold for $820?

31. CONCERT PARKING A concert promoter gets $33\frac{1}{3}\%$ of the revenue the arena receives from its parking concession the night of the performance. How much can the promoter make if 6,000 cars are anticipated and parking is $6 a car?

32. KITCHENWARE PARTY A homemaker invited her neighbors to a kitchenware party to show off cookware and utensils. As party hostess, she received 12% of the total sales. How much was purchased if she received $41.76 for hosting the party?

33. WATCH SALE See Illustration 6. What is the regular price and the rate of discount for the watch that is on sale?

ILLUSTRATION 6

34. STEREO SALE See Illustration 7. What is the regular price and the rate of discount for the stereo system that is on sale?

ILLUSTRATION 7

35. RING SALE What does a ring regularly sell for if it has been discounted 20% and is on sale for $149.99?

36. BLINDS SALE What do vinyl blinds regularly sell for if they have been discounted 55% and are on sale for $49.50?

37. VCR SALE What is the sale price and the discount rate for a VCR with remote that regularly sells for $399.97 and is being discounted $50?

38. CAMCORDER SALE What is the sale price and the discount rate for a camcorder that regularly sells for $559.97 and is being discounted $80?

39. REBATE See Illustration 8. Find the discount, the discount rate, and the reduced price for a case of motor oil if a shopper receives the manufacturer's rebate mentioned in the ad.

ILLUSTRATION 8

40. DOUBLE COUPONS See Illustration 9. Find the discount, the discount rate, and the reduced price for a box of cereal that normally sells for $3.29 if a shopper presents the coupon at a store that doubles the value of the coupon.

ILLUSTRATION 9

WRITING *Write a paragraph using your own words.*

41. List the pros and cons of working on commission.

43. In Example 6, explain why you get the correct answer by finding 75% of the regular price.

42. Explain the difference between a tax and a tax rate.

44. Explain how to find the sale price of an item if you know the discount rate.

REVIEW

45. Multiply: $-5(-5)(-2)$.

47. Evaluate: $\left(\frac{1}{2}\right)^2 + \left(\frac{1}{3}\right)^2$.

49. Divide: $0.2\overline{)34.68}$.

51. Evaluate: $|-5 - 8|$.

46. Divide: $\frac{3}{5} \div \frac{3}{10}$.

48. Multiply: $0.45 \cdot 675$.

50. Evaluate: $-4 - (-7)$.

52. Evaluate: $\sqrt{25} - \sqrt{16}$.

Percent

Here we discuss some estimation methods that can be used when working with percent. To begin, we consider a way to find 10% of a number quickly. Recall that 10% of a number is found by multiplying the number by 10% or 0.1. When multiplying a number by 0.1, we simply move the decimal point one place to the left to find the result.

EXAMPLE 1 *10% of a number.* Find 10% of 234.

Solution To find 10% of 234, move the decimal point 1 place to the left.

$$234 = 23.4.0$$

Thus, 10% of 234 is 23.4, or approximately 23.

To find 15% of a number, first find 10% of the number. Then find half of that to obtain the other 5%. Finally, add the two results.

EXAMPLE 2 *Estimating 15% of a number.* Estimate 15% of 78.

Solution 10% of 78 is 7.8, or about 8. ⟶   8
Add half of 8 to get the other 5%. ⟶ + 4
                                            12

Thus, 15% of 78 is approximately 12.

To find 20% of a number, first find 10% of it and then double that result. A similar procedure can be used when working with any multiple of 10%.

EXAMPLE 3 *Estimating 20% of a number.* Estimate 20% of 3,234.15.

Solution 10% of 3,234.15 is 323.415 or about 323. To find 20%, double that.

Thus, 20% of 3,234.15 is approximately 646.

To find 50% of a number means to find $\frac{1}{2}$ of the number. To find 25% of a number, first find 50% of it, then divide that result by 2.

EXAMPLE 4   *Estimating 25% of a number.*   Estimate 25% of 16,813.

Solution   16,813 is about 16,800. Half of that is 8,400. Thus, 50% of 16,813 is approximately 8,400.

To estimate 25% of 16,813, divide 8,400 by 2. Thus, 25% of 16,813 is approximately 4,200.

■

100% of a number is the number itself. To find 200% of a number, double the number.

EXAMPLE 5   *Estimating 200% of a number.*   Estimate 200% of 65.198.

Solution   65.198 is about 65. To find 200% of 65, double it. Thus, 200% of 65.198 is approximately 130.

■

STUDY SET   *In Exercises 1–6, estimate the answer to each problem.*

1. 20% of the 815 students attending a small college were enrolled in a science course. How many students is this?

2. In the grocery store, a 65-ounce bottle of window cleaner was marked "25% free." How many ounces are free?

3. By how much is the price of a VCR discounted if the regular price of $196.88 is reduced by 30%?

4. A restaurant tip is normally 15% of the cost of the meal. Find the tip on a dinner costing $38.64.

5. An insurance company paid 50% of the $107,809 it cost to rebuild a home that was destroyed by fire. How much did the insurance company pay?

6. Of the 2,580 vehicles inspected at a safety checkpoint, 10% had code violations. How many cars had code violations?

In Exercises 7–10, approximate the percent and then estimate the answer to each problem.

7. The attendance at a seminar was only 31% of what the organizers had anticipated. If 68 people were expected, how many actually attended the seminar?

8. Of the 900 students in a school, 16% were on the principal's honor roll. How many students were on the honor roll?

9. On election day, 48% of the 6,200 workers at the polls were volunteers. How many volunteers helped with the election?

10. Each department at a college was asked to cut its budget by 21%. By how much money should the mathematics department budget be reduced if it is currently $4,515?

Interest

In this section, you will learn about

- **Simple interest**
- **Compound interest**

INTRODUCTION. When money is borrowed, the lender expects to be paid back the amount of the loan plus an additional charge for the use of the money. The additional charge is called **interest.** When money is deposited in a bank, the depositor is paid for the use of the money. The money the deposit earns is also called interest. In general, interest is money that is paid for the use of money.

Simple interest

Interest is calculated in one of two ways: either as **simple interest** or as **compound interest.** We will begin by discussing simple interest. First, we need some key terms associated with borrowing or lending money.

Principal: the amount of money that is invested, deposited, or borrowed.

Interest rate: a percent that is used to calculate the amount of interest to be paid. It is usually expressed as an annual (yearly) rate.

Time: the length of time (usually in years) that the money is invested, deposited, or borrowed.

The amount of interest depends on the principal, the rate, and the time. That is why all three are usually mentioned in advertisements for bank accounts, investments, and loans. (See Figure 5-9.)

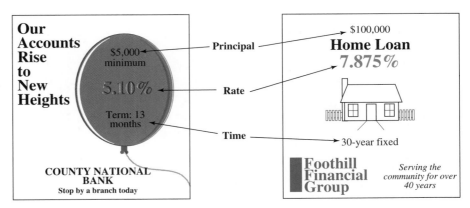

FIGURE 5-9

Simple interest is interest earned on the original principal. It is found by using a formula.

Simple interest formula	Interest = principal · rate · time
	or
	$I = Prt$
	where the rate r is expressed as an annual rate and the time t is expressed in years.

EXAMPLE 1 *Finding the interest earned.* $3,000 is invested for 1 year at a rate of 5%. How much interest is earned?

Solution

We will use the formula $I = Prt$ to calculate the interest earned. The principal is $3,000, the interest rate is 5% (or 0.05), and the time is 1 year.

$$P = 3,000 \qquad r = 5\% = 0.05 \qquad t = 1 \text{ year}$$

$I = Prt$	Write the interest formula.
$I = 3,000 \cdot 0.05 \cdot 1$	Substitute the values for P, r, and t.
$I = 150$	Do the multiplication.

The interest earned in 1 year is $150.

Self Check

If $4,200 is invested for 2 years at a rate of 4% annual interest, how much interest is earned?

Answer: $336

When using the formula $I = Prt$, the time must be expressed in years. If the time is given in days or months, we rewrite it as a fractional part of a year. For example, a 30-day investment lasts $\frac{30}{365}$ of a year, since there are 365 days in a year. For a 6-month loan, we express the time as $\frac{6}{12}$ or $\frac{1}{2}$ of a year.

EXAMPLE 2 *Paying off a loan.* To start a carpet-cleaning business, a couple borrows $5,500 to purchase equipment and supplies. If their loan has a 14% interest rate, how much must they repay at the end of the 90-day period?

Solution

First, we find the amount of interest paid on the loan. We must rewrite the time as a fractional part of a year.

$$P = 5,500 \qquad r = 14\% = 0.14 \qquad t = \frac{90}{365}$$

$I = Prt$	Write the interest formula.
$I = 5,500 \cdot 0.14 \cdot \dfrac{90}{365}$	Substitute the values for P, r, and t.
$I = \dfrac{5,500}{1} \cdot \dfrac{0.14}{1} \cdot \dfrac{90}{365}$	Write 5,500 and 0.14 as fractions.
$I = \dfrac{69,300}{365}$	Use a calculator to multiply the numerators. Multiply the denominators.
$I \approx 189.86$	Use a calculator to do the division. Round to the nearest cent.

The interest on the loan is $189.86. To find how much they must pay back, we add the principal and the interest.

$$5,500 + 189.86 = 5,689.86$$

The couple must pay back $5,689.86 at the end of 90 days.

Self Check

How much must be repaid if $3,200 is borrowed at a rate of 15% for 120 days?

Answer: $3,357.81

Compound interest

Most savings accounts pay **compound interest** rather than simple interest. Compound interest is interest paid on accumulated interest. To illustrate this concept, suppose that $2,000 is deposited in a savings account at a rate of 5% for 1 year. We can use the formula $I = Prt$ to calculate the interest earned at the end of 1 year.

$$I = Prt$$
$$I = 2,000 \cdot 0.05 \cdot 1 \quad \text{Substitute for } P, r, \text{ and } t.$$
$$I = 100 \quad \quad \quad \text{Do the multiplication.}$$

Interest of $100 was earned. At the end of the first year, the account contains the interest ($100) plus the original principal ($2,000) for a balance of $2,100.

Suppose that the money remains in the savings account for another year at the same interest rate. For the second year, interest will be paid on a principal of $2,100. That is, during the second year, we earn *interest on the interest* as well as on the original $2,000 principal. Using $I = Prt$, we can find the interest earned in the second year.

$$I = Prt$$
$$I = 2,100 \cdot 0.05 \cdot 1 \quad \text{Substitute for } P, r, \text{ and } t.$$
$$I = 105 \quad \quad \quad \text{Do the multiplication.}$$

In the second year, $105 of interest is earned. The account now contains that interest plus the $2,100 principal, for a total of $2,205.

As Figure 5-10 shows, we calculated the simple interest two times to find the compound interest.

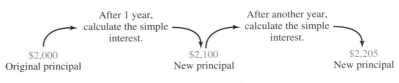

FIGURE 5-10

If we compute the *simple interest* on $2,000, at 5% for 2 years, the interest earned is $I = 2,000 \cdot 0.05 \cdot 2 = 200$. Thus, the account balance would be $2,200. Comparing the balances, the account earning compound interest will contain $5 more than the account earning simple interest.

In the previous example, the interest was calculated at the end of each year, or **annually.** When compounding, we can compute the interest in other time increments, such as **semiannually** (twice a year), **quarterly** (four times a year), or even **daily.**

EXAMPLE 3 *Finding compound interest.* As a gift for her newborn granddaughter, a grandmother opens a $1,000 savings account in the baby's name. The interest rate is 4.2% compounded quarterly. Find the amount of money the child will have in the bank on her first birthday.

Solution If the interest is compounded quarterly, the interest will be computed four times in one year. In this case, we must express the time as a fractional part of a year.

First quarter

$$P = 1,000 \quad \quad r = 4.2\% = 0.042 \quad \quad t = \frac{1}{4}$$

$$I = 1,000 \cdot 0.042 \cdot \frac{1}{4}$$

$$I = \$10.50$$

The interest earned in the first quarter is $10.50. This now becomes part of the principal for the second quarter:

$1,000 + $10.50 = $1,010.50

Second quarter

$P = 1,010.50 \qquad r = 0.042 \qquad t = \dfrac{1}{4}$

$I = 1,010.50 \cdot 0.042 \cdot \dfrac{1}{4}$

$I \approx \$10.61$ (Rounded)

The interest earned in the second quarter is $10.61. This becomes part of the principal for the third quarter:

$1,010.50 + $10.61 = $1,021.11

Third quarter

$P = 1,021.11 \qquad r = 0.042 \qquad t = \dfrac{1}{4}$

$I = 1,021.11 \cdot 0.042 \cdot \dfrac{1}{4}$

$I \approx \$10.72$ (Rounded)

The interest earned in the third quarter is $10.72. This now becomes part of the principal for the fourth quarter:

$1,021.11 + $10.72 = $1,031.83

Fourth quarter

$P = 1,031.83 \qquad r = 0.042 \qquad t = \dfrac{1}{4}$

$I = 1,031.83 \cdot 0.042 \cdot \dfrac{1}{4}$

$I \approx \$10.83$ (Rounded)

The interest earned the fourth quarter is $10.83. Adding this to the existing principal, we get:

$1,031.83 + $10.83 = $1,042.66

The amount that has accumulated in the account after four quarters, or 1 year, is $1,042.66. ■

Computing compound interest by hand is tedious. The **compound interest formula** can be used to find the total amount of money that an account will contain at the end of the term.

Compound interest formula	The total amount A in an account can be found using the formula $$A = P\left(1 + \dfrac{r}{n}\right)^{nt}$$ where P is the principal, r is the annual interest rate expressed as a decimal, t is the length of time in years, and n is the number of compoundings in one year.

A businessman invests $9,250 at 7.6% interest to be compounded monthly. To find what the investment will be worth in 3 years, we use the compound interest formula with the following values.

$$P = \$9,250 \quad r = 7.6\% = 0.076 \quad t = 3 \text{ years} \quad n = 12 \text{ times a year (monthly)}$$

We apply the compound interest formula:

$$A = P\left(1 + \frac{r}{n}\right)^{nt} \qquad \text{Write the compound interest formula.}$$

$$A = 9,250\left(1 + \frac{0.076}{12}\right)^{12(3)} \qquad \text{Substitute the values of } P, r, t, \text{ and } n.$$

$$A = 9,250\left(1 + \frac{0.076}{12}\right)^{36} \qquad \text{Simplify the exponent: } 12(3) = 36.$$

Then we use a calculator to evaluate.

Keystrokes: $\boxed{9}\,\boxed{2}\,\boxed{5}\,\boxed{0}\,\boxed{\times}\,\boxed{(}\,\boxed{(}\,\boxed{1}\,\boxed{+}\,\boxed{.}\,\boxed{0}\,\boxed{7}\,\boxed{6}\,\boxed{\div}\,\boxed{1}\,\boxed{2}\,\boxed{)}\,\boxed{)}\,\boxed{y^x}$
$\boxed{3}\,\boxed{6}\,\boxed{=}$ `11610.43875`

Rounded to the nearest cent, the amount in the account after 3 years will be $11,610.44.

If your calculator does not have parenthesis keys, calculate the sum inside the parentheses first. Then find the power. Finally, multiply by 9,250.

EXAMPLE 4   *Compound interest.* A man deposited $50,000 in a long-term account at 6.8% interest compounded daily. How much money will he be able to withdraw in 7 years if the principal is to remain in the bank?

Solution

"Compounded daily" means that compounding will be done 365 times in a year.

$$P = \$50,000 \qquad r = 6.8\% = 0.068 \qquad t = 7 \text{ years} \qquad n = 365 \text{ times a year}$$

$$A = P\left(1 + \frac{r}{n}\right)^{nt} \qquad \text{Write the compound interest formula.}$$

$$A = 50,000\left(1 + \frac{0.068}{365}\right)^{365(7)} \qquad \text{Substitute the values of } P, r, t, \text{ and } n.$$

$$A = 50,000\left(1 + \frac{0.068}{365}\right)^{2,555} \qquad 365(7) = 2,555.$$

$$A \approx 80,477.58 \qquad \text{Use a calculator. Round to the nearest cent.}$$

The account will contain $80,477.58 at the end of 7 years. To find the amount the man can withdraw, we subtract.

$$80,477.58 - 50,000 = 30,477.58$$

The man can withdraw $30,477.58 without having to touch the $50,000 principal.

Self Check

Find the amount of interest $25,000 will earn in 10 years if it is deposited in an account at 5.99% interest compounded daily.

Answer: $20,505.20

VOCABULARY *In Exercises 1–6, fill in the blanks to make a true statement.*

1. In banking, the original amount of money borrowed or deposited is known as the _____.

2. Borrowers pay _____ to lenders for the use of their money.

3. The percent that is used to calculate the amount of interest to be paid is called the _____ rate.

4. _____ interest is interest paid on accumulated interest.

5. Interest computed only on the original principal is called _____ interest.

6. Percent means per _____.

CONCEPTS

7. Tell how many times a year the interest on a savings account is calculated if the interest is compounded
   a. semiannually   **b.** quarterly
   c. daily   **d.** monthly

8. Express each of the following as a fraction of a year. Simplify the fraction.
   a. 6 months   **b.** 90 days
   c. 120 days   **d.** 1 month

9. When doing calculations with percents, they must be changed to decimals or fractions. Change each percent to a decimal.

   a. 7%   **b.** 9.8%   **c.** $6\frac{1}{4}\%$

10. When doing calculations with percents, they must be changed to decimals or fractions. Change each percent to a fraction.

   a. $33\frac{1}{3}\%$ **b.** $66\frac{2}{3}\%$ **c.** $16\frac{2}{3}\%$

11.

⌐1st qtr⌐	⌐2nd qtr⌐	⌐3rd qtr⌐	⌐4th qtr⌐

$1,000 $1,050 $1,102.50 $1,157.63 $1,215.51

ILLUSTRATION 1

   a. What concept studied in this section is illustrated by the diagram in Illustration 1?

   b. What was the original principal?
   c. How many times was the interest found?
   d. How much interest was earned on the first compounding?
   e. For how long was the money invested?

12. $3,000 is deposited in a savings account that earns 10% interest compounded annually. Complete the series of calculations in Illustration 2 to find how much money will be in the account at the end of 2 years.

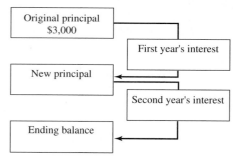

ILLUSTRATION 2

NOTATION

13. In the formula $I = Prt$, what operations are indicated by Prt?

14. In the formula $A = P\left(1 + \dfrac{r}{n}\right)^{nt}$, how many operations must be performed to find A?

APPLICATIONS *In Exercises 15–26, use simple interest.*

15. RETIREMENT INCOME   A retiree invests $5,000 in a savings plan that pays 6% per year. What will the account balance be at the end of the first year?

16. INVESTMENT   A developer promised a return of 8% annual interest on an investment of $15,000 in her company. How much could an investor expect to make in the first year?

17. REMODELING   A homeowner borrows $8,000 to pay for a kitchen remodeling project. The terms of the loan are 9.2% annual interest and repayment in 2 years. How much interest will be paid on the loan?

18. CREDIT UNION   A farmer borrowed $7,000 from a credit union. The money was loaned at 8.8% annual interest for 18 months. How much money did the credit union charge him for the use of the money?

19. MEETING A PAYROLL   In order to meet end-of-the-month payroll obligations, a small business had to borrow $4,200 for 30 days. How much did the business have to repay if the interest rate was 18%?

20. CAR LOAN   To purchase a car, a man takes out a loan for $2,000. If the interest rate is 9% per year, how much interest will he have to pay at the end of the 120-day loan period?

21. SAVINGS ACCOUNT   Find the interest earned on $10,000 at $7\frac{1}{4}$% for 2 years. Use the chart in Illustration 3 to organize your work.

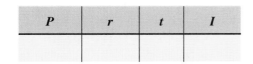

P	r	t	I

ILLUSTRATION 3

22. TUITION   A student borrows $300 from an educational fund to pay for books for spring semester. If the loan is for 45 days at $3\frac{1}{2}$% annual interest, what will the student owe at the end of the loan period?

23. LOAN APPLICATION   Complete the loan application form in Illustration 4.

24. LOAN APPLICATION   Complete the loan application form in Illustration 5.

25. LOW-INTEREST LOAN   An underdeveloped country receives a low-interest loan from a bank to finance the construction of a water treatment plant. What must the country pay back at the end of 2 years if the loan is for $18 million at 2.3%?

Loan Application Worksheet

1. Amount of loan (principal) ___$1,200.00___

2. Length of loan (time) ___2 YEARS___

3. Annual percentage rate ___8%___

4. Interest charged _____

5. Total amount to be repaid _____

6. Check method of repayment:
   ☐ 1 lump sum   ☑ monthly payments

   Borrower agrees to pay ___24___ equal payments of _____ to repay loan.

ILLUSTRATION 4

Loan Application Worksheet

1. Amount of loan (principal) ___$810.00___

2. Length of loan (time) ___9 mos.___

3. Annual percentage rate ___12%___

4. Interest charged _____

5. Total amount to be repaid _____

6. Check method of repayment:
   ☐ 1 lump sum   ☑ monthly payments

   Borrower agrees to pay ___9___ equal payments of _____ to repay loan.

ILLUSTRATION 5

26. REDEVELOPMENT   A city is awarded a low-interest loan to help renovate the downtown business district. The $40 million loan, at 1.75%, must be repaid in $2\frac{1}{2}$ years. How much interest will the city have to pay?

▦   *In Exercises 27–34, use a calculator.*

27. COMPOUNDING ANNUALLY   If $600 is invested in an account that earns 8% compounded annually, what will the account balance be after 3 years?

28. COMPOUNDING SEMIANNUALLY   If $600 is invested in an account that earns annual interest of 8% compounded semiannually, what will the account balance be at the end of 3 years?

29. COLLEGE FUND   A ninth-grade student opens a savings account that locks her money in for 4 years at an annual rate of 6% compounded daily. If the initial de-

posit is $1,000, how much money will be in the account when she begins college in four years?

30. CERTIFICATE OF DEPOSIT   A 3-year certificate of deposit pays an annual rate of 5% compounded daily. The maximum allowable deposit is $90,000. What is the most interest a depositor can earn from the CD?

31. TAX REFUND   A couple deposits an income tax refund check of $545 in an account paying an annual rate of 4.6% compounded daily. What will the size of the account be at the end of 1 year?

32. INHERITANCE After receiving an inheritance of $11,000, a man deposits the money in an account paying an annual rate of 7.2% compounded daily. How much money will be in the account at the end of 1 year?

33. LOTTERY Suppose you won $500,000 in the lottery and deposited the money in a savings account that paid an annual rate of 6% interest compounded daily. How much interest would you earn each year?

34. CASH GIFT After receiving a $250,000 cash gift, a university decides to deposit the money in an account paying an annual rate of 5.88% compounded quarterly. How much money will the account contain in 5 years?

WRITING *Use a paragraph to explain in your own words.*

35. What is the difference between simple and compound interest?

36. Explain: *Interest is the amount of money paid for the use of money.*

37. On some accounts, banks charge a penalty if the depositor withdraws the money before the end of the term. Why would a bank do this?

38. Explain why it is better for a depositor to open a savings account that pays 5% interest compounded daily than one that pays 5% interest compounded monthly.

REVIEW

39. Simplify: $\sqrt{\dfrac{1}{4}}$.

40. Find: $\left(\dfrac{1}{4}\right)^2$.

41. Add: $\dfrac{3}{7} + \dfrac{2}{5}$.

42. Subtract: $\dfrac{3}{7} - \dfrac{2}{5}$.

43. Multiply: $2\dfrac{1}{2} \cdot 3\dfrac{1}{3}$.

44. Divide: $-12\dfrac{1}{2} \div 5$.

45. Find 6% of 200.

46. Evaluate: $(0.2)^2 - (0.3)^2$.

Percent

Since the word *percent* means *per hundred*, we can think of a percent as the numerator of a fraction that has a denominator of 100.

To change a percent to a fraction, we drop the % sign and write the given number over 100. Finish each conversion.

$$67\% = \frac{}{100} \qquad 56\% = \frac{56}{} = \frac{14}{25} \qquad .05\% = \frac{}{100} = \frac{5}{10,000} = \frac{1}{}$$

To change a percent to a decimal, we drop the percent sign and move the decimal point two places to the left. Finish each conversion.

$$67\% = \qquad\qquad 56\% = \qquad\qquad 0.05\% =$$

To change a fraction to a percent, we write the fraction as a decimal and then move the decimal point two places to the right and insert a % sign. Finish each conversion.

$$\frac{3}{4} = 0.75 = \qquad\qquad \frac{4}{5} = \qquad = 80\%$$

$$\frac{5}{8} = \qquad = 62.5\% \qquad \frac{25}{4} = 6.25 =$$

To solve problems involving percent, we use the percent formula.

$$\text{Amount} = \text{percent} \cdot \text{base}$$

Solve each problem.

1. Find 32% of 620.

2. 300 is what percent of 500?

3. 25 is 40% of what number?

4. Find 125% of 850.

5. 106.25 is what percent of 625?

6. 163.84 is 32% of what number?

Percents are often used to compute interest. If I is the interest, P is the principal, r is the annual rate (or percent), and t is the length of time in years, the formula for simple interest is

$$I = Prt$$

7. Find the amount of interest that will be earned if $10,000 is invested for 5 years at 6% annual interest.

If A is the amount, P is the principal, r is the annual rate of interest, t is the length of time in years, and n is the number of compoundings in one year, the formula for compound interest is

$$A = P\left(1 + \frac{r}{n}\right)^{nt}$$

8. Find the amount of interest that will be earned of $10,000 is invested for 5 years at 6% annual interest, compounded quarterly.

| SECTION 5.1 | *Percents, Decimals, and Fractions* |

CONCEPTS

Percent means per one hundred.

To change a percent to a fraction, drop the % sign and put the given number over 100.

To change a percent to a decimal, drop the % sign and move the decimal point 2 places to the left.

To change a decimal to a percent, move the decimal point 2 places to the right and insert a % sign.

To change a fraction to a percent, write the fraction as a decimal by dividing its numerator by its denominator. Then move the decimal point 2 places to the right and insert a % sign.

REVIEW EXERCISES

1. Express the amount of each figure that is shaded as a percent, as a decimal, and as a fraction. Each set of squares represents 100%.

a. 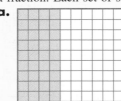   b.    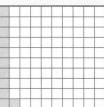

2. In Problem 1, part a, what percent of the figure is not shaded?

3. Change each percent to a fraction.
   a. 15%   **b.** 120%   **c.** $9\frac{1}{4}\%$ **d.** 0.1%

4. Change each percent to a decimal.
   a. 27%   **b.** 8%   **c.** 155%   **d.** $1\frac{4}{5}\%$

5. Change each decimal to a percent.
   a. 0.83   **b.** 0.625   **c.** 0.051   **d.** 6

6. Change each fraction to a percent.
   a. $\frac{1}{2}$ **b.** $\frac{4}{5}$ **c.** $\frac{7}{8}$ **d.** $\frac{1}{16}$

7. Find the exact percent equivalent for each fraction.
   a. $\frac{1}{3}$ **b.** $\frac{5}{6}$

8. Change each fraction to a percent. Round to the nearest hundredth.
   a. $\frac{5}{9}$ **b.** $\frac{8}{3}$

9. BILL OF RIGHTS   There are 27 amendments to the Constitution of the United States. The first ten are known as the Bill of Rights. What percent of the amendments were adopted after the Bill of Rights? (Round to the nearest one percent.)

The percent formula:
   Amount = percent · base

10. Identify the amount, the base, and the percent in the statement "15 is $33\frac{1}{3}$% of 45."

11. Use the percent formula and the question "What number is 32% of 96?" to form an equation. Let the amount be represented by *A*.

_____

12. Solve each problem.
   a. What number is 40% of 500?   **b.** 16% of what number is 20?

   c. 1.4 is what percent of 80?   **d.** $66\frac{2}{3}$% of 3,150 is what number?

   e. Find 220% of 55.   **f.** What is 0.05% of 60,000?

13. RACING   The nitro–methane fuel mixture used to power some experimental cars is 96% nitro and 4% methane. How many gallons of each fuel component are needed to fill a 15-gallon fuel tank?

14. HOME SALES   After the first day on the market, 51 homes in a new subdivision had already sold. This was 75% of the total number of homes available. How many homes were originally for sale?

15. HURRICANE DAMAGE   96 of the 110 trailers in a mobile home park were either damaged or destroyed by hurricane winds. What percent is this? (Round to the nearest one percent.)

16. TIPPING   The cost of dinner for a family of five at a restaurant was $36.20. Find the amount of the tip if it should be 15% of the cost of dinner.

A *circle graph* is a way of presenting data for comparison. The sizes of the segments of the circle indicate the percents of the whole represented by each category.

17. AIR POLLUTION
Complete Illustration 1
(a circle graph)
to show the given data.

Source of Carbon Monoxide Air Pollution	
Transportation vehicles	63%
Fuel combustion in homes, offices, electrical plants	12%
Industrial processes	8%
Solid-waste disposal	3%
Miscellaneous	14%

ILLUSTRATION 1

18. EARTH'S SURFACE The surface of the earth is approximately 196,800,000 square miles. Use the information in Illustration 2 to determine the number of square miles of the earth's surface that are covered with water.

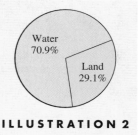

Applications of Percent

To find the total price of an item:
    Total price = purchase price + sales tax

19. SALES RECEIPT Complete the sales receipt in Illustration 3.

CAMERA CENTER	
35mm Canon Camera	$59.99
SUBTOTAL	$59.99
SALES TAX @ 5.5%	
TOTAL	

ILLUSTRATION 3

20. SALES TAX RATE Find the sales tax rate if the sales tax on the purchase of an automobile priced at $12,300 was $492.

Commission is based on a percent of the total dollar amount of the goods or services sold.

21. COMMISSION If the commission rate is 6%, find the commission earned by an appliance salesperson who sells a washing machine for $369.97 and a dryer for $299.97.

To find *percent of increase or decrease:*
1. Subtract the smaller number from the larger to find the amount of increase or decrease.
2. Find what percent the difference is of the original amount.

22. TROOP SIZE The size of a peace-keeping force was increased from 10,000 to 12,500 troops. What percent of increase is this?

23. GAS MILEAGE Experimenting with a new brand of gasoline in her truck, a woman found that the gas mileage fell from 18.8 to 17.0 miles per gallon. What percent of decrease is this? (Round to the nearest tenth of a percent.)

The difference between the original price and the sale price of an item is called the *discount.*

To find the *sale price:*
    Sale price = original price − discount

24. TOOL CHEST See Illustration 4. Use the information in the advertisement to find the discount, the original price, and the discount rate on the tool chest.

Sale price $139.99
Save $50!
Tool Chest Professional quality

ILLUSTRATION 4

Simple interest is interest earned on the original principal and is found using the formula

$$I = Prt$$

where P is the principal, r is the annual interest rate, and t is the length of time in years.

25. Find the interest earned on $6,000 invested at 8% per year for 2 years. Use the following chart to organize your work.

P	r	t	I

26. CODE VIOLATIONS   A business was ordered to correct safety code violations in a production plant. To pay for the needed corrections, the company borrowed $10,000 at 12.5% for 90 days. Find the total amount that must be paid after 90 days.

Compound interest is interest earned on interest.

The compound interest formula:

$$A = P\left(1 + \frac{r}{n}\right)^{nt}$$

where A is the amount in the account, P is the principal, r is the annual interest rate, n is the number of compoundings in one year, and t is the length of time in years.

27. MONTHLY PAYMENTS   A couple borrows $1,500 for 1 year at $7\frac{3}{4}$% and decides to repay the loan by making 12 equal monthly payments. How much will each monthly payment be?

28. Find the amount of money that will be in a savings account at the end of 1 year if $2,000 is the initial deposit and the annual interest rate of 7% is compounded semiannually. (*Hint:* find the simple interest twice.)

29. Find the amount that will be in a savings account at the end of 3 years if $5,000 is deposited at an annual rate of $6\frac{1}{2}$% compounded daily.

30. CASH GRANT   Each year a cash grant is given to a deserving college student. The grant consists of the interest earned that year on a $500,000 savings account. What is the cash award for the year if the money is invested at an annual rate of 8.3% compounded daily?

Chapter 5   Test

1. In Illustration 1, each set of 100 squares represents 100%. Express the amount of the figure that is shaded as a percent. Then express that percent as a fraction and as a decimal.

2. See Illustration 2. Express the amount of the figure that is shaded as a percent, a fraction, and a decimal.

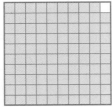

ILLUSTRATION 1

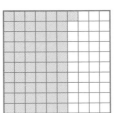

ILLUSTRATION 2

3. Change each percent to a decimal.
   a. 67%   **b.** 12.3%   **c.** $9\frac{3}{4}\%$

4. Change each fraction to a percent.
   a. $\frac{1}{4}$ **b.** $\frac{5}{8}$ **c.** $\frac{3}{25}$

5. Change each decimal to a percent.
   a. 0.19   **b.** 3.47   **c.** 0.005

6. Change each percent to a fraction.
   a. 55%   **b.** 0.01%   **c.** 125%

7. Change $\frac{7}{30}$ to a percent. Round to the nearest hundredth of a percent.

8. WEATHER REPORT   A weatherman states that there is a 40% chance of rain. What are the chances that it will not rain?

9. Find the exact percent equivalent for the fraction $\frac{2}{3}$.

10. Find the exact percent equivalent for the fraction $\frac{1}{4}$.

11. SHRINKAGE   See Illustration 3, a label on a new pair of jeans.
   a. How much length will be lost due to shrinkage?

   b. What will be the resulting length?

12. 65 is what percent of 1,000?

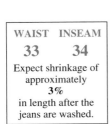

```
WAIST   INSEAM
 33       34
Expect shrinkage of
 approximately
    3%
in length after the
jeans are washed.
```

ILLUSTRATION 3

13. TIPPING   Find the amount of a 15% tip on a meal costing $25.40.

14. CABLE CHANNELS   Of the 50 television channels carried by a cable service, only 15 are included in the basic service package. What percent is this?

15. SWIMMING WORKOUT   A swimmer was able to complete 18 laps before a shoulder injury forced him to stop. This was only 20% of a typical workout. How many laps does he normally complete during a workout?

16. COLLEGE EMPLOYEES   The 700 employees at a community college fall into three major categories, as shown in Illustration 4. How many employees are in administration?

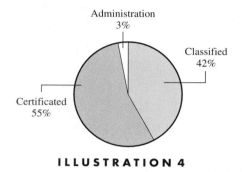

ILLUSTRATION 4

17. What number is 24% of 600?

18. PURCHASING A REFRIGERATOR   Illustration 5 shows the number of hours the average production worker took to earn enough money to buy a refrigerator in 1940 and 1990. Find the percent of decrease, to the nearest one percent.

ILLUSTRATION 5

19. HOMEOWNER'S INSURANCE   An insurance salesperson receives a 4% commission on the annual premium of any policy she sells. What is her commission on a homeowner's policy if the premium is $898?

20. COST-OF-LIVING INCREASE   A teacher earning $40,000 just received a cost-of-living increase of 3.6%. What is the teacher's new salary?

21. CAR WAX SALE   A car waxing kit, regularly priced at $14.95, is on sale for $3 off. What is the sale price, the discount, and the rate?

22. POPULATION INCREASE   After a new freeway was completed, the population of a city it passed through increased from 12,808 to 15,565 in two years. What percent of increase is this? (Round to the nearest one percent.)

23. Find the interest on a loan of $3,000 at 5% per year for 1 year.

24. Find the amount of interest earned on an investment of $24,000 paying an annual rate of 6.4% interest compounded daily for 3 years.

Ratio, Proportion, and Measurement

6

RATIOS AND PROPORTIONS CAN BE USED TO SOLVE PROBLEMS INVOLVING PRICING, TRAVEL, MIXTURES, AND MANUFACTURING. THEY ARE ALSO USED TO CONVERT FROM ONE UNIT OF MEASUREMENT TO ANOTHER.

6.1 Ratio

In this section, you will learn about

- **Ratios**
- **Unit costs**
- **Rates**

INTRODUCTION.   The concept of *ratio* occurs often in real-life situations. For example,

- To prepare fuel for an outboard marine engine, gasoline must be mixed with oil in the ratio of 50 to 1.
- To make 14-karat jewelry, gold is mixed with other metals in the ratio of 14 to 10.
- In the stock market, losing stocks might outnumber winning stocks in the ratio of 5 to 3.
- At Rock Valley College, the ratio of faculty to students is 1 to 18.

In this section, we will discuss ratios and use them to solve problems.

Ratios

Ratios give us a way to compare numerical quantities.

Ratio   A **ratio** is the comparison of two numbers by their indicated quotient.

Some examples of ratios are

$$\frac{5}{7}, \qquad \frac{25}{35}, \qquad \text{and} \qquad \frac{131}{229}$$

- The fraction $\frac{5}{7}$ can be read as "the ratio of 5 to 7."

- The fraction $\frac{25}{35}$ can be read as "the ratio of 25 to 35."

- The ratio $\frac{131}{229}$ can be read as "the ratio of 131 to 229."

Because the fractions $\frac{5}{7}$ and $\frac{25}{35}$ represent equal numbers, they are **equal ratios.**

EXAMPLE 1   *Writing ratios.*   Express each phrase as a ratio in lowest terms.   **a.** the ratio of 15 to 12   and   **b.** the ratio of 0.3 to 1.2.

Solution

a. The ratio of 15 to 12 can be written as the fraction $\frac{15}{12}$. After simplifying, the ratio is $\frac{5}{4}$.

b. The ratio of 0.3 to 1.2 can be written as the fraction $\frac{0.3}{1.2}$. We can simplify this fraction as follows:

$$\frac{0.3}{1.2} = \frac{0.3 \cdot 10}{1.2 \cdot 10} \quad \text{To clear the decimal, multiply both the numerator and the denominator by 10.}$$

$$= \frac{3}{12} \quad \text{Multiply: } 0.3 \cdot 10 = 3 \text{ and } 1.2 \cdot 10 = 12.$$

$$= \frac{1}{4} \quad \text{Simplify the fraction: } \frac{3}{12} = \frac{\overset{1}{\cancel{3}} \cdot 1}{\underset{1}{\cancel{3}} \cdot 4} = \frac{1}{4}.$$

Self Check

Express each ratio in lowest terms:

a. The ratio of 8 to 12

b. The ratio of 3.2 to 16

Answer: **a.** $\frac{2}{3}$,   **b.** $\frac{1}{5}$ ■

EXAMPLE 2   *Writing ratios.*   Express each phrase as a ratio in lowest terms: **a.** the ratio of 3 inches to 7 inches   and   **b.** the ratio of 6 ounces to 1 pound.

Solution

a. The ratio of 3 inches to 7 inches can be written as the fraction $\frac{3 \text{ inches}}{7 \text{ inches}}$, or just $\frac{3}{7}$.

b. When possible, we should express ratios in the same units. Since there are 16 ounces in 1 pound, the proper ratio is $\frac{6 \text{ ounces}}{16 \text{ ounces}}$, which simplifies to $\frac{3}{8}$.

Self Check

Express each ratio in lowest terms:

a. The ratio of 8 ounces to 2 pounds

b. The ratio of 2 feet to 1 yard. (*Hint:* 3 feet = 1 yard.)

Answer: **a.** $\frac{1}{4}$,   **b.** $\frac{2}{3}$ ■

EXAMPLE 3   *Student-to-faculty ratio.*   At a college, there are 2,772 students and 154 faculty members. Write a fraction in simplified form that expresses the ratio of students per faculty member.

Solution

The ratio of students to faculty is 2,772 to 154. We can write this ratio as the fraction $\frac{2,772}{154}$ and simplify it.

$$\frac{2,772}{154} = \frac{18 \cdot \overset{1}{\cancel{154}}}{1 \cdot \underset{1}{\cancel{154}}} \quad \text{Factor 2,772 as } 18 \cdot 154.$$

$$= \frac{18}{1} \quad \text{Divide out the common factor of 154.}$$

The ratio of students to faculty is 18 to 1.

Self Check

In a college, 224 members of a graduating class of 632 went on to graduate school. Write a fraction in simplified form that expresses the ratio of number of students going on to the number in the graduating class.

Answer: $\frac{28}{79}$ ■

Unit costs

The *unit cost* of an item is the ratio of its cost to its quantity. For example, the unit cost (the cost per pound) of 5 pounds of coffee priced at $18.75 is given by the ratio

$$\frac{\$18.75}{5 \text{ lb}} = \frac{\$1,875}{500 \text{ lb}}$$ To eliminate the decimal, multiply the numerator and the denominator by 100.

$$= \$3.75 \text{ per lb}$$ Do the division.

The unit cost is $3.75 per pound.

EXAMPLE 4

Comparison shopping. Olives come packaged in a 10-ounce jar, which sells for $2.49, or in a 6-ounce jar, which sells for $1.53. Which is the better buy?

Solution

To find the better buy, we must find each unit cost.

$$\frac{\$2.49}{10 \text{ oz}} = \frac{249\text{¢}}{10 \text{ oz}}$$ Change $2.49 to 249 cents.

$$= 24.9\text{¢ per oz}$$ Divide 249 by 10.

$$\frac{\$1.53}{6 \text{ oz}} = \frac{153\text{¢}}{6 \text{ oz}}$$ Change $1.53 to 153 cents.

$$= 25.5\text{¢ per oz}$$ Do the division.

The unit cost is less when olives are packaged in 10-ounce jars, so that is the better buy.

Self Check

A fast-food restaurant sells a 12-ounce cola for 72¢ and a 16-ounce cola for 99¢. Which is the better buy?

Answer: the 12-oz cola

Rates

Ratios can be used to compare quantities with different units. Such ratios are called *rates*. For example, if we drive 372 miles in 6 hours, the average rate of speed is the ratio of the miles driven to the length of time of the trip.

$$\text{Average rate of speed} = \frac{372 \text{ miles}}{6 \text{ hours}} = \frac{62 \text{ miles}}{1 \text{ hour}}$$
$$\frac{372}{6} = \frac{\overset{1}{\cancel{6}} \cdot 62}{\cancel{6} \cdot 1} = \frac{62}{1}$$

The ratio $\frac{62 \text{ miles}}{1 \text{ hour}}$ can be expressed in any of the following forms:

$$62\frac{\text{miles}}{\text{hour}}, \quad 62 \text{ miles per hour}, \quad 62 \text{ miles/hour}, \quad \text{or} \quad 62 \text{ mph}$$

EXAMPLE 5

Finding hourly rates of pay. A student earns $152 for working 16 hours. Find his hourly rate of pay.

Solution

We can write the rate of pay as the ratio

$$\text{Rate of pay} = \frac{\$152}{16 \text{ hr}}$$

and simplify by dividing 152 by 16.

$$\text{Rate of pay} = 9.50\frac{\$}{\text{hr}}$$

The rate of pay is $9.50 per hour.

Self Check

Joan earns $316 per 40-hour week managing a dress shop. Set up a ratio and find her hourly rate of pay.

Answer: $7.90 per hour

Energy consumption. One household used 795 kilowatt hours (kwh) of electricity during a 30-day period. Find the rate of energy consumption in kilowatt hours per day.

Solution

We can write the rate of energy consumption as the ratio

$$\text{Rate of energy consumption} = \frac{795 \text{ kwh}}{30 \text{ days}}$$

and simplify by dividing 795 by 30.

$$\text{Rate of energy consumption} = 26.5 \frac{\text{kwh}}{\text{day}}$$

The rate of energy consumption was 26.5 kilowatt hours per day.

Self Check

To heat a house for 30 days, a furnace burned 69 therms of natural gas. Find the rate of gas consumption in therms per day.

Answer: 2.3 therms per day ■

Tax rate. A textbook costs $44.52, including sales tax. If the tax was $2.52, find the sales tax rate.

Solution

Since the tax was $2.52, the cost of the book alone was

$$\$44.52 - \$2.52 = \$42.00$$

We can write the sales tax rate as the ratio

$$\text{Sales tax rate} = \frac{\text{amount of sales tax}}{\text{cost of the book, without tax}}$$

$$= \frac{\$2.52}{\$42}$$

and simplify by dividing 2.52 by 42.

$$\text{Sales tax rate} = 0.06$$

The sales tax rate is 0.06, or 6%.

Self Check

A sport coat costs $160.50, including sales tax. If the cost of the coat without tax is $150, find the sales tax rate.

Answer: 7% ■

Accent on Technology ***Computing gas mileage***

A man drove a total of 775 miles. Along the way, he stopped for gas three times, pumping 10.5, 11.3, and 8.75 gallons of gas. He started with the tank half full and ended with the tank half full. To find how many miles he got per gallon, we need to divide the total distance by the total number of gallons of gas consumed.

$$\frac{775}{10.5 + 11.3 + 8.75} \quad \longleftarrow \text{Total distance}$$
$$\longleftarrow \text{Total number of gallons consumed}$$

We can make this calculation by pressing these keys on a scientific calculator.

Evaluate: $\dfrac{775}{10.5 + 11.3 + 8.75}$

Keystrokes: 7 7 5 ÷ (1 0 . 5 + 1 1 . 3 +
8 . 7 5) = | 25.368249 |

To the nearest hundredth, he got 25.37 mpg.

VOCABULARY *Fill in the blanks to make a true statement.*

1. A ratio is a _____ of two numbers by their indicated _____.

2. The _____ of an item is the ratio of its cost to its quantity.

3. The ratios $\frac{2}{3}$ and $\frac{4}{6}$ are _____ ratios.

4. The ratio $\frac{500 \text{ miles}}{15 \text{ hours}}$ is called a _____.

CONCEPTS

5. Give three examples of ratios that you have encountered in the past week.

6. Suppose that a basketball player made 8 free throws out of 12 tries. The ratio of $\frac{8}{12}$ can be simplified as $\frac{2}{3}$. Interpret this result.

NOTATION

7. Write the ratio 2 is to 5 as a fraction.

8. Express the ratio $\frac{440 \text{ miles}}{8 \text{ hours}}$ in two different ways.

PRACTICE *In Exercises 9–24, express each phrase as a ratio in lowest terms.*

9. 5 to 7

10. 3 to 5

11. 17 to 34

12. 19 to 38

13. 22 to 33

14. 14 to 21

15. 7 to 24.5

16. 0.65 to 0.15

17. 4 ounces to 12 ounces

18. 3 inches to 15 inches

19. 12 minutes to 1 hour

20. 8 ounces to 1 pound

21. 3 days to 1 week

22. 4 inches to 2 yards

23. 18 months to 2 years

24. 8 feet to 4 yards

In Exercises 25–28, refer to the monthly budget shown in Illustration 1. Give each ratio in lowest terms.

25. Find the total amount of the budget.

26. Find the ratio of the amount budgeted for rent to the total budget.

27. Find the ratio of the amount budgeted for food to the total budget.

28. Find the ratio of the amount budgeted for the phone to the total budget.

Item	Amount
Rent	$800
Food	$600
Gas and electric	$180
Phone	$100
Entertainment	$120

ILLUSTRATION 1

In Exercises 29–32, refer to the list of tax deductions shown in Illustration 2. Give each ratio in lowest terms.

29. Find the total amount of the deductions.

30. Find the ratio of the real estate tax deduction to the total deductions.

31. Find the ratio of the charitable contributions to the total deductions.

32. Find the ratio of the mortgage interest deduction to the union dues deduction.

Item	Amount
Medical expenses	$ 995
Real estate taxes	$1,245
Charitable contributions	$1,680
Mortgage interest	$4,580
Union dues	$ 225

ILLUSTRATION 2

In Exercises 33–50, find each ratio and express it in lowest terms. Use a calculator when it is helpful.

33. FACULTY-TO-STUDENT RATIO   At a college, there are 125 faculty members and 2,000 students. Find the faculty-to-student ratio.

34. RATIO OF MEN TO WOMEN   In a state senate, there are 94 men and 24 women. Find the ratio of men to women.

35. UNIT COST OF GASOLINE   A driver pumped 17 gallons of gasoline into his tank at a cost of $21.59. Write a ratio of dollars to gallons, and give the unit cost of the gasoline.

36. UNIT COST OF GRASS SEED   A 50-pound bag of grass seed costs $222.50. Write a ratio of dollars to pounds, and give the unit cost of grass seed.

37. UNIT COST OF CRANBERRY JUICE   A 12-ounce can of cranberry juice sells for 84¢. Give the unit cost in cents per ounce.

38. UNIT COST OF BEANS   A 24-ounce package of green beans sells for $1.29. Give the unit cost in cents per ounce.

39. COMPARISON SHOPPING   A 6-ounce can of orange juice sells for 89¢, and an 8-ounce can sells for $1.19. Which is the better buy?

40. COMPARING SPEEDS   A car travels 345 miles in 6 hours, and a truck travels 376 miles in 6.2 hours. Which vehicle is going faster?

41. COMPARING READING SPEEDS   One seventh grader read a 54-page book in 40 minutes, and another read an 80-page book in 62 minutes. If the books were equally difficult, which student read faster?

42. COMPARISON SHOPPING   A 30-pound bag of fertilizer costs $12.25, and an 80-pound bag costs $30.25. Which is the better buy?

43. EMPTYING A TANK   A 11,880-gallon tank can be emptied in 27 minutes. Write a ratio of gallons to minutes, and give the rate of flow in gallons per minute.

44. RATE OF PAY   Ricardo worked for 27 hours to help insulate a hockey arena. For his work, he received $337.50. Write a ratio of dollars to hours, and find his hourly rate of pay.

45. SALES TAX   A sweater cost $36.75 after sales tax had been added. Find the tax rate as a percent if the sweater retailed for $35.

46. REAL ESTATE TAX   The real estate taxes on a summer home assessed at $75,000 were $1,500. Find the tax rate as a percent.

47. AUTO TRAVEL   A car's odometer reads 34,746 at the beginning of a trip. At the end five hours later, it reads 35,071. How far has the car traveled? What is the average rate of speed?

48. RATE OF SPEED   An airplane travels from Chicago to San Francisco, a distance of 1,883 miles, in 3.5 hours. Find the rate of speed of the plane.

49. COMPARING GAS MILEAGE   One car went 1,235 miles on 51.3 gallons of gasoline, and another went 1,456 miles on 55.78 gallons. Which car got the better gas mileage?

50. COMPARING ELECTRIC RATES   In one community, a bill for 575 kilowatt hours of electricity is $38.81. In a second community, a bill for 831 kwh is $58.10. In which community is electricity cheaper?

WRITING   Write a paragraph using your own words.

51. Some people think that the word *ratio* comes from the words *rational number.* Explain why this may be true.

52. In this section, ratios were used in three ways. Describe each of these ways.

REVIEW   In Exercises 53–58, do each operation.

53. $3.05 + 17.17 + 25.317$

54. $3.5\overline{)157.85}$

55. $13.2 + 25.07 \cdot 7.16$

56. $\dfrac{4}{3} - \dfrac{1}{4}$

57. $5 - 3\dfrac{1}{4}$

58. $\dfrac{5}{3} \div 1\dfrac{2}{3}$

Proportion

In this section, you will learn about

- **Proportions**
- **Means and extremes of a proportion**
- **Solving proportions**
- **Problem solving**

Gallons	Cost
2	$ 2.64
5	$ 6.60
8	$10.56
12	$15.84
20	$26.40

TABLE 6-1

INTRODUCTION. Ratios are often equal. For example, consider Table 6-1, in which we are given the costs of various numbers of gallons of gasoline.

If we find the ratios of the costs to the number of gallons purchased, we will see that they are equal.

$$\frac{\$2.64}{2} = \$1.32, \qquad \frac{\$6.60}{5} = \$1.32, \qquad \frac{\$10.56}{8} = \$1.32,$$

$$\frac{\$15.84}{12} = \$1.32, \qquad \text{and} \qquad \frac{\$26.40}{20} = \$1.32$$

Each ratio represents the cost of 1 gallon of gasoline. When two ratios such as $\frac{\$2.64}{2}$ and $\frac{\$6.60}{5}$ are equal, they form a *proportion*.

Proportions

Proportion

A **proportion** is a statement that two ratios are equal.

Some examples of proportions are

$$\frac{1}{2} = \frac{3}{6}, \qquad \frac{3}{7} = \frac{9}{21}, \qquad \frac{8}{1} = \frac{40}{5}, \qquad \text{and} \qquad \frac{5}{6} = \frac{10}{12}$$

- The proportion $\frac{1}{2} = \frac{3}{6}$ can be read as "1 is to 2 as 3 is to 6."
- The proportion $\frac{3}{7} = \frac{9}{21}$ can be read as "3 is to 7 as 9 is to 21."
- The proportion $\frac{8}{1} = \frac{40}{5}$ can be read as "8 is to 1 as 40 is to 5."
- The proportion $\frac{5}{6} = \frac{10}{12}$ can be read as "5 is to 6 as 10 is to 12."

The terms in a proportion are numbered as follows:

First term ⟶ $\dfrac{5}{6}$ = $\dfrac{10}{12}$ ⟵ Third term
Second term ⟶ ⟵ Fourth term

Means and extremes of a proportion

In any proportion, the first and fourth terms are called the **extremes.** The second and third terms are called the **means.**

In the proportion $\frac{1}{2} = \frac{3}{6}$, 1 and 6 are the **extremes,** and 2 and 3 are the **means.**

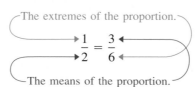

The extremes of the proportion.

$$\frac{1}{2} = \frac{3}{6}$$

The means of the proportion.

In this proportion, the product of the extremes is equal to the product of the means.

$$1 \cdot 6 = 6 \quad \text{and} \quad 2 \cdot 3 = 6$$

This example illustrates a fundamental property of proportions.

Fundamental property of proportions	In any proportion, the product of the extremes is equal to the product of the means.

As another example, we consider the proportion

$$\frac{5}{6} = \frac{10}{12}$$

- The product of the extremes is $5 \cdot 12 = 60$.
- The product of the means is $6 \cdot 10 = 60$.

Again we see that in a proportion, the product of the extremes (60) is equal to the product of the means (60).

To determine whether an equation is a proportion, we can check to see whether the product of the extremes is equal to the product of the means.

EXAMPLE 1 *Determining whether an equation is a proportion.*
Determine whether each equation is a proportion:

a. $\dfrac{3}{7} = \dfrac{9}{21}$ and **b.** $\dfrac{8}{3} = \dfrac{13}{5}$.

Solution
In each case, we check to see whether the product of the extremes is equal to the product of the means.

a. The product of the extremes is $3 \cdot 21 = 63$. The product of the means is $7 \cdot 9 = 63$. Since the products are equal, the equation is a proportion: $\frac{3}{7} = \frac{9}{21}$.

b. The product of the extremes is $8 \cdot 5 = 40$. The product of the means is $3 \cdot 13 = 39$. Since the products are not equal, the equation is not a proportion: $\frac{8}{3} \neq \frac{13}{5}$.

Self Check
Determine whether the equation is a proportion.

$$\frac{6}{13} = \frac{18}{39}$$

Answer: yes

When two pairs of numbers such as 2, 3 and 8, 12 form a proportion, we say that they are **proportional.** To show that 2, 3 and 8, 12 are proportional, we check to see whether the equation

$$\frac{2}{3} = \frac{8}{12}$$

is a proportion. To do so, we find the product of the extremes and the product of the means:

$$2 \cdot 12 = 24 \qquad 3 \cdot 8 = 24$$

Since the products are equal, the equation is a proportion, and the numbers are proportional.

EXAMPLE 2

EXAMPLE 2   *Determining whether numbers are proportional.*
Determine whether 3, 7 and 36, 91 are proportional.

Solution

We check to see whether $\frac{3}{7} = \frac{36}{91}$ is a proportion by finding two products:

$3 \cdot 91 = 273$ The product of the extremes.

$7 \cdot 36 = 252$ The product of the means.

Since the products are not equal, the numbers are not proportional.

Solving proportions

Suppose that we know three terms in the proportion

$$\frac{?}{5} = \frac{24}{20}$$

To find the missing term, we represent it by x, multiply the extremes and multiply the means, set them equal, and then find x as follows:

$$\frac{x}{5} = \frac{24}{20}$$

$20 \cdot x = 5 \cdot 24$ In a proportion, the product of the extremes is equal to the product of the means.

$20 \cdot x = 120$ Multiply: $5 \cdot 24 = 120$.

$\dfrac{20 \cdot x}{20} = \dfrac{120}{20}$ To undo the multiplication by 20, divide both sides by 20.

$x = 6$ Simplify: $\frac{20}{20} = 1$ and $\frac{120}{20} = 6$.

The first term is 6.

The second line of the solution of the previous problem contains the expression $20 \cdot x$. It is common to write expressions such as this in a more abbreviated form: $20x$, in this case.

EXAMPLE 3   *Solving proportions.*   Solve the proportion $\dfrac{12}{18} = \dfrac{3}{x}$.

Solution

$$\frac{12}{18} = \frac{3}{x}$$

$12 \cdot x = 18 \cdot 3$ In a proportion, the product of the extremes equals the product of the means.

$12x = 54$ Write $12 \cdot x$ as $12x$. Multiply: $18 \cdot 3 = 54$.

$\dfrac{12x}{12} = \dfrac{54}{12}$ To undo the multiplication by 12, divide both sides by 12.

$x = \dfrac{9}{2}$ Simplify: $\frac{12}{12} = 1$ and $\frac{54}{12} = \frac{9}{2}$.

Thus, $x = \frac{9}{2}$.

EXAMPLE 4 *Solving proportions.* Find the third term of the proportion $\dfrac{3.5}{7.2} = \dfrac{x}{15.84}$.

Self Check
Find the second term of the proportion $\dfrac{6.7}{x} = \dfrac{33.5}{38}$.

Solution

$$\frac{3.5}{7.2} = \frac{x}{15.84}$$

$3.5(15.84) = 7.2 \cdot x$ In a proportion, the product of the extremes equals the product of the means.

$55.44 = 7.2x$ Multiply: $3.5 \cdot 15.84 = 55.44$. Write $7.2 \cdot x$ as $7.2x$.

$\dfrac{55.44}{7.2} = \dfrac{7.2x}{7.2}$ To undo the multiplication by 7.2, divide both sides by 7.2.

$7.7 = x$ Simplify: $\frac{55.44}{7.2} = 7.7$ and $\frac{7.2}{7.2} = 1$.

The third term is 7.7.

Answer: 7.6

Accent on Technology **Solving equations with a calculator**

To solve the equation in Example 4 with a calculator, we can proceed as follows.

$$\frac{3.5}{7.2} = \frac{x}{15.84}$$

$$\frac{3.5(15.84)}{7.2} = x \qquad \text{Multiply both sides by 15.84.}$$

We can find x by pressing these keys on a scientific calculator.

Evaluate: $\dfrac{3.5(15.84)}{7.2}$

Keystrokes: $\boxed{3}\ \boxed{.}\ \boxed{5}\ \boxed{\times}\ \boxed{1}\ \boxed{5}\ \boxed{.}\ \boxed{8}\ \boxed{4}\ \boxed{\div}\ \boxed{7}\ \boxed{.}\ \boxed{2}\ \boxed{=}\ \boxed{\qquad\qquad 7.7}$

Thus, $x = 7.7$.

Problem solving

We can use proportions to solve many problems.

EXAMPLE 5 *Grocery shopping.* If 5 apples cost $1.15, how much will 16 apples cost?

Self Check
If 9 tickets to a concert cost $112.50, how much will 15 tickets cost?

Solution
Let c represent the cost of 16 apples. The ratios of the numbers of apples to their costs are equal.

5 apples is to $1.15 as 16 apples is to c.

5 apples $\longrightarrow$ $\dfrac{5}{1.15} = \dfrac{16}{c}$ $\longleftarrow$ 16 apples
Cost of 5 apples $\longrightarrow$ $\qquad\qquad$ $\longleftarrow$ Cost of 16 apples

$5 \cdot c = 1.15(16)$ In a proportion, the product of the extremes is equal to the product of the means.

$5c = 18.4$ Do the multiplication: $1.15(16) = 18.4$.

$$\frac{5c}{5} = \frac{18.4}{5}$$ To undo the multiplication by 5, divide both sides by 5.

$$c = 3.68$$ Simplify: $\frac{5}{5} = 1$ and $\frac{18.4}{5} = 3.68$.

Sixteen apples will cost $3.68.

EXAMPLE 6 *Mixing solutions.* A solution contains 2 quarts of antifreeze and 5 quarts of water. How many quarts of antifreeze must be mixed with 18 quarts of water to have the same concentration?

Solution

Let q represent the number of quarts of antifreeze to be mixed with the water. The ratios of the quarts of antifreeze to the quarts of water are equal.

2 qt antifreeze is to 5 qt water as q qt antifreeze is to 18 qt water.

2 qt antifreeze → $\quad \frac{2}{5} = \frac{q}{18} \quad$ ← q qt of antifreeze
5 qt of water → $\qquad\qquad\qquad$ ← 18 qt of water

$$2 \cdot 18 = 5q$$ In a proportion, the product of the extremes is equal to the product of the means.

$$36 = 5q$$ Do the multiplication: $2 \cdot 18 = 36$.

$$\frac{36}{5} = \frac{5q}{5}$$ To undo the multiplication by 5, divide both sides by 5.

$$\frac{36}{5} = q$$ Simplify: $\frac{5}{5} = 1$.

The mixture should contain $\dfrac{36}{5}$ or 7.2 quarts of antifreeze.

EXAMPLE 7 *Baking.* A recipe for rhubarb cake calls for $1\frac{1}{4}$ cups of sugar for every $2\frac{1}{2}$ cups of flour. How many cups of flour are needed if the baker intends to use 3 cups of sugar?

Solution

Let f represent the number of cups of flour to be mixed with the sugar. The ratios of the cups of sugar to the cups of flour are equal.

$1\frac{1}{4}$ cups sugar is to $2\frac{1}{2}$ cups flour as 3 cups sugar is to f cups flour.

$1\frac{1}{4}$ cups sugar → $\dfrac{1\frac{1}{4}}{2\frac{1}{2}} = \dfrac{3}{f}$ ← 3 cups sugar
$2\frac{1}{2}$ cups flour → $\qquad\qquad$ ← f cups flour

$$\frac{1.25}{2.5} = \frac{3}{f}$$ Change the fractions to decimals.

$$1.25f = 2.5 \cdot 3$$ In a proportion, the product of the extremes is equal to the product of the means.

$$1.25f = 7.5$$ Do the multiplication: $2.5 \cdot 3 = 7.5$.

$$\frac{1.25f}{1.25} = \frac{7.5}{1.25}$$ To undo the multiplication by 1.25, divide both sides by 1.25.

$$f = 6$$ Divide: $\frac{1.25}{1.25} = 1$ and $\frac{7.5}{1.25} = 6$.

The baker should use 6 cups of flour.

EXAMPLE 8 *Manufacturing.* In a manufacturing process, 15 parts out of 90 were found to be defective. How many defective parts will be expected in a run of 120 parts?

Self Check
How many defective parts will be expected in a run of 3,000 parts?

Solution

Let d represent the expected number of defective parts. In each run, the ratio of the defective parts to the total number of parts should be the same.

15 defective parts is to 90 as d defective parts is to 120.

15 defective parts $\longrightarrow$ $\dfrac{15}{90} = \dfrac{d}{120}$ $\longleftarrow$ d defective parts
90 parts $\longrightarrow$ $\longleftarrow$ 120 parts

$15 \cdot 120 = 90d$ In a proportion, the product of the extremes is equal to the product of the means.

$1{,}800 = 90d$ Do the multiplication: $15 \cdot 120 = 1{,}800$.

$\dfrac{1{,}800}{90} = \dfrac{90d}{90}$ To undo the multiplication by 90, divide both sides by 90.

$20 = d$ Divide: $\frac{1{,}800}{90} = 20$ and $\frac{90}{90} = 1$.

The expected number of defective parts is 20.

Answer: 500

STUDY SET   Section 6.2

VOCABULARY *Fill in the blanks to make a true statement.*

1. A _____ is a statement that two _____ are equal.

2. The first and fourth terms of a proportion are called the _____ of the proportion.

3. The second and third terms of a proportion are called the _____ of the proportion.

4. When two pairs of numbers form a proportion, we say that the numbers are _____.

CONCEPTS *Fill in the blanks to make a true statement.*

5. The equation $\frac{3}{7} = \frac{9}{21}$ will be a proportion if the product ____ is equal to the product ____.

6. Since $3 \cdot 10 = 5 \cdot 6$, _____ is a proportion.

NOTATION *Solve each proportion.*

7. $\dfrac{12}{18} = \dfrac{x}{24}$

$12 \cdot 24 =$ ____   The product of the extremes equals the product of the means.

____ $= 18x$ Multiply.

$\dfrac{288}{} = \dfrac{18x}{}$ Divide both sides by 18.

$16 = x$ Simplify.

8. $\dfrac{14}{x} = \dfrac{49}{17.5}$

_____ $= 49x$ The product of the extremes equals the product of the means.

____ $= 49x$ Multiply.

$\dfrac{245}{} = \dfrac{49x}{}$ Divide both sides by 49.

$5 = x$ Simplify.

PRACTICE *In Exercises 9–16, tell whether each statement is a proportion.*

9. $\dfrac{9}{7} = \dfrac{81}{70}$

10. $\dfrac{5}{2} = \dfrac{20}{8}$

11. $\dfrac{7}{3} = \dfrac{14}{6}$

12. $\dfrac{13}{19} = \dfrac{65}{95}$

13. $\dfrac{9}{19} = \dfrac{38}{80}$

14. $\dfrac{40}{29} = \dfrac{29}{22}$

15. $\dfrac{10.4}{3.6} = \dfrac{41.6}{14.4}$

16. $\dfrac{13.23}{3.45} = \dfrac{39.96}{11.35}$

In Exercises 17–32, solve each proportion.

17. $\dfrac{2}{3} = \dfrac{x}{6}$

18. $\dfrac{3}{6} = \dfrac{x}{8}$

19. $\dfrac{5}{10} = \dfrac{3}{x}$

20. $\dfrac{7}{14} = \dfrac{2}{x}$

21. $\dfrac{6}{x} = \dfrac{8}{4}$

22. $\dfrac{4}{x} = \dfrac{2}{8}$

23. $\dfrac{x}{3} = \dfrac{9}{3}$

24. $\dfrac{x}{2} = \dfrac{18}{6}$

25. $\dfrac{x}{2.5} = \dfrac{3.7}{9.25}$

26. $\dfrac{8.5}{x} = \dfrac{4.25}{1.7}$

27. $\dfrac{-8}{20} = \dfrac{x}{5}$

28. $\dfrac{9}{3} = \dfrac{-6}{x}$

29. $-\dfrac{4}{12} = \dfrac{-6}{x}$

30. $-\dfrac{5}{x} = -\dfrac{20}{44}$

31. $\dfrac{x}{5.2} = \dfrac{-4.65}{7.8}$

32. $-\dfrac{8.6}{2.4} = \dfrac{x}{6}$

APPLICATIONS ▦ *In Exercises 33–54, set up and solve a proportion. Use a calculator when it is helpful.*

33. GROCERY SHOPPING   If 3 pints of yogurt cost $1, how much will 51 pints cost?

34. SHOPPING FOR CLOTHES   If shirts are on sale at two for $25, how much will 5 shirts cost?

35. GARDENING   Garden seeds are on sale at 3 packets for 50¢. How much will 39 packets cost?

36. COOKING   A recipe for spaghetti sauce requires four 16-ounce bottles of ketchup to make two gallons of sauce. How many bottles of ketchup are needed to make 10 gallons of sauce?

37. BUSINESS PERFORMANCE   The bar graph in Illustration 1 shows the yearly costs incurred and the revenue received by a business. How do the ratios of costs to revenue for 1995 and 1996 compare?

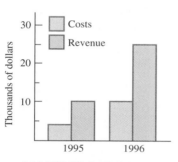

ILLUSTRATION 1

38. RAMP STEEPNESS   In Illustration 2, find the ratio of the rise to the run for each ramp. Is one ramp steeper than the other?

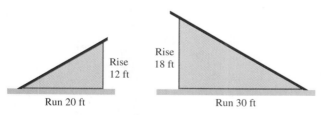

ILLUSTRATION 2

39. MIXING PERFUME   A perfume is to be mixed in the ratio of 3 drops of pure essence to 7 drops of alcohol.

How many drops of pure essence should be mixed with 56 drops of alcohol?

40. MAKING COLOGNE   A cologne can be made by mixing 2 drops of pure essence with 5 drops of distilled water. How much water should be used with 15 drops of pure essence?

41. MAKING COOKIES   A recipe for chocolate chip cookies calls for $1\frac{1}{4}$ cups of flour and 1 cup of sugar. The recipe will make $3\frac{1}{2}$ dozen cookies. How many cups of flour will be needed to make 12 dozen cookies?

42. MAKING BROWNIES   A recipe for brownies calls for 4 eggs and $1\frac{1}{2}$ cups of flour. If the recipe makes 15 brownies, how many cups of flour will be needed to make 130 brownies?

43. QUALITY CONTROL   In a manufacturing process, 95% of the parts made are to be within specifications. How many defective parts would be expected in a run of 940 pieces?

44. QUALITY CONTROL   Out of a sample of 500 men's shirts, 17 were rejected because of crooked collars. How many crooked collars would you expect to find in a run of 15,000 shirts?

45. GAS CONSUMPTION   If a car can travel 42 miles on 1 gallon of gas, how much gas is needed to travel 315 miles?

46. GAS CONSUMPTION   If a truck gets 12 miles per gallon of gas, how far can it go on 17 gallons?

47. PAYCHECK   Bill earns $412 for a 40-hour week. If he missed 10 hours of work last week, how much did he get paid?

48. MODEL RAILROADING   An HO-scale model railroad engine is 9 inches long. If HO scale is 87 feet to 1 foot, how long is a real engine?

49. MODEL RAILROADING   An N-scale model railroad caboose is 3.5 inches long. If N scale is 169 feet to 1 foot, how long is a real caboose?

50. MODEL HOUSES   A model house is built to a scale of 1 inch to 8 inches. If the model house is 36 inches wide, how wide is the real house?

51. STAFFING The school board has determined that there should be 3 teachers for every 50 students. Complete Illustration 3 by filling in the number of teachers for each school.

	Glenwood High	Goddard Junior High	Sellers Elementary
Enrollment	2,700	1,900	850
Teachers			

ILLUSTRATION 3

52. DRAFTING In a scale drawing, a 280-foot antenna tower is drawn 7 inches high. The building next to it is drawn 2 inches high. How tall is the actual building?

53. MIXING FUEL The instructions on a can of oil intended to be added to lawnmower gasoline read as shown in Illustration 4. Are these instructions correct? (*Hint:* There are 128 ounces in 1 gallon.)

Recommended	Gasoline	Oil
50 to 1	6 gal	16 oz

ILLUSTRATION 4

54. MIXING FUEL How much oil should be mixed with 28 gallons of gas? (See Exercise 53.)

WRITING *Write a paragraph using your own words.*

55. Explain the difference between a ratio and a proportion.

56. Explain how to tell whether the equation $\frac{3.2}{3.7} = \frac{5.44}{6.29}$ is a proportion.

REVIEW

57. Change $\frac{9}{10}$ to a percent.

58. Change $\frac{7}{8}$ to a percent.

59. Change $33\frac{1}{3}\%$ to a fraction.

60. Change 75% to a fraction.

61. Find 30% of 1,600.

62. Find $\frac{1}{2}\%$ of 520.

63. SHOPPING Maria bought a dress for 25% off the original price of $98. How much did the dress cost?

64. SHOPPING Bill purchased a shirt on sale for $17.50. Find the original cost of the shirt if it was marked down 30%.

65. SHOPPING Ricardo bought a pair of shoes for 30% off the regular price of $59, and a pair of boots for 40% off the regular price of $79. What did he have to pay?

66. SHOPPING Anita purchased a purse for 25% off the regular price of $39.95, and a pair of gloves for 50% off the regular price of $6.95. What did she have to pay?

6.3 *American Units of Measurement*

In this section, you will learn about
- **American units of length**
- **Converting units of length**
- **American units of weight**
- **American units of capacity**
- **Units of time**

INTRODUCTION. In this chapter, we will discuss the measurement of distance, weight, capacity, and time. Two common systems of measurement are the American (or English) system and the metric system. We will discuss American units in this section and metric units in the next section. Some common American units are *inches, feet, miles, ounces, pounds, tons, cups, pints, quarts,* and *gallons.* In the United States, American units are used for many nonscientific purposes:

- A newborn baby is 20 inches long.
- The distance from Rockford to Chicago is 80 miles.
- First-class postage for a letter that weighs less than 1 ounce is 32¢.
- A football player weighs 285 pounds.
- We buy milk in quarts or gallons.

American units of length

A ruler is one of the most common devices used for measuring distances. Figure 6-1 shows only a portion of a ruler; most rulers are 12 inches or 1 foot long. Since 12 inches = 1 foot, a ruler is divided into 12 equal distances of 1 inch. Each inch is divided into halves of an inch, quarters of an inch, eighths of an inch, and sixteenths of an inch. Several distances are measured using the ruler shown in Figure 6-1.

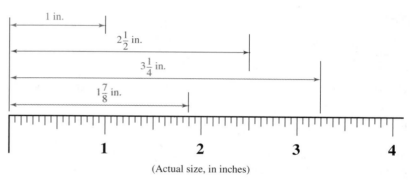

(Actual size, in inches)

FIGURE 6-1

EXAMPLE 1   ***Measuring the length of a nail.***   To the nearest $\frac{1}{4}$ inch, find the length of the nail in Figure 6-2.

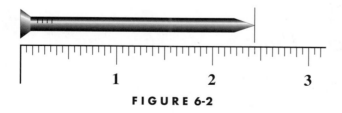

FIGURE 6-2

Solution

We place the zero end of the ruler by one end of the nail and note that the other end of the nail is closer to the $2\frac{1}{2}$-inch mark than to the $2\frac{1}{4}$-inch mark on the ruler. To the nearest quarter-inch, the nail is $2\frac{1}{2}$ inches long.

Self Check

To the nearest $\frac{1}{4}$ inch, find the width of the circle below.

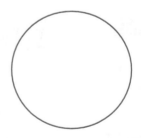

Answer: $1\frac{1}{4}$ in.   ■

EXAMPLE 2 *Measuring the length of a paper clip.* To the nearest $\frac{1}{8}$ inch, find the length of the paper clip in Figure 6-3.

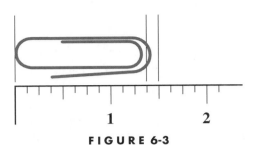

FIGURE 6-3

Self Check

To the nearest $\frac{1}{8}$ inch, find the length of the jumbo paper clip below.

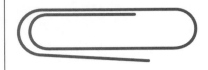

Answer: $1\frac{7}{8}$ in.

Solution

We place the zero end of the ruler by one end of the paper clip and note that the other end is closer to the $1\frac{3}{8}$-inch mark than to the $1\frac{1}{2}$-inch mark on the ruler. To the nearest eighth of an inch, the paper clip is $1\frac{3}{8}$ inches long.

Each point on a ruler, like each point on a number line, has a number associated with it: the distance between the point and 0. As Example 3 illustrates, the distance between any two points on a ruler (or on a number line) is the difference of the numbers associated with those points.

EXAMPLE 3 *Measuring the length of a ticket.* Find the length of the tear-off part of the concert ticket in Figure 6-4.

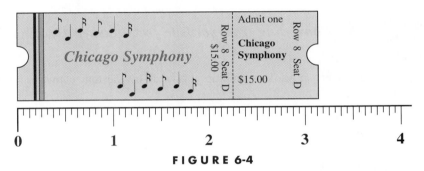

FIGURE 6-4

Solution After placing the zero end of the ruler by one end of the ticket, we note that the length of the entire ticket is $3\frac{1}{8}$ inches and that the length of the longer part is $2\frac{1}{4}$ inches. Since the length of the tear-off part of the ticket is the *difference* between these two lengths, we must subtract $2\frac{1}{4}$ from $3\frac{1}{8}$.

$$3\frac{1}{8} - 2\frac{1}{4} = \frac{25}{8} - \frac{9}{4} \qquad \text{Change the mixed numbers to improper fractions.}$$

$$= \frac{25}{8} - \frac{9 \cdot 2}{4 \cdot 2} \qquad \text{Write the fractions with a common denominator.}$$

$$= \frac{25}{8} - \frac{18}{8} \qquad \text{Do the multiplication in the numerator and in the denominator.}$$

$$= \frac{25 - 18}{8} \qquad \text{Subtract the fractions.}$$

$$= \frac{7}{8} \qquad \text{Simplify.}$$

The length of the tear-off portion of the ticket is $\frac{7}{8}$ inch.

Converting units of length

American units of length are related in the following ways.

American units of length	12 inches (in.) = 1 foot (ft) 36 inches = 1 yard (yd)
	3 feet = 1 yard 5,280 feet = 1 mile (mi)

To convert from one unit to another, we use *unit conversion factors.* To find the unit conversion factor between yards and feet, we begin with this fact:

$$3 \text{ ft} = 1 \text{ yd}$$

If we divide both sides of this equation by 1 yard, we get

$$\frac{3 \text{ ft}}{1 \text{ yd}} = \frac{1 \text{ yd}}{1 \text{ yd}}$$

$$\frac{3 \text{ ft}}{1 \text{ yd}} = 1 \qquad \text{A number divided by itself is 1: } \frac{1 \text{ yd}}{1 \text{ yd}} = 1.$$

The fraction $\frac{3 \text{ ft}}{1 \text{ yd}}$ is called a **unit conversion factor,** because its value is 1. It can be read as "3 feet per yard." Since this fraction is equal to 1, multiplying a length by this fraction does not change its measure; it only changes the units of measure.

EXAMPLE 4   *Converting from yards to feet.*   Convert 7 yards to feet.

Self Check

Convert 9 yards to feet.

Solution

Since there are 3 feet per yard, we multiply 7 yards by the unit conversion factor $\frac{3 \text{ ft}}{1 \text{ yd}}$ to get

$$7 \text{ yd} = 7 \text{ yd} \cdot \frac{3 \text{ ft}}{1 \text{ yd}} \qquad \frac{3 \text{ ft}}{1 \text{ yd}} = 1.$$

$$= 7 \overset{1}{\text{ yd}} \cdot \frac{3 \text{ ft}}{\underset{1}{1 \text{ yd}}} \qquad \text{The units of yards divide out.}$$

$$= 7 \cdot 3 \text{ ft}$$

$$= 21 \text{ ft} \qquad \text{Multiply: } 7 \cdot 3 = 21.$$

Seven yards is equal to 21 feet.

Answer: 27 ft

Notice that in Example 4, we eliminated the units of yards and introduced the units of feet by multiplying by the appropriate unit conversion factor. Since there are 12 inches per foot, the unit conversion factor between inches and feet is $\frac{12 \text{ in.}}{1 \text{ ft}}$.

EXAMPLE 5   *Converting from feet to inches.*   Convert $1\frac{3}{4}$ feet to inches.

Self Check

Convert 1.5 feet to inches.

Solution

Since there are 12 inches per foot, we multiply $1\frac{3}{4}$ feet by the unit conversion factor $\frac{12 \text{ in.}}{1 \text{ ft}}$ to get

$$1\frac{3}{4}\text{ ft} = \frac{7}{4}\text{ ft} \cdot \frac{12\text{ in.}}{1\text{ ft}} \qquad 1\frac{3}{4} = \frac{7}{4} \text{ and } \frac{12\text{ in.}}{1\text{ ft}} = 1.$$

$$= \frac{7}{4}\overset{1}{\cancel{\text{ft}}} \cdot \frac{12\text{ in.}}{1\underset{1}{\cancel{\text{ft}}}} \qquad \text{The units of feet divide out.}$$

$$= \frac{7 \cdot 12}{4 \cdot 1}\text{ in.} \qquad \text{Multiply the fractions.}$$

$$= 21\text{ in.} \qquad \text{Simplify: } \frac{7 \cdot 12}{4 \cdot 1} = \frac{7 \cdot 3 \cdot \overset{1}{\cancel{4}}}{\underset{1}{\cancel{4}} \cdot 1} = 7 \cdot 3 = 21.$$

Thus, $1\frac{3}{4}$ feet is equal to 21 inches.

Answer: 18 in.

Accent on Technology   *Finding the length of a football field in miles*

A football field is 100 yards long (excluding the end zones). To find this distance in miles, we set up the problem so that the correct units divide out and leave us with units of miles. Since there are 3 feet per yard and 5,280 feet per mile, we multiply 100 yards by $\frac{3\text{ ft}}{1\text{ yd}}$ and $\frac{1\text{ mi}}{5,280\text{ ft}}$.

$$100\text{ yd} = 100\text{ yd} \cdot \frac{3\text{ ft}}{1\text{ yd}} \cdot \frac{1\text{ mi}}{5,280\text{ ft}} \qquad \frac{3\text{ ft}}{1\text{ yd}} = 1 \text{ and } \frac{1\text{ mi}}{5,280\text{ ft}} = 1.$$

$$= 100\text{ yd} \cdot \frac{3\overset{1}{\cancel{\text{ft}}}}{1\text{ yd}} \cdot \frac{1\text{ mi}}{5,280\underset{1}{\cancel{\text{ft}}}} \qquad \text{Divide out the units of yards and feet.}$$

$$= \frac{100 \cdot 3}{5,280}\text{ mi} \qquad \text{Multiply the fractions.}$$

We can do this arithmetic using a scientific calculator by pressing these keys.

Evaluate: $\dfrac{100 \cdot 3}{5,280}$

Keystrokes: $\boxed{1}\ \boxed{0}\ \boxed{0}\ \boxed{\times}\ \boxed{3}\ \boxed{\div}\ \boxed{5}\ \boxed{2}\ \boxed{8}\ \boxed{0}\ \boxed{=}$ $\boxed{0.0568181}$

To the nearest hundredth, a football field is 0.06 mile long.

American units of weight

American units of weight are related in the following ways.

American units of weight	16 ounces (oz) = 1 pound (lb)
	2,000 pounds = 1 ton

To convert units of weight, we use the following unit conversion factors.

To convert from	use the unit conversion factor	To convert from	use the unit conversion factor
pounds to ounces	$\frac{16\text{ oz}}{1\text{ lb}}$	ounces to pounds	$\frac{1\text{ lb}}{16\text{ oz}}$
tons to pounds	$\frac{2,000\text{ lb}}{1\text{ ton}}$	pounds to tons	$\frac{1\text{ ton}}{2,000\text{ lb}}$

EXAMPLE 6 *Converting from ounces to pounds.* Convert 40 ounces to pounds.

Solution

Since there is 1 pound per 16 ounces, we multiply 40 ounces by the unit conversion factor $\frac{1\text{ lb}}{16\text{ oz}}$ to get

$$40 \text{ oz} = 40 \text{ oz} \cdot \frac{1\text{ lb}}{16\text{ oz}} \qquad \frac{1\text{ lb}}{16\text{ oz}} = 1.$$

$$= 40 \overset{1}{\text{ oz}} \cdot \frac{1\text{ lb}}{16 \underset{1}{\text{ oz}}} \qquad \text{The units of ounces divide out.}$$

$$= \frac{40}{16} \text{ lb}$$

$$= 2.5 \text{ lb} \qquad \text{Divide: } \frac{40}{16} = 2.5.$$

Forty ounces is equal to 2.5 pounds.

Self Check

Convert 60 ounces to pounds.

Answer: 3.75 lb

EXAMPLE 7 *Converting from pounds to ounces.* Convert 25 pounds to ounces.

Solution

Since there are 16 ounces per pound, we multiply 25 pounds by the unit conversion factor $\frac{16\text{ oz}}{1\text{ lb}}$ to get

$$25 \text{ lb} = 25 \text{ lb} \cdot \frac{16\text{ oz}}{1\text{ lb}} \qquad \frac{16\text{ oz}}{1\text{ lb}} = 1.$$

$$= 25 \overset{1}{\text{ lb}} \cdot \frac{16\text{ oz}}{1 \underset{1}{\text{ lb}}} \qquad \text{The units of pounds divide out.}$$

$$= 25 \cdot 16 \text{ oz}$$

$$= 400 \text{ oz} \qquad \text{Multiply: } 25 \cdot 16 = 400.$$

Twenty-five pounds is equal to 400 ounces.

Self Check

Convert 60 pounds to ounces.

Answer: 960 oz

Accent on Technology   *Finding the weight of a car in pounds*

A car weighs 1.7 tons. To find its weight in pounds, we set up the problem so that the correct units divide out and leave us with pounds. Since there are 2,000 pounds per ton, we multiply by $\frac{2,000\text{ lb}}{1\text{ ton}}$.

$$1.7 \text{ tons} = 1.7 \text{ tons} \cdot \frac{2,000\text{ lb}}{1\text{ ton}} \qquad \frac{2,000\text{ lb}}{1\text{ ton}} = 1.$$

$$= 1.7 \overset{1}{\text{ tons}} \cdot \frac{2,000\text{ lb}}{1 \underset{1}{\text{ ton}}} \qquad \text{Divide out the units of tons.}$$

$$= 1.7 \cdot 2,000 \text{ lb}$$

We can do this multiplication using a scientific calculator by pressing these keys.

Evaluate: $1.7 \cdot 2,000$

Keystrokes: $\boxed{1}\ \boxed{.}\ \boxed{7}\ \boxed{\times}\ \boxed{2}\ \boxed{0}\ \boxed{0}\ \boxed{0}\ \boxed{=}$ $\boxed{3400}$

The car weighs 3,400 pounds.

American units of capacity

American units of capacity are related as follows.

American units of capacity	
1 cup (c) = 8 fluid ounces (fl oz)	1 pint (pt) = 2 cups (c)
1 quart (qt) = 2 pints (pt)	1 gallon (gal) = 4 quarts (qt)

To convert units of capacity, we use the following unit conversion factors.

To convert from	use the unit conversion factor	To convert from	use the unit conversion factor
cups to ounces	$\frac{8 \text{ fl oz}}{1 \text{ c}}$	ounces to cups	$\frac{1 \text{ c}}{8 \text{ fl oz}}$
pints to cups	$\frac{2 \text{ c}}{1 \text{ pt}}$	cups to pints	$\frac{1 \text{ pt}}{2 \text{ c}}$
quarts to pints	$\frac{2 \text{ pt}}{1 \text{ qt}}$	pints to quarts	$\frac{1 \text{ qt}}{2 \text{ pt}}$
gallons to quarts	$\frac{4 \text{ qt}}{1 \text{ gal}}$	quarts to gallons	$\frac{1 \text{ gal}}{4 \text{ qt}}$

EXAMPLE 8   *Changing pints to ounces.*   If a recipe calls for 3 pints of milk, how many fluid ounces of milk should be used?

Solution

Since there are 2 cups per pint and 8 fluid ounces per cup, we multiply 3 pints by unit conversion factors of $\frac{2 \text{ c}}{1 \text{ pt}}$ and $\frac{8 \text{ fl oz}}{1 \text{ c}}$.

$$3 \text{ pt} = 3 \text{ pt} \cdot \frac{2 \text{ c}}{1 \text{ pt}} \cdot \frac{8 \text{ fl oz}}{1 \text{ c}} \qquad \frac{2 \text{ c}}{1 \text{ pt}} = 1 \text{ and } \frac{8 \text{ fl oz}}{1 \text{ c}} = 1.$$

$$= 3 \text{ pt} \cdot \frac{\overset{1}{\cancel{2 \text{ c}}}}{\underset{1}{1 \text{ pt}}} \cdot \frac{8 \text{ fl oz}}{\underset{1}{\cancel{1 \text{ c}}}} \qquad \text{Divide out the units of pints and cups.}$$

$$= 3 \cdot 2 \cdot 8 \text{ fl oz}$$

$$= 48 \text{ fl oz}$$

Since 3 pints is equal to 48 fluid ounces, 48 fluid ounces of milk should be used.

Self Check

How many pints are in 1 gallon?

Answer: 8 pt

Units of time

Units of time are related in the following ways.

Units of time	
1 minute (min) = 60 seconds (sec)	1 hour (hr) = 60 minutes
1 day = 24 hours	

EXAMPLE 9 *Converting from minutes to seconds.*
Convert 50 minutes to seconds.

Solution
Since there are 60 seconds per minute, we multiply 50 minutes by the unit conversion factor $\frac{60 \text{ sec}}{1 \text{ min}}$ to get

$$50 \text{ min} = 50 \text{ min} \cdot \frac{60 \text{ sec}}{1 \text{ min}} \qquad \frac{60 \text{ sec}}{1 \text{ min}} = 1.$$

$$= 50 \overset{1}{\cancel{\text{min}}} \cdot \frac{60 \text{ sec}}{\underset{1}{\cancel{\text{min}}}} \qquad \text{The units of minutes divide out.}$$

$$= 50 \cdot 60 \text{ sec}$$

$$= 3,000 \text{ sec} \qquad \text{Multiply: } 50 \cdot 60 = 3,000.$$

Fifty minutes is equal to 3,000 seconds.

Self Check
Convert 180 hours to days.

Answer: 7.5 days

STUDY SET Section 6.3

VOCABULARY *Fill in the blanks to make a true statement.*

1. Units of inches, feet, and miles are examples of _____ units of length.

2. A ruler is subdivided into 12 _____.

3. The value of any unit conversion factor is ___.

4. _____, pounds, and _____ are examples of American units of weight.

5. Some examples of American units of capacity are _____, _____, _____, and gallons.

6. Some units of time are _____, _____, and days.

CONCEPTS *Fill in the blanks to make a true statement.*

7. 12 in. = ___ ft

8. ___ ft = 1 yd

9. 1 mi = ____ ft

10. 1 yd = ___ in.

11. ___ ounces = 1 pound

12. ____ pounds = 1 ton

13. 1 cup = ___ fluid ounces

14. 1 pint = ___ cups

15. 2 pints = ___ quart(s)

16. 4 quarts = _____

17. 1 day = ___ hours

18. 2 hours = ___ minutes

NOTATION *Complete each solution.*

19. Convert 12 yards to inches.

$$12 \text{ yd} = 12 \text{ yd} \cdot \frac{\underline{\quad} \text{ in.}}{1 \text{ yd}}$$

$$= 12 \cdot \underline{\quad} \text{ in.}$$

$$= 432 \text{ in.} \qquad \text{Do the multiplication.}$$

20. Convert 1 ton to ounces.

$$1 \text{ ton} = 1 \text{ ton} \cdot \frac{\underline{\quad} \text{ lb}}{1 \text{ ton}} \cdot \frac{\underline{\quad} \text{ oz}}{1 \text{ lb}}$$

$$= 1 \cdot 2,000 \cdot 16 \text{ oz}$$

$$= \underline{\quad} \text{ oz} \qquad \text{Do the multiplication.}$$

21. Convert 12 pints to gallons.

$$12 \text{ pt} = 12 \text{ pt} \cdot \frac{1 \text{ qt}}{\text{pt}} \cdot \frac{1 \text{ gal}}{\text{qt}}$$

$$= \underline{\quad} \cdot \frac{1}{2} \cdot \frac{1}{4} \text{ gal}$$

$$= 1.5 \text{ gal} \qquad \textit{Do the multiplication.}$$

22. Convert 37,440 minutes to days.

$$37,440 \text{ min} = 37,440 \text{ min} \cdot \frac{1 \text{ hr}}{\text{min}} \cdot \frac{1 \text{ day}}{\text{hr}}$$

$$= \frac{\quad}{60 \cdot 24} \text{ days}$$

$$= 26 \text{ days} \qquad \textit{Do the arithmetic.}$$

PRACTICE *In Exercises 23–30, use a ruler with a scale in inches to measure each object to the nearest $\frac{1}{8}$ inch.*

23. The width of a dollar bill

24. The length of a dollar bill

25. The diameter of a quarter

26. The diameter of a penny

27. The length of a sheet of typing paper

28. The width of a sheet of typing paper

29. The height of a Coca-Cola can

30. The distance between the Q and P keys on a standard keyboard, measured center to center

In Exercises 31–66, do each conversion.

31. 4 feet to inches

32. 7 feet to inches

33. $3\frac{1}{2}$ feet to inches

34. $2\frac{2}{3}$ feet to inches

35. 24 inches to feet

36. 54 inches to feet

37. 8 yards to inches

38. 288 inches to yards

39. 90 inches to yards

40. 12 yards to inches

41. 56 inches to feet

42. 44 inches to feet

43. 5 yards to feet

44. 21 feet to yards

45. 7 feet to yards

46. $4\frac{2}{3}$ yards to feet

47. 15,840 feet to miles

48. 2 miles to feet

49. $\frac{1}{2}$ mile to feet

50. 1,320 feet to miles

51. 80 ounces to pounds

52. 8 pounds to ounces

53. 7,000 pounds to tons

54. 2.5 tons to ounces

55. 12.4 tons to pounds

56. 48,000 ounces to tons

57. 3 quarts to pints

58. 20 quarts to gallons

59. 16 pints to gallons

60. 3 gal to fluid ounces

61. 32 fluid ounces to pints

62. 2 quarts to fluid ounces

63. 240 minutes to hours

64. 2,400 seconds to hours

65. 7,200 minutes to days

66. 691,200 seconds to days

APPLICATIONS ▦ *In Exercises 67–80, use a calculator when it is helpful.*

67. HEIGHT OF THE GREAT PYRAMID The Great Pyramid in Egypt is about 450 feet high. Express this distance in yards.

68. LENGTH OF THE SPHINX The great Sphinx of Egypt is 240 feet long. Express this in inches.

69. HEIGHT OF THE HOOVER DAM The Hoover Dam in Colorado is 726 feet high. Express this distance in miles.

70. HEIGHT OF THE SEARS TOWER The Sears Tower in Chicago has 110 stories and is 1,454 feet tall. To the nearest hundredth, express this height in miles.

71. WEIGHT OF WATER One gallon of water weighs about 8 pounds. Express this weight in ounces.

72. WEIGHT OF A BABY A newborn baby weighed 136 ounces. Express this weight in pounds.

73. WEIGHT OF STONES The Great Pyramid contains 2,300,000 blocks with an average weight of 5,000 lb. Express the weight of the average block in tons.

74. WEIGHT OF FOOTBALL PLAYERS The offensive line of the Dallas Cowboys football team averaged 320 pounds. Express this weight in ounces.

75. BUYING PAINT A painter estimates that he will need 17 gallons of paint for a job. To take advantage of a closeout sale on quart cans, he decides to buy the paint in quarts. How many cans will he need to buy?

76. BAKING PIES A recipe for pumpkin pie calls for $\frac{3}{4}$ cup of milk. How many pies can be made with 1 gallon of milk?

77. SCHOOL LUNCHES Each student attending Eagle River elementary school receives one pint of milk for lunch each day. If 575 students attend the school, how many gallons of milk are used each day?

78. RADIATOR CAPACITY The radiator capacity of a piece of earth-moving equipment is 39 quarts. If the radiator is drained and new coolant put in, how many gallons of new coolant will be used?

79. TAKING A HIKE A college student walks 11 miles in 155 minutes. To the nearest tenth, how many hours does he walk?

80. SPACE FLIGHT Some astronauts were in space for 186 hours. How many days did the flight take?

WRITING *Write a paragraph using your own words.*

81. Explain how to find the unit conversion factor that will convert feet to inches.

82. Explain how to find the unit conversion factor that will convert pints to gallons.

REVIEW *In Exercises 83–90, round each number as indicated.*

83. 3,673.263; nearest hundred

84. 3,673.263; nearest ten

85. 3,673.263; nearest hundredth

86. 3,673.263; nearest tenth

87. 0.100602; nearest thousandth

88. 0.100602; nearest hundredth

89. 0.09999; nearest tenth

90. 0.09999; nearest one

6.4 *Metric Units of Measurement*

In this section, you will learn about

- **Metric units of length**
- **Converting units of length**
- **Metric units of mass**
- **Metric units of capacity**
- **Applications**

INTRODUCTION. In the United States, metric units are used for scientific purposes:

- A tumor is 7 centimeters wide.
- The speed of sound is approximately 330 meters per second.
- The mass of an object is 3 grams.
- A beaker holds 1 liter.

Almost every country except the United States uses the metric system exclusively.

Metric units of length

The basic metric unit of length is the **meter** (m). One meter is approximately 39 inches, slightly more than 1 yard. Larger and smaller units are designated by using prefixes in front of this basic unit.

deka means tens   *deci* means tenths
hecto means hundreds   *centi* means hundredths
kilo means thousands   *milli* means thousandths

Metric units of length	
1 dekameter (dam) = 10 meters. 1 dam is a little less than 11 yards.	**1 decimeter** (dm) = $\frac{1}{10}$ of 1 meter. 1 dm is about the length of your palm.
1 hectometer (hm) = 100 meters. 1 hm is about 1 football field long, plus one endzone.	**1 centimeter** (cm) = $\frac{1}{100}$ of 1 meter. 1 cm is about as wide as the nail of your little finger.
1 kilometer (km) = 1,000 meters. 1 km is about $\frac{3}{5}$ mile.	**1 millimeter** (mm) = $\frac{1}{1,000}$ of 1 meter. 1 mm is about the thickness of a dime.

Some rulers use metric units to measure distances. Figure 6-5 shows a portion of a metric ruler and some distances measured on it.

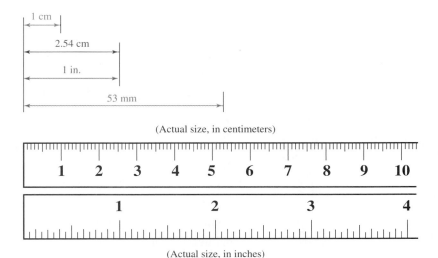

(Actual size, in centimeters)

(Actual size, in inches)

FIGURE 6-5

EXAMPLE 1   *Measuring the length of a nail.*   To the nearest centimeter, find the length of the nail in Figure 6-6.

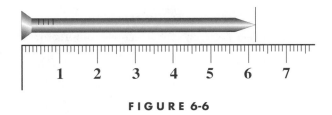

FIGURE 6-6

Solution
We place the zero end of the ruler by one end of the nail and note that the other end of the nail is closer to the 6-cm mark than to the 7-cm mark on the ruler. To the nearest centimeter, the nail is 6 cm long.

Self Check
To the nearest centimeter, find the width of the circle below.

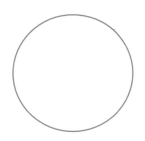

Answer: 3 cm

EXAMPLE 2   *Measuring the length of a paper clip.*   To the nearest millimeter, find the length of the paper clip in Figure 6-7.

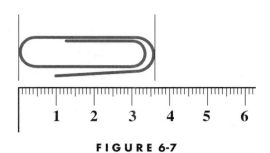

FIGURE 6-7

Solution

We place the zero end of the ruler by one end of the paper clip and note that the other end is closer to the 36-mm mark than to the 35-mm mark on the ruler. To the nearest millimeter, the paper clip is 36 mm long.

Self Check

To the nearest millimeter, find the length of the jumbo paper clip below.

Answer: 47 mm

Converting units of length

Metric units of length are related in the following ways.

Metric units of length			
1 kilometer (km) = 1,000 meters	or	1 meter = $\frac{1}{1,000}$ kilometer
1 hectometer (hm) = 100 meters	or	1 meter = $\frac{1}{100}$ hectometer
1 dekameter (dam) = 10 meters	or	1 meter = $\frac{1}{10}$ dekameter
1 decimeter (dm) = $\frac{1}{10}$ meter	or	1 meter = 10 decimeters
1 centimeter (cm) = $\frac{1}{100}$ meter	or	1 meter = 100 centimeters
1 millimeter (mm) = $\frac{1}{1,000}$ meter	or	1 meter = 1,000 millimeters

One advantage of the metric system is that multiplying or dividing by a unit conversion factor involves multiplying or dividing by a power of 10. The following table shows the unit conversion factors for the metric system.

To convert between	use the unit conversion factor		
kilometers and meters	$\frac{1,000 \text{ m}}{1 \text{ km}}$	or	$\frac{1 \text{ km}}{1,000 \text{ m}}$
hectometers and meters	$\frac{100 \text{ m}}{1 \text{ hm}}$	or	$\frac{1 \text{ hm}}{100 \text{ m}}$
dekameters and meters	$\frac{10 \text{ m}}{1 \text{ dam}}$	or	$\frac{1 \text{ dam}}{10 \text{ m}}$
decimeters and meters	$\frac{1 \text{ m}}{10 \text{ dm}}$	or	$\frac{10 \text{ dm}}{1 \text{ m}}$
centimeters and meters	$\frac{1 \text{ m}}{100 \text{ cm}}$	or	$\frac{100 \text{ cm}}{1 \text{ m}}$
millimeters and meters	$\frac{1 \text{ m}}{1,000 \text{ mm}}$	or	$\frac{1,000 \text{ mm}}{1 \text{ m}}$

TABLE 6-2

EXAMPLE 3	*Changing centimeters to meters.* Convert 350 centimeters to meters.

Solution

Since there is 1 meter per 100 centimeters, we multiply 350 centimeters by the unit conversion factor $\frac{1 \text{ m}}{100 \text{ cm}}$ to get

$$350 \text{ cm} = 350 \text{ cm} \cdot \frac{1 \text{ m}}{100 \text{ cm}} \qquad \frac{1 \text{ m}}{100 \text{ cm}} = 1.$$

$$= 350 \overset{1}{\cancel{\text{cm}}} \cdot \frac{1 \text{ m}}{100 \underset{1}{\cancel{\text{cm}}}} \qquad \text{The units of centimeters divide out.}$$

$$= \frac{350}{100} \text{ m}$$

$$= 3.5 \text{ m} \qquad \text{Divide by 100 by moving the decimal point 2 places to the left.}$$

Thus, 350 centimeters = 3.5 meters.

EXAMPLE 4	*Changing meters to millimeters.* Convert 2.4 meters to millimeters.

Solution

Since there are 1,000 millimeters per meter, we multiply 2.4 meters by the unit conversion factor $\frac{1,000 \text{ mm}}{1 \text{ m}}$ to get

$$2.4 \text{ m} = 2.4 \text{ m} \cdot \frac{1,000 \text{ mm}}{1 \text{ m}} \qquad \frac{1,000 \text{ mm}}{1 \text{ m}} = 1.$$

$$= 2.4 \overset{1}{\cancel{\text{m}}} \cdot \frac{1,000 \text{ mm}}{1 \underset{1}{\cancel{\text{m}}}} \qquad \text{The units of meters divide out.}$$

$$= 2.4 \cdot 1,000 \text{ mm}$$

$$= 2,400 \text{ mm} \qquad \text{Multiply by 1,000 by moving the decimal point 3 places to the right.}$$

Thus, 2.4 meters = 2,400 millimeters.

EXAMPLE 5	*Changing kilometers to dekameters.* Convert 3.2 kilometers to dekameters.

Solution

To convert to dekameters, we set up the problem so that the correct units divide out and leave us with units of dekameters. Since there are 1,000 meters per kilometer and 1 dekameter per 10 meters, we multiply 3.2 kilometers by $\frac{1,000 \text{ m}}{1 \text{ km}}$ and $\frac{1 \text{ dam}}{10 \text{ m}}$.

$$3.2 \text{ km} = 3.2 \overset{1}{\cancel{\text{km}}} \cdot \frac{1,000 \overset{1}{\cancel{\text{m}}}}{1 \underset{1}{\cancel{\text{km}}}} \cdot \frac{1 \text{ dam}}{10 \underset{1}{\cancel{\text{m}}}} \qquad \frac{1 \text{ km}}{1 \text{ km}} = 1 \text{ and } \frac{1 \text{ m}}{1 \text{ m}} = 1.$$

$$= \frac{3.2 \cdot 1,000}{10} \text{ dam} \qquad \text{Multiply.}$$

$$= 320 \text{ dam} \qquad \text{Simplify.}$$

Thus, 3.2 kilometers = 320 dekameters.

Metric units of mass

The **mass** of an object is a measure of the amount of material in the object. When an object is moved about in space, its mass does not change. One basic unit of mass in the metric system is the **gram.**

The **weight** of an object is determined by the earth's gravitational pull on the object. Since gravitational pull on an object decreases as the object gets farther from earth, the object weighs less as it gets farther from the earth's surface. This is why astronauts experience weightlessness in space. However, since most of us remain near the earth's surface, we will use the words *mass* and *weight* interchangeably. Thus, a mass of 30 grams is said to weigh 30 grams.

Metric units of weight are related in the following ways.

Metric units of weight	1 gram (g) = 1,000 milligrams (mg)	or	1 milligram = $\frac{1}{1,000}$ gram
	1 gram = 100 centigrams (cg)	or	1 centigram = $\frac{1}{100}$ gram
	1 kilogram (kg) = 1,000 g	or	1 gram = $\frac{1}{1,000}$ kilogram
	1 metric ton (t) = 1,000 kg	or	1 kilogram = $\frac{1}{1,000}$ metric ton

Table 6-3 shows the unit conversion factors for the metric system.

To convert between	use the unit conversion factor		
grams and milligrams	$\frac{1,000 \text{ mg}}{1 \text{ g}}$	or	$\frac{1 \text{ g}}{1,000 \text{ mg}}$
grams and centigrams	$\frac{100 \text{ cg}}{1 \text{ g}}$	or	$\frac{1 \text{ g}}{100 \text{ cg}}$
grams and kilograms	$\frac{1,000 \text{ g}}{1 \text{ kg}}$	or	$\frac{1 \text{ kg}}{1,000 \text{ g}}$
kilograms and metric tons	$\frac{1,000 \text{ kg}}{1 \text{ t}}$	or	$\frac{1 \text{ t}}{1,000 \text{ kg}}$

TABLE 6-3

EXAMPLE 6   *Changing centigrams to milligrams.*   Convert 2.5 centigrams to milligrams.

Solution

To convert to milligrams, we set up the problem so that the correct units divide out and leave us with units of milligrams. Since there is 1 gram per 100 centigrams and 1,000 mg to 1 gram, we multiply 2.5 centigrams by $\frac{1 \text{ g}}{100 \text{ cg}}$ and $\frac{1,000 \text{ mg}}{1 \text{ g}}$.

$$2.5 \text{ cg} = 2.5 \text{ cg} \cdot \frac{\overset{1}{\cancel{1 \text{ g}}}}{100 \text{ cg}} \cdot \frac{1,000 \text{ mg}}{\underset{1}{\cancel{1 \text{ g}}}} \qquad \tfrac{1 \text{ cg}}{1 \text{ cg}} = 1 \text{ and } \tfrac{1 \text{ g}}{1 \text{ g}} = 1.$$

$$= \frac{2.5 \cdot 1,000}{100} \text{ mg} \qquad \text{Multiply.}$$

$$= 25 \text{ mg} \qquad \text{Simplify.}$$

Thus, 2.5 centigrams = 25 milligrams.

Self Check

Convert 7.3 centigrams to milligrams.

Answer: 73 mg

EXAMPLE 7   *Changing grams to kilograms.*   Convert 3,200 grams to kilograms.

Solution

To convert to kilograms, we set up the problem so that the correct units divide out and leave us with units of kilograms. Since there is 1 kilogram per 1,000 grams, we multiply 3,200 grams by $\frac{1\,kg}{1,000\,g}$.

$$3,200\ g = 3,200\ \overset{1}{\cancel{g}} \cdot \frac{1\ kg}{\underset{1}{1,000\ \cancel{g}}} \qquad \frac{1\,g}{1\,g} = 1.$$

$$= \frac{3,200}{1,000}\ kg \qquad \text{Multiply the fractions.}$$

$$= 3.2\ kg \qquad \begin{array}{l}\text{Do the division by moving the decimal point in 3,200}\\ \text{3 places to the left.}\end{array}$$

Thus, 3,200 grams = 3.2 kilograms.

Self Check

Convert 5,000 grams to kilograms.

Answer: 5 kg   ■

Metric units of capacity

Some metric units of capacity are milliliters, liters, hectometers, and kiloliters. They are related as follows.

Metric units of capacity	
1 liter (L) = 1,000 cubic centimeters	1 deciliter (dL) = $\frac{1}{10}$ liter
1 hectoliter (hL) = 100 liters	1 centiliter (cL) = $\frac{1}{100}$ liter
1 kiloliter (kL) = 1,000 liters	1 milliliter (mL) = $\frac{1}{1,000}$ liter

Table 6-4 shows the corresponding unit conversion factors.

To convert between	use the unit conversion factor	
liters and hectoliters	$\frac{100\ L}{1\ hL}$	or $\frac{1\ hL}{100\ L}$
liters and kiloliters	$\frac{1,000\ L}{1\ kL}$	or $\frac{1\ kL}{1,000\ L}$
liters and deciliters	$\frac{1\ L}{10\ dL}$	or $\frac{10\ dL}{1\ L}$
liters and centiliters	$\frac{1\ L}{100\ cL}$	or $\frac{100\ cL}{1\ L}$
liters and milliliters	$\frac{1\ L}{1,000\ mL}$	or $\frac{1,000\ mL}{1\ L}$

TABLE 6-4

Applications

EXAMPLE 8   *Changing from liters to centiliters.*   How many centiliters are there in three 1-liter bottles of cola?

Solution

To convert to centiliters, we set up the problem so that the correct units divide out and leave us with units of centiliters. Since there are 100 centiliters per 1 liter, we multiply 3 liters by the unit conversion factor $\frac{100\ cL}{1\ L}$.

Self Check

How many milliliters are in two 1-liter bottles of cola?

$$3 \text{ L} = 3 \text{ L} \cdot \frac{100 \text{ cL}}{1 \text{ L}} \qquad \tfrac{100 \text{ cL}}{1 \text{ L}} = 1.$$

$$= 3 \overset{1}{\cancel{\text{L}}} \cdot \frac{100 \text{ cL}}{1 \underset{1}{\cancel{\text{L}}}} \qquad \text{The units of liters divide out.}$$

$$= 3 \cdot 100 \text{ cL}$$

$$= 300 \text{ cL}$$

Thus, there are 300 centiliters in 3 liters of cola.

Answer: 2,000 mL

EXAMPLE 9 *Changing from milligrams to centigrams.* A bottle of Verapamil, a drug taken for high blood pressure, contains 30 tablets. If each tablet contains 180 mg of active ingredient, how many centigrams of active ingredient are in the bottle?

Self Check

One brand name for Verapamil is Isoptin. If a bottle of Isoptin contains 90 tablets, each containing 200 mg of active ingredient, how many centigrams of active ingredient are in the bottle?

Solution

Since there are 30 tablets and each one contains 180 mg of active ingredient, there are

$$30 \cdot 180 \text{ mg} = 5{,}400 \text{ mg}$$

of active ingredient in the bottle.

To convert milligrams to centigrams, we multiply 5,400 milligrams by $\frac{1 \text{ g}}{1{,}000 \text{ mg}}$ and $\frac{100 \text{ cg}}{1 \text{ g}}$ to get

$$5{,}400 \text{ mg} = 5{,}400 \overset{1}{\cancel{\text{mg}}} \cdot \frac{1 \overset{1}{\cancel{\text{g}}}}{1{,}000 \underset{1}{\cancel{\text{mg}}}} \cdot \frac{100 \text{ cg}}{1 \underset{1}{\cancel{\text{g}}}}$$

$$= \frac{5400 \cdot 100}{1{,}000} \text{ cg}$$

$$= 540 \text{ cg}$$

There are 540 centigrams of active ingredient in the bottle.

Answer: 1,800 cg

STUDY SET   Section 6.4

VOCABULARY *Fill in the blanks to make a true statement.*

1. *Deka* means _____.

2. *Hecto* means _____.

3. *Kilo* means _____.

4. *Deci* means _____.

5. *Centi* means _____.

6. *Milli* means _____.

7. The amount of material in an object is called its _____.

8. The _____ of an object is determined by the earth's gravitational pull on the object.

CONCEPTS *Fill in the blanks to make a true statement.*

9. 1 dekameter = _____ meters

10. 1 decimeter = _____ meter

11. 1 centimeter = _____ meter

12. 1 kilometer = _____ meters

13. 1 millimeter = _____ meter

14. 1 hectometer = _____ meters

15. 1 gram = _____ milligrams

16. 100 centigrams = _____ gram

17. 1 kilogram = _____ grams

18. 1,000 kilograms = _____ metric ton

19. 1 liter = _____ cubic centimeters

20. 1 kiloliter = _____ liters

21. 1 centiliter = _____ liter

22. 1 milliliter = _____ liter

23. 100 liters = _____ hectoliter

24. 10 deciliters = _____ liter

NOTATION   *Complete each solution.*

25. Convert 20 centimeters to meters.

$$20 \text{ cm} = 20 \text{ cm} \cdot \frac{\text{m}}{100 \text{ cm}}$$

$$= \frac{20}{\quad} \text{ m}$$

$$= 0.2 \text{ m} \qquad \text{Do the division.}$$

26. Convert 300 centigrams to grams.

$$300 \text{ cg} = 300 \text{ cg} \cdot \frac{\text{g}}{100 \text{ cg}}$$

$$= \frac{\quad}{100} \text{ g}$$

$$= 3 \text{ g} \qquad \text{Do the division.}$$

27. Convert 2 kilometers to decimeters.

$$2 \text{ km} = 2 \text{ km} \cdot \frac{\text{m}}{1 \text{ km}} \cdot \frac{10 \text{ dm}}{\text{m}}$$

$$= 2 \cdot \quad \cdot 10 \text{ dm}$$

$$= 20,000 \text{ dm} \qquad \text{Do the multiplication.}$$

28. Convert 3 deciliters to milliliters.

$$3 \text{ dL} = 3 \text{ dL} \cdot \frac{1 \text{ L}}{\text{dL}} \cdot \frac{\text{mL}}{1 \text{ L}}$$

$$= \frac{\quad \cdot 1,000}{10} \text{ mL}$$

$$= 300 \text{ mL} \qquad \text{Do the arithmetic.}$$

PRACTICE   *In Exercises 29–32, use a metric ruler to measure each object to the nearest millimeter.*

29. The length of a dollar bill

30. The width of a dollar bill

31. The diameter of a nickel

32. The diameter of a quarter

In Exercises 33–36, use a metric ruler to measure each object to the nearest centimeter.

33. The length of a sheet of typing paper

34. The width of a sheet of typing paper

35. The length of one octave on a piano keyboard

36. The distance between the Q and P keys on a standard keyboard, measured center to center

In Exercises 37–84, convert each measurement between the given metric units.

37. 3 m = _____ cm

38. 5 m = _____ cm

39. 5.7 m = _____ cm

40. 7.36 km = _____ dam

41. 0.31 dm = _____ cm

42. 73.2 m = _____ dm

43. 76.8 hm = _____ mm

44. 165.7 km = _____ m

45. 4.72 cm = _____ dm

46. 0.593 cm = _____ dam

47. 453.2 cm = _____ m

48. 675.3 cm = _____ m

49. 0.325 dm = _____ m

50. 0.0034 mm = _____ m

51. 3.75 cm = _____ mm

52. 0.074 cm = _____ mm

53. 0.125 m = _____ mm

54. 134 m = _____ hm

55. 675 dam = _____ cm

56. 0.00777 cm = _____ dam

57. 638.3 m = _____ hm

58. 6.77 cm = _____ m

59. 6.3 mm = _____ cm

60. 6.77 mm = _____ cm

61. 695 dm = _____ m

62. 6,789 cm = _____ dm

63. 5,689 m = _____ km

64. 0.0579 km = _____ mm

65. 576.2 mm = _____ dm

66. 65.78 km = _____ dam

67. 6.45 dm = _____ km

68. 6.57 cm = _____ mm

69. 658.23 m = _____ km

70. 0.0068 hm = _____ km

71. 3 g = _____ mg

72. 5 g = _____ cg

73. 2 kg = _____ g

74. 1 t = _____ kg

75. 1 t = _____ g

76. 2 kg = _____ cg

77. 500 mg = _____ g

78. 500 mg = _____ cg

79. 3 kL = _____ L

80. 500 mL = _____ L

81. 500 cL = _____ mL

82. 400 L = _____ hL

83. 2 hL = _____ cL

84. 5 kL = _____ hL

In Exercises 85–94, solve each problem.

85. LENGTH OF A SWIMMING POOL   A swimming pool is 50 meters long. Give this length in decimeters.

86. HEIGHT OF THE HANCOCK CENTER   The John Hancock Center has 100 stories and is 343 meters high. Give this height in hectometers.

87. WEIGHT OF A BABY   A baby weighs 4 kilograms. Give this weight in centigrams.

88. WEIGHT OF A GOLD CHAIN   A gold chain weighs 1,500 milligrams. Give this weight in grams.

89. VOLUME OF ROOT BEER   How many deciliters of root beer are there in two 1-liter bottles?

90. VOLUME OF WINE   How many liters of wine are in a 750-mL bottle?

91. BUYING OLIVES   The net weight of a bottle of olives is 284 grams. Find the smallest number of bottles that must be purchased to have at least 1 kilogram of olives.

92. BUYING COFFEE   A can of Cafe Vienna has a net weight of 133 grams. Find the smallest number of cans that must be packaged to have at least 1 metric ton of coffee.

93. MEDICINE   A bottle of hydrochlorothiazine contains 60 tablets. If each tablet contains 50 milligrams of active ingredient, how many grams of active ingredient are in the bottle?

94. MEDICINE   A bottle of acetaminophen contains 30 tablets. If each tablet contains 30 milligrams of active ingredient, how many centigrams of active ingredient are in the bottle?

WRITING   *Write a paragraph using your own words.*

95. To change 3.452 kilometers to meters, we can just move the decimal point in 3.452 three places to the right to get 3,452 meters. Explain why.

96. To change 7,532 grams to kilograms, we can just move the decimal point in 7,532 three places to the left to get 7.532 kilograms. Explain why.

REVIEW

97. Find 7% of $342.72.

98. $17.23 is what percent of $344.60?

99. $32.16 is 8% of what amount?

100. Multiply: $3\frac{1}{7} \cdot 2\frac{1}{2}$.

101. Divide: $3\frac{1}{7} \div 2\frac{1}{2}$.

102. Simplify: $3\frac{1}{7} + 2\frac{1}{2} \cdot 3\frac{1}{3}$.

103. Solve the proportion: $\dfrac{16}{18} = \dfrac{24}{x}$.

104. Solve the proportion: $\dfrac{-3.3}{x} = \dfrac{5.5}{10.5}$.

6.5   *Converting between American and Metric Units*

In this section, you will learn about

- **Converting between American and metric units**
- **Comparing American and metric units of temperature**

Introduction.   It is often necessary to convert between American units and metric units. For example, we must convert units to answer the following questions:

- Is a building that is 40 feet high taller than a building that is 13 meters high?
- Does a 2-pound package of butter contain more butter than a 1-kilogram package?
- Is a quart of soda pop more or less than a liter of soda pop?

In this section, we will discuss how to make such decisions.

Converting between American and metric units

We can convert between American and metric units of length by using the accompanying table.

Equivalent Lengths	
American to metric	**Metric to American**
1 in. = 2.54 cm	1 cm = 0.3937 in.
1 ft = 0.3048 m	1 m = 3.2808 ft
1 yd = 0.9144 m	1 m = 1.0936 yd
1 mi = 1.6093 km	1 km = 0.6214 mi

EXAMPLE 1 *Changing yards to meters.* A football field is 100 yards long. Express this distance in meters.

Solution

Since there are 0.9144 meters in 1 yard, we can substitute 0.9144 meters for 1 yard.

$$100 \text{ yd} = 100(\textbf{1 yd})$$
$$= 100(\textbf{0.9144 m}) \quad \text{Substitute 0.9144 meters for 1 yard.}$$
$$= 91.44 \text{ m} \quad \text{Do the multiplication.}$$

To the nearest tenth, a football field is 91.4 meters long.

Self Check

A swimming pool is 50 meters long. Express this distance in feet.

Answer: 164.04 ft ■

EXAMPLE 2 *Comparing heights.* Is a building that is 40 feet high taller than a building that is 13 meters high?

Solution

To decide which building is taller, we can change 40 feet to meters.

$$40 \text{ ft} = 40(\textbf{1 ft})$$
$$= 40(\textbf{0.3048 m}) \quad \text{Substitute 0.3048 meters for 1 foot.}$$
$$= 12.192 \text{ m} \quad \text{Do the multiplication.}$$

Since a 40-foot building is only 12.192 meters high, the 13-meter building is taller. The answer is no.

Self Check

Which race is longer?

a. A 500-meter race

b. A 550-yard race

Answer: the 550-yard race ■

We can convert between American units of weight and metric units of mass by using the accompanying table.

Equivalent Weights and Masses	
American to metric	**Metric to American**
1 oz = 28.35 g	1 g = 0.035 oz
1 lb = 0.454 kg	1 kg = 2.2 lb

EXAMPLE 3 *Changing pounds to grams.* Change 50 pounds to grams.

Solution

$$50 \text{ lb} = 50(\textbf{1 lb})$$
$$= 50(\textbf{16 oz}) \quad \text{Substitute 16 ounces for 1 pound.}$$
$$= 50(16)(1 \text{ oz})$$
$$= 50(16)(\textbf{28.35 g}) \quad \text{Substitute 28.35 grams for 1 ounce.}$$
$$= 22,680 \text{ g} \quad \text{Do the multiplication.}$$

Thus, 50 pounds is equal to 22,680 grams.

Self Check

Change 20 kilograms to pounds.

Answer: 44 lb ■

EXAMPLE 4	*Comparing weights.* Does a 2-pound package of butter contain more butter than a 1-kilogram package?

Solution

To decide which contains more butter, we can change 2 pounds to kilograms.

$$2 \text{ lb} = 2(\mathbf{1 \text{ lb}})$$
$$= 2(\mathbf{0.454 \text{ kg}}) \qquad \text{Substitute 0.454 kilograms for 1 pound.}$$
$$= 0.908 \text{ kg} \qquad \text{Do the multiplication.}$$

Since a 2-pound package weighs only 0.908 kilogram, the 1-kilogram package contains more butter.

Self Check

Which person weighs more?

a. A person who weighs 165 pounds

b. A person who weighs 76 kilograms

Answer: the person who weighs 76 kg

We can convert between American and metric units of capacity by using the accompanying table.

Equivalent Capacities	
American to metric	**Metric to American**
1 fl oz = 0.030 L	1 L = 33.8 fl oz
1 pt = 0.473 L	1 L = 2.1 pt
1 qt = 0.946 L	1 L = 1.06 qt
1 gal = 3.785 L	1 L = 0.264 gal

EXAMPLE 5	*Changing from milliliters to quarts.* A bottle of SevenUp contains 750 milliliters. Convert this measure to quarts.

Solution

We convert milliliters to liters and then liters to quarts.

$$750 \text{ mL} = 750 \text{ mL} \cdot \frac{1 \text{ L}}{1,000 \text{ mL}} \qquad \tfrac{1 \text{ L}}{1,000 \text{ mL}} = 1.$$
$$= \frac{750}{1,000} \text{ L} \qquad \text{The units of mL divide out.}$$
$$= \frac{3}{4} \text{ L} \qquad \tfrac{750}{1,000} = \tfrac{3 \cdot 250}{4 \cdot 250} = \tfrac{3}{4}.$$
$$= \frac{3}{4}(1.06 \text{ qt}) \qquad \text{Substitute 1.06 quart for 1 liter.}$$
$$= 0.795 \text{ qt} \qquad \text{Do the multiplication.}$$

The bottle contains 0.795 quart.

Self Check

A student bought a 355-mL can of cola. How many ounces did he buy?

Answer: 12 oz

From the table of equivalent capacities, we see that 1 liter is equal to 1.06 quarts. Thus, one liter of soda pop is more than one quart of soda pop.

EXAMPLE 6	*Comparison shopping.* A two-quart bottle of soda pop is priced at $1.89, and a one-liter bottle is priced at 97¢. Which is the better buy?

Solution

We can convert 2 quarts to liters and find the price per liter of the two-quart bottle.

Self Check

Thirty-four fluid ounces of aged vinegar costs $3.49. A one-liter bottle of the same vinegar costs $3.17. Which is the better buy?

$$2 \text{ qt} = 2(\mathbf{1 \text{ qt}})$$
$$= 2(\mathbf{0.946 \text{ L}}) \quad \text{Substitute } 0.946 \text{ liter for } 1 \text{ quart.}$$
$$= 1.892 \text{ L} \quad \text{Do the multiplication.}$$

Thus, the two-quart bottle contains 1.892 liters. To find the price per liter of the two-quart bottle, we divide $\frac{\$1.89}{1.892}$ to get

$$\frac{\$1.89}{1.892} = \$0.998942917$$

Since the price per liter of the two-quart bottle is a little more than 99¢, the one-liter bottle priced at 97¢ is the better buy.

Answer: the one-liter bottle ■

Comparing American and metric units of temperature

In the American system, we measure temperature using **degrees Fahrenheit** (°F). In the metric system, we measure temperature using **degrees Celsius** (°C). These two scales are shown on the thermometers in Figure 6-8.

From Figure 6-8, we can see that

- 212° F = 100° C   Water boils.
- 32° F = 0° C   Water freezes.
- 5° F = −15° C   A cold winter day.
- 95° F = 35° C   A hot summer day.

As we have seen, there is a formula that enables us to convert from degrees Fahrenheit to degrees Celsius. There is also a formula to convert from degrees Celsius to degrees Fahrenheit.

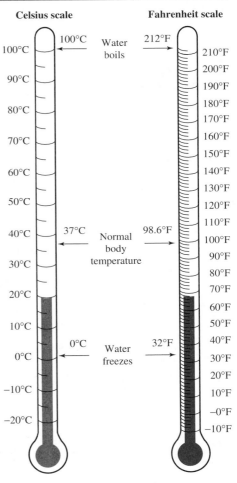

FIGURE 6-8

Conversion formulas for temperature	If F is the temperature in degrees Fahrenheit and C is the corresponding temperature in degrees Celsius, then $$C = \frac{5(F-32)}{9} \quad \text{and} \quad F = \frac{9}{5}C + 32$$

EXAMPLE 7 *Converting from degrees Fahrenheit to degrees Celsius.* Warm bath water is 90° F. Find the equivalent temperature in degrees Celsius.

Self Check

Hot coffee is 110° F. To the nearest tenth of a degree, express this temperature in degrees Celsius.

Solution

We substitute 90 for F in the formula $C = \dfrac{5F - 160}{9}$ and simplify.

$$C = \frac{5F - 160}{9}$$

$$= \frac{5(90) - 160}{9} \quad \text{Substitute 90 for } F.$$

$$= \frac{450 - 160}{9} \quad \text{Multiply: } 5(90) = 450.$$

$$= 32.222222 \quad \text{Do the arithmetic.}$$

To the nearest tenth of a degree, the equivalent temperature is 32.2° C.

Answer: 43.3° C

EXAMPLE 8 *Converting from degrees Celsius to degrees Fahrenheit.* A dishwasher manufacturer recommends that dishes be rinsed in hot water with a temperature of 60° C. Express this temperature in degrees Fahrenheit.

Self Check

To see whether a baby has a fever, her mother takes her temperature with a Celsius thermometer. If the reading is 38.8° C, does the baby have a fever? (*Hint:* Normal body temperature is 98.6° F.)

Solution

We substitute 60 for C in the formula $F = \dfrac{9}{5}C + 32$ and simplify.

$$F = \frac{9}{5}C + 32$$

$$= \frac{9}{5}(60) + 32 \quad \text{Substitute 60 for } C.$$

$$= \frac{540}{5} + 32 \quad \text{Multiply: } \tfrac{9}{5}(60) = \tfrac{540}{5}.$$

$$= 108 + 32 \quad \text{Do the division.}$$

$$= 140 \quad \text{Do the addition.}$$

The manufacturer recommends that dishes be rinsed in 140° F water.

Answer: yes

STUDY SET Section 6.5

VOCABULARY *Fill in the blanks to make a true statement.*

1. In the American system, temperatures are measured in degrees _____.

2. In the metric system, temperatures are measured in degrees _____.

3. The formula used for changing degrees Celsius to degrees Fahrenheit is _____ .

4. A formula used for changing degrees Fahrenheit to degrees Celsius is _____ .

NOTATION *Complete each solution.*

5. Change 4,500 feet to kilometers.

4,500 ft = 4,500 (_____ m)	Substitute 0.3048 m for 1 foot.
= _____ m	Multiply.
= 1.3716 km	Change 1,371.6 m to kilometers.

7. Change 8 liters to gallons.

8 L = 8(_____ gal)	Substitute 0.264 gallons for 1 liter.
= 2.112 gal	Multiply.

6. Change 3 kilograms to ounces.

3 kg = 3(___ lb)	Substitute 2.2 pounds for 1 kilogram.
= 3(2.2)(___ oz)	Substitute 16 ounces for 1 pound.
= 105.6 oz	Multiply.

8. Change 70° C to degrees Fahrenheit.

$$F = \frac{9}{5}C + 32$$

$= \dfrac{9}{5}(\_\_\_) + 32$	Substitute 70 for C.
$= \_\_\_ + 32$	Multiply $\frac{9}{5}$ and 70.
$= 158° \text{ F}$	Add.

PRACTICE *In Exercises 9–40, make each conversion. Since most conversions are approximate, answers will vary slightly depending on the method used.*

9. 3 ft = _____ cm **10.** 7.5 yd = _____ m **11.** 3.75 m = _____ in. **12.** 2.4 km = _____ mi

13. 12 km = _____ ft **14.** 3,212 cm = _____ ft **15.** 5,000 in. = _____ m **16.** 25 mi = _____ km

17. 37 oz = _____ kg **18.** 10 lb = _____ kg **19.** 25 lb = _____ g **20.** 7.5 oz = _____ g

21. 0.5 kg = _____ oz **22.** 35 g = _____ lb **23.** 17 g = _____ oz **24.** 100 kg = _____ lb

25. 3 fl oz = _____ L **26.** 2.5 pt = _____ L **27.** 7.2 L = _____ fl oz **28.** 5 L = _____ qt

29. 0.75 qt = _____ mL **30.** 3 pt = _____ mL **31.** 500 mL = _____ qt **32.** 2,000 mL = _____ gal

33. 50° F = _____ C **34.** 67.7° F = _____ C **35.** 50° C = _____ F **36.** 36.2° C = _____ F

37. −10° C = _____ F **38.** −22.5° C = _____ F **39.** −5° F = _____ C **40.** −10° F = _____ C

APPLICATIONS *Solve each problem.*

41. DISTANCE BETWEEN JERUSALEM AND BETHLEHEM The distance between Jerusalem and Bethlehem is 8 kilometers. To the nearest mile, give this distance in miles.

42. LENGTH OF THE DEAD SEA The Dead Sea is 80 kilometers long. To the nearest mile, give this distance in miles.

43. DISTANCE BETWEEN ROCKFORD AND CHICAGO The distance between Rockford, IL and Chicago is 90 miles. To the nearest tenth, give this distance in kilometers.

44. HEIGHT OF MOUNT WASHINGTON The highest peak of the White Mountains of New Hampshire is Mount Washington, at 6,288 feet. To the nearest tenth, give this height in kilometers.

45. SPEED OF A CHEETAH A cheetah can run 112 kilometers per hour. How fast is this in mph?

46. SPEED OF A LION A lion can run 50 mph. How fast is this in kilometers per hour?

47. TRACK AND FIELD A shot-put weighs 7.264 kilograms. Give this weight in pounds.

48. WEIGHTLIFTING A weightlifter pressed 125 kilograms. Give this weight in pounds.

49. COMPARING WEIGHTS One box weighs 30 pounds, and another weighs 13 kilograms. Which is heavier?

50. POSTAL REGULATIONS You can mail a package weighing up to 70 pounds via priority mail. Can you mail a package that weighs 32 kilograms by priority mail?

51. COMPARISON SHOPPING Which is the better buy, 3 quarts of root beer for $4.50 or 2 liters of root beer for $3.60?

52. COMPARISON SHOPPING Which is the better buy, 3 gallons of antifreeze for $10.35 or 12 liters of antifreeze for $10.50?

53. TAKING A SHOWER When you take a shower, which water temperature would you choose: 15° C, 28° C, or 50° C?

54. DRINKING WATER To get a cold drink of water, which temperature would you choose: −2° C, 10° C, or 25° C?

55. SNOWY WEATHER At which temperature might it snow: −5° C, 0° C, or 10° C?

56. RUNNING THE AIR CONDITIONER At which temperature would you be likely to run the air conditioner: 15° C, 20° C, or 30° C?

WRITING *Write a paragraph using your own words.*

57. Explain how to change kilometers to miles.

58. Explain how to change 50° C to degrees Fahrenheit.

REVIEW *In Exercises 59–66, do each operation.*

59. $\dfrac{3}{5} + \dfrac{4}{3}$

60. $\dfrac{3}{5} - \dfrac{4}{3}$

61. $\dfrac{3}{5} \cdot \dfrac{4}{3}$

62. $\dfrac{3}{5} \div \dfrac{4}{3}$

63. $3.25 + 4.8$

64. $3.25 - 4.8$

65. $3.25 \cdot 4.8$

66. $4.8\overline{)15.6}$

Proportions

A **proportion** is a statement that two ratios are equal.

Fill in the blanks as we set up a proportion to solve a problem.

1. TEACHER'S AIDES   For every 15 children on the playground, a child care center is required to have 2 teacher's aides supervising. How many teacher's aides will be needed to supervise 75 children?

Step 1: Let x = the number of _____.

The ratios of the number of children to the number of teacher's aides must be equal.

15 children are to ___ aides as ___ children are to ___ aides.

Expressing this as a proportion, we have $\dfrac{15}{\quad} = \dfrac{75}{\quad}$.

In the proportion $\frac{15}{2} = \frac{75}{x}$, 15 and x are the *extremes* and 2 and 75 are the *means*. After setting up the proportion, we solve it using the fact that the product of the extremes is equal to the product of the means.

Step 2: Solve the proportion: $\dfrac{15}{2} = \dfrac{75}{x}$.

___ $\cdot\, x = 2 \cdot$ ___	The product of the extremes equals the product of the means.
$15x =$ ___	Do the multiplication.
$\dfrac{15x}{\quad} = \dfrac{150}{\quad}$	Divide both sides by 15.
$x =$ ___	Simplify.

Ten teacher's aides are needed to supervise 75 children.

Set up and solve each problem using a proportion. A calculator will be helpful.

2. PARKING   A city code requires that companies provide 10 parking spaces for every 12 employees. How many spaces will be needed if a company employs 450 people?

3. MOTION PICTURES   Every 2 seconds, 3 feet of motion picture film pass through the projector. How many feet of film are there in a movie that runs for 120 minutes?

CHAPTER REVIEW

| SECTION 6.1 | *Ratio* |

CONCEPTS

A *ratio* is a comparison of two numbers by their indicated quotient.

A *unit cost* is the ratio of the cost of an item to its quantity.

A *rate* is a ratio used to compare quantities with different units.

REVIEW EXERCISES

1. Express each phrase as a ratio in lowest terms.
   a. The ratio of 4 inches to 12 inches
   b. The ratio of 8 ounces to 2 pounds

2. Mixed nuts come packaged in a 12-ounce can, which sells for $4.95, or an 8-ounce can, which sells for $3.25. Which is the better buy?

3. Find the hourly rate of pay for a student who earned $333.25 for working 43 hours.

4. An airplane flew from Chicago to San Francisco, a distance of 1,850 miles, in 3 hours and 45 minutes. Find the plane's average rate of speed.

| SECTION 6.2 | *Proportion* |

A *proportion* is a statement that two ratios are equal.

In any proportion, the product of the *extremes* is equal to the product of the *means*.

When two pairs of numbers form a proportion, we say that the numbers are proportional.

If three terms of a proportion are known, we can solve for the missing term.

5. Consider the proportion $\frac{5}{15} = \frac{25}{75}$.
   a. Which term is the fourth term?
   b. Which term is the second term?

6. Determine whether each of the following statements is a proportion.
   a. $\frac{15}{29} = \frac{105}{204}$
   b. $\frac{17}{7} = \frac{204}{84}$

7. Determine whether the numbers are proportional.
   a. 5, 9 and 20, 36
   b. 7, 13 and 29, 54

8. Solve each proportion.
   a. $\frac{12}{18} = \frac{3}{x}$
   b. $\frac{4}{x} = \frac{2}{8}$
   c. $\frac{x}{7.5} = \frac{6.9}{22.5}$
   d. $\frac{4.8}{6.6} = \frac{x}{9.9}$
   e. $\frac{x}{15} = \frac{-24}{45}$
   f. $\frac{-0.08}{x} = \frac{0.04}{0.06}$

9. If a truck can go 35 miles on 2 gallons of gas, how far can it go on 11 gallons?

10. In a manufacturing process, 12 parts out of 66 were found to be defective. How many defective parts will be expected in a run of 1,650 parts?

American Units of Measurement

Common American units of length are *inches, feet, yards,* and *miles.*

11. Use a ruler to measure the length of the computer mouse in Illustration 1 to the nearest quarter of an inch.

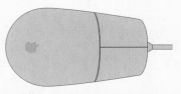

ILLUSTRATION 1

12 in. = 1 ft
3 ft = 1 yd
36 in. = 1 yd
5,280 ft = 1 mi

12. Make each conversion.
   a. 5 yards to feet
   c. 66 inches to feet
   e. 9,240 feet to miles
   b. 6 yards to inches
   d. 25.5 feet to inches
   f. 1 mile to yards

Common American units of weight are *ounces, pounds,* and *tons.*

13. Make each conversion.
   a. 32 ounces to pounds
   c. 3 tons to ounces
   b. 17.2 pounds to ounces
   d. 4,500 pounds to tons

16 oz = 1 lb
2,000 lb = 1 ton

Common American units of capacity are *fluid ounces, cups, pints, quarts,* and *gallons.*

14. Make each conversion.
   a. 5 pints to fluid ounces
   c. 17 quarts to cups
   e. 5 gallons to pints
   b. 8 cups to gallons
   d. 176 fluid ounces to quarts
   f. 3.5 gallons to cups

1 c = 8 fl oz
1 pt = 2 c
1 qt = 2 pt
1 gal = 4 qt

Units of time are *seconds, minutes, hours,* and *days.*

15. Make each conversion.
   a. 20 minutes to seconds
   c. 200 hours to days
   e. 4.5 days to hours
   b. 900 seconds to minutes
   d. 6 hours to minutes
   f. 1 day to seconds

1 min = 60 sec
1 hr = 60 min
1 day = 24 hr

16. The Sears Tower in Chicago is 1,454 feet high. Express this distance in yards.

17. A magnum is a two-quart bottle used for measuring wine. How many magnums will be needed to hold 50 gallons of wine?

Metric Units of Measurement

Common metric units of length are *millimeter, centimeter, decimeter, meter, dekameter, hectometer,* and *kilometer.*

18. Use a metric ruler to measure the length of the computer mouse in Illustration 2 to the nearest centimeter.

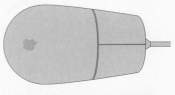

ILLUSTRATION 2

$1 \text{ mm} = \frac{1}{1,000} \text{ m}$

$1 \text{ cm} = \frac{1}{100} \text{ m}$

$1 \text{ dm} = \frac{1}{10} \text{ m}$

$1 \text{ dam} = 10 \text{ m}$

$1 \text{ hm} = 100 \text{ m}$

$1 \text{ km} = 1,000 \text{ m}$

Common metric units of mass are *milligrams, centigrams, grams, kilograms,* and *metric tons.*

$1 \text{ mg} = \frac{1}{1,000} \text{ g}$

$1 \text{ cg} = \frac{1}{100} \text{ g}$

$1 \text{ g} = \frac{1}{1,000} \text{ kg}$

$1 \text{ kg} = \frac{1}{1,000} \text{ t}$

Common metric units of capacity are *milliliters, centiliters, deciliters, liters, hectoliters,* and *kiloliters.*

$1 \text{ mL} = \frac{1}{1,000} \text{ L}$

$1 \text{ cL} = \frac{1}{100} \text{ L}$

$1 \text{ dL} = \frac{1}{10} \text{ L}$

$1 \text{ L} = 1,000 \text{ cc}$

$1 \text{ hL} = 100 \text{ L}$

$1 \text{ kL} = 1,000 \text{ L}$

19. Make each conversion.
  a. 475 centimeters to meters
  b. 8 meters to millimeters
  c. 3 dekameters to kilometers
  d. 2 hectometers to decimeters
  e. 5 kilometers to hectometers
  f. 2,500 meters to hectometers

20. Make each conversion.
  a. 7 centigrams to milligrams
  b. 800 centigrams to grams
  c. 5,425 grams to kilograms
  d. 5,425 grams to metric tons
  e. 7,500 milligrams to grams
  f. 5,000 centigrams to kilograms

21. A bottle of Extra Strength Tylenol® contains 100 caplets, 500 milligrams each. How many grams of Tylenol are in the bottle?

22. Make each conversion.
  a. 150 centiliters to liters
  b. 3,250 liters to kiloliters
  c. 1 hectoliter to deciliters
  d. 400 milliliters to centiliters
  e. 2 kiloliters to hectoliters
  f. 4 deciliters to milliliters

23. How many centiliters are there in a 2-liter bottle of root beer?

| SECTION 6.5 | *Converting between American and Metric Units* |

$1 \text{ in.} = 2.54 \text{ cm}$

$1 \text{ ft} = 0.3048 \text{ m}$

$1 \text{ yd} = 0.9144 \text{ m}$

$1 \text{ mi} = 1.6093 \text{ km}$

$1 \text{ cm} = 0.3937 \text{ in.}$

$1 \text{ m} = 3.2808 \text{ ft}$

$1 \text{ m} = 1.0936 \text{ yd}$

$1 \text{ km} = 0.6214 \text{ mi}$

24. A swimming pool is 25 meters long. Express this distance in feet.

25. The World Trade Center is 419 meters high, and the Empire State Building is 1,250 feet high. Which building is taller?

26. The distance between Rockford and Eagle River is 300 miles. Express this distance in kilometers.

27. Michael Jordan is 6 feet, 6 inches tall. Express his height in centimeters.

$1 \text{ oz} = 28.35 \text{ g}$

$1 \text{ lb} = 0.454 \text{ kg}$

$1 \text{ g} = 0.035 \text{ oz}$

$1 \text{ kg} = 2.2 \text{ lb}$

28. Make each conversion.
  a. 30 ounces to grams
  b. 15 kilograms to pounds
  c. 25 pounds to grams. (Round to the nearest thousand.)
  d. 2,000 pounds to kilograms. (Round to the nearest ten.)

29. A man weighs 84 kilograms. Find his weight in pounds.

1 fl oz = 0.030 L
1 pt = 0.473 L
1 qt = 0.946 L
1 gal = 3.785 L
1 L = 33.8 fl oz
1 L = 2.1 pt
1 L = 1.06 qt
1 L = 0.264 gal

Two units used to measure
temperature are degrees
Fahrenheit and degrees Celsius.

$$C = \frac{5(F - 32)}{9}$$

$$F = \frac{9}{5}C + 32$$

30. LaCroix® bottled water can be purchased in bottles containing 17 fluid ounces. Mountain Valley® water can be purchased in half-liter bottles. Which bottle contains more water?

31. One gallon of bleach costs $1.39. A 5-liter economy bottle costs $1.80. Which is the better buy?

32. A 2-quart bottle of root beer costs $1.18, and a 1.5-liter bottle costs 95¢. Which is the better buy?

33. Change 77° F to degrees Celsius.

34. Change 40° C to degrees Fahrenheit.

35. Which temperature of water would you like to swim in: 10° C, 30° C, 50° C, or 70° C?

36. At which temperature would you be likely to run the furnace: 10° C, 30° C, 50° C, or 70° C?

In Problems 1–2, write each phrase as a ratio in lowest terms.

1. The ratio of 6 feet to 8 feet

2. The ratio of 8 ounces to 3 pounds

3. Two pounds of coffee can be purchased for $3.38, and a 5-pound can can be purchased for $8.50. Which is the better buy?

4. A household used 675 kilowatt hours of electricity during a 30-day month. Find the rate of electric consumption in kilowatt hours per day.

In Problems 5–6, tell whether each expression is a proportion.

5. $\dfrac{25}{33} = \dfrac{350}{460}$

6. $\dfrac{22}{35} = \dfrac{176}{280}$

7. Write the third term of the proportion $\dfrac{3}{5} = \dfrac{6}{10}$.

8. Are the numbers 7, 15 and 245, 525 proportional?

In Problems 9–12, solve each proportion.

9. $\dfrac{x}{3} = \dfrac{35}{7}$

10. $\dfrac{15.3}{x} = \dfrac{3}{12.4}$

11. $\dfrac{0.07}{-0.5} = \dfrac{x}{1.5}$

12. $\dfrac{25}{0.1} = \dfrac{50}{x}$

13. If 13 ounces of coffee costs $2.79, how much would you expect to pay for 16 ounces?

14. A recipe calls for $\frac{2}{3}$ cup of sugar and 2 cups of flour. How much sugar should be used with 5 cups of flour?

15. Convert 180 inches to feet.

16. Convert 18 feet to yards.

17. Convert 10 pounds to ounces.

18. A car weighs 1.6 tons. Find its weight in pounds.

19. How many fluid ounces are in a 1-gallon carton of milk?

20. How many minutes are there in a seven-day week?

21. How many centimeters are in 5 meters?

22. How many millimeters are in 0.5 meter?

23. How many meters are in 10 kilometers?

24. Convert 8,000 centigrams to kilograms. (*Hint:* 100 cg = 1 g.)

25. Convert 70 liters to hectoliters.

26. A bottle contains 50 tablets, each one of which contains 150 mg of medicine. How many grams of medicine does the bottle contain?

27. Which is the longer distance: a 100-yard race or an 80-meter race?

28. Which person is heavier: Jim, who weighs 160 pounds, or Ricardo, who weighs 72 kilograms?

29. A two-quart bottle of soda pop costs $1.73, and one-liter bottle costs 89¢. Which is the better buy? (*Hint:* 1 quart = 0.946 liter.)

30. Change 40° C to degrees Fahrenheit. (*Hint:* $F = \frac{9}{5}C + 32$.)

31. A contractor told his clients that as they increased the number of square feet in their home remodeling project, the cost of the project would increase proportionally. Explain this statement.

32. Explain how larger and smaller units of length in the metric system are designated by using prefixes in front of the basic unit of meter.

In Exercises 1–2, consider the number 6,245,867.

1. Round to the nearest thousand.

2. Round to the nearest million.

In Exercises 3–4, find the perimeter of each figure.

3. A rectangle that is 8 meters long and 3 meters wide.

4. A square with sides that are 13 inches long.

In Exercises 5–6, find the area of each figure.

5. A rectangle that is 8 meters long and 3 meters wide.

6. A square with sides that are 13 inches long.

In Exercises 7–8, find the prime factorization of each number.

7. 120

8. 600

In Exercises 9–12, do the operations and simplify.

9. $\dfrac{5}{10} \cdot \dfrac{2}{15}$

10. $\dfrac{6}{9} \div \dfrac{15}{7}$

11. $\dfrac{1}{2} + \dfrac{3}{5}$

12. $\dfrac{5}{8} - \dfrac{2}{5}$

In Exercises 13–14, consider the decimal 57.544.

13. Round to the nearest tenth.

14. Round to the nearest hundredth.

In Exercises 15–18, do the operations.

15. $29.703 + 321.35$

16. $287.23 - 179.97$

17. $7.89 \cdot 0.27$

18. $3.8\overline{)17.746}$

In Exercises 19–20, change each fraction into a decimal fraction.

19. $\dfrac{3}{5}$

20. $\dfrac{35}{99}$

In Exercises 21–22, write each number as a decimal rounded to the nearest tenth.

21. $5\dfrac{5}{8}$

22. $-4\dfrac{7}{9}$

In Exercises 23–26, simplify each expression.

23. $\sqrt{121}$

24. $\sqrt{\dfrac{81}{4}}$

25. $\sqrt{0.25}$

26. $3\sqrt{144} - \sqrt{49}$

In Exercises 27–30, solve each problem.

27. Find 87% of 900.

28. 64.8 is what percent of 540?

29. 32.4 is 45% of what number?

30. Find 16% of 600.

31. How much interest will be earned if $25,000 is invested at an annual rate of 7% for 8 years? (Assume simple interest.)

32. How much interest will be earned in Exercise 31 if interest is compounded quarterly?

In Exercises 33–34, solve each proportion.

33. $\dfrac{x}{5} = \dfrac{14}{35}$

34. $\dfrac{9}{11} = \dfrac{x}{66}$

35. A student earns $131.25 for working 25 hours. Find her hourly rate of pay.

36. 3 pounds of coffee sells for $6.75. Find the unit cost.

In Exercises 37–44, do each conversion.

37. 5 feet to inches

38. 80 inches to feet

39. 26 quarts to gallons

40. 100 ounces to pounds

41. 4 meters to centimeters

42. 650 centimeters to meters

43. 4 kilograms to grams

44. 8.5 millimeters to centimeters

Descriptive Statistics

7

NEWSPAPERS AND MAGAZINES OFTEN PRESENT INFORMATION IN THE FORM OF GRAPHS AND TABLES. IN THIS CHAPTER, WE WILL SHOW HOW INFORMATION CAN BE OBTAINED BY READING MANY TYPES OF GRAPHS. WE WILL THEN DISCUSS THREE MEASURES OF CENTRAL TENDENCY: THE MEAN, THE MEDIAN, AND THE MODE.

7.1 Reading Graphs and Tables

In this section, you will learn about

- **Reading data from tables**
- **Reading bar graphs**
- **Reading pictographs**
- **Reading pie graphs**
- **Reading line graphs**
- **Reading histograms and frequency polygons**

INTRODUCTION. It is often said that a picture is worth a thousand words. In this section, we will show how to read information from mathematical pictures called *graphs*.

Reading data from tables

The **table** in Figure 7-1(a), the **bar graph** in Figure 7-1(b), and the **pie graph** in Figure 7-1(c) all show the results of a survey of viewers' opinions. In the bar graph, the length of each bar represents the percent of responses in each category. In the pie

Ratings of Prime-Time News Coverage

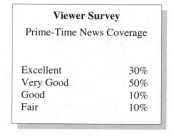

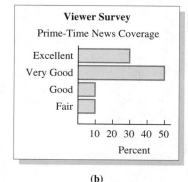

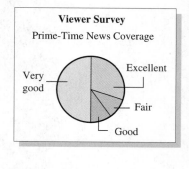

| (a) | (b) | (c) |

FIGURE 7-1

graph, the size of each region represents the percent of response. The two graphs tell the story more quickly and more clearly than the table of numbers.

It is easy to see from either graph that the largest percent of those surveyed rated the programming *very good,* and that the responses *good* and *fair* were tied for last. The same information is available in the table of Figure 7-1(a), but it is not as easy to see at a glance.

Data are often presented in tables, with information organized in rows and columns. To read a table, we must find the intersection of the row and the column that contains the needed information.

Postal rates (in 1998) for priority mail appear in Figure 7-2. To find the cost of mailing an $8\frac{1}{2}$-pound package by priority mail to postal zone 4, we find the *row* of the postage table for a package that does not exceed 9 pounds. We then find the *column* for zone 4. At the intersection of this row and this column, we read the number 9.35. This means that it will cost $9.35 to mail the package.

Priority Mail

Weight not over (lb)	Zone 1, 2, & 3	4	5	6	7	8
1	$3.00	3.00	3.00	3.00	3.00	3.00
2	3.00	3.00	3.00	3.00	3.00	3.00
3	4.00	4.00	4.00	4.00	4.00	4.00
4	5.00	5.00	5.00	5.00	5.00	5.00
5	6.00	6.00	6.00	6.00	6.00	6.00
6	6.35	6.90	7.10	7.20	7.80	8.00
7	6.65	7.80	8.10	8.40	9.20	9.80
8	6.95	8.70	9.05	9.50	10.40	11.60
9	7.40	9.35	10.00	10.60	11.30	13.00
10	7.85	10.00	10.75	11.40	12.15	14.05
11	8.25	10.65	11.45	12.20	13.00	15.10
12	8.70	11.30	12.20	13.00	13.90	16.50

FIGURE 7-2

Reading bar graphs

EXAMPLE 1 *Reading bar graphs.* The bar graph in Figure 7-3 shows the total income generated by three sectors of the economy in each of three years. The height of each bar, representing income in billions of dollars, is measured on the scale on the vertical *axis.* The years appear on the horizontal axis. Read the graph to answer the following questions.

a. What income was generated by retail sales in 1980?

b. Which sector of the economy consistently generated the most income?

c. By what amount did income from the wholesale sector increase between the years 1970 and 1990?

Solution

a. The second group of bars indicates income in 1980, and the middle bar of that group shows sales in the retail sector. The height of that bar is approximately 75,

Self Check

What income was generated by the service sector in 1990?

which represents $75 billion. The retail income generated in 1980 was about $75 billion.

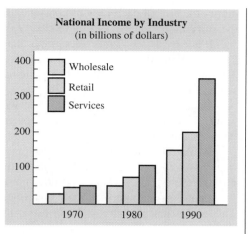

FIGURE 7-3

b. In each group, the rightmost bar is the tallest. That bar, according to the key, represents income from the service sector of the economy. Therefore, services consistently generated the most income.

c. According to the key, the leftmost bar in each group shows income from the wholesale sector. That sector generated about $30 billion in 1970 and $150 billion in 1990. The amount of increase in income is the difference of these two quantities:

$150 billion − $30 billion = $120 billion

Wholesale income increased by $120 billion between 1970 and 1990.

Answer: $350 billion

EXAMPLE 2   *Reading bar graphs.* The bar graph in Figure 7-4 shows the number of cars of various models purchased in Dale County for two consecutive years.

Self Check
Which model has shown the greatest decrease in sales?

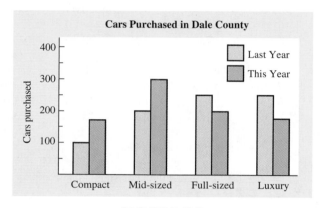

FIGURE 7-4

a. Which models have shown a decrease in sales?

b. Which model showed the greatest increase in sales?

Solution

a. In each pair, the bar on the left gives last year's sales. The bar on the right represents this year's sales. Only for full-sized and luxury cars is the left bar taller than the right bar. This means that full-sized and luxury cars have decreased in sales.

b. Sales of compact and mid-sized cars have increased over last year, because for these models the right bar is taller than the left bar. The difference in the heights of the bars represents the amount of increase. That increase is greater for mid-sized cars. Of all models, mid-sized cars have shown the greatest increase in sales.

Answer: luxury cars

Reading pictographs

A **pictograph** is like a bar graph, but the bars are composed of pictures, where each picture represents a quantity. In Figure 7-5, each picture represents 50 pizzas ordered during exam week. The top bar contains three complete pizzas and one partial pizza, indicating that the men in the men's residence hall ordered 3 · 50, or 150 pizzas, plus about $\frac{1}{4}$ of 50, or 13 pizzas. This totals 163 pizzas. The women in the women's residence hall ordered $4\frac{1}{2}$ · 50, or 225 pizzas.

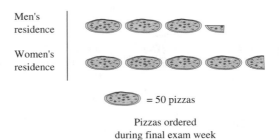

Men's residence

Women's residence

= 50 pizzas

Pizzas ordered during final exam week

FIGURE 7-5

Reading pie graphs

EXAMPLE 3   *Reading pie graphs.*   The pie graph in Figure 7-6 gives information about world gold production. The entire circle represents the world's total production, and the sizes of the segments of the circle represent the parts of that total contributed by various nations and regions. Use the graph to answer the following questions.

a. What percent of the total was the combined production of the United States and Canada?

b. What percent of the total production came from sources other than those listed?

c. If the world's total production of gold was 56.3 million ounces during the year of the survey, how many ounces did Australia produce?

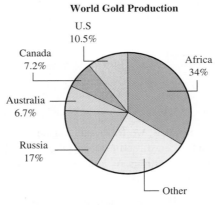

World Gold Production

U.S 10.5%

Canada 7.2%

Australia 6.7%

Russia 17%

Africa 34%

Other

FIGURE 7-6

Self Check
To the nearest tenth of a million, how many ounces of gold did Russia produce?

Solution

a. According to the graph, the United States produced 10.5% and Canada produced 7.2% of the total. Together, they produced (10.5 + 7.2)%, or 17.7% of the total.

b. To find the percent of gold produced by countries that are not listed, we add the contributions of all the listed sources and subtract that total from 100%.

$$100\% - (34\% + 10.5\% + 7.2\% + 6.7\% + 17\%) = 100\% - 75.4\%$$
$$= 24.6\%$$

The countries that are not listed produced 24.6% of the world's total production of gold.

c. From the graph, we see that Australia produced 6.7% of the world's gold. Since the world total was 56.3 million ounces, Australia's share (in millions of ounces) was

$$6.7\% \text{ of } 56.3 = (0.067)(56.3)$$
$$= 3.7721$$

Rounded to the nearest tenth of a million, Australia produced 3.8 million ounces of gold.

Answer: 9.6 million ounces

Reading line graphs

Another graph, called a **line graph,** is used to show how quantities change with time. From such a graph, we can determine when a quantity is increasing and when it is decreasing.

EXAMPLE 4   *Reading line graphs.*   The line graph in Figure 7-7 shows how U.S. automobile production has changed since 1900. Look at the graph and answer the following questions.

a. How many automobiles were manufactured in 1940?

b. How many were manufactured in 1950?

c. Over which 20-year span did automobile production increase most rapidly?

d. When did production decrease?

e. Why is a broken line used for a portion of the graph?

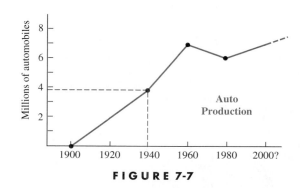

FIGURE 7-7

Solution

a. To find the number of autos produced in 1940, we follow the dashed line from the label 1940 straight up to the graph, and then directly over to the scale. There, we read about 3.8. Since the scale indicates millions of automobiles, approximately 3.8 million autos were produced in 1940.

b. To find the number of autos produced in 1950, we find the point halfway between 1940 and 1960. From there, we move up to the graph and then sideways to the scale, where we read about 5. Approximately 5 million autos were manufactured in 1950.

c. Since the upward tilt of the graph is the greatest between 1940 and 1960, auto production increased most rapidly in those years.

d. Between 1960 and 1980, the graph drops, indicating that auto production decreased during that period.

e. Because beyond the year 2000 is still in the future, production levels are a *projection,* and the broken line indicates that the numbers are only estimates.

EXAMPLE 5   *Reading line graphs.*   The graph in Figure 7-8 shows the movements of two trains. The horizontal axis represents time, and the vertical axis represents the distance that the trains have traveled.

a. How are the trains moving at time A?

b. At what time (A, B, C, D, or E) are both trains stopped?

c. At what times have both trains gone the same distance?

Self Check

How many more autos were produced in 1960 than in 1940?

Answer: about 3.2 million

Self Check

In Figure 7-8, what is train 1 doing at time D?

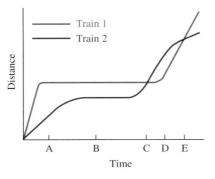

FIGURE 7-8

Solution

The movement of train 1 is represented by the red line, and that of train 2 is represented by the blue line.

a. At time A, the blue line is rising. This shows that the distance traveled by train 2 is increasing: At time A, train 2 is moving.

At time A, the red line is horizontal. This indicates that the distance traveled by train 1 is not changing: At time A, train 1 is stopped.

b. To find the time at which both trains are stopped, we find the time at which both the red and the blue lines are horizontal. At time B, both trains are stopped.

c. At any time, the height of a line gives the distance a train has traveled. Both trains have traveled the same distance whenever the two lines are the same height—that is, at any time when the lines intersect. This occurs at times C and E.

Answer: Train 1, which had been stopped, is beginning to move. ■

Reading histograms and frequency polygons

A pharmaceutical company is sponsoring a series of reruns of old Westerns. The marketing department must choose from three advertisements:

1. Children talking about Chipmunk Vitamins

2. A college student catching a quick breakfast and a TurboPill Vitamin

3. A grandmother talking about Seniors Vitamins

A survey of the viewing audience records the age of each viewer, counting the number in the 6-to-15-year-old age group, the 16-to-25-year-old age group, and so on. The graph of the data is the **histogram** shown in Figure 7-9. The vertical axis, labeled *frequency,* indicates the number of viewers in each age group. For example, the histogram shows that 105 viewers are in the 36-to-45-year-old age group.

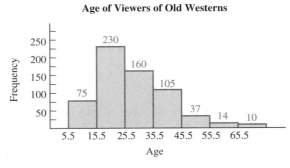

FIGURE 7-9

A histogram is a bar graph with several important features:

1. The bars of a histogram touch.

2. Data values never fall at the edge of a bar.

3. The width of each bar represents a numeric value.

The width of each bar in Figure 7-9 represents an age span of 10 years. Since most viewers are in the 16-to-25-year-old age group, the marketing department decides to advertise TurboPills in commercials appealing to active young adults.

EXAMPLE 6   ***Carry-on luggage.***   An airline weighs the carry-on luggage of 2,260 passengers. See the histogram in Figure 7-10.

a. How many passengers carried luggage in the 8-to-11-pound range?

b. How many carried luggage in the 12-to-19-pound range?

Solution   **a.** The second bar, with edges at 7.5 and 11.5 pounds, corresponds to the 8-to-11-pound range. Use the height of the bar (or the number written there) to determine that 430 passengers carried such luggage.

b. The 12-to-19-pound range is covered by two bars. The total number of passengers with luggage in this range is 970 + 540, or 1,510.

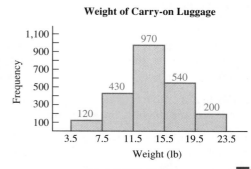

FIGURE 7-10

A special line graph, called a **frequency polygon**, can be constructed from the histogram in Figure 7-10 by joining the center points at the top of each bar. See Figure 7-11. On the horizontal axis, we write the coordinate of the middle value of each bar. After erasing the bars, we get the frequency polygon shown in Figure 7-12.

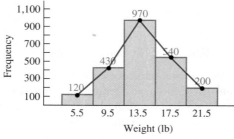

FIGURE 7-11

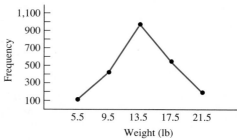

FIGURE 7-12

STUDY SET   Section 7.1

VOCABULARY   *In Exercises 1–6, fill in the blanks to make a true statement.*

1. In Illustration 1, figure ___ is a bar graph.

2. In Illustration 1, figure ___ is a pie graph.

3. In Illustration 1, figure ___ is a pictograph.

4. In Illustration 1, figure ___ is a line graph.

5. In Illustration 1, figure ___ is a histogram.

6. In Illustration 1, figure ___ is a frequency polygon.

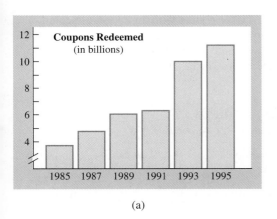

(a)

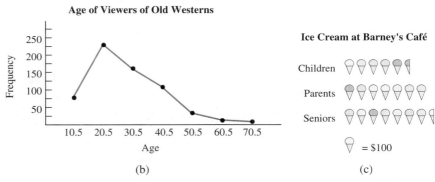

Age of Viewers of Old Westerns

(b)

Ice Cream at Barney's Café

(c)

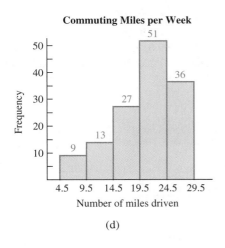

(d)

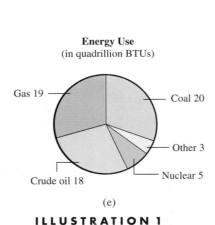

Energy Use
(in quadrillion BTUs)

Gas 19   Coal 20

Other 3

Crude oil 18   Nuclear 5

(e)

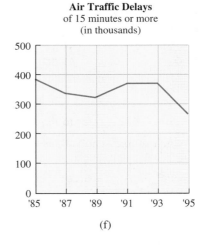

Air Traffic Delays
of 15 minutes or more
(in thousands)

(f)

ILLUSTRATION 1

CONCEPTS   *In Exercises 7–8, fill in the blanks to make the statements true.*

7. The bars of a _____ touch.

8. The width of each bar of a histogram represents a _____ value.

APPLICATIONS   *In Exercises 9–12, refer to the postal rate table in Figure 7-2 on page 337.*

9. PRIORITY MAIL   Find the cost of using priority mail to send a package weighing $7\frac{1}{4}$ pounds to zone 3.

10. PRIORITY MAIL   Find the cost of sending (priority mail) a package weighing $2\frac{1}{4}$ pounds to zone 5.

11. COMPARING POSTAGE   Juan wants to send a 6-pound 1-ounce package to a friend living in zone 2. Fourth-class postage would be $1.79. How much could he save by sending the package fourth class instead of priority mail?

12. SENDING TWO PACKAGES   Jenny wants to send a birthday gift and an anniversary gift to her brother, who lives in zone 6. One package weighs 2 pounds 9 ounces, and the other weighs 3 pounds 8 ounces. If she uses priority mail, how much will she save by sending both gifts as one package instead of two? (*Hint:* 16 ounces = 1 pound.)

In Exercises 13–16, refer to the federal income tax tables in Illustration 2 on the next page.

13. FILING A JOINT RETURN   Raul has an adjusted income of $57,100, is married, and files jointly. Compute his tax.

14. FILING A SINGLE RETURN   Herb is single and has an adjusted income of $79,250. Compute his tax.

15. TAX-SAVING STRATEGIES Angelina is single and has an adjusted income of $53,000. If she gets married, she will gain other deductions that will reduce her income by $2,000, and she can file a joint return. How much will she save in tax by getting married?

16. FILING STATUS A man with an adjusted income of $53,000 married a woman with an adjusted income of $75,000. They filed a joint return. Would they have saved on their taxes if they had both stayed single?

Single—Schedule X			
If adjusted income is:		**The tax is:**	
Over	**But not over**		**Of the amount over**
$0	$23,350	15%	$0
23,350	56,550	$3,502.50 + 28%	23,350
56,550	117,950	12,798.50 + 31%	56,550
117,950	256,500	31,832.50 + 36%	117,950
256,500	———	81,710.50 + 39.6%	256,500

Married Filing Jointly or Qualifying Widow(er)—Schedule Y-1			
If adjusted income is:		**The tax is:**	
Over	**But not over**		**Of the amount over**
$0	$39,000	15%	$0
39,000	94,250	$5,850.00 + 28%	39,000
94,250	143,600	21,320.00 + 31%	94,250
143,600	256,500	36,618.50 + 36%	143,600
256,500	———	77,262.50 + 39.6%	256,500

ILLUSTRATION 2

In Exercises 17–22, refer to the bar graph in Illustration 3.

17. Which source supplied the least energy in 1975?

18. Which energy source remained unchanged between 1975 and 1995?

19. What percent of electrical energy was produced by oil in 1975?

20. Which source provided about 8% of electrical energy in 1995?

21. What was the approximate percent of increase in the use of energy from coal?

22. What was the approximate percent of increase in the use of nuclear power between 1975 and 1995?

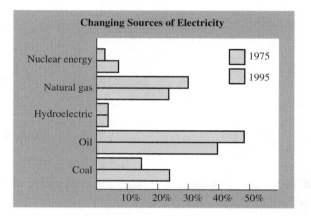

ILLUSTRATION 3

In Exercises 23–28, refer to the bar graph in Illustration 4.

23. The world production of lead in 1970 was approximately equal to the production of zinc in another year. In what other year was that?

24. The world production of zinc in 1990 was approximately equal to the production of lead in another year. In what other year was that?

25. In what year was the production of zinc less than one-half that of lead?

26. In what year was the production of zinc more than twice that of lead?

27. By how many metric tons did the production of zinc increase between 1970 and 1980?

28. By how many metric tons did the production of lead decrease between 1980 and 1990?

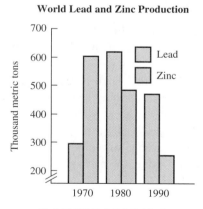

ILLUSTRATION 4

In Exercises 29–32, refer to the bar graph in Illustration 5.

29. In which categories of moving violation have arrests decreased since last month?

30. Last month, which violation occurred most often?

31. This month, which violation occurred least often?

32. Which violation has shown the greatest decrease in number of arrests since last month?

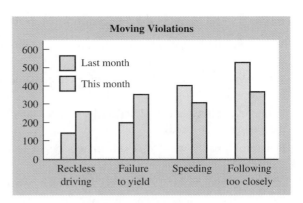

ILLUSTRATION 5

In Exercises 33–36, refer to the pictograph in Illustration 6.

33. Which group (children, parents, or seniors) spent the most money on ice cream at Barney's Café?

34. How much money did parents spend on ice cream?

35. How much more money did seniors spend than parents?

36. How much more money did seniors spend than children?

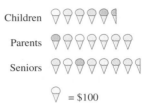

ILLUSTRATION 6

In Exercises 37–42, refer to the pie graph in Illustration 7.

37. Two of the seven languages considered are spoken by groups of about the same size. Which languages are they?

38. Of the languages in the graph, which is spoken by the greatest number of people?

39. Do more people speak Russian or English?

40. What percent of the world's population speak Russian or English?

41. What percent of the world's population speak a language other than these seven?

42. What percent of the world's population do not speak either French or German?

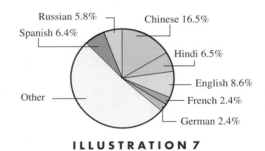

ILLUSTRATION 7

In Exercises 43–48, refer to the pie graph in Illustration 8.

43. What percent of total energy sources is nuclear energy?

44. What percent of total energy sources are represented by coal and crude oil combined?

45. By what percent does energy derived from coal exceed that derived from crude oil?

46. By what percent does energy derived from coal exceed that derived from nuclear sources?

47. Solar energy accounts for less than what percent of total energy sources?

48. If production of nuclear energy tripled in the next 10 years and other sources remained the same, what percent of total energy sources would nuclear energy be?

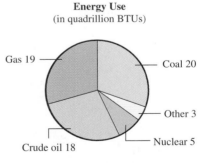

ILLUSTRATION 8

In Exercises 49–54, refer to the line graph in Illustration 9.

49. What were the average weekly earnings in mining for the year 1975?

50. What were the average weekly earnings in construction for the year 1980?

51. In the period between 1982 and 1984, which salary was increasing most rapidly?

52. In approximately what year did miners begin to earn more than construction workers?

53. In the period from 1970 to 1995, which workers received the greatest increase in wages?

54. In what five-year interval did wages in mining increase most rapidly?

In Exercises 55–60, refer to the line graph in Illustration 10.

55. Which runner ran faster at the start of the race?

56. Which runner stopped to rest first?

57. Which runner dropped the baton and had to go back to get it?

58. At what times (A, B, C, or D) was runner 1 stopped and runner 2 running?

59. Describe what was happening at time D.

60. Which runner won the race?

ILLUSTRATION 9

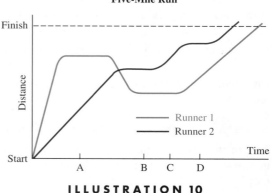

ILLUSTRATION 10

61. COMMUTING MILES   An insurance company has collected data on the number of miles its employees drive to and from work. The data are presented in the histogram in Illustration 11. How many employees commute between 14.5 and 19.5 miles each week?

62. COMMUTING MILES   How many employees commute 14 miles or less each week? (See Illustration 11.)

63. NIGHT SHIFT STAFFING   A hospital administrator surveyed the medical staff to determine the number of room calls during the night. She constructed the frequency polygon in Illustration 12. On how many nights were there about 30 room calls?

64. NIGHT SHIFT STAFFING   How many times did the staff handle the greatest number of room calls? (See Illustration 12.)

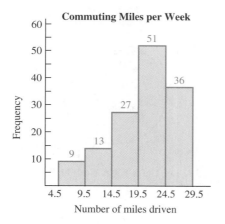

ILLUSTRATION 11

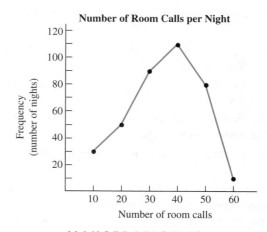

ILLUSTRATION 12

65. MAKING A BAR GRAPH Use the data in Illustration 13 to make a bar graph showing the number of U.S. farms in the years 1950 to 1990.

66. MAKING A LINE GRAPH Use the data in Illustration 13 to make a line graph showing the average acreage of U.S. farms for the years 1950 to 1990.

U.S. Farms 1950–1990		
Year	Number of U.S. farms (in millions)	Average size of U.S. farms (acres)
1950	5.3	215
1960	3.8	305
1970	2.8	390
1980	2.1	447
1990	2.0	470

ILLUSTRATION 13

SAVE! On purchases of	
$10	$100–$250
$25	$250–$500
$50	over $500

ILLUSTRATION 14

67. MAKING A LINE GRAPH The coupon in Illustration 14 provides savings for shoppers. Make a line graph that relates the original price (in dollars, on the horizontal axis) to the sale price (on the vertical axis).

68. MAKING A HISTOGRAM To study the effect of fluoride in preventing tooth decay, researchers counted the number of fillings in the teeth of 28 patients and recorded these results:

3, 7, 11, 21, 16, 22, 18, 8, 12, 3, 7, 2, 8, 19, 12, 19, 12, 10, 13, 10, 14, 15, 14, 14, 9, 10, 12, 13

Tally the results by completing the table in Illustration 15. Then make a histogram. The first bar extends from 0.5 to 5.5, the second bar from 5.5 to 10.5, and so on.

Number of fillings	Frequency
1–5	
6–10	
11–15	
16–20	
21–25	

ILLUSTRATION 15

WRITING *Write a paragraph using your own words.*

69. What kind of presentation (table, bar graph, line graph, pie chart, pictograph, or histogram) is most appropriate for displaying each type of information?

- The percent of students, classified by major
- The percent of Biology majors each year since 1970
- The number of hours spent studying for finals
- Various ethnic populations of the ten largest cities
- Average annual salary of corporate executives for ten major industries

Explain your choices.

70. A histogram is a special type of bar graph. Explain.

REVIEW *In Exercises 71–74, do the operations.*

71. $5 - 3 \cdot 4$

72. $3(6 - 9) + 4$

73. $\left(\dfrac{1}{2} + \dfrac{1}{3} \right)^2$

74. $5^2 + |6 - 10|$

75. Write the prime numbers between 10 and 30.

76. Write the first ten composite numbers.

77. Write the even numbers less than 6 that are not prime.

78. Write the prime numbers between 0 and 10.

Mean, Median, and Mode

In this section, you will learn about

- **The mean (the arithmetic average)**
- **The median**
- **The mode**

INTRODUCTION. Graphs are not the only way of describing lists of numbers compactly. We can often find *one* number that is typical of all of the numbers in a list. We have already seen one such typical number: the *mean,* or the *average.* There are two others: the *median,* and the *mode.* In this section, we will discuss these three measures of central tendency.

The mean (the arithmetic average)

A student has taken five tests this semester, scoring 87, 73, 89, 92, and 84. To find out how well she is doing, she calculates the **mean,** or the **arithmetic average,** of these grades, by finding the sum of the grades and then dividing by 5:

$$\text{Mean score} = \frac{87 + 73 + 89 + 92 + 84}{5}$$

$$= \frac{425}{5}$$

$$= 85$$

The mean score is 85. Some exams were better and some were worse, but 85 is a good indication of her performance in the class.

The mean (arithmetic average)	The **mean,** or the **arithmetic average,** of several values is given by the formula $$\text{Mean (or average)} = \frac{\text{sum of the values}}{\text{number of values}}$$

EXAMPLE 1 **Department store sales.** The week's sales in three departments of the Tog Shoppe are given in the table in Figure 7-13. Find the mean of the daily sales in the women's department for this week.

Self Check

In Example 1, find the mean daily sales in all departments on Wednesday.

	Men's department	Women's department	Children's department
Monday	$2,315	$3,135	$1,110
Tuesday	2,020	2,310	890
Wednesday	1,100	3,206	1,020
Thursday	2,000	2,115	880
Friday	955	1,570	1,010
Saturday	850	2,100	1,000

FIGURE 7-13

Solution

Use a calculator to add the sales in the women's department for the week. Then divide the sum of those six values by 6.

$$\text{Mean sales in the women's department} = \frac{3,135 + 2,310 + 3,206 + 2,115 + 1,570 + 2,100}{6}$$

$$= \frac{14,436}{6}$$

$$= 2,406$$

The mean of the week's daily sales in the women's department is $2,406.

Answer: $1,775.33

EXAMPLE 2　*Miles driven per day.*　In January, Bob drove a total of 4,805 miles. On the average, how many miles did he drive each day?

Self Check

If he drove 3,360 miles in February, 1999, how many miles did he drive each day, on average?

Solution

To find the average number of miles driven each day, we divide the total number of miles by the number of days. Because there are 31 days in January, we divide 4,805 by 31:

$$\text{Average number of miles per day} = \frac{\text{total miles driven}}{\text{number of days}}$$

$$= \frac{4,805}{31}$$

$$= 155$$

On average, Bob drove 155 miles each day.

Answer:　120

The median

The mean is not always representative of the values in a list. For example, suppose that the weekly earnings of four workers in a small business are $280, $300, $380, and $240, and the owner of the company pays himself $5,000.

The mean salary is

$$\text{Mean salary} = \frac{280 + 300 + 380 + 240 + 5,000}{5}$$

$$= \frac{6,200}{5}$$

$$= 1,240$$

The owner could say, "Our employees earn an average of $1,240 per week." Clearly, the mean does not fairly represent the typical worker's salary.

A better measure of the company's typical salary is the *median:* the salary in the middle when all the numbers are arranged by size.

$$240 \quad 280 \quad 300 \quad 380 \quad 5{,}000$$

$\uparrow$
The middle salary.

The typical worker earns $300 per week, far less than the mean salary.

If there is an even number of values in a list, there is no middle value. In that case, the median is the mean of the two numbers closest to the middle. For example, there is no middle number in the list 2, 5, 6, 8, 13, 17. The two numbers closest to the middle are 6 and 8. The median is the mean of 6 and 8, which is $\frac{6+8}{2}$, or 7.

The median	The **median** of several values is the middle value. To find the median:
	1. Arrange the values in increasing order.
	2. If there is an odd number of values, the median is the value in the middle.
	3. If there is an even number of values, the median is the average of the two values that are closest to the middle.

EXAMPLE 3   *Finding a median score.*   On an exam, there were three scores of 59, four scores of 77, and scores of 43, 47, 53, 60, 68, 82, and 97. Find the median score.

Solution

We arrange the 14 scores in increasing order:

43 47 53 59 59 59 **60** **68** 77 77 77 77 82 97

Since there is an even number of scores, the median is the mean of the two scores closest to the middle: the 60 and the 68.

The median is $\dfrac{60 + 68}{2}$, or 64.

Self Check
Find five numbers that have the same mean and median.

Answer: One answer is 1, 2, 3, 4, and 5 (mean = median = 3).

The mode

A hardware store displays 20 outdoor thermometers. Twelve of them read 68°, and the other eight have different readings. To choose an accurate thermometer, should we choose one with a reading that is closest to the *mean* of all 20, or to their *median?* Neither. Instead, we should choose one of the 12 that all read the same, figuring that any of those that agree will likely be correct.

By choosing that temperature that appears most often, we have chosen the *mode* of the 20 numbers.

The mode	The **mode** of several values is the single value that occurs most often. The mode of several values is also called the **modal value.**

EXAMPLE 4 *Finding the mode.* Find the mode of these values: 3, 6, 5, 7, 3, 7, 2, 4, 3, 5, 3, 7, 8, 7, 3, 7, 6, 3, 4.

Solution
To find the mode of the numbers in the list, we make a chart of the distinct numbers that appear and make tally marks to record the number of times they occur.

2	3	4	5	6	7	8
/	////// /	//	//	//	/////	/

Because 3 occurs more times than any other number, it is the mode.

Self Check
Find the mode of these values:
2, 3, 4, 6, 2, 4, 3, 4, 3, 4, 2, 5

Answer: 4

STUDY SET Section 7.2

VOCABULARY *In Exercises 1-4, fill in the blanks to make a true statement.*

1. The sum of the values in a distribution divided by the number of values in the distribution is called the _____ of the distribution.

2. The value that appears most often in a distribution is called the _____ of the distribution.

3. The middle value in a distribution is called the _____ of the distribution.

4. The modal value of a distribution is the value in the distribution that appears _____ often.

CONCEPTS *In Exercises 5-6, fill in the blanks to make a true statement.*

5. The mean of several values is given by

Mean = $\dfrac{\text{the sum of the values}}{\rule{3cm}{0.4pt}}$

6. Complete the formula.

$\dfrac{\text{Average number of miles}}{\text{driven per day}} = \dfrac{\rule{3cm}{0.4pt}}{\text{number of days}}$

PRACTICE *In Exercises 7–12, find the mean of the numbers.*

7. 3, 4, 7, 7, 8, 11, 16

8. 13, 15, 17, 17, 15, 13

9. 5, 9, 12, 35, 37, 45, 60, 77

10. 0, 0, 3, 4, 7, 9, 12

11. 15, 7, 12, 19, 27, 17, 19, 35, 20

12. 45, 67, 42, 35, 86, 52, 91, 102

In Exercises 13–18, find the median of the numbers.

13. 2, 5, 9, 9, 9, 17, 29

14. 16, 18, 27, 29, 35, 47

15. 4, 7, 2, 11, 5, 4, 9, 17

16. 0, 0, 3, 4, 0, 0, 3, 4, 5

17. 18, 17, 2, 9, 21, 23, 21, 2

18. 5, 13, 5, 23, 43, 56, 32, 45

In Exercises 19–24, find the mode (if any) of the numbers.

19. 3, 5, 7, 3, 5, 4, 6, 7, 2, 3, 1, 4

20. 12, 12, 17, 17, 12, 13, 17, 12

21. 5, 9, 12, 35, 37, 45, 60

22. 0, 3, 0, 2, 7, 0, 6, 0, 3, 4, 2, 0

23. 23.1, 22.7, 23.5, 22.7, 34.2, 22.7

24. $\dfrac{1}{2}, \dfrac{1}{3}, \dfrac{1}{3}, 2, \dfrac{1}{2}, 2, \dfrac{1}{5}, \dfrac{1}{2}, 5, \dfrac{1}{3}$

APPLICATIONS

25. SOFT DRINK PRICES A survey of soft-drink machines indicates the following prices for a can (in cents): 50, 60, 50, 50, 70, 75, 50, 45, 50, 50, 65, 75, 60, 75, 100, 50, 80, 75. Find the mean price of a soft drink.

26. COMPUTER SUPPLIES Several computer stores reported differing prices for toner cartridges for a laser printer (in dollars): 51, 55, 73, 75, 72, 70, 53, 59, 75. Find the mean price of a toner cartridge.

27. SOFT DRINK PRICES   Find the median price for a soft drink. (See Exercise 25.)

28. COMPUTER SUPPLIES   Find the median price for a toner cartridge. (See Exercise 26.)

29. SOFT DRINK PRICES   Find the modal price for a soft drink. (See Exercise 25.)

30. COMPUTER SUPPLIES   Find the mode of the prices for a toner cartridge. (See Exercise 26.)

31. CHANGING TEMPERATURES   Temperatures are recorded at hourly intervals, as in Illustration 1. Find the average temperature of the period from midnight to 11:00 A.M.

Time	Temperature	Time	Temperature
12:00 A.M.	53	12:00 noon	71
1:00	53	1:00 P.M.	75
2:00	57	2:00	77
3:00	58	3:00	77
4:00	59	4:00	79
5:00	59	5:00	72
6:00	60	6:00	70
7:00	62	7:00	64
8:00	64	8:00	61
9:00	66	9:00	59
10:00	68	10:00	53
11:00	70	11:00	51

ILLUSTRATION 1

32. SEMESTER GRADE   Frank's algebra grade is based on the average of four exams, which will count equally. His grades are 75, 80, 90, and 85. Find his average.

33. AVERAGE TEMPERATURE   Find the average temperature for the 24-hour period recorded in Illustration 1.

34. WEIGHTED FINAL   If Frank's professor decided to count the fourth examination double, what would Frank's average be? (See Exercise 32.)

35. FLEET MILEAGE   An insurance company's sales force uses 37 cars. Last June, those cars logged a total of 98,790 miles. On the average, how many miles did each car travel that month?

36. BUDGETING FOR GROCERIES   The Hinrichs family spent $519 on groceries last April. On the average, how much did they spend each day?

37. DAILY MILEAGE   Find the average number of miles driven daily for each car in Exercise 35.

38. GROCERY COSTS   See Exercise 36. The Hinrichs family has five members. What is the average spent for groceries for one family member for one day?

39. EXAM AVERAGES   Roberto received the same score on each of five exams, and his mean score is 85. Find his median score and his modal score.

40. BETTER THAN AVERAGE   The scores on the first exam of the students in a history class were 57, 59, 61, 63, 63, 63, 87, 89, 95, 99, and 100. Kia got a score of 70 and claims that "70 is better than average." Which of the three measures of central tendency is she better than: the mean, the median, or the mode?

41. COMPARING GRADES   A student received scores of 37, 53, and 78 on three quizzes. His sister received scores of 53, 57, and 58. Who had the better average? Whose grades were more consistent?

42. What is the average of all of the integers from -100 to 100, inclusive?

WRITING   *Write a paragraph using your own words.*

43. Explain how to find the mean, the median, and the mode of several numbers.

44. Explain why a set of numbers might have no modal value.

REVIEW

45. Find the prime factorization of 81.

46. Find the LCD for two fractions whose denominators are 36 and 81.

In Exercises 47–52, do each operation.

47. $\dfrac{3}{4} \cdot \dfrac{2}{9}$

48. $\dfrac{2}{15} \div \dfrac{4}{5}$

49. $\dfrac{18}{5} + \dfrac{12}{5}$

50. $\dfrac{7}{12} - \dfrac{5}{12}$

51. $\dfrac{8}{5} + \dfrac{3}{10}$

52. $\dfrac{5}{6} - \dfrac{1}{12}$

Mean, Median, and Mode

To indicate the center of a distribution, we can use the mean, the median, or the mode.

- The **mean** of a distribution is the sum of the values in the distribution divided by the number of values in the distribution:

$$\text{Mean} = \frac{\text{sum of the values in the distribution}}{\text{number of values in the distribution}}$$

- The **median** is the middle value in a distribution. Just as many scores are above the median as are below it. If there is an even number of values in the distribution, the median is the mean of the two values that are closest to the middle.

- The **mode** of a distribution is the value that occurs most often.

Consider the following distribution: 3, 7, 4, 12, 15, 23, 17, 21, 15, 20.

1. Calculate the mean.

2. Find the median.

3. Find the mode.

4. Are the mean, the median, and the mode the same number?

Consider the distribution 2, 4, 6, 6, 8, 10.

5. Calculate the mean.

6. Find the median.

7. Find the mode.

8. Are the mean, the median, and the mode the same number?

Construct a distribution with the following characteristics.

9. The mean is greater than the mode.

10. The mean is less than the median.

11. The mode is less than the median.

12. The mode is greater than the median.

<table>
<tr><td></td><td>*Reading Graphs and Tables*</td></tr>
</table>

CONCEPTS

Numerical information can be presented in the form of tables, bar graphs, pictographs, pie graphs, and line graphs.

REVIEW EXERCISES

In Exercises 1–2, refer to the table in Illustration 1.

1. WIND-CHILL TEMPERATURE   Find the wind-chill temperature on a 10° F day when a 15-mph wind is blowing.

2. WIND SPEED   The wind-chill temperature is −25° F, and the outdoor temperature is 15° F. How fast is the wind blowing?

Determining the Wind-Chill Temperature

Wind speed	Actual temperature												
	35° F	30° F	25° F	20° F	15° F	10° F	5° F	0° F	−5° F	−10° F	−15° F	−20° F	−25° F	−30° F
5 mph	33°	27°	21°	16°	12°	7°	0°	−5°	−10°	−15°	−21°	−26°	−31°	−36°
10 mph	22	16	10	3	−3	−9	−15	−22	−27	−34	−40	−46	−52	−58
15 mph	16	9	−2	−5	−11	−18	−25	−31	−38	−45	−51	−58	−65	−72
20 mph	12	4	−3	−10	−17	−24	−31	−39	−46	−53	−60	−67	−74	−81
25 mph	8	1	−7	−15	−22	−29	−36	−44	−51	−59	−66	−74	−81	−88
30 mph	6	−2	−10	−18	−25	−33	−41	−49	−56	−64	−71	−79	−86	−93
35 mph	4	−4	−12	−20	−27	−35	−43	−52	−58	−67	−74	−82	−89	−97
40 mph	3	−5	−13	−21	−29	−37	−45	−53	−60	−69	−76	−84	−92	−100
45 mph	2	−6	−14	−22	−30	−38	−46	−54	−62	−70	−78	−85	−93	−102

ILLUSTRATION 1

3. Refer to Illustration 2 to answer each question.
   a. How many coupons were redeemed in 1987?
   b. Between what years did the number of redeemed coupons remain essentially unchanged?
   c. In what two-year period did the number of redeemed coupons increase the most?
   d. What was the percent of the greatest two-year increase?

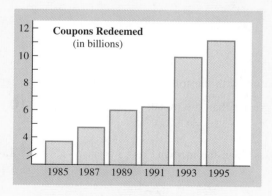

ILLUSTRATION 2

4. Refer to Illustration 3.
  a. How many eggs were produced in Wisconsin in 1985?
  b. How many eggs were produced in Nebraska in 1987?
  c. In what year was the egg production of Wisconsin equal to that of Nebraska?

  d. What was the total egg production of Wisconsin and Nebraska in 1988?

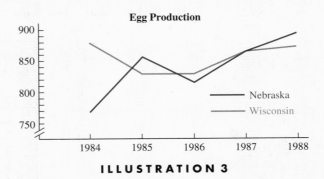

ILLUSTRATION 3

In Exercises 5–6, refer to Illustration 4.

5. A survey of the television viewing habits of 320 households produced the histogram in Illustration 4. How many households watch between 6 and 15 hours of TV each week?

6. How many households watch 11 hours or more each week?

A *histogram* is a bar graph with these features:
1. The bars of the histogram touch.
2. Data values never fall at the edge of a bar.
3. The width of each bar represents a numeric value.

A *frequency polygon* is a special line graph formed from a histogram.

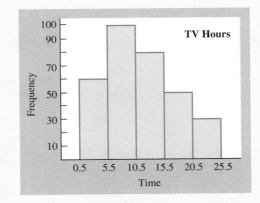

ILLUSTRATION 4

Mean, Median, and Mode

The *mean* (or average) is given by the formula

$$\text{Mean} = \frac{\text{sum of the values}}{\text{number of values}}$$

7. EARNING AN A   Jose worked hard this semester, earning grades of 87, 92, 97, 100, 100, 98, 90, and 98. If he needs a 95 average to earn an A in the class, did he make it?

To find the *median* of several values:

1. Arrange the values in increasing order.
2. If there is an odd number of values, the median is the value in the middle.
3. If there is an even number of values, the median is the average of the two values that are closest to the middle.

The *mode* of several numbers is the single value that occurs most often.

8. GRADE SUMMARY   The students in a mathematics class had final averages of 43, 83, 40, 100, 40, 36, 75, 39, and 100. When asked how well her students did, their teacher answers, "43 was typical." What measure is the teacher using?

9. PRETZEL PACKAGING   Samples of SnacPak pretzels were weighed to find out whether the package claim "Net weight 1.2 ounces" was accurate. The tally appears in Illustration 5. Find the modal weight.

10. Find the mean weight of the samples in Exercise 9.

Weights of SnacPak Pretzels	
Ounces	**Number**
0.9	1
1.0	6
1.1	18
1.2	23
1.3	2
1.4	0

ILLUSTRATION 5

In Problems 1–4, refer to Illustration 1. Keeping one prisoner for one month costs $2,266.

1. How much money is spent monthly, per prisoner, to pay the prison staff?

2. How much money is spent monthly, per prisoner, on office costs?

3. What percent of the monthly allotment is spent on one prisoner's food?

4. What percent of the monthly allotment is spent on one prisoner's recreation and training?

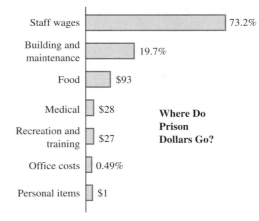

Staff wages — 73.2%
Building and maintenance — 19.7%
Food — $93
Medical — $28
Recreation and training — $27
Office costs — 0.49%
Personal items — $1

Where Do Prison Dollars Go?

ILLUSTRATION 1

In Problems 5–6, refer to Illustration 2.

5. Approximately what percent of all employees are in the food and clothing industries?

6. Among workers in food and clothing, 2.4 million are in food, and the rest are in clothing. What percent of all workers are in clothing?

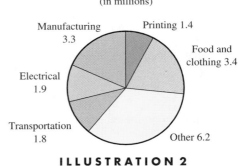

Employees in Industry
(in millions)

Manufacturing 3.3
Printing 1.4
Food and clothing 3.4
Electrical 1.9
Transportation 1.8
Other 6.2

ILLUSTRATION 2

In Problems 7–10, use the information given in Illustration 3.

7. How many air traffic delays occurred in 1995?

8. How many air traffic delays in 1991 were due to the weather?

9. Which year was worst for air traffic delays?

10. The percent for "Other" causes for delays appears smudged. What should the value be?

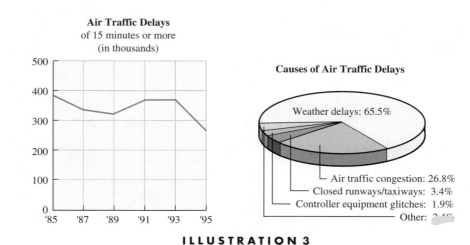

Air Traffic Delays
of 15 minutes or more
(in thousands)

Causes of Air Traffic Delays

Weather delays: 65.5%

Air traffic congestion: 26.8%
Closed runways/taxiways: 3.4%
Controller equipment glitches: 1.9%
Other: 2.4%

ILLUSTRATION 3

In Problems 11–14, refer to Illustration 4 and choose the best answer from the following statements.

A. Both bicyclists are moving, and bicyclist 1 is faster than 2.
B. Both bicyclists are moving, and bicyclist 2 is faster than 1.
C. Bicyclist 1 is stopped, and bicyclist 2 is not.
D. Bicyclist 2 is stopped, and bicyclist 1 is not.
E. Both bicyclists are stopped.

11. Indicate what is happening at time *A*.

12. Indicate what is happening at time *B*.

13. Indicate what is happening at time *C*.

14. Which bicyclist won the race?

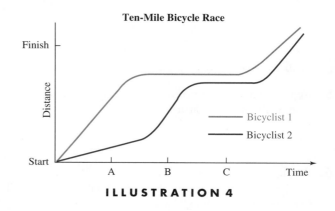

Ten-Mile Bicycle Race

Finish

Distance

Bicyclist 1
Bicyclist 2

Start

A   B   C   Time

ILLUSTRATION 4

In Problems 15–18, refer to this information: The hours served last month by the individual volunteers at a homeless shelter were 4, 6, 8, 2, 8, 10, 11, 9, 5, 12, 5, 18, 7, 5, 1, and 9.

15. Find the median of the hours of volunteer service.

16. Find the mean of the hours of volunteer service.

17. Find the mode of the hours of volunteer service.

18. If the one value of 18 were removed from the list, which would be most affected: the mean or the median?

An Introduction to Algebra

8

ALGEBRA IS THE LANGUAGE OF MATHEMATICS. IN THIS CHAPTER, YOU WILL LEARN MORE ABOUT THINKING AND WRITING IN THIS LANGUAGE, USING ITS MOST IMPORTANT COMPONENT—A VARIABLE.

8.1 Solving Equations by Addition and Subtraction

In this section, you will learn about

- **Equations**
- **Checking solutions**
- **Solving equations**
- **Problem solving**

INTRODUCTION. The language of mathematics is *algebra*. The word *algebra* comes from the title of a book written by the Arabian mathematician Al-Khowarazmi around A.D. 800. Its title, *Ihm aljabr wa'l muqabalah,* means restoration and reduction, a process then used to solve equations. In this section, we will begin discussing equations, one of the most powerful ideas in algebra.

Equations

An **equation** is a statement indicating that two expressions are equal. Equations contain an equals sign ($=$). Some examples of equations are

$$x + 5 = 21, \qquad 16 + 5 = 21, \qquad \text{and} \qquad 10 + 5 = 21$$

In the equation $x + 5 = 21$, the expression $x + 5$ is called the **left-hand side,** and 21 is called the **right-hand side.** The letter x is the **variable** (or the **unknown**).

An equation can be true or false. For example, $16 + 5 = 21$ is a true equation, whereas $10 + 5 = 21$ is a false equation. An equation containing a variable can be true or false, depending upon the value of the variable. If $x = 16$, the equation $x + 5 = 21$ is true, because

$$\mathbf{16} + 5 = 21 \qquad \text{Substitute 16 for } x.$$

However, this equation is false for all other values of x.

Any number that makes an equation true when substituted for its variable is said to *satisfy* the equation. Such numbers are called **solutions** or **roots.** Because 16 is the only number that satisfies $x + 5 = 21$, it is the only solution of the equation.

Checking solutions

Checking a solution. Verify that 8 is a solution of the equation $x - 3 = 5$.

Self Check
Is 8 a solution of $x + 17 = 25$?

Solution

We substitute 8 for the variable x in the equation and verify that both sides of the equation are equal.

$$x - 3 = 5$$
$$8 - 3 \stackrel{?}{=} 5 \quad \text{Substitute 8 for } x. \text{ Read } \stackrel{?}{=} \text{ as "is possibly equal to."}$$
$$5 = 5 \quad \text{Do the subtraction.}$$

Since both sides of the equation equal 5, 8 is a solution.

Answer: yes

Checking a solution. Is 23 a solution of the equation $12 = y - 10$?

Self Check
Is 5 a solution of $20 = 16 + y$?

Solution

We substitute 23 for y and simplify.

$$12 = y - 10$$
$$12 \stackrel{?}{=} 23 - 10 \quad \text{Substitute 23 for } y.$$
$$12 \neq 13 \quad \text{Do the subtraction.}$$

Because the left-hand and right-hand sides are not equal, 23 is not a solution.

Answer: no

Solving equations

Normally, the solution of an equation is not given—you must find it yourself. This process is called *solving the equation*. To develop an understanding of the properties and procedures used to solve an equation, we will examine $x + 2 = 5$ and make some observations as we solve it in a practical way.

We can think of the scales shown in Figure 8-1(a) as representing the equation $x + 2 = 5$. The weight (in grams) on the left-hand side of the scale is $x + 2$, and the weight (in grams) on the right-hand side is 5. Because these weights are equal, the scale is in balance. To find x, we need to isolate it. That can be accomplished by removing 2 grams from the left-hand side of the scale. Common sense tells us that we must also remove 2 grams from the right-hand side if the scales are to remain in balance. In Figure 8-1(b), we can see that x grams will be balanced by 3 grams. We say that we have *solved* the equation and that the *solution* is 3.

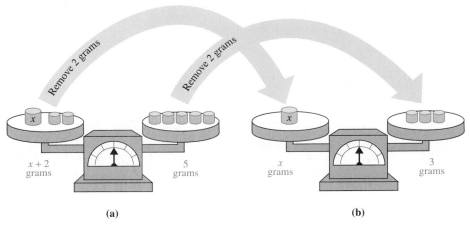

(a)   (b)

FIGURE 8-1

From this example, we can draw some observations about solving an equation.

Observations

1. To find the value of x, we needed to isolate it on the left-hand side of the scales.
2. To isolate x, we had to undo the addition of 2 grams. This was accomplished by subtracting 2 grams from the left-hand side.
3. We wanted the scales to remain in balance. When we subtracted 2 grams from the left-hand side, we subtracted the same amount from the right-hand side.

The observations suggest a property of equality: *If the same quantity is subtracted from equal quantities, the results will be equal quantities.* We can express this property, called the **subtraction property of equality,** using the variables a, b, and c.

Subtraction property of equality	Let a, b, and c be any three numbers. If $a = b$, then $a - c = b - c$.

When we use this property, the resulting equation will be equivalent to the original equation.

Equivalent equations	Two equations are **equivalent equations** when they have the same solutions.

In the previous example, we found that $x + 2 = 5$ is equivalent to $x = 3$.

EXAMPLE 3 *Solving an equation.* Solve the equation $x + 2 = 5$.

Solution

To isolate x on one side of the equation, we undo the addition of 2 by subtracting 2 from both sides of the equation.

$$x + 2 = 5 \qquad \text{The equation to solve.}$$
$$x + 2 - 2 = 5 - 2 \qquad \text{Subtract 2 from both sides.}$$
$$x = 3 \qquad \text{Do the subtractions: } 2 - 2 = 0 \text{ and } 5 - 2 = 3.$$

The solution is 3. We check by substituting 3 for x in the original equation and simplifying.

$$x + 2 = 5$$
$$3 + 2 \stackrel{?}{=} 5 \qquad \text{Substitute 3 for } x.$$
$$5 = 5 \qquad \text{Do the addition: } 3 + 2 = 5.$$

The solution checks.

Self Check

Solve the equation $x + 7 = 14$ and then check the result.

Answer: 7

A second property that we will use to solve equations involves addition. It is based on the following idea: *If the same quantity is added to equal quantities, the results will be equal quantities.* Using variables, we have the following property.

Addition property of equality	Let a, b, and c be any three numbers. If $a = b$, then $a + c = b + c$.

We can think of the scales shown in Figure 8-2(a) as representing the equation $x - 2 = 3$. To find x, we need to use the addition property of equality and add 2 grams of weight to each side. The scales will remain in balance. From the scales in Figure 8-2(b), we can see that x grams will be balanced by 5 grams.

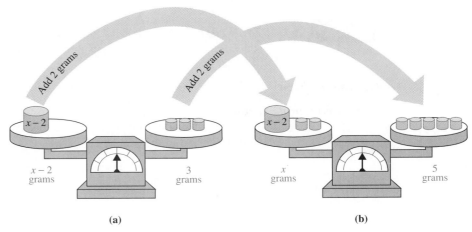

FIGURE 8-2

EXAMPLE 4 *Solving an equation.* Solve the equation $3 = x - 2$.	**Self Check**

Self Check
Solve the equation $12 = x - 8$ and then check the result.

Solution

To isolate x on one side of the equation, we undo the subtraction of 2 by adding 2 to both sides of the equation.

$$3 = x - 2 \qquad \text{The equation to solve.}$$
$$3 + 2 = x + 2 - 2 \qquad \text{Add 2 to both sides.}$$
$$5 = x \qquad \text{Do the operations: } 3 + 2 = 5 \text{ and } 2 - 2 = 0.$$
$$x = 5 \qquad \text{Since } 5 = x, x = 5.$$

The solution is 5. We check by substituting 5 for x in the original equation and simplifying.

$$3 = x - 2$$
$$3 \stackrel{?}{=} 5 - 2 \qquad \text{Substitute 5 for } x.$$
$$3 = 3 \qquad \text{Do the subtraction: } 5 - 2 = 3.$$

The solution checks.

Answer: 20 ■

Problem solving

The key to problem solving is to understand the problem and then to devise a plan for solving it. The following list of steps provides a good strategy to follow.

Strategy for problem solving	1. **Analyze the problem** by reading it carefully to understand the given facts. What information is given? What vocabulary is given? What are you asked to find? Often a diagram will help you visualize the facts of a problem.
	2. **Form an equation** by picking a variable to represent the quantity to be found. Then express all other unknown quantities as expressions involving that variable. Finally, write an equation expressing a quantity in two different ways.
	3. **Solve the equation.**
	4. **State the conclusion.**
	5. **Check the result** in the words of the problem.

We will now use this five-step strategy to solve problems. The purpose of the following examples is to help you learn the strategy, even though you can solve the examples without it.

EXAMPLE 5

Total earnings. Table 8-1 shows the income a person earned from various sources during one year. Find the total income.

Analyze the problem   We are asked to find the total income.

Form an equation   Let n = the total income.
To form an equation involving n, we look for a key word or phrase in the problem.

Source	Amount
Wages	$21,000
Savings	$ 1,750
Stocks	$ 2,275
Rentals	$ 7,275

TABLE 8-1

Key word: *total*   **Translation:** *addition*

We can write the total income in two ways.

Total income	is	the income from wages	plus	the income from savings	plus	the income from stocks	plus	the income from rentals.
n	=	21,000	+	1,750	+	2,275	+	7,275

Solve the equation $n = 21,000 + 1,750 + 2,275 + 7,275$ The variable n is isolated on the left.
$n = 32,300$ Do the additions.

State the conclusion   The person received a total income of $32,300.

Check the result   We can add the numbers again or check the result by estimation. To estimate, we round each amount to the nearest thousand.

$$21,000 + 2,000 + 2,000 + 7,000 = 32,000$$

The answer, 32,300, is reasonable.

EXAMPLE 6

Small business. Last year a hairdresser lost 17 customers who moved away. If he now has 73 customers, how many did he have originally?

Analyze the problem   We are asked to find the number of customers the hairdresser had before any moved away. We know that the hairdresser started with some unknown number of customers, and after 17 moved away, he had 73 left.

Form an equation   We can let c = the original number of customers.
To form an equation involving c, we need to look for a key word or phrase in the problem.

Key phrase: *moved away*   **Translation:** *subtract*

We can write the remaining number of customers in two ways.

The original number of customers	minus	17	is	the remaining number of customers.
c	−	17	=	73

Solve the equation $c - 17 = 73$
$c + 17 - 17 = 73 + 17$ To undo the subtraction of 17 and isolate c, add 17 to both sides.
$c = 90$ Simplify: $17 - 17 = 0$ and $73 + 17 = 90$.

State the conclusion	He originally had 90 customers.
Check the result	The hairdresser had 90 customers. After losing 17, he has $90 - 17$, or 73 left. The solution checks. ■

EXAMPLE 7   *Buying a house.*   Sue wants to buy a house that costs $87,000. Since she has only $15,000 for a down payment, she will have to take a mortgage. How much will she have to borrow?

Analyze the problem   We want to find the size of the mortgage. To buy the house, she will have to use the $15,000 for a down payment and borrow some additional money.

Form an equation   We can let $x =$ the money that she needs to borrow.
To form an equation involving x, we need to look for a key word or phrase in the problem or analysis.

Key phrase: *borrow some additional money*   **Translation:** *addition*

We can write the total cost of the house in two ways.

The amount Sue has	plus	the amount she needs	is	the total cost of the house.
15,000	+	x	=	87,000

Solve the equation

$$15,000 + x = 87,000$$
$$15,000 + x - \mathbf{15,000} = 87,000 - \mathbf{15,000}$$

To undo the addition of 15,000 and isolate x, subtract 15,000 from both sides.

$$x = 72,000$$

Do the subtraction:
$87,000 - 15,000 = 72,000$.

State the conclusion   She needs to borrow $72,000.

Check the result   With a $72,000 mortgage, she will have $15,000 + $72,000, which is the $87,000 that is necessary to buy the house. The solution checks. ■

STUDY SET   Section 8.1

VOCABULARY   *Fill in the blanks to make a true statement.*

1. An equation is a statement that two expressions are _____.

2. A _____ of an equation is a number that satisfies the equation.

3. The answer to an equation is called a _____ or a _____.

4. A letter that is used to represent a number is called a _____.

5. _____ equations have exactly the same solutions.

6. To solve an equation, we _____ the variable on one side of the equals sign.

CONCEPTS   *In Exercises 7–8, complete the following rules.*

7. If $x = y$ and c is any number, then $x + c =$ _____.

8. If $x = y$ and c is any number, then $x - c =$ _____.

9. In the equation $x + 6 = 10$, what operation is performed on the variable? How do we undo that operation to isolate the variable?

10. In $9 = y - 5$, what operation is performed on the variable? How do we undo that operation to isolate the variable?

11. Solve $x + 8 = 24$.

$\quad\quad x + 8 = 24$ The equation to solve.

$\quad x + 8 - \underline{\quad} = 24 - \underline{\quad}$ Subtract 8 from both sides.

$\quad\quad\quad\quad x = 16$ Simplify.

12. Solve $x - 8 = 24$.

$\quad\quad x - 8 = 24$ The equation to solve.

$\quad x + \underline{\quad} - 8 = 24 + \underline{\quad}$ Add 8 to both sides.

$\quad\quad\quad\quad x = 32$ Simplify.

PRACTICE *In Exercises 13–20, tell whether each statement is an equation.*

13. $x = 2$

14. $y - 3$

15. $7x < 8$

16. $7 + x = 2$

17. $x + y = 0$

18. $3 - 3y > 2$

19. $1 + 1 = 3$

20. $5 = a + 2$

In Exercises 21–32, tell whether the number is a solution of the equation.

21. $x + 2 = 3$; 1

22. $x - 2 = 4$; 6

23. $a - 7 = 0$; 7

24. $x + 4 = 4$; 0

25. $8 - y = y$; 5

26. $10 - c = c$; 5

27. $x + 32 = 0$; 16

28. $x - 1 = 0$; 4

29. $z + 7 = z$; 7

30. $n - 9 = n$; 9

31. $x = x$; 0

32. $x = 2$; 0

In Exercises 33–72, use the addition or subtraction property of equality to solve each equation. Check all solutions.

33. $x - 7 = 3$

34. $y - 11 = 7$

35. $a - 2 = 5$

36. $z - 3 = 9$

37. $1 = b - 2$

38. $0 = t - 1$

39. $x - 4 = 0$

40. $c - 3 = 0$

41. $y - 7 = 6$

42. $a - 2 = 4$

43. $70 = x - 5$

44. $66 = b - 6$

45. $312 = x - 428$

46. $x - 307 = 113$

47. $x - 117 = 222$

48. $y - 27 = 317$

49. $x + 9 = 12$

50. $x + 3 = 9$

51. $y + 7 = 12$

52. $c + 11 = 22$

53. $t + 19 = 28$

54. $s + 45 = 84$

55. $23 + x = 33$

56. $34 + y = 34$

57. $5 = 4 + c$

58. $41 = 23 + x$

59. $99 = r + 43$

60. $92 = r + 37$

61. $512 = x + 428$

62. $x + 307 = 513$

63. $x + 117 = 222$

64. $y + 38 = 321$

65. $3 + x = 7$

66. $b - 4 = 8$

67. $y - 5 = 7$

68. $z + 9 = 23$

69. $4 + a = 12$

70. $5 + x = 13$

71. $x - 13 = 34$

72. $x - 23 = 19$

APPLICATIONS *In Exercises 73–74, solve each problem by completing the outline of the five-step problem-solving strategy.*

73. ARCHAEOLOGY A 1,700-year-old manuscript is 425 years older than the clay jar in which it was found. How old is the jar?

Analyze the problem
What are you asked to find? _____ How old is the manuscript? _____ How much older is the manuscript than the jar? _____

Form an equation
Since we want to find the age of the jar, we can let $x =$ _____. To set up an equation involving x, we look for a key word or phrase in the problem.

Key phrase: _____
Translation: _____

We can write the age of the manuscript in two ways.

The age of the manuscript	is		plus	the age of the jar
____	=	425	+	____

Solve the equation

$\quad\quad\quad \underline{\quad} = 425 + x$

$1,700 - \underline{\quad} = 425 + x - \underline{\quad}$ Subtract 425 from both sides.

$\quad\quad\quad \underline{\quad} = x$ Do the subtractions.

State the conclusion
_____

Check the result
The jar is 1,275 years old. The manuscript, at age 1,700, is 425 years older than the 1,275-year-old jar. The solution checks.

74. BANKING   After a student wrote a $1,500 check to pay for a car, he had a balance of $750 in his account. How much did he have in the account before he wrote the check?

Analyze the problem
What are you asked to find? _____
____. How large a check did he write? _____ What was his balance after he wrote the check? _____

Form an equation
Since we want to find his balance before he wrote the check, we let $x =$ _____
To set up an equation involving x, we look for a key word or phrase in the problem.

Key phrase: _____
Translation: _____

We can write the balance now in the account in two ways.

	minus	1,500	equals	
x	$-$	____	$=$	750

Solve the equation

$$\underline{} - 1,500 = 750$$
$$x + \underline{} - 1,500 = 750 + \underline{} \quad \text{Add 1,500 to both sides.}$$
$$x = \underline{} \quad \text{Simplify.}$$

State the conclusion
_____

Check the result
The original balance was $2,250. After writing a check, his balance was $2,250 − $1,500, or $750. The solution checks.

In Exercises 75–82, use the five-step problem-solving strategy to solve each problem.

75. PARTY INVITATIONS   Three of Mia's party invitations were lost in the mail, but 59 were delivered. How many invitations did she send?

76. COST OF A CONDOMINIUM   The price of a condominium is $57,500 less than the cost of a house. The house costs $102,700. Find the price of the condominium.

77. COST SHARING   Kim and Mary shared the costs of driving to Denver. Kim paid $12 more than Mary. If Mary paid $68, how much did Kim pay?

78. BUYING GOLF CLUBS   A man needs $345 for a new set of golf clubs. How much more money does he need if he now has $317?

79. BUYING CLOTHES   Heather paid $48 for a sweater. Holly bought the same sweater on sale for $9 less. How much did Holly pay for the sweater?

80. COMPARING WAGES   Last week, Sally made $543 more than Juan. How much money did Sally make if Juan made $123?

81. AUTO REPAIR   A woman paid $29 less to have her car fixed at a muffler shop than she would have paid at a gas station. At the gas station, she should have paid $219. How much did she pay to have her car fixed?

82. BUS RIDER   A man had to wait 20 minutes for a bus today. Three days ago, he had to wait 15 minutes longer than he did today, because four buses passed by without stopping. How long did he wait three days ago?

WRITING   *Write a paragraph using your own words.*

83. Explain what it means for a number to satisfy an equation.

84. Explain how to tell whether a number is a solution of an equation.

85. Explain what Figure 8-1 is trying to show.

86. Explain what Figure 8-2 is trying to show.

REVIEW   *In Exercises 87–88, round 325,784 to the specified place.*

87. nearest ten

88. nearest thousand

89. Evaluate: $2 \cdot 3^2 \cdot 5$.

90. Represent $4 + 4 + 4 + 4 + 4 + 4 + 4$ as a multiplication.

91. Evaluate: $8 - 2(3) + 1^3$.

92. Write 1,055 in words.

Solving Equations by Division and Multiplication

In this section, you will learn about

- **The division property of equality**
- **The multiplication property of equality**
- **Problem solving**

INTRODUCTION. In the previous section, we learned how to solve equations of the forms

$$x - 4 = 10 \qquad \text{and} \qquad x + 5 = 16$$

by using the addition and subtraction properties of equality. In this section, we will learn how to solve equations of the forms

$$2x = 14 \qquad \text{and} \qquad \frac{x}{5} = 25$$

by using the division and multiplication properties of equality.

The division property of equality

To solve many equations, we must divide both sides of the equation by the same non-zero number. The resulting equation will be equivalent to the original one. This idea is summed up in the division property of equality: *If equal quantities are divided by the same nonzero quantity, the results will be equal quantities.*

Division property of equality	If a, b, and c are any three numbers, then If $a = b$, then $\dfrac{a}{c} = \dfrac{b}{c}$. $\qquad (c \neq 0)$

We will now consider how to solve the equation $2x = 8$. You will recall that $2x$ means $2 \cdot x$. Therefore, the given equation can be rewritten as $2 \cdot x = 8$. We can think of the scales in Figure 8-3(a) as representing the equation $2 \cdot x = 8$. The weight on the left-hand side of the scale is $2 \cdot x$ grams and the weight on the right-hand side is 8 grams. Because these weights are equal, the scale is in balance. To find x, we use the division property of equality to remove half of the weight from each side. The scales will remain in balance. From the scales shown in Figure 8-3(b), we see that x grams will be balanced by 4 grams.

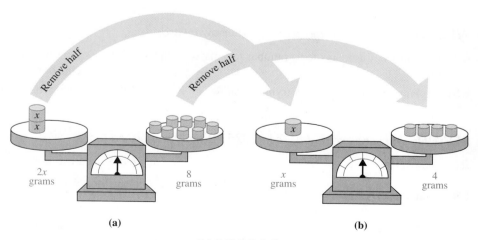

(a)   (b)

FIGURE 8-3

EXAMPLE 1   *Solving equations.*   Solve the equation $2x = 8$.

Self Check
Solve the equation $17x = 153$ and
then check the result.

Solution

Recall that $2x = 8$ means $2 \cdot x = 8$. To isolate x on the left-hand side of the equation, we undo the multiplication by 2 by dividing both sides of the equation by 2.

$$2x = 8 \quad \text{The equation to solve.}$$

$$\frac{2x}{2} = \frac{8}{2} \quad \text{To undo the multiplication by 2, divide both sides by 2.}$$

$$x = 4 \quad \text{Do the division: } 8 \div 2 = 4.$$

The solution is 4. Check it as follows:

$$2x = 8$$

$$2 \cdot 4 \overset{?}{=} 8 \quad \text{Substitute 4 for } x.$$

$$8 = 8 \quad \text{Do the multiplication: } 2 \cdot 4 = 8.$$

Because the final equation is a true statement, the solution checks.

Answer: 9

The multiplication property of equality

We can also multiply both sides of an equation by the same nonzero number to get an equivalent equation. This idea is summed up in the multiplication property of equality: *If equal quantities are multiplied by the same nonzero quantity, the results will be equal quantities.*

Multiplication property of equality	If a, b, and c are any three numbers, then If $a = b$, then $c \cdot a = c \cdot b$. $(c \neq 0)$

We can think of the scales shown in Figure 8-4(a) as representing the equation $\frac{x}{3} = 25$. The weight on the left-hand side of the scale is $\frac{x}{3}$ grams, and the weight on the right-hand side is 25 grams. Because these weights are equal, the scale is in balance. To find x, we can use the multiplication property of equality to triple (or multiply by 3) the weight on each side. The scales will remain in balance. From the scales shown in Figure 8-4(b), we can see that x grams will be balanced by 75 grams.

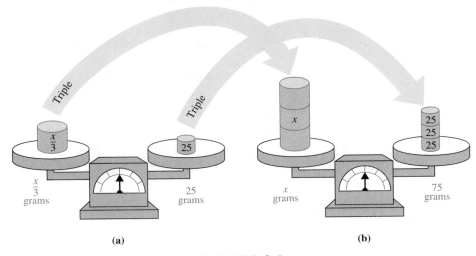

(a)   **(b)**

FIGURE 8-4

EXAMPLE 2 *Solving equations.* Solve the equation $\frac{x}{3} = 25$.

Solution

To isolate x on the left-hand side of the equation, we undo the division of the variable by 3 by multiplying both sides by 3.

$$\frac{x}{3} = 25 \qquad \text{The equation to solve.}$$

$$3 \cdot \frac{x}{3} = 3 \cdot 25 \qquad \text{Multiply both sides by 3.}$$

$$x = 75 \qquad \text{Do the multiplication: } 3 \cdot 25 = 75.$$

Check:

$$\frac{x}{3} = 25$$

$$\frac{75}{3} \overset{?}{=} 25 \qquad \text{Substitute 75 for } x.$$

$$25 = 25 \qquad \text{Do the division: } 75 \div 3 = 25.$$

Self Check

Solve the equation $\frac{x}{12} = 24$ and then check the result.

Answer: 288

Problem solving

As before, we can use equations to solve problems, applying the five-step strategy for problem solving. Remember that the purpose of these early examples is to help you learn the strategy, even though you can solve the examples without it.

EXAMPLE 3 *Buying electronics.* Find the total cost of six Direct TV® satellite dishes that are on sale for $499 each.

Analyze the problem   We are asked to find the total cost. We can either use 499 as an addend six times or multiply 499 by 6. Since it is easier, we will multiply.

Form an equation   We can let c = the total cost of the satellite dishes.
To form an equation involving c, we need to look for a key word or phrase in the problem.

Key phrase: *total of six dishes, each costing $499*   **Translation:** *multiplication*

We can write the total cost in two ways.

The total number of dishes	multiplied by	the cost of each dish	equals	the total cost.
6	$\cdot$	499	=	c

Solve the equation
$6 \cdot 499 = c$ The variable c is isolated on the right.
$2{,}994 = c$ Do the multiplication: $6 \cdot 499 = 2{,}994$.

State the conclusion	The total cost will be $2,994.
Check the result	We can check by estimation. Since each dish costs a little less than $500, we would expect the total cost to be a little less than 6 · $500, or $3,000. An answer of $2,994 is reasonable. ■

EXAMPLE 4

	Splitting an inheritance. Seven brothers will inherit $343,000. If they split the money evenly, how much will each brother get?
Analyze the problem	We are asked to find how much each brother will get if they split $343,000 evenly.
Form an equation	We can let g = the number of dollars each brother will get. To form an equation involving g, we need to look for a key word or phrase in the problem.

Key phrase: *split the money evenly*   **Translation:** *division*

We can write the share each brother will get in two ways.

The total amount of the inheritance	divided by	the number of brothers	equals	the share each brother will get.
343,000	÷	7	=	g

Solve the equation	$343,000 \div 7 = g$ The variable g is isolated on the right.
	$\dfrac{343,000}{7} = g$ $343,000 \div 7$ means the same as $\frac{343,000}{7}$.
	$49,000 = g$ Do the division: $\frac{343,000}{7} = 49,000$.
State the conclusion	Each brother will get $49,000.
Check the result	If we multiply $49,000 by 7, we get a total of $343,000. ■

EXAMPLE 5

	Buying paper. A writer estimates that she will need 520 sheets of high-quality paper to make eight copies of a manuscript. If the paper comes packaged at 65 sheets per pack, how many packs will she need to buy?
Analyze the problem	We are asked to find how many packages of paper the writer needs. Since each pack contains 65 sheets and the writer will buy some unknown number of packs, the product of 65 and the number of packs must equal 520 sheets.
Form an equation	We can let p = the number of packs the writer should buy. To form an equation involving p, we need to look for a key word or phrase. In this case, we find it in the analysis of the problem.

Key word: *product*   **Translation:** *multiplication*

We can write the number of sheets needed in two ways.

The number of sheets in one pack	times	the number of packs	equals	the number of sheets needed.
65	·	p	=	520

Solve the equation	$65p = 520$
	$\dfrac{65p}{65} = \dfrac{520}{65}$ To undo the multiplication by 65 and isolate p, divide both sides by 65.
	$p = 8$ Simplify: $520 \div 65 = 8$.
State the conclusion	The writer needs to buy 8 packs.
Check the result	If we multiply 65 by 8, we get 520.

EXAMPLE 6

Cost of a band. A five-piece band worked on New Year's Eve. If each player earned \$120, how much did the band cost?

Analyze the problem We are asked to find how much the band cost. We know that the total cost of the band divided by the number of players will give each person's share.

Form an equation We can let c = the total cost of the band.
To form an equation involving c, we need to look for a key word or phrase. In this case, we find it in the analysis of the problem.

Key phrase: *divided by*   **Translation:** *division*

We can write the wages for each player in two ways.

$$\underbrace{\frac{\text{The total cost of the band}}{\text{the number in the band}}}_{} \quad \text{equals} \quad \underbrace{\text{the wages for each player.}}_{}$$

$$\frac{c}{5} \qquad = \qquad 120$$

Solve the equation

$$\frac{c}{5} = 120$$

$$5 \cdot \frac{c}{5} = 5 \cdot 120 \quad \text{To undo the division by 5 and isolate } c, \text{ multiply both sides by 5.}$$

$$c = 600 \quad \text{Do the multiplication: } 5 \cdot 120 = 600.$$

State the conclusion The band cost \$600.

Check the result If we divide \$600 by 5, we get the individual wages of \$120.

STUDY SET   Section 8.2

VOCABULARY *Fill in the blanks to make a true statement.*

1. The statement "If equal quantities are divided by the same nonzero quantity, the results will be equal quantities" expresses the _____ property of equality.

2. The statement "If equal quantities are multiplied by the same nonzero quantity, the results will be equal quantities" expresses the _____ property of equality.

CONCEPTS *Fill in the blanks to make a true statement.*

3. To undo a multiplication by 6, we _____ both sides of the equation by ___.

4. To undo a division by 8, we _____ both sides of the equation by ___.

5. In the equation $4t = 40$, what operation is being performed on the variable? How do we undo it?

6. In the equation $\frac{t}{15} = 1$, what operation is being performed on the variable? How do we undo it?

7. Name the first step in solving each of the following equations.
  a. $x + 5 = 10$ **b.** $x - 5 = 10$

  c. $5x = 10$ **d.** $\frac{x}{5} = 10$

8. For each of the following equations, check the given answer.
  a. $16 = t - 8$; $t = 33$
  b. $16 = t + 8$; $t = 8$
  c. $16 = 8t$; $t = 128$
  d. $16 = \frac{t}{8}$; $t = 2$

9. Complete the following statement of the division property of equality. Assume that both sides are to be divided by z, which is not zero.

  If $x = y$, then _____ .

10. Complete the following statement of the multiplication property of equality. Assume that both sides are to be multiplied by z, which is not zero.

  If $x = y$, then _____ .

NOTATION *In Exercises 11–12, complete the solution of each equation.*

11. Solve: $3x = 12$.

  $3x = 12$

  $\dfrac{3x}{} = \dfrac{12}{}$ Divide both sides by 3.

  $x = 4$ Do the division.

12. Solve: $\frac{x}{5} = 9$.

  $\dfrac{x}{5} = 9$

  $\underline{} \cdot \dfrac{x}{5} = \underline{} \cdot 9$ Multiply both sides by 5.

  $x = 45$ Do the multiplications.

PRACTICE *In Exercises 13–20, use the division property of equality to solve each equation. Check each solution.*

13. $3x = 3$ **14.** $5x = 5$ **15.** $32x = 192$ **16.** $24x = 120$

17. $17y = 51$ **18.** $19y = 76$ **19.** $34y = 204$ **20.** $18y = 90$

In Exercises 21–28, use the multiplication property of equality to solve each equation. Check each solution.

21. $\dfrac{x}{7} = 2$ **22.** $\dfrac{x}{12} = 4$ **23.** $\dfrac{y}{14} = 3$ **24.** $\dfrac{y}{13} = 5$

25. $\dfrac{a}{15} = 15$ **26.** $\dfrac{b}{25} = 25$ **27.** $\dfrac{c}{13} = 13$ **28.** $\dfrac{d}{11} = 11$

In Exercises 29–40, solve each equation. Check each solution.

29. $9z = 9$ **30.** $3z = 6$ **31.** $7x = 21$ **32.** $13x = 52$

33. $43t = 86$ **34.** $96t = 288$ **35.** $21s = 21$ **36.** $31x = 155$

37. $\dfrac{d}{20} = 2$ **38.** $\dfrac{x}{16} = 4$ **39.** $400 = \dfrac{t}{3}$ **40.** $250 = \dfrac{y}{2}$

41. INVESTING   An unlucky investor has forgotten how much he invested in ABC company, but he knows that his stock is worth only $\frac{1}{3}$ of his original purchase price. If the current value of his stock is $12,500, how much did he pay?

Analyze the problem
What are you asked to find? _____
How much is the stock worth now? _____

Form an equation
Since we want to find how much he paid, we can let $x =$ _____. To set up an equation involving x, we look for a key word or phrase in the problem.

Key phrase: _____
Translation: _____
We can now form the equation.

The original cost of the stock	equals	the current value of the stock.

$$\frac{x}{3} = \underline{\qquad}$$

Solve the equation
$$\frac{x}{3} = \underline{\quad}$$

$$\underline{\quad} \cdot \frac{x}{3} = \underline{\quad} \cdot 12{,}500 \qquad \text{To undo the division, multiply both sides by 3.}$$

$$x = \underline{\qquad\qquad} \qquad \text{Do the multiplication.}$$

State the conclusion
_____

Check the result
The original cost of the stock was $37,500. Since its current value is only $\frac{1}{3}$ of its original value, its current value is $37,500 divided by 3, which is $12,500. The solution checks.

42. INVESTING   A lucky investor has watched the value of his portfolio double in the last 12 months. If the current value of his portfolio is $274,552, what was its value one year ago?

Analyze the problem
What are you asked to find? _____
_____ How much is the portfolio worth now? _____

Form an equation
We can let $x =$ _____.
To set up an equation involving x, we look for a key word or phrase in the problem.

Key phrase: _____
Translation: _____
We can now form the equation.

times	the value of the portfolio one year ago	equals	the current value of the portfolio.
2 $\cdot$	____	=	$274,552

Solve the equation
$$2x = \underline{\qquad}$$

$$\frac{2x}{\underline{\ }} = \frac{274{,}552}{\underline{\ }} \qquad \text{To undo the multiplication, divide both sides by 2.}$$

$$x = \underline{\qquad\qquad} \qquad \text{Do the division.}$$

State the conclusion
_____

Check the result
If the value of the portfolio one year ago was $137,276 and it doubled, its current value would be $274,552. The solution checks.

In Exercises 43–48, use the five-step problem-solving strategy to solve each problem.

43. SPEED READING   An advertisement for a speed reading program claimed that successful completion of the course could triple a person's reading rate. If Alicia can currently read 130 words a minute, at what rate can she expect to read after taking the classes?

44. COST OVERRUN   Lengthy delays and skyrocketing costs caused a rapid-transit construction project to go over budget by a factor of 10. The final audit showed the project costing $540 million. What was the initial cost estimate?

45. LOTTO WINNERS   The grocery store employees listed in Illustration 1 pooled their money to buy $120 worth of lottery tickets each week, with the understanding they would split the prize equally if they happened to win. One week they did have the winning ticket and won $480,000. What was each employee's share of the winnings?

ILLUSTRATION 1

46. ANIMAL SHELTER   The number of phone calls to an animal shelter quadrupled after the evening news showed a 3-minute segment explaining the services the shelter offered. Prior to the publicity, the shelter received 8 calls a day. How many calls did the shelter receive each day after being featured on the news?

47. OPEN HOUSE   The attendance at an elementary school open house was only $\frac{1}{4}$ of what the principal had expected. If 120 people visited the school that evening, how many had she expected to attend?

48. INFOMERCIAL   The number of orders received each week by a company selling a skin care product increased fivefold after a well-known Hollywood celebrity was added to the company's television infomercial. After adding the celebrity, the company received about 175 orders each week. How many orders were received each week before adding the celebrity to their advertisement?

WRITING   *Write a paragraph using your own words.*

49. Explain what Figure 8-3 is trying to show.

50. Explain what Figure 8-4 is trying to show.

REVIEW   *In Exercises 51–58, solve each problem.*

51. Find the perimeter of a rectangle with sides measuring 8 cm and 16 cm.

52. Find the area of a rectangle with sides measuring 23 inches and 37 inches.

53. Find the prime factorization of 120.

54. Find the prime factorization of 150.

55. Evaluate: $3^2 \cdot 2^3$.

56. Evaluate: $5 + 6 \cdot 3$.

57. On the golf team, John had the scores shown in Illustration 2 for his first five matches. Find his match average.

58. Solve the equation $x - 4 = 20$.

Match	Score
1	76
2	80
3	74
4	83
5	72

ILLUSTRATION 2

8.3   *Algebraic Expressions and Formulas*

In this section, you will learn about

- **Algebraic expressions**
- **Evaluating algebraic expressions**
- **Formulas**
- **Formulas from business**
- **Formulas from science**

INTRODUCTION.   An algebraic expression is a combination of variables and numbers with the operation symbols of addition, subtraction, multiplication, and division. In this section, we will be replacing the variables in these expressions with specific numbers. Then, using the rule for order of operations, we will evaluate each expression.

We will also study formulas. Like algebraic expressions, formulas involve variables.

Algebraic expressions

In the equation $1,700 = 425 + x$, the expression $425 + x$ is called an **algebraic expression**. Algebraic expressions are the building blocks of equations.

Algebraic expressions	Variables and numbers (called **constants**) can be combined with the operations of addition, subtraction, multiplication, and division to create **algebraic expressions.**

Here are some examples of algebraic expressions.

$5(2a)$	This algebraic expression is a combination of the constants 5 and 2, the variable a, and the operation of multiplication.
$x + 2x + 3x$	This algebraic expression involves the variable x, the constants 2 and 3, and the operations of addition and multiplication.
$t^2 \cdot 3t^4$	This algebraic expression contains powers of the variable t, the constants 3, 2, and 4, and the operation of multiplication.
$5(r - 6)$	This algebraic expression is a combination of the constants 5 and 6, the variable r, and the operations of subtraction and multiplication.

Evaluating algebraic expressions

The manufacturer's instructions for installing a kitchen garbage disposal include the diagram in Figure 8-5. Word phrases are used to describe the lengths of the pieces of pipe that are necessary to connect the disposal to the drain line.

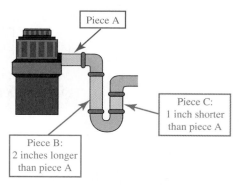

Piece A

Piece C:
1 inch shorter
than piece A

Piece B:
2 inches longer
than piece A

FIGURE 8-5

The instructions tell us that the lengths of pieces B and C are related to the length of piece A. If we let $x =$ the length of piece A, then the lengths of the two other pieces of pipe can be expressed using algebraic expressions, as shown in Figure 8-6.

Since piece B is 2 inches longer than piece A,

$x + 2 =$ length of piece B

Since piece C is 1 inch shorter than piece A,

$x - 1 =$ length of piece C

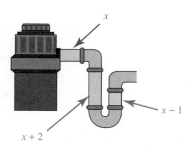

x

$x - 1$

$x + 2$

FIGURE 8-6

Model	Length of piece A
#101	2 inches
#201	3 inches
#301	4 inches

TABLE 8-2

See Table 8-2, which shows part of the manufacturer's instruction sheet. Suppose that model #201 is being installed. The table tells us that piece A should be 3 inches long. We then find the lengths of the other two pieces of pipe by replacing x in each of the algebraic expressions with 3.

To find the length of piece B:

Replace x with 3.

$$x + 2 = 3 + 2$$
$$= 5$$

Piece B should be 5 inches long.

To find the length of piece C:

Replace x with 3.

$$x - 1 = 3 - 1$$
$$= 2$$

Piece C should be 2 inches long.

When we replace the variable in an algebraic expression with a specific number and then apply the rule for order of operations, we are **evaluating the algebraic expression.** In the previous example, we say that we *substituted* 3 for x to find the lengths of the two other pieces of pipe.

In the next three examples, we will replace the variables with positive or negative numbers. It is a good idea to write parentheses around the number when it is inserted into the algebraic expression in place of the variable.

EXAMPLE 1 *Evaluating algebraic expressions.* Evaluate each algebraic expression for $x = 3$: **a.** $2x - 1$ and **b.** $\dfrac{-x - 15}{6}$.

Solution

a. $2x - 1 = 2(3) - 1$ Replace x with 3. Use parentheses.

$\qquad\quad = 6 - 1$ Do the multiplication first: $2(3) = 6$.

$\qquad\quad = 5$ Do the subtraction.

b. $\dfrac{-x - 15}{6} = \dfrac{-(3) - 15}{6}$ Substitute 3 for x. Use parentheses.

$\qquad\quad = \dfrac{-3 - 15}{6}$ The opposite of 3 is -3.

$\qquad\quad = \dfrac{-3 + (-15)}{6}$ Add the opposite of 15.

$\qquad\quad = \dfrac{-18}{6}$ Do the addition: $-3 + (-15) = -18$.

$\qquad\quad = -3$ Do the division.

Self Check

Evaluate each algebraic expression for $y = 5$:

a. $5y - 4$

b. $\dfrac{-y - 15}{5}$

Answer: **a.** 21, **b.** -4

EXAMPLE 2 *Evaluating algebraic expressions.* Evaluate each algebraic expression for $a = -2$: **a.** $-3a + 4a^2$ and **b.** $2 + 3(a + 1)$.

Solution

a. $-3a + 4a^2 = -3(-2) + 4(-2)^2$ Replace a with -2.

$\qquad\qquad = -3(-2) + 4(4)$ Find the power: $(-2)^2 = 4$.

$\qquad\qquad = 6 + 16$ Do the multiplications: $-3(-2) = 6$ and $4(4) = 16$.

$\qquad\qquad = 22$ Do the addition.

b. $2 + 3(a + 1) = 2 + 3[(-2) + 1]$ Substitute -2 for a.

$\qquad\qquad\quad = 2 + 3(-1)$ Do the addition inside the brackets: $-2 + 1 = -1$.

$\qquad\qquad\quad = 2 + (-3)$ Do the multiplication: $3(-1) = -3$.

$\qquad\qquad\quad = -1$ Do the addition.

Self Check

Evaluate each algebraic expression for $t = -3$:

a. $-2t + 4t^2$

b. $-t + 2(t + 1)$

Answer: **a.** 42, **b.** -1

When evaluating algebraic expressions containing two or more variables, we need to know the value of each variable.

EXAMPLE 3 *Expressions involving two variables.*
Evaluate the expression $(8hg + 6g)^2$ for $h = -1$ and $g = 5$.

Self Check
Evaluate the expression
$(5rs + 4s)^2$ for $r = -1$ and $s = 5$.

Solution

$$(8hg + 6g)^2 = [8(-1)(5) + 6(5)]^2 \quad \text{Replace } h \text{ with } -1 \text{ and } g \text{ with } 5.$$

$$= (-40 + 30)^2 \quad \text{Do the multiplications inside the brackets:} \\ 8(-1)(5) = -40 \text{ and } 6(5) = 30.$$

$$= (-10)^2 \quad \text{Do the addition inside the parentheses:} \\ -40 + 30 = -10.$$

$$= 100 \quad \text{Find the power: } (-10)^2 = 100.$$

Answer: 25

Formulas

A **formula** is a mathematical expression used to state a known relationship between two or more variables. Formulas are used in many fields: economics, physical education, anthropology, biology, automotive repair, and nursing, just to name a few. In this section, we will consider seven formulas from business, science, and mathematics.

Formulas from business

A Formula to Find the Sale Price
If a car that normally sells for $12,000 is discounted $1,500, you can find the sale price using the formula

| Sale price | = | original price | − | discount |

Using variables to represent the sale price (s), the original price (p), and the discount (d), this formula can be written

$$s = p - d$$

If we substitute 12,000 for p, 1,500 for d, and simplify, we can find the sale price of the car.

$$s = p - d$$
$$= 12,000 - 1,500 \quad \text{Substitute 12,000 for } p \text{ and 1,500 for } d.$$
$$= 10,500 \quad \text{Do the subtraction.}$$

The sale price of the car is $10,500.

A Formula to Find the Retail Price
To make a profit, a merchant must sell a product for more than he paid for it. The price at which he sells the product, called the *retail price,* is the *sum* of what the item cost him and the markup.

| Retail price | = | cost | + | markup |

Using the variables r to represent the retail price, c the cost, and m the markup, we can write this formula as

$$\boxed{r = c + m}$$

As an example, suppose that a store buys a lamp for $35 and then marks up the cost $20 before selling it. We can find the retail price of the lamp using this formula.

$$r = c + m$$
$$= 35 + 20 \quad \text{Substitute 35 for } c \text{ and 20 for } m.$$
$$= 55 \quad\quad\;\; \text{Do the addition.}$$

Thus, the retail price of the lamp is $55.

A Formula to Find Profit

The profit a business makes is the *difference* of the revenue (the money it takes in) and the costs.

$$\text{Profit} \quad = \quad \text{revenue} \quad - \quad \text{costs}$$

Using variables to write the formula, we have

$$\boxed{p = r - c}$$

As an example, suppose that a charity telethon took in $14 million in donations but had expenses totaling $2 million. We can find the profit made by the charity by subtracting the expenses (costs) from the donations (revenue).

$$p = r - c$$
$$= 14 - 2 \quad \text{Substitute 14 for } r \text{ and 2 for } c.$$
$$= 12 \quad\quad\;\; \text{Do the subtraction.}$$

The charity made a profit of $12 million from the telethon.

Formulas from science

A Formula to Find the Distance Traveled

If we know the rate (speed) at which we are traveling (say, 55 miles per hour) and the time we will be driving at that rate (say, 3 hours), we can find the distance traveled using the formula

$$\text{Distance} \quad = \quad \text{rate} \quad \cdot \quad \text{time}$$

Writing the formula using variables, we have

$$\boxed{d = rt}$$

If we replace the rate r with 55 and the time t with 3, we can use this formula to find the distance traveled.

$$d = rt$$
$$d = 55(3) \quad \text{Substitute 55 for } r \text{ and 3 for } t.$$
$$= 165 \quad\quad\; \text{Do the multiplication.}$$

The distanced traveled in 3 hours at a rate of 55 miles per hour is 165 miles.

WARNING! When using this formula for distance, make sure that the units are similar. For example, if the rate is given in miles per *hour*, the time must be expressed in *hours*.

A Formula for Converting Degrees Fahrenheit to Degrees Celsius

The electronic message board in front of a bank flashes two temperature readings. This is because temperature can be measured using the Fahrenheit or the Celsius scale. The Fahrenheit scale is used in the American system of measurement and the Celsius scale in the metric system. The two scales are shown on the thermometers in Figure 8-7. This should help you to see how the two scales are related. There is a formula to convert a Fahrenheit reading to a Celsius reading:

$$C = \frac{5(F - 32)}{9}$$

Later we will see that there is a formula to convert a Celsius reading to a Fahrenheit reading.

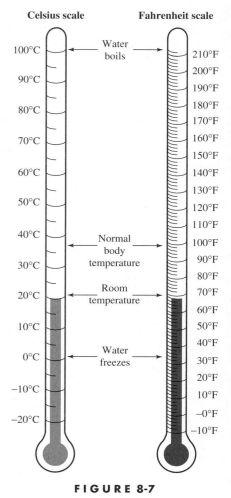

FIGURE 8-7

EXAMPLE 4 *Converting from degrees Fahrenheit to degrees Celsius.* The thermostat in an office building was set at 77° F. Convert this setting to degrees Celsius.

Solution

$$C = \frac{5(F - 32)}{9}$$

$$= \frac{5(77 - 32)}{9} \quad \text{Replace } F \text{ with 77.}$$

$$= \frac{5(45)}{9} \quad \text{Do the subtraction: } 77 - 32 = 45.$$

$$= \frac{225}{9} \quad \text{Do the multiplication: } 5(45) = 225.$$

$$= 25 \quad \text{Do the division.}$$

The thermostat is set at 25° C.

Self Check

The temperature on a hot July day was 104° F. Convert this temperature to degrees Celsius.

Answer: 40° C

A Formula to Find the Distance an Object Falls

The distance an object falls (in feet) when it is dropped from a height is related to the time (in seconds) that it has been falling by the formula

$$\text{Distance fallen} = 16 \cdot (\text{time})^2$$

Writing this relationship using variables, we have

$$d = 16t^2$$

EXAMPLE 5 *Finding the distance an object falls.* Find the distance a camera would fall in 6 seconds if it was dropped by a vacationer taking a hot-air balloon ride.

Self Check

Find the distance a rock would fall in 3 seconds if it was dropped over the edge of the Grand Canyon.

Solution

We will use the formula $d = 16t^2$.

$$d = 16t^2$$
$$d = 16(6)^2 \quad \text{Replace } t \text{ with 6.}$$
$$= 16(36) \quad \text{Find the power: } 6^2 = 36.$$
$$= 576 \quad \text{Do the multiplication.}$$

The camera would fall 576 feet.

Answer: 144 ft

STUDY SET   Section 8.3

VOCABULARY   *In Exercises 1–4, fill in the blanks to make a true statement.*

1. A _____ is a mathematical expression that states a known relationship between two or more variables.

2. An _____ is a combination of variables, numbers, and the operation symbols for addition, subtraction, multiplication, and division.

3. To evaluate an algebraic expression, we _____ specific numbers for the variables in the expression and apply the rules for order of operations.

4. A _____ is a letter that stands for a number.

CONCEPTS

5. Show the misunderstanding that occurs if we don't write parentheses around -8 when evaluating the expression $2x + 10$ for $x = -8$.

6. a. Which of the formulas studied in this section involve a *difference* of two quantities?

b. Which of the formulas studied in this section involve a *product*?

7. The plans for building a children's swing set are shown in Illustration 1. The builder can choose a size that is appropriate for the children who will use it.
a. Describe the lengths of parts 1, 2, and 3 using a variable and algebraic expressions.

b. If the builder chooses to make part 1 60 inches long, how long should parts 2 and 3 be?

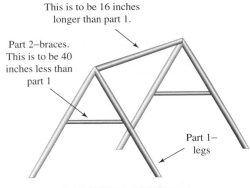

Part 3–crossbar.
This is to be 16 inches longer than part 1.

Part 2–braces.
This is to be 40 inches less than part 1

Part 1–legs

ILLUSTRATION 1

8. A television studio art department plans to construct a series of set decorations out of plywood, using the plan shown in Illustration 2.

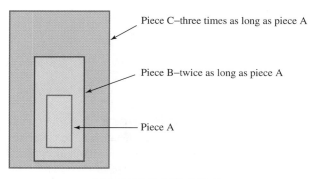

Piece C–three times as long as piece A

Piece B–twice as long as piece A

Piece A

ILLUSTRATION 2

a. Describe the lengths of pieces A, B, and C using a variable and algebraic expressions.

b. For the foreground, designers will make piece A 15 inches long. How long should pieces B and C be?

c. For the background, piece A will be 30 inches long. How long should pieces B and C be?

9. A ticket outlet adds a service charge of $2 to the price of every ticket it sells. Complete the pricing chart in Illustration 3.

Ticket price	Service charge	Total cost
$20	$2	
$25	$2	
$p	$2	
$10p	$2	

ILLUSTRATION 3

10. A tire shop charges $5 to balance each new tire it sells. Complete the pricing chart in Illustration 4.

Price of new tire	Balancing charge	Total cost
$45	$5	
$65	$5	
$t	$5	
$5t	$5	

ILLUSTRATION 4

11. Complete the chart in Illustration 5 by finding the distance traveled in each instance. Be sure to give the units for each of your answers.

	Rate (mph)	Time (hr)	Distance traveled
Bus	25	2	
Bike	12	4	
Walking	3	t	
Running	5	1	
Car	x	3	

ILLUSTRATION 5

12. Each semester, a college charges its students a campus service fee of $10, a parking fee of $12, and an enrollment fee of $13 per unit. Complete the fee chart in Illustration 6.

Units taken	Total charges units	Campus service fee	Parking fee	Total charges
4				
10				
12				
u				

ILLUSTRATION 6

13. Illustration 7 shows a dashboard. Explain what each instrument measures. How are these measurements related?

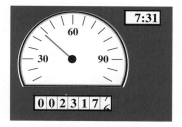

ILLUSTRATION 7

14. What occupation might use a formula that finds
   a. target heart rate after a workout

   b. gas mileage of a car
   c. age of a fossil
   d. equity in a home
   e. dosage to administer
   f. cost-of-living index

15. A car travels at a rate of 65 miles per hour for 15 minutes. What is wrong with the following thinking?

$$d = rt$$
$$d = 65(15)$$
$$d = 975$$

The car travels 975 miles in 15 minutes.

NOTATION

17. Use variables to write the formula that relates each of the quantities listed below.
   a. Rate, distance, time
   b. Centigrade temperature, Fahrenheit temperature

   c. Time, the distance an object falls when dropped

16. SPREADSHEET   The data from a chemistry experiment is shown in Illustration 8. To obtain a result, the chemist must use the formula

   SUM(A1:D1) − MINIMUM(D1:D3)

   If A1:D1 means from cell A1 up to and including cell D1, find the result.

	A	**B**	**C**	**D**
1	12	−5	6	−2
2	15	4	5	−4
3	6	4	−2	8

ILLUSTRATION 8

18. Use variables to write the formula that relates each of the quantities listed below.
   a. Original price, sale price, discount
   b. Number of values, average, sum of values
   c. Cost, profit, revenue
   d. Markup, retail price, cost

PRACTICE   *In Exercises 19–38, evaluate each expression for the given value of the variable.*

19. $3x + 5$ for $x = 4$

20. $1 + 7a$ for $a = 2$

21. $-p$ for $p = -4$

22. $-j$ for $j = -9$

23. $-4t$ for $t = -10$

24. $-12m$ for $m = -6$

25. $\dfrac{x - 8}{2}$ for $x = -4$

26. $\dfrac{-10 + y}{-4}$ for $y = -6$

27. $2(p + 9)$ for $p = -12$

28. $3(r - 20)$ for $r = 15$

29. $x^2 + 14$ for $x = 3$

30. $16 - d^2$ for $d = 5$

31. $8s - s^3$ for $s = -2$

32. $5r + r^3$ for $r = 1$

33. $4x^2$ for $x = 5$

34. $3f^2$ for $f = 3$

35. $3b - b^2$ for $b = -4$

36. $5a - a^2$ for $a = -3$

37. $\dfrac{24 + k}{3k}$ for $k = 3$

38. $\dfrac{4 - h}{h - 4}$ for $h = -1$

In Exercises 39–54, evaluate each algebraic expression for the given values of the variables.

39. $\dfrac{x}{y}$ for $x = 30$ and $y = -10$

40. $\dfrac{e}{3f}$ for $e = 24$ and $f = -8$

41. $-x - y$ for $x = -1$ and $y = 8$

42. $-a - 5b$ for $a = -9$ and $b = 6$

43. $x(5h - 1)$ for $x = -2$ and $h = 2$

44. $c(2k - 7)$ for $c = -3$ and $k = 4$

45. $b^2 - 4ac$ for $b = -3$, $a = 4$ and $c = -1$

46. $3r^2h$ for $r = 4$ and $h = 2$

47. $x^2 - y^2$ for $x = 5$ and $y = -2$

48. $x^3 - y^3$ for $x = -1$ and $y = 2$

49. $\dfrac{50 - 6s}{-t}$ for $s = 5$ and $t = 4$

50. $\dfrac{7v - 5r}{-r}$ for $v = 8$ and $r = 4$

51. $-5abc + 1$ for $a = -2$, $b = -1$, and $c = 3$

52. $-rst + 2t$ for $r = -3$, $s = -1$, and $t = -2$

53. $5s^2t$ for $s = -3$ and $t = -1$

54. $-3k^2t$ for $k = -2$ and $t = -3$

In Exercises 55–68, use the appropriate formula to answer each question.

55. It costs a snack bar owner 20 cents to make a snow cone. If the markup is 50 cents, what is the price of a snow cone?

56. Find the distance covered by a jet if it travels for 3 hours at 550 miles per hour.

57. A school carnival brought in revenues of $13,500 and had costs of $5,300. What was the profit?

58. For the month of June, a florist's cost of doing business was $3,795. If June revenues totaled $5,115, what was her profit for the month?

59. A jewelry store buys bracelets for $18 and marks them up $5. What is the retail price of a bracelet?

60. A shopkeeper marks up the cost of every item she carries by the amount she paid for the item. If a fan costs her $27, what does she charge for the fan?

61. Find the distance covered by a car traveling 60 miles per hour for 5 hours.

62. Find the sale price of a pair of skis that normally sells for $200 but is discounted $35.

63. Find the Celsius temperature reading if the Fahrenheit reading is 14°.

64. Find the Celsius temperature reading if the Fahrenheit reading is 113°.

65. The area of a trapezoid is given by the formula $A = \frac{1}{2}h(b_1 + b_2)$, where h is the height and b_1 and b_2 are the bases. Find the area of a trapezoid whose bases measure 32 meters and 16 meters and whose height is 10 meters.

66. On its first night of business, a pizza parlor brought in $445. The owner estimated his costs that night to be $295. What was the profit?

67. Find the distance a ball has fallen 2 seconds after being dropped from a tall building.

68. A store owner buys a pair of pants for $25 and marks them up $15 for sale. What is the retail price of the pants?

APPLICATIONS

69. PROFITS   Illustration 9 shows the revenue and the costs of doing business for a car rental company over a three-year period.

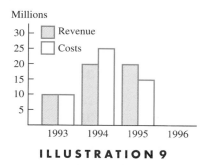

ILLUSTRATION 9

a. Find the profit or loss for each of the 3 years shown.

b. For 1996, revenue was $25 million and costs were $15 million. Enter this data on the chart and then find the profit.

70. THERMOMETER SCALE A thermometer manufacturer wishes to scale a thermometer in both degrees Celsius and degrees Fahrenheit. Find the missing Celsius degree measures in Illustration 10.

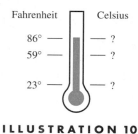

ILLUSTRATION 10

71. SPREADSHEET   A store manager wants to use a spreadsheet to post the prices of sale items for the checkers at the cash registers. (See Illustration 11.) If column B lists the regular price and column C lists the discount, write a formula using column names to have the computer find the sale price. Then fill in column D.

	A	B	C	D
1	Bath towel set	$25	$5	
2	Pillows	$15	$3	
3	Comforter	$53	$11	

ILLUSTRATION 11

72. DEALER MARKUP   A car dealer marks up the cars he sells $500 above factory invoice (that is, $500 over what it costs him to purchase the car from the factory). Complete the pricing chart in Illustration 12.

Model	Factory invoice	Markup	Price
Minivan	$15,600		
Pickup	$13,200		
Convertible	$x		

ILLUSTRATION 12

73. FOOTBALL   The results of each carry that a running back had during a game are recorded in Illustration 13. Find his average yards gained per carry for the game.

Play	Result
Sweep	Gain of 16 yd
Pitch	Gain of 10 yd
Dive	Gain of 4 yd
Dive	Loss of 2 yd
Dive	No gain
Sweep	Loss of 4 yd

ILLUSTRATION 13

74. QUARTERLY REPORT   The financial performance of a computer company in each of the four quarters of the year is shown in Illustration 14. Find the company's quarterly financial average. The figures are in millions of dollars.

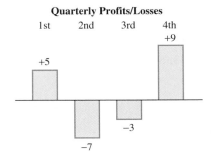

ILLUSTRATION 14

CALCULATOR   *In Exercises 75–78, use a calculator to answer each question.*

75. SHIPPING   Illustration 15 shows the number of ships using the docking facilities of a port one week in July. Find the average number of ships per day for that week.

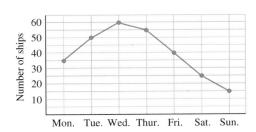

ILLUSTRATION 15

Time falling	Distance traveled	Time intervals
1 sec		Distance traveled 0 sec to 1 sec
2 sec		Distance traveled 1 sec to 2 sec
3 sec		Distance traveled 2 sec to 3 sec
4 sec		Distance traveled 3 sec to 4 sec

ILLUSTRATION 16

76. FALLING OBJECT   See Illustration 16. First, find the distance in feet traveled by a falling object in 1, 2, 3, and 4 seconds. Enter the results in the middle column. Then find the distance the object traveled over each time interval and enter it in the right-hand column.

77. CUSTOMER SATISFACTION SURVEY   As customers were leaving a restaurant, they were asked to rate the service they had received. Good service was rated with a 5, fair service with a 3, and poor service with a 1. The tally sheet compiled by the questioner is shown in Illustration 17. What was the restaurant's average score on this survey?

Type of service	Point value	Number
Good	5	53
Fair	3	26
Poor	1	9

ILLUSTRATION 17

78. DISTANCE TRAVELED
   a. When in orbit, the space shuttle travels at a rate of approximately 17,250 miles per hour. How far does it travel in one day?
   b. The speed of light is approximately 186,000 miles per second. How far will light travel in one minute?
   c. The speed of a sound wave in air is about 1,100 feet per second at normal temperatures. How far does it travel in half a minute?

79. Explain the process of evaluating an algebraic expression.

80. Explain how we can use a stopwatch to find the distance traveled by a falling object.

81. Write a definition for each of these business words: *revenue, markup,* and *profit.*

82. What is a formula?

REVIEW

83. Which of these are prime numbers?
9, 15, 17, 33, 37, 41

84. How can this repeated multiplication be rewritten in simpler form? $2 \cdot 2 \cdot 2 \cdot 2 \cdot 2$

85. Evaluate: $|-2 + (-5)|$.

86. Multiply: $-3(-2)(4)$.

87. In the equation $\frac{x}{3} = -4$, what operation is performed on the variable?

88. Express this situation using an inequality symbol and a variable: City code requires the height of the homes on the coastline to be under 20 feet.

89. Subtract: $-3 - (-6)$.

90. Which is undefined: division of zero or division by zero?

8.4 *Simplifying Algebraic Expressions and the Distributive Property*

In this section, you will learn about

- **Simplifying algebraic expressions involving multiplication**
- **The distributive property**
- **Extending the distributive property**

INTRODUCTION. In mathematics, it is almost always useful to replace something complicated with something that is equivalent and *simpler* in form. In this section, we will simplify algebraic expressions, and the result will be an equivalent but simpler expression.

Simplifying algebraic expressions involving multiplication

To **simplify an algebraic expression,** we use properties of algebra to write the given expression in a more usable form. Two of the properties used to simplify algebraic expressions are the associative and the commutative properties of multiplication. Recall that the associative property allows us to change the grouping of the factors involved in a multiplication. The commutative property allows us to change the order of the factors.

Consider $6(5x)$, an algebraic expression involving multiplication. We want to write it in a simpler form. Since $5x = 5 \cdot x$, we can rewrite $6(5x)$ as $6 \cdot (5 \cdot x)$.

$$6(5x) = 6 \cdot (5 \cdot x) \quad \text{\small $5x = 5 \cdot x$.}$$
$$= (6 \cdot 5)x \quad \text{\small Apply the associative property of multiplication. Instead of grouping the 5 with the x, we group it with the 6.}$$
$$= 30x \quad \text{\small Do the operation inside the parentheses first: $6 \cdot 5 = 30$.}$$

We say that $6(5x)$ simplifies to $30x$. That is, $6(5x) = 30x$.

In the previous example, we used the associative property to regroup the factors of the expression so that the numbers were separated from the variable. After multiplying the numbers, that result was then multiplied by the variable.

EXAMPLE 1   *Simplifying an algebraic expression involving multiplication.*   Simplify each algebraic expression: **a.** $-2(7x)$ and   **b.** $-12t(-6)$.

Solution

a. $-2(7x) = (-2 \cdot 7)x$ Apply the associative property of multiplication to regroup the factors.

         $= -14x$ Do the operation inside the parentheses first: $-2 \cdot 7 = -14$.

b. $-12t(-6) = -12(-6)t$ Apply the commutative property of multiplication. Change the order of the factors.

         $= [-12(-6)]t$ Apply the associative property of multiplication to group the numbers together.

         $= 72t$ Do the operation in the brackets first: $-12(-6) = 72$.

Self Check

Simplify each algebraic expression:

a. $4 \cdot 8r$

b. $-3y(-5)$

Answer: **a.** $32r$,   **b.** $15y$ ■

In the next example, we will work with algebraic expressions that involve two variables.

EXAMPLE 2   *Simplifying an algebraic expression involving two variables.*   Simplify each algebraic expression: **a.** $-4m(-5n)$ and   **b.** $2(6y)(-4z)$.

Solution

a. $-4m(-5n) = [-4(-5)](m \cdot n)$ Group the numbers and variables separately, using the commutative and associative properties of multiplication.

         $= 20mn$ Do the multiplication inside the brackets: $-4(-5) = 20$. Write $m \cdot n$ as mn.

b. $2(6y)(-4z) = [2(6)(-4)](y \cdot z)$ Use the commutative and associative properties to change the order and regroup the factors.

         $= (-48)(yz)$ Do the multiplication inside the brackets: $2(6)(-4) = -48$. Write $y \cdot z$ as yz.

         $= -48yz$ Show the multiplication without parentheses.

Self Check

Simplify each algebraic expression:

a. $-7k(-5t)$

b. $2(4a)(-3d)$

Answer: **a.** $35kt$,   **b.** $-24ad$ ■

The distributive property

Another property of algebra that is used to simplify algebraic expressions is the **distributive property.** To introduce this property, we will examine the expression $2(5 + 3)$, which can be evaluated in two ways.

Method 1   Rule for Order of Operations

Because of the grouping symbols contained in $2(5 + 3)$, the rule for order of operations states that we must compute the *sum* inside the parentheses first.

       $2(5 + 3) = 2(8)$ Do the addition inside the parentheses first: $5 + 3 = 8$.

              $= 16$ Do the multiplication.

Method 2 The Distributive Property

The distributive property allows us to evaluate the numerical expression $2(5 + 3)$ in another way. We can "distribute" the factor of 2 across to the 5 and across to the 3. We then find each of those products separately and add the results. This process is called *removing parentheses*.

"Distribute" the 2.

$$2(5 + 3) = 2(5 + 3)$$

Apply the distributive property. Each term inside the parentheses is multiplied by the factor outside the parentheses.

First product ⎯⎯⎯⎯⎯   Second product

$$= 2(5) + 2(3)$$

$$= 10 + 6$$ Apply the rules for order of operations. Do the multiplications first: $2(5) = 10$ and $2(3) = 6$.

$$= 16$$ Do the addition.

Notice that the result in each method is 16.

We now state the distributive property in symbols.

The distributive property	If a, b, and c are any numbers, then $a(b + c) = ab + ac$

Since subtraction is the same as adding the opposite, the distributive property also holds for subtraction.

The distributive property	If a, b, and c are any numbers, then $a(b - c) = ab - ac$

EXAMPLE 3 *Applying the distributive property.* Apply the distributive property to remove parentheses.
  a. $3(s + 7)$ and   **b.** $6(x - 1)$

Solution

a. $3(s + 7) = 3s + 3(7)$ Apply the distributive property.

$$= 3s + 21$$ Do the multiplication.

b. $6(x - 1) = 6x - 6(1)$ Apply the distributive property.

$$= 6x - 6$$ Do the multiplication.

Self Check

Apply the distributive property to remove parentheses:

a. $5(h + 4)$

b. $9(a - 3)$

Answer: **a.** $5h + 20$,
b. $9a - 27$

 WARNING! If an expression contains parentheses, that does not automatically mean that the distributive property can be applied to simplify it. For example, the distributive property does not apply to the expressions $5(4x)$ or $5(4 \cdot x)$, where a product is multiplied by 5. The distributive property does apply to $5(4 + x)$ or $5(4 - x)$, where a sum or difference is multiplied by 5.

EXAMPLE 4 *Applying the distributive property.* Multiply:
  a. $-3(4x + 2)$ and   **b.** $-9(3 - 2t)$.

Solution

a. $-3(4x + 2) = -3(4x) + (-3)(2)$ Apply the distributive property.

$$= -12x + (-6)$$ Do the multiplications.

$$= -12x - 6$$ Subtraction is the same as addition of the opposite.

Self Check

Multiply:

a. $-4(6y + 8)$

b. $-7(2 - 8m)$

b. $-9(3 - 2t) = -9(3) - (-9)(2t)$ Apply the distributive property.

$\qquad\qquad = -27 - (-18t)$ Do the multiplications.

$\qquad\qquad = -27 + [-(-18t)]$ Subtraction is the same as addition of the opposite.

$\qquad\qquad = -27 + 18t$ Apply the double negative rule: $-(-18t) = 18t$.

Answer: **a.** $-24y - 32$,   **b.** $-14 + 56m$ ∎

Extending the distributive property

The distributive property can be extended to situations where there are more than two terms within parentheses.

The extended distributive property	If a, b, c, and d are any numbers, then $$a(b + c + d) = ab + ac + ad$$

EXAMPLE 5   *Applying the extended distributive property.*   Apply the extended distributive property to remove parentheses: $-6(-3x - 6y + 8)$.

Solution   We distribute the factor -6 over the three terms inside the parentheses.

$-6(-3x - 6y + 8) = -6(-3x) - (-6)(6y) + (-6)(8)$ Apply the distributive property.

$\qquad\qquad\qquad\qquad = 18x - (-36y) + (-48)$ Do the multiplications.

$\qquad\qquad\qquad\qquad = 18x + [-(-36y)] + (-48)$ Write the subtraction as addition of the opposite of $-36y$.

$\qquad\qquad\qquad\qquad = 18x + 36y - 48$ Apply the double negative rule: $-(-36y) = 36y$. ∎

Since multiplication is commutative, we can write the distributive property in either of the following forms:

$$(b + c)a = ba + ca$$
$$(b - c)a = ba - ca$$

EXAMPLE 6   *Applying the distributive property.*   Multiply: **a.** $(5 + 3r)7$ and   **b.** $(4 - x)2$.

Solution

a. $(5 + 3r)7 = (5)7 + (3r)7$ Apply the distributive property.

$\qquad\qquad = 35 + 21r$ Do the multiplications.

b. $(4 - x)2 = (4)2 - (x)2$ Apply the distributive property.

$\qquad\qquad = 8 - 2x$ Do the multiplications.

Self Check

Multiply:

a. $(8 + 7x)5$

b. $(5 - c)3$

Answer: **a.** $40 + 35x$,   **b.** $15 - 3c$ ∎

At first glance, the expression $-(x + 8)$ doesn't appear to be in the proper form to be simplified using the distributive property; the number in front of the parentheses is missing. But the negative sign in front of the parentheses represents the number -1.

The negative sign represents -1.

$-(x + 8) = -1(x + 8)$

$\qquad\qquad = -1(x) + (-1)(8)$ Apply the distributive property. Distribute -1.

$\qquad\qquad = -x + (-8)$ Do the multiplications.

$\qquad\qquad = -x - 8$ Subtraction is the same as adding the opposite.

EXAMPLE 7

"Distributing" a negative sign. Simplify the expression: $-(-6 - 2e)$.

Solution

$-(-6 - 2e) = -1(-6 - 2e)$	Rewrite the negative sign in front of the parentheses as -1.
$= -1(-6) - (-1)(2e)$	Apply the distributive property. Distribute the -1.
$= 6 - (-2e)$	Do the multiplications.
$= 6 + [-(-2e)]$	Write the subtraction as addition of the opposite.
$= 6 + 2e$	Apply the double negative rule: $-(-2e) = 2e$.

Self Check

Simplify the expression:
$-(-2t + 4)$.

Answer: $2t - 4$

After working several problems like Example 7, you will notice that it is not necessary to show each of the steps. The result can be obtained very quickly by changing the sign of each term inside the parentheses and dropping the parentheses.

STUDY SET   Section 8.4

VOCABULARY   *In Exercises 1–4, fill in the blanks to make a true statement.*

1. The _____ tells us how to multiply $5(7 + x)$. After doing the multiplication to obtain $35 + 5x$, we say that the parentheses were _____.

2. To _____ an algebraic expression means to use algebraic properties to write it in simpler form.

3. When an algebraic expression is simplified, the result is an _____ expression.

4. A _____ is a letter that stands for a number.

CONCEPTS

5. State the distributive property using the variables x, y, and z.

6. Use the variables r, s, and t to state the distributive property of subtraction.

7. The following expressions are examples of the *right* and *left* distributive properties:

$5(w + 7)$ and $(w + 7)5$

Which of the two do you think would be termed the right distributive property?

8. For each of the following expressions, tell whether the distributive property applies.
   a. $2(5t)$ **b.** $-2(5 - t)$
   c. $5(-2 \cdot t)$ **d.** $(-2t)5$
   e. $(2)(-t)5$ **f.** $(5 + t)(-2)$

9. The distributive property can be demonstrated using Illustration 1. Fill in the blanks below.
   Two groups of 6 plus three groups of 6 is ___ groups of 6. Therefore,
   ___ $\cdot 2 +$ ___ $\cdot 3 = 6($___ $+$ ___$)$

10. The distributive property of subtraction can be demonstrated using Illustration 2. Fill in the blanks below.
   Six groups of 5 minus four groups of 5 leaves ___ groups of 5. Therefore,
   ___ $\cdot 6 -$ ___ $\cdot 4 = 5($___ $-$ ___$)$

ILLUSTRATION 1

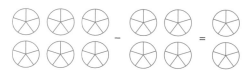

ILLUSTRATION 2

11. Write the expression $-(y + 9)$ without parentheses.

12. When we write $6x$, what operation is indicated?

13. Explain what the arrows are illustrating.

$$-9(y - 7)$$

14. Simplify each expression.
a. $-(-5)$
b. $-(-x)$

NOTATION *In Exercises 15–18, complete each solution.*

15. Multiply $-5(7n)$.

$-5(7n) = (\underline{\quad} \cdot 7)n$ Apply the associative property of multiplication.

$= -35n$ Do the operation inside the parentheses.

16. Multiply $6y(-9)$.

$6y(-9) = 6(\underline{\quad})y$ Apply the commutative property of multiplication.

$= [6(\underline{\quad})]y$ Apply the associative property of multiplication.

$= -54y$ Do the multiplication in the brackets.

17. Multiply $-9(-4 - 5y)$.

$-9(-4 - 5y) = (\underline{\quad})(-4) - (\underline{\quad})(5y)$ Apply the distributive property.

$= 36 - (\underline{\quad})$ Do the multiplications.

$= 36 + 45y$ Add the opposite.

18. Multiply $4(2a + b - 1)$.

$4(2a + b - 1) = 4(\underline{\quad}) + 4(\underline{\quad}) - \underline{\quad}(1)$ Apply the distributive property.

$= 8a + 4b - 4$ Do the multiplications.

PRACTICE *In Exercises 19–42, simplify each algebraic expression.*

19. $2(6x)$
20. $4(7b)$
21. $-5(6y)$
22. $-12(6t)$
23. $-10(-10t)$
24. $-8(-6k)$
25. $(4s)3$
26. $(9j)7$
27. $2c \cdot 7$
28. $11f \cdot 9$
29. $-5 \cdot 8h$
30. $-8 \cdot 4d$
31. $-7x(6y)$
32. $13a(-2b)$
33. $4r \cdot 4s$
34. $7x \cdot 7y$
35. $2x(5y)(3)$
36. $4(3z)(4)$
37. $5r(2)(-3b)$
38. $4d(5)(-3e)$
39. $5 \cdot 8c \cdot 2$
40. $3 \cdot 6j \cdot 2$
41. $(-1)(-2e)(-4)$
42. $(-1)(-5t)(-1)$

In Exercises 43–62, use the distributive property to remove parentheses.

43. $4(x + 1)$
44. $5(3 + y)$
45. $4(4 - x)$
46. $5(7 + k)$
47. $-2(3e + 3)$
48. $-5(7t + 2)$
49. $-8(2q - 6)$
50. $-5(3p - 8)$

51. $-4(-3 - 5s)$
52. $-6(-1 - 3d)$
53. $(7 + 4d)6$
54. $(8r + 2)7$

55. $(5r - 6)(-5)$
56. $(3z - 7)(-8)$
57. $(-4 - 3d)6$
58. $(-4 - 2j)5$

59. $3(3x - 7y + 2)$
60. $5(4 - 5r + 8s)$
61. $-3(-3z - 3x - 5y)$
62. $-10(5e + 4a + 6t)$

In Exercises 63–70, write each expression without using parentheses.

63. $-(x + 3)$
64. $-(5 + y)$
65. $-(4t + 5)$
66. $-(8x + 4)$
67. $-(-3w - 4)$
68. $-(-6 - 4y)$
69. $-(5x - 4y + 1)$
70. $-(6r - 5f + 1)$

In Exercises 71–78, each expression is the result of an application of the distributive property. What was the original algebraic expression?

71. $2(5) + 2(4x)$
72. $3(3y) + 3(7)$
73. $-4(5) - 3x(5)$
74. $-8(7) - (4s)(7)$

75. $-3(4y) - (-3)(2)$
76. $-5(11s) - (-5)(11t)$
77. $3(4) - 3(7t) - 3(5s)$
78. $2(7y) + 2(8x) - 2(4)$

79. Explain what it means to simplify an algebraic expression. Give an example.

81. Explain how to apply the distributive property.

80. Explain the commutative and associative properties of multiplication.

82. Explain why the distributive property applies to $2(3 + x)$ but does not apply to $2(3x)$.

REVIEW

83. Write the following steps of the problem-solving strategy in proper order: check the result, form an equation, state the conclusion, analyze the problem, solve the equation.

85. Identify the operation associated with each word: product, quotient, difference, sum.

87. Insert the proper inequality symbol: -6 ___ -7.

89. Which of the following involve area: carpeting a room, fencing a yard, walking around a lake, and painting a wall?

84. Find $-1 - (-4)$.

86. What are the steps used to find the mean (average) of a set of scores?

88. Fill in the blank to make a true statement: To factor a number means to express it as the _____ of other whole numbers.

90. Write seven squared and seven cubed.

8.5 *Combining Like Terms*

In this section, you will learn about

- **Terms of an algebraic expression**
- **Coefficients of a term**
- **Terms and factors**
- **Like terms**
- **Combining like terms**
- **Perimeter formulas**

INTRODUCTION. In this section, we will see how the distributive property can be used to simplify algebraic expressions that involve addition and subtraction. We will also review the concept of perimeter and write the formulas for the perimeter of a rectangle and a square using variables.

Terms of an algebraic expression

Addition signs break algebraic expressions into smaller parts called **terms.** The expression $3x + 8$ contains two terms, $3x$ and 8.

$$3x + 8$$

Term ⟍   ⟋ Term

The addition sign breaks the expression
into two terms.

| **Term** | A **term** is a number or a product of a number and one or more variables. |

If an algebraic expression involves subtraction, the subtraction can be expressed as addition of the opposite. For example, $5x - 6$ can be written in the equivalent form $5x + (-6)$. We can then see that $5x - 6$ contains two terms, $5x$ and -6.

EXAMPLE 1 *Identifying terms.* List the terms of each algebraic expression: **a.** $-3a + 5 + 8a$, **b.** $-24rst$, and **c.** $x - 5 - 3x + 10$.

Solution

a. $-3a + 5 + 8a$ contains three terms: $-3a$, 5, and $8a$.

b. $-24rst$ is one term.

c. $x - 5 - 3x + 10$ can be written as $x + (-5) + (-3x) + 10$. It contains four terms: x, -5, $-3x$, and 10.

Self Check

List the terms of each algebraic expression.

a. $-12y + 5y + 10$

b. $-4abc$

c. $9 - m - 6n + 12$

Answer: **a.** $-12y$, $5y$, 10, **b.** $-4abc$, **c.** 9, $-m$, $-6n$, 12

Coefficients of a term

A term of an algebraic expression can consist of a single number, a single variable, or a product of numbers and variables.

Numerical coefficient | In a term that is the product of a number and one or more variables, the number factor is called the **numerical coefficient** of the term.

In the expression $3x$, 3 is the *numerical coefficient,* and x is the *variable* part. Some more examples are shown in Table 8-3.

Term	Numerical coefficient	Variable part
$6b^2$	6	b^2
$-5xy$	-5	xy
$9rst$	9	rst
x	1	x
$-y$	-1	y

TABLE 8-3

Notice that when there is no number in front of a variable, the coefficient is understood to be 1. For example, the coefficient of the term x is 1. If there is only a negative (or opposite) sign in front of the variable, the coefficient is understood to be -1. Therefore, $-y$ can be thought of as $-1y$.

EXAMPLE 2 *Identifying coefficients of terms.* Identify the coefficient of each term in the expression $-5x^2 + x - 15y$.

Solution

We can write $-5x^2 + x - 15y$ as $-5x^2 + x + (-15y)$.

Term	Coefficient
$-5x^2$	-5
x	1
$-15y$	-15

Self Check

Identify the coefficient of each term in the expression $6y^3 - y + 7x$.

Answer: 6, -1, 7

Terms and factors

It is important to be able to distinguish between the *terms* of an algebraic expression and the *factors* of a term. Terms are separated by + signs, and factors are numbers that are multiplied together.

Consider the algebraic expression $-2x + 3xy$, which contains two terms.

The first term.   The second term.

$$-2x + 3xy$$

The first term contains two factors: -2 and x.

The second term contains three factors: 3, x, and y.

EXAMPLE 3   Determine whether the variable y is a factor or a term of each expression: **a.** $8 + y$,   and   **b.** $8y$.

Solution

a. y is a term of $8 + y$, since it is separated from 8 by a $+$ sign.

b. y is a factor of $8y$, since it is multiplied by 8.

Self Check

Determine whether the variable x is a factor or a term of each expression: **a.** $-5x$ and **b.** $x + 5$.

Answer: a. factor, b. term ■

Like terms

The expression $5t + 7s - 6t + 10$ contains four terms. The variable part of two of the terms, $5t$ and $-6t$, are identical. We say those two terms are **like** or **similar terms**.

Like terms (similar terms)   **Like terms,** or **similar terms,** are terms with exactly the same variables and exponents.

Like terms	Unlike terms
$2x$ and $3x$	$2x$ and $2t$
Identical variables	Different variables
$-3a^2b$, $16a^2b$ and $125a^2b$	$8xy - 14xy^2$
Identical variables to identical powers	Different exponent on the variable y

 WARNING! When looking for like terms, don't look at the coefficients of the terms. Consider only their variable parts.

EXAMPLE 4   *Identifying like terms.*   List the like terms in each of the following expressions.

a. $5a + 6 + 3a$

b. $x^2 - 3x^5 + 2$

c. $-5x + 3 + 1 + x$

Solution

a. $5a + 6 + 3a$ contains three terms. The like terms are $5a$ and $3a$.

b. $x^2 - 3x^5 + 2$ does not contain any like terms.

c. $-5x + 3 + 1 + x$ contains two pairs of like terms: $-5x$ and x are like terms and the constants 3 and 1 are like terms.

Self Check

List the like terms in each of the following expressions:

a. $10c + 8 + 7$

b. $1 + 3x + x^2 + x^3$

c. $4r + 8t - 3r + t$

Answer: a. 8 and 7,   b. none,   c. $4r$ and $-3r$, $8t$ and t ■

Combining like terms

If we are to add or subtract objects, they must be similar. For example, fractions that are to be added must have a common denominator. When adding decimals, we align columns to be sure that we add tenths to tenths, hundredths to hundredths, and so on. The same is true when working with terms of an algebraic expression. They can be added or subtracted only if they are like or similar terms.

This expression cannot be simplified because its terms are unlike.

$$3x + 4y$$

Unlike terms.
The variable parts are not identical.

This expression can be simplified, because it contains like terms.

$$3x + 4x$$

Like terms.
The variable parts are identical.

To simplify an expression containing like terms, we use the distributive property. For example, we can simplify $3x + 4x$ as follows:

$3x + 4x = (3 + 4)x$ Apply the distributive property.

$\qquad = 7x$ Do the addition in the parentheses: $3 + 4 = 7$.

We say that we have *simplified* the expression $3x + 4x$ by combining like terms. The result is the equivalent expression $7x$.

EXAMPLE 5 ***Simplifying algebraic expressions.*** Simplify each expression by combining like terms: **a.** $-3x + 7x$ and **b.** $6yz - 4yz$.

Self Check

Simplify each expression by combining like terms:

a. $-5b + 10b$

b. $12c - 9c$

Solution

a. $-3x + 7x = (-3 + 7)x$ Apply the distributive property.

$\qquad = 4x$ Do the addition inside the parentheses: $-3 + 7 = 4$.

b. $6yz - 4yz = (6 - 4)yz$ Apply the distributive property.

$\qquad = 2yz$ Do the subtraction inside the parentheses: $6 - 4 = 2$.

Answer: a. $5b$, **b.** $3c$ ■

The results of Example 5 suggest the following rule.

Combining like terms   To add or subtract like terms, combine their numerical coefficients and keep the same variables and exponents.

EXAMPLE 6 ***Combining like terms.*** Simplify by combining like terms: **a.** $-7x + (-9x)$ and **b.** $6mp - 3mp$.

Self Check

Simplify by combining like terms:

a. $-5n + (-2n)$

b. $15r - 4r$

Solution

a. $-7x + (-9x) = -16x$ Add the coefficients of the like terms: $-7 + (-9) = -16$. Keep the variable x.

b. $6mp - 3mp = 3mp$ Subtract: $6 - 3 = 3$. Keep the variables mp.

Answer: a. $-7n$, **b.** $11r$ ■

EXAMPLE 7 *Expressions involving subtraction.* Simplify by combining like terms: **a.** $16h - 24h$ and **b.** $-4d - (-5d)$.

Solution

a. $16h - 24h = 16h + (-24h)$ Add the opposite of $24h$.

$\qquad\qquad = -8h$ Add the coefficients of the like terms: $16 + (-24) = -8$. Keep the variable h.

b. $-4d - (-5d) = -4d + [-(-5d)]$ Add the opposite of $-5d$. Brackets are used to show this.

$\qquad\qquad = -4d + 5d$ $-(-5d) = 5d$.

$\qquad\qquad = 1d$ Add the coefficients of the like terms: $-4 + 5 = 1$. Keep the variable d.

$\qquad\qquad = d$ $1d = d$.

EXAMPLE 8 *Combining like terms.* Simplify $8s - 8S - 5s + S$ by combining like terms.

Solution The lowercase s and the capital S are different variables. We will first write each subtraction as addition of the opposite and then proceed as follows.

$8s - 8S - 5s + S = 8s + (-8S) + (-5s) + S$ Rewrite the subtraction as addition of the opposite.

$\qquad\qquad = 8s + (-5s) + (-8S) + S$ Use the commutative property of addition to get the like terms together.

$\qquad\qquad = 3s + (-7S)$ Combine like terms: $8s + (-5s) = 3s$ and $(-8S) + S = -7S$.

$\qquad\qquad = 3s - 7S$ Write as a subtraction.

EXAMPLE 9 *Simplifying an algebraic expression.* Simplify $4(x + 3) - 2(x - 1)$.

Solution

$4(x + 3) - 2(x - 1) = 4x + 12 - 2x + 2$ Apply the distributive property twice.

$\qquad\qquad = 4x - 2x + 12 + 2$ Use the commutative property of addition to get like terms together.

$\qquad\qquad = 2x + 14$ Combine like terms.

The expressions in Examples 8 and 9 contained two sets of like terms. In each solution, we rearranged terms so that like terms were next to each other. However, with practice you will find that you can combine like terms without having to write them next to each other.

EXAMPLE 10 *Combining without rearranging.* Simplify $-5(x - 2) + 8x - 6$.

Solution

$-5(x - 2) + 8x - 6 = -5x + 10 + 8x - 6$ Distribute -5.

$\qquad\qquad = 3x + 4$ Combine like terms: $-5x + 8x = 3x$ and $10 - 6 = 4$.

Perimeter formulas

FIGURE 8-8

To develop the formula for the perimeter of a rectangle, we let l = length of the rectangle and w = width of the rectangle. (See Figure 8-8.) Then

$P = l + w + l + w$ The perimeter is the distance around the rectangle.

$= 2l + 2w$ Combine like terms.

The perimeter of a rectangle	The perimeter P of a rectangle with length l and width w is given by $$P = 2l + 2w$$

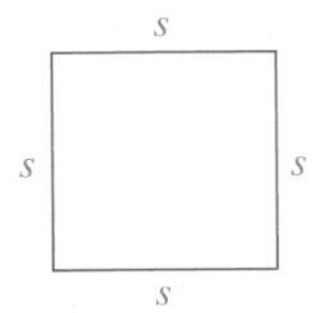

To develop the formula for the perimeter of a square, we let s = the length of a side of the square. (See Figure 8-9.) Then

$P = s + s + s + s$ Add the lengths of the four sides.

$= 4s$ Combine like terms.

FIGURE 8-9

The perimeter of a square	The perimeter of a square with sides of length s is given by $$P = 4s$$

EXAMPLE 11   ***Energy conservation.***   See Figure 8-10. Find the cost to weatherstrip the front door and window of the house if the material costs 20 cents a foot.

FIGURE 8-10

Analyze the problem   To find the cost of material, we must find the perimeter of the door and the window. The door is in the shape of a rectangle, and the window is in the shape of a square.

Form an equation   Let P represent the total perimeter and translate the words of the problem into an equation. We will express the total perimeter in two ways.

The total perimeter	is	the perimeter of the door	plus	the perimeter of the window.
P	$=$	$2l + 2w$	$+$	$4s$

Write the formulas for the perimeter of a rectangle and a square.

Solve the equation $P = 2(7) + 2(3) + 4(3)$ Replace l with 7, w with 3, and s with 3.

$= 14 + 6 + 12$ Do the multiplications.

$= 32$ Do the additions.

State the conclusion   The total perimeter is 32 feet. At 20 cents a foot, the total cost will be $(32 \cdot 20)$ cents. This is 640 cents or $6.40.

Check the result   We can check the results by estimation. The perimeter is approximately 30 feet, and $30 \cdot 20 = 600$ cents, which is $6. The answer of $6.40 seems reasonable.   ■

VOCABULARY *In Exercises 1–8, fill in the blanks to make a true statement.*

1. A _____ is a number or a product of a number and one or more variables.

2. In the term $5t$, 5 is called the numerical _____.

3. The _____ of a geometric figure is the distance around it.

4. A _____ is a general rule that mathematically describes a relationship between two or more variables.

5. $2(x + 3) = 2x + 2 \cdot 3$ is an example of the _____ property.

6. Terms with exactly the same variables and exponents are called _____.

7. When an algebraic expression contains like terms, an _____ expression can be obtained by _____ like terms.

8. The numbers multiplied together to form a product are called _____.

CONCEPTS

9. Tell whether x is used as a factor or as a term.
   a. $12 + x$ **b.** $7x$
   c. $12y + 12x - 6$ **d.** $-36xy$

10. Tell whether $6y$ is used as a factor or as a term.
   a. $6yx$ **b.** $10 + 6y$
   c. $9xy + 6y$ **d.** $6y - 18$

11. Use the variables a, b, and c to state the distributive property.

12. Use the distributive property to complete this expression.

$(w + y + z)x = ?$

13. What is the numerical coefficient of each term?
   a. $11x$ **b.** $8t$
   c. $-4x^2$ **d.** a
   e. $-x$ **f.** $102xy$

14. What is the numerical coefficient of the second term of each expression?
   a. $5x^2 + 6x + 7$ **b.** $xy - x + y + 10$
   c. $9y^2 + y + 8$ **d.** $5x^3 - 4x^2 + 3x + 1$

15. When simplifying an algebraic expression, some students use underlining.

$\underline{\underline{3y}} - \underline{4x} + \underline{\underline{5y}} + \underline{4x}$

What purpose does the underlining serve?

16. Tell whether each statement is true or false.
   a. $x = 1x$
   b. $2x = 2 \cdot x$
   c. $x - y = x + (-y)$
   d. $-y = -1y$

17. Illustration 1 shows the distance (in miles) that two men are from the office. Find the total distance traveled by the men from home to office.

ILLUSTRATION 1

18. The heights of two trees are shown in Illustration 2. Find the sum of their heights.

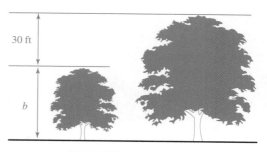

ILLUSTRATION 2

19. What does this diagram illustrate?

$9x + 5x = 14x$

20. What does this diagram illustrate?

$12k - 4k = 8k$

NOTATION *In Exercises 21–26, complete each solution.*

21. Simplify $5x + 7x$.

$\quad 5x + 7x = (5 + \underline{\quad})x \quad$ Apply the distributive property.

$\quad\quad\quad = 12x \quad\quad\quad\quad$ Do the addition inside the parentheses.

22. Simplify $12w - 16w$.

$\quad 12w - 16w = (\underline{\quad} - 16)w \quad$ Apply the distributive property.

$\quad\quad\quad = -4w \quad\quad\quad\quad$ Do the subtraction inside the parentheses.

23. Simplify $2a - 3b + 5a$.

$\quad 2a - 3b + 5a = 2a + 5a - \underline{\quad} \quad$ Rearrange the terms.

$\quad\quad\quad = (2 + 5)a - \underline{\quad} \quad$ Apply the distributive property.

$\quad\quad\quad = 7a - 3b \quad\quad\quad$ Do the addition.

24. Simplify $5 - 3h - 7$.

$\quad 5 - 3h - 7 = 5 - \underline{\quad} - 3h \quad$ Rearrange the terms.

$\quad\quad\quad = 5 + (\underline{\quad}) - 3h \quad$ Write subtraction as addition of the opposite.

$\quad\quad\quad = -2 - 3h \quad\quad\quad$ Do the addition.

25. Simplify $2(x - y) + 3x$.

$\quad 2(x - y) + 3x = 2x - \underline{\quad} + 3x \quad$ Apply the distributive property.

$\quad\quad\quad = 5x - 2y \quad\quad\quad$ Combine like terms.

26. Simplify $-3(y - b) - b$.

$\quad -3(y - b) - b = \underline{\quad} + \underline{\quad} - b \quad$ Apply the distributive property.

$\quad\quad\quad = -3y + 2b \quad\quad\quad$ Combine like terms.

PRACTICE *In Exercises 27–42, simplify by combining like terms, if possible.*

27. $6t + 9t$

28. $7r + 5r$

29. $5s - s$

30. $8y - y$

31. $-5x + 6x$

32. $-8m + 6m$

33. $-5d + 9d$

34. $-4a + 12a$

35. $3e - 7e$

36. $2s - 4s$

37. $-3x - 4x$

38. $-7y - 9y$

39. $4z - 10z$

40. $3w - 18w$

41. $h - 7b$

42. $j - 8s$

In Exercises 43–46, identify the terms of each expression.

43. $3x^2 - 5x + 4$

44. $6 + 12y - y^2$

45. $5 + 5t - 8t + 4$

46. $3x - y - 5x + y$

In Exercises 47–50, what exponent must appear in each box to make the terms like terms?

47. $3x^{\boxed{}}y, \ -6x^2y$

48. $7a^3b^4, \ 21a^{\boxed{}}b^4$

49. $-8h^5c^{\boxed{}}, \ -5h^{\boxed{}}c^2$

50. $25mn^3, \ -15m^{\boxed{}}n^{\boxed{}}$

In Exercises 51–60, simplify by combining like terms.

51. $6t + 9 + 5t + 3$

52. $5x + 3 + 5x + 4$

53. $3w - 4 - w - 1$

54. $6y + 6 - y - 1$

55. $-4r + 8R + 2R - 3r + R$

56. $12a - A - a - 8A - a$

57. $-45d - 12a - 5d + 12a$

58. $-m - n - 8m + n$

59. $4x - 3y - 7 + 4x - 2 - y$

60. $2a + 8 - b - 5 + 5a - 9b$

In Exercises 61–72, simplify each expression.

61. $4(x + 1) + 5(6 + x)$

62. $7(1 + y) + 8(2y + 3)$

63. $5(3 - 2s) + 4(2 - 3s)$

64. $6(t - 3) + 9(2 - t)$

65. $-4(6 - 4e) + 3(e + 1)$

66. $-5(7 - 4t) + 3(2 + 5t)$

67. $3t - (t - 8)$

68. $6n - (4n + 1)$

69. $-2(2 - 3x) - 3(x - 4)$

70. $-3(1 - y) - 5(2y - 6)$

71. $-4(-4y + 5) - 6(y + 2)$

72. $-3(-6y - 8) - 4(5 - y)$

73. MOBILE HOME DESIGN The design of a mobile home calls for a six-inch wide strip of stained pine around the outside of *each* exterior wall, as shown in Illustration 3. Find the number of feet of pine needed to complete this design.

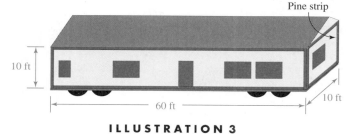

Pine strip

10 ft

60 ft

10 ft

ILLUSTRATION 3

74. LANDSCAPING DESIGN A landscape architect has designed a planter surrounding two birch trees, as shown in Illustration 4. The planter is to be outlined with redwood edging in the shape of a rectangle and two squares. How much edging will be needed to complete the job?

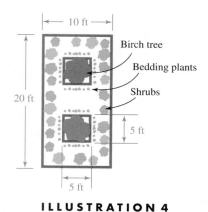

10 ft

Birch tree

Bedding plants

Shrubs

20 ft

5 ft

5 ft

ILLUSTRATION 4

WRITING *Write a paragraph using your own words.*

75. Explain what it means for two terms to be like terms.

76. Explain what it means to say that the coefficient of x is an understood 1.

77. Explain the difference between a term and a factor. Give some examples.

78. The formula for the perimeter of a rectangle is $P = 2l + 2w$. Explain why we can write this formula in the form $P = 2(l + w)$.

REVIEW

79. Solve: $-4t - 3 = -11$.

80. Evaluate: $(-1)(-1)(-1)$.

81. Find the prime factorization of 100.

82. List the integers.

83. Fill in the blank to make a true statement. The _____ of a number is the distance between it and 0 on the number line.

84. State the division property of equality in words.

8.6 *Simplifying Expressions to Solve Equations*

In this section, you will learn about

- **Solutions to equations**
- **Combining like terms**
- **Variables on both sides of the equation**
- **Removing parentheses**
- **A strategy for solving equations**

INTRODUCTION. We must often simplify algebraic expressions to solve equations. Sometimes it will be necessary to combine like terms in order to isolate the variable on one side of the equation. At other times, it will be necessary to apply the distributive property to write an equation in a form that can be solved. In this section, we will discuss both of these situations.

Solutions to equations

To solve an equation means to find all values of the variable that make the equation a true statement.

EXAMPLE 1 *Checking a solution.* Is $x = -3$ a solution of
$3x - 1 + 2x = 6x + 8$?

Solution

$$3x - 1 + 2x = 6x + 8$$
$$3(-3) - 1 + 2(-3) \stackrel{?}{=} 6(-3) + 8 \quad \text{Replace } x \text{ with } -3.$$
$$-9 + (-1) + (-6) \stackrel{?}{=} -18 + 8 \quad \text{Do the multiplications. Write subtraction as addition of the opposite.}$$
$$-16 = -10 \quad \text{Simplify the left and right sides of the equation.}$$

Since $-16 \neq -10$, -3 is not a solution.

Self Check

Is $w = -4$ a solution of
$-4w - 6 = 2w - 2 - 5w$?

Answer: yes

Combining like terms

When solving equations, we must often combine like terms before applying any properties of equality.

EXAMPLE 2 *Combining like terms.* Solve $5x - 4 - 7x = -4 - 10$. Check the solution.

Solution

$$5x - 4 - 7x = -4 - 10$$
$$-2x - 4 = -14 \quad \text{Combine like terms.}$$
$$-2x - 4 + 4 = -14 + 4 \quad \text{To undo the subtraction of 4, add 4 to both sides.}$$
$$-2x = -10 \quad \text{Simplify.}$$
$$\frac{-2x}{-2} = \frac{-10}{-2} \quad \text{To undo the multiplication by } -2, \text{ divide both sides by } -2.$$
$$x = 5 \quad \text{Do the divisions.}$$

We check the solution of 5 by substituting 5 for x in the original equation and simplifying.

$$5x - 4 - 7x = -4 - 10$$
$$5(5) - 4 - 7(5) \stackrel{?}{=} -4 - 10 \quad \text{Replace } x \text{ with 5.}$$
$$25 - 4 - 35 \stackrel{?}{=} -14 \quad \text{Do the multiplications on the left side. Do the subtraction on the right side.}$$
$$21 - 35 \stackrel{?}{=} -14 \quad 25 - 4 = 21.$$
$$-14 = -14 \quad 21 - 35 = -14.$$

The solution checks.

Self Check

Solve $8 + 2r - 7 = 6 - 11$.
Check the solution.

Answer: $r = -3$

Variables on both sides of the equation

When solving equations, we want to isolate the variable on one side of the $=$ sign. If variables appear on both sides, we can use the addition (or subtraction) property of equality to get all variable terms on one side and all constant terms on the other.

EXAMPLE 3 *Variables on both sides of an equation.* Solve $6y - 4 = 12 + 2y$.

Solution

$$6y - 4 = 12 + 2y$$

$6y - 4 - 2y = 12 + 2y - 2y$ To eliminate $2y$ from the right-hand side, subtract $2y$ from both sides.

$4y - 4 = 12$ Combine like terms.

$4y - 4 + 4 = 12 + 4$ To eliminate -4 from the left-hand side, add 4 to both sides.

$4y = 16$ Combine like terms.

$\dfrac{4y}{4} = \dfrac{16}{4}$ To undo the multiplication by 4, divide both sides by 4.

$y = 4$ Do the divisions.

Self Check

Solve $5B + 6 = -18 - 3B$.

Answer: $B = -3$

EXAMPLE 4 *Variables on both sides of an equation.* Solve $5t + 1 = 3(2t) + 9$.

Solution

$$5t + 1 = 3(2t) + 9$$

$5t + 1 = 6t + 9$ Do the multiplication: $3(2t) = 6t$.

$5t + 1 - 5t = 6t + 9 - 5t$ To eliminate $5t$ from the left-hand side, subtract $5t$ from both sides.

$1 = t + 9$ Combine like terms.

$1 - 9 = t + 9 - 9$ To eliminate 9 from the right-hand side, subtract 9 from both sides.

$-8 = t$ Simplify.

$t = -8$ Interchange the sides of the equation.

Self Check

Solve $(6d)2 - 3 = 13d + 9$.

Answer: $d = -12$

Removing parentheses

At times, you will have to apply the distributive property to solve an equation.

EXAMPLE 5 *Removing parentheses to solve an equation.* Solve $-3(x - 4) = 2(2x + 6)$.

Solution

$$-3(x - 4) = 2(2x + 6)$$

$-3x - (-3)(4) = 2(2x) + 2(6)$ Apply the distributive property to remove parentheses.

$-3x + 12 = 4x + 12$ Do the multiplications.

$-3x + 12 + 3x = 4x + 12 + 3x$ To eliminate $-3x$ from the left-hand side, add $3x$ to both sides.

$12 = 7x + 12$ Combine like terms.

$12 - 12 = 7x + 12 - 12$ To eliminate 12 from the right-hand side, subtract 12 from both sides.

$0 = 7x$ Simplify.

$\dfrac{0}{7} = \dfrac{7x}{7}$ To undo the multiplication by 7, divide both sides by 7.

$0 = x$ Simplify.

$x = 0$ Interchange the sides of the equation.

Self Check

Solve $7(4 - t) = -2(3t - 3)$

Answer: $t = 22$

A strategy for solving equations

To summarize, when solving an equation we must isolate the variable on one side of the = sign. At times, this requires that we remove parentheses and/or combine like terms. The following steps should be applied, in order, when solving an equation.

Strategy for solving equations	1. Use the distributive property to remove any parentheses.
	2. Combine like terms on either side of the equation.
	3. Apply the addition or subtraction properties of equality to get the variables on one side of the = sign and the constants on the other.
	4. Continue to combine like terms when necessary.
	5. Undo the operations of multiplication and division to isolate the variable.

You won't always have to use all five steps to solve a given equation. If a step doesn't apply, skip it and go to the next step.

EXAMPLE 6 *Removing parentheses.* Solve $7 + 2x = 2 - (4x + 7)$.

Self Check
Solve $7 + 4x = -13 - (3x + 1)$.

Solution

$$7 + 2x = 2 - (4x + 7)$$

$7 + 2x = 2 - 4x - 7$ Apply the distributive property to remove parentheses.

$7 + 2x = -5 - 4x$ Combine like terms.

$7 + 2x + 4x = -5 - 4x + 4x$ To eliminate $-4x$ from the right-hand side, add $4x$ to both sides.

$7 + 6x = -5$ Combine like terms.

$7 + 6x - 7 = -5 - 7$ To eliminate 7 from the left-hand side, subtract 7 from both sides.

$6x = -12$ Combine like terms.

$\dfrac{6x}{6} = \dfrac{-12}{6}$ Divide both sides by 6.

$x = -2$ Do the divisions.

Answer: $x = -3$

STUDY SET   Section 8.6

VOCABULARY *In Exercises 1–6, fill in the blanks to make a true statement.*

1. To _____ an equation means to find all values of the variable that make the equation a true statement.

2. To _____ a solution means to substitute that value into the original equation to see whether a true statement results.

3. In $2(x + 4)$, to remove parentheses means to apply the _____ property.

4. Algebraic _____ are simplified, and _____ are solved.

5. The phrase *combine like terms* refers to the operations of _____ and _____.

6. A _____ is a letter that represents a number. A _____ is a number that is fixed and does not change value.

CONCEPTS

7. What property is illustrated here?

Let x, y, and z be any numbers.

If $x = y$, then $x + z = y + z$.

8. What property is illustrated here?

Let x, y, and z be any numbers with $z \neq 0$.

If $x = y$, then $\dfrac{x}{z} = \dfrac{y}{z}$.

9. What property is illustrated here?

Let x, y, and z be any numbers.

If $x = y$, then $x - z = y - z$.

10. What property is illustrated here?

Let x, y, and z be any numbers, with $z \neq 0$.

If $x = y$, then $z \cdot x = z \cdot y$.

11. a. Evaluate $2(x + 1)$ for $x = -4$.
  b. Multiply $2(x + 1)$.
  c. Solve $2(x + 1) = -4$.

12. List, in order, the five steps of the strategy for solving equations.

13. a. Simplify $3x + 4 - x$.
  b. Solve $3x + 4 - x = 8$.

14. a. Solve $6x + 1 - 5x = 13$.
  b. Evaluate $6x + 1 - 5x$ for $x = 3$.

NOTATION *In Exercises 15–18, complete each solution.*

15. Solve $4x + 5 - 2x = -15$.

$$4x + 5 - 2x = -15$$
$$\underline{} + 5 = -15 \qquad \text{Combine like terms.}$$
$$2x + 5 - \underline{} = -15 - \underline{} \qquad \text{Subtract 5 from both sides.}$$
$$2x = -20 \qquad \text{Combine like terms.}$$
$$\frac{2x}{} = \frac{-20}{} \qquad \text{Divide both sides by 2.}$$
$$x = -10 \qquad \text{Do the divisions.}$$

16. Solve $8y - 6 = -2 + 10y$.

$$8y - 6 = -2 + 10y$$
$$8y - 6 - \underline{} = -2 + 10y - \underline{} \qquad \text{Subtract } 8y \text{ from both sides.}$$
$$-6 = -2 + \underline{} \qquad \text{Combine like terms.}$$
$$-6 + \underline{} = -2 + 2y + \underline{} \qquad \text{Add 2 to both sides.}$$
$$-4 = \underline{} \qquad \text{Simplify.}$$
$$\frac{-4}{} = \frac{2y}{} \qquad \text{Divide both sides by 2.}$$
$$-2 = y \qquad \text{Do the divisions.}$$

17. Solve $5(x - 9) = 5$.

$$5(x - 9) = 5$$
$$5x - 5(\underline{}) = 5 \qquad \text{Apply the distributive property.}$$
$$5x - \underline{} = 5 \qquad \text{Do the multiplication.}$$
$$5x - 45 + \underline{} = 5 + \underline{} \qquad \text{Add 45 to both sides.}$$
$$\underline{} = 50 \qquad \text{Combine like terms.}$$
$$\frac{5x}{} = \frac{50}{} \qquad \text{Divide both sides by 5.}$$
$$x = 10 \qquad \text{Do the divisions.}$$

18. Solve $-2(-1 - 3x) = 4(4 + x)$.

$$-2(-1 - 3x) = 4(4 + x)$$
$$2 + \underline{} = 16 + 4x \qquad \text{Apply the distributive property.}$$
$$2 + \underline{} - 4x = 16 + 4x - 4x \qquad \text{Subtract } 4x \text{ from both sides.}$$
$$2 + \underline{} = 16 \qquad \text{Simplify.}$$
$$2 + 2x - \underline{} = 16 - \underline{} \qquad \text{Subtract 2 from both sides.}$$
$$2x = \underline{} \qquad \text{Combine like terms.}$$
$$\frac{2x}{2} = \frac{}{2} \qquad \text{Divide both sides by 2.}$$
$$x = 7 \qquad \text{Do the divisions.}$$

PRACTICE *In Exercises 19–42, solve each equation.*

19. $5y - 7 + 1 = 10 - 1$

20. $3x - 2 = 12 + (-5)$

21. $5a - 4 + a = -28$

22. $4 = 7t + 8 - 5t$

23. $-3x - 5 = -7 - 16$

24. $7 - 4y + 1 = 0$

25. $4(x - 2) = 0$

26. $10(4z - 4) = 0$

27. $3s + 1 = 4s - 7$

28. $-4x - 7 = 3x + 21$

29. $38 - 5w = -10 + 7w$

30. $-x - 2 = 3x + 2$

31. $2(x + 6) = 4$

32. $9(y - 1) = 27$

33. $-(c - 4) = 3$

34. $-(6 - 2x) = -8$

35. $-3(2w - 3) = 9$ **36.** $-4(5t + 2) = -8$ **37.** $-16 = 2(t + 4)$ **38.** $-10 = 5(y - 7)$

39. $\dfrac{x}{2} - 2 - 1 = 4$ **40.** $4 + \dfrac{x}{3} = -3 - 3$ **41.** $1 - 3 - 5 + \dfrac{c}{-4} = 0$ **42.** $10 - 13 = 5 + \dfrac{t}{-5}$

In Exercises 43–56, solve each equation.

43. $2(x + 6) - 4 = 2$ **44.** $5(2r - 1) + 3 = -12$

45. $2(4y + 8) = 3(2y - 2)$ **46.** $3(7 - y) = 3(2y + 1)$

47. $5 - (7 - y) = -5$ **48.** $10 - (x - 5) = 40$

49. $4(r - 1) - 5(2r + 6) = -4$ **50.** $7(2q - 1) - 2(6q + 3) = -23$

51. $-(6x + 3) + 9(2x + 2) = 3$ **52.** $-(5x - 4) + 6(2x - 7) = -3$

53. $-3(4 - 2x) - 5(2x + 1) = -1$ **54.** $-4(2s + 5) - 4(2s - 5) = 32$

55. $35p + 2 - (15p + 3) = 58 + 1$ **56.** $45K - 6 - (35K + 12) = 11 + 11$

In Exercises 57–60, check whether the given number is a solution of the equation.

57. $4x + 2(x - 1) = -33, x = -5$ **58.** $5 - (4t + 9) = 10, t = -2$

59. $6f + 8 - f = 11 + 4f, f = 3$ **60.** $4d + 8 - d = 5d - 2, d = 5$

WRITING   *Write a paragraph using your own words.*

61. Explain the difference between an algebraic expression and an equation.

62. What does it mean to solve an equation?

63. When solving an equation, what does it mean to isolate the variable on one side of the equation?

64. Explain what it means to remove parentheses. When and how is it done?

REVIEW

65. Subtract: $-7 - 9$.

66. Which numbers are *not* factors of 28: 4, 6, 7, 8?

67. Evaluate: $\dfrac{-8 + 2}{-2 + 4}$.

68. What is wrong with the design of this number line?

69. Find: $-(-5)$.

70. Using 2 and 3, illustrate the commutative property of addition.

71. What is the sign of the product of two negative integers?

72. Evaluate: $3 + 4[-4 - 3(-2)]$.

8.7   *Exponents*

In this section, you will learn about

- **The product rule for exponents**
- **The power rule for exponents**
- **The power rule for products**

INTRODUCTION. In this chapter, we have applied the commutative, associative, and distributive properties to simplify algebraic expressions. We will now discuss three properties (or rules) that are used to simplify expressions that involve exponents.

The product rule for exponents

We have seen that exponents are used to represent repeated multiplication. For example, x^5 is an exponential expression with base x and an exponent of 5. It is called a *power of x*. Applying the definition of an exponent, we see that

$$x^5 = \underbrace{x \cdot x \cdot x \cdot x \cdot x}_{5 \text{ factors of } x}$$

EXAMPLE 1 *Writing expressions without exponents.* Write each expression without exponents: **a.** n^3, **b.** $(3h)^2$, and **c.** $(-6b)^4$.

Solution

a. For n^3, the base is n and the exponent is 3. Therefore,

$$n^3 = n \cdot n \cdot n$$

b. For $(3h)^2$, the base is $3h$ and the exponent is 2. Therefore,

$$(3h)^2 = 3h \cdot 3h$$

c. For $(-6b)^4$, the base is $-6b$ and the exponent is 4. Therefore,

$$(-6b)^4 = (-6b)(-6b)(-6b)(-6b)$$

Self Check

Write each expression without exponents:

a. t^4

b. $(12f)^3$

c. $(-2x)^2$

Answer: **a.** $t \cdot t \cdot t \cdot t$, **b.** $12f \cdot 12f \cdot 12f$, **c.** $(-2x)(-2x)$

The expression $x^3 \cdot x^5$ is the *product* of two powers of x. To develop a rule for multiplying them, we will use the fact that an exponent indicates repeated multiplication.

$$x^3 \cdot x^5 = \underbrace{(x \cdot x \cdot x)}_{3 \text{ factors of } x}\underbrace{(x \cdot x \cdot x \cdot x \cdot x)}_{5 \text{ factors of } x}$$

x^3 means to write x as a factor 3 times.
x^5 means to write x as a factor 5 times.

$$= \underbrace{x \cdot x \cdot x \cdot x \cdot x \cdot x \cdot x \cdot x}_{8 \text{ factors of } x}$$

Do the multiplication to get 8 factors of x.

$$= x^8$$

Since x is used as a factor 8 times, we can write the product as x^8.

Notice that the exponent of the result is the *sum* of the exponents in $x^3 \cdot x^5$.

$$\overset{\text{Sum of the exponents}}{x^3 \cdot x^5 = x^{3+5} = x^8}$$

This observation suggests the following rule.

The product rule for exponents	For any number x and any positive integers m and n, $$x^m \cdot x^n = x^{m+n}$$ To multiply two exponential expressions with the same base, add the exponents and keep the common base.

EXAMPLE 2 *Simplifying exponential expressions.* Simplify each product: **a.** $3^4 \cdot 3^7$, **b.** $(y^2)(y^4)$, and **c.** $x^2 x^4 x^9$.

Solution

a. $3^4 \cdot 3^7 = 3^{4+7}$ Since the bases are the same, we add the exponents and keep the common base.

$= 3^{11}$ Do the addition: $4 + 7 = 11$.

Self Check

Simplify each product:

a. $5^3 \cdot 5^6$

b. $m^2 \cdot m^3$

c. $a^3 a^8 a^2$

b. $(y^2)(y^4) = y^{2+4}$ Since the bases are the same, add the exponents and keep the common base.

       $= y^6$ Do the addition: $2 + 4 = 6$.

c. $x^2 x^4 x^9 = x^{2+4+9}$ Since the bases are the same, add the exponents and keep the common base.

       $= x^{15}$ Do the addition: $2 + 4 + 9 = 15$.

Answer: **a.** 5^9,   **b.** m^5,   **c.** a^{13} ■

> **WARNING!** We cannot use the product rule for exponents to simplify an expression such as $x^4 + x^3$, because it is not a product. Neither can we use it to simplify $x^4 \cdot y^3$, because the bases are not the same.

The product rule for exponents can be used to simplify more complicated algebraic expressions involving multiplication.

EXAMPLE 3   *Using the product rule.*   Simplify each product:
   a. $3a(5a)$ and   **b.** $-2t^2 \cdot 6t^6$.

Self Check

Simplify each product:

a. $4m \cdot 6m$

b. $8r^3(-5r^2)$

Solution

a. $3a(5a) = (3 \cdot 5)(a \cdot a)$ Apply the commutative and associative properties to change the order and regroup the factors.

      $= (3 \cdot 5)(a^1 \cdot a^1)$ Recall that $a = a^1$.

      $= 15a^{1+1}$ Do the multiplication. Add the exponents and keep the common base.

      $= 15a^2$ Do the addition: $1 + 1 = 2$.

b. $-2t^2 \cdot 6t^6 = (-2 \cdot 6)(t^2 \cdot t^6)$ Change the order of the factors and regroup them.

      $= -12t^{2+6}$ Do the multiplication. Add the exponents and keep the common base.

      $= -12t^8$ Do the addition: $2 + 6 = 8$.

Answer: **a.** $24m^2$,   **b.** $-40r^5$ ■

Exponential expressions often contain more than one variable.

EXAMPLE 4   *Working with two variables.*   Simplify the following:
   a. $n^2 m \cdot n^8 m^3$ and   **b.** $4xy^2(-3x^2 y^3)$.

Self Check

Simplify the following:

a. $c^3 d^2 \cdot cd^5$

b. $-7a^2 b^3(8a^4 b^5)$

Solution

a. $n^2 m \cdot n^8 m^3 = (n^2 \cdot n^8)(m \cdot m^3)$ Change the order and group the factors with like bases.

      $= n^{2+8} \cdot m^{1+3}$ Add the exponents of the like bases. Recall that $m = m^1$.

      $= n^{10} m^4$ Do the additions.

b. $4xy^2(-3x^2 y^3) = [4(-3)](x \cdot x^2)(y^2 \cdot y^3)$ Group the factors with like bases.

      $= -12 \cdot x^{1+2} \cdot y^{2+3}$ Do the multiplication. Add the exponents of the like bases.

      $= -12x^3 y^5$ Do the additions.

Answer: **a.** $c^4 d^7$,   **b.** $-56a^6 b^8$ ■

The power rule for exponents

To develop the power rule for exponents, we will consider the expression $(x^2)^5$. Notice that the base, x^2, is itself a power. Therefore, we are working with a power of a power. We will again rely on the definition of an exponent to find a rule for simplifying this exponential expression.

$$(x^2)^5 = x^2 \cdot x^2 \cdot x^2 \cdot x^2 \cdot x^2 \quad \text{The exponent 5 tells us to write the base } x^2 \text{ five times.}$$

$$= x^{2+2+2+2+2} \quad \text{Since the bases are alike, add the exponents and keep the common base.}$$

$$= x^{10} \quad \text{Do the addition: } 2 + 2 + 2 + 2 + 2 = 10.$$

Notice that the exponent of the result is the *product* of the exponents in $(x^2)^5$.

Product of the exponents

$$(x^2)^5 = x^{2 \cdot 5} = x^{10}$$

This observation suggests the following rule.

The power rule for exponents	For any number x and any positive integers m and n, $$(x^m)^n = x^{m \cdot n}$$ To raise an exponential expression to a power, keep the base and multiply the exponents.

EXAMPLE 5   ***The power rule for exponents.***   Simplify each expression: **a.** $(2^3)^7$ and   **b.** $(b^5)^3$.

Solution

a. $(2^3)^7 = 2^{3 \cdot 7}$ Apply the power rule for exponents by keeping the base and multiplying the exponents.

$$= 2^{21} \quad \text{Do the multiplication: } 3 \cdot 7 = 21.$$

b. $(b^5)^3 = b^{5 \cdot 3}$ Keep the base and multiply the exponents.

$$= b^{15} \quad \text{Do the multiplication: } 5 \cdot 3 = 15.$$

Self Check
Simplify each expression:

a. $(4^2)^6$

b. $(y^6)^4$

Answer: **a.** 4^{12},   **b.** y^{24} ◼

In some cases, when simplifying algebraic expressions involving exponents, two properties of exponents must be applied.

EXAMPLE 6   ***Applying two properties.***   Simplify: **a.** $(n^3)^4(n^2)^5$ and   **b.** $(n^2n^3)^5$.

Solution

a. $(n^3)^4(n^2)^5 = n^{3 \cdot 4} \cdot n^{2 \cdot 5}$ Keep each base and multiply their exponents.

$$= n^{12} \cdot n^{10} \quad \text{Do the multiplications: } 3 \cdot 4 = 12 \text{ and } 2 \cdot 5 = 10.$$

$$= n^{12+10} \quad \text{Since the bases are alike, keep the base and add the exponents.}$$

$$= n^{22} \quad \text{Do the addition: } 12 + 10 = 22.$$

b. $(n^2n^3)^5 = (n^{2+3})^5$ Work inside the parentheses first. Since the bases are alike, keep the base and add the exponents.

$$= (n^5)^5 \quad \text{Do the addition: } 2 + 3 = 5.$$

$$= n^{5 \cdot 5} \quad \text{Keep the base and multiply the exponents.}$$

$$= n^{25} \quad \text{Do the multiplication: } 5 \cdot 5 = 25.$$

Self Check
Simplify:

a. $(x^4)^2(x^3)^3$

b. $(x^4x^2)^3$

Answer: **a.** x^{17},   **b.** x^{18} ◼

The power rule for products

The exponential expression $(2x)^4$ has an exponent of 4 and a base of $2x$. The base $2x$ is a product, since $2x = 2 \cdot x$. Therefore, $(2x)^4$ is a power of a product. To find a rule to simplify it, we will again use the definition of exponent.

$$(2x)^4 = 2x \cdot 2x \cdot 2x \cdot 2x \qquad \text{Write the base, } 2x, \text{ as a factor 4 times.}$$

$$= (2 \cdot 2 \cdot 2 \cdot 2)(x \cdot x \cdot x \cdot x) \qquad \text{Apply the commutative and associative properties of multiplication to change the order and group like factors.}$$

$$= 2^4 x^4 \qquad \text{The factor 2 and the factor } x \text{ are both repeated 4 times. Apply the definition of an exponent.}$$

The result has factors of 2 and x. In the original problem, they were inside the parentheses. Each is now raised to the fourth power.

Each factor inside the parentheses ends up being raised to the 4th power.

$$(2x)^4 = 2^4 x^4$$

This observation suggests the following rule.

The power rule for products	For any numbers x and y, and any positive integer m, $$(xy)^m = x^m y^m$$ To raise a product to a power, raise each factor of the product to that power.

EXAMPLE 7   *The power rule for products.*   Simplify each expression: **a.** $(8a)^2$ and   **b.** $(2bx)^3$.

Solution

a. $(8a)^2 = 8^2 a^2$ To raise $8a$ to a power, raise each factor of the product to that power.

$\qquad = 64a^2$ Find the power: $8^2 = 64$.

b. $(2bx)^3 = 2^3 b^3 x^3$ To raise $2bx$ to a power, raise each factor of the product to that power.

$\qquad = 8b^3 x^3$ Find the power: $2^3 = 8$.

EXAMPLE 8   *Applying two properties.*   Simplify each expression: **a.** $(10a^2)^3$ and   **b.** $(3c^5 d^3)^4$.

Solution

a. $(10a^2)^3 = 10^3 (a^2)^3$ To raise $10a^2$ to a power, raise each factor of the product to that power.

$\qquad = 10^3 a^{2 \cdot 3}$ To raise a^2 to a power, keep the base and multiply the exponents.

$\qquad = 10^3 a^6$ Do the multiplication: $2 \cdot 3 = 6$.

$\qquad = 1{,}000 a^6$ Find the power: $10^3 = 1{,}000$.

b. $(3c^5 d^3)^4 = 3^4 (c^5)^4 (d^3)^4$ To raise $3c^5 d^3$ to a power, raise each factor of the product to that power.

$\qquad = 3^4 c^{5 \cdot 4} d^{3 \cdot 4}$ To raise c^5 and d^3 to powers, keep the bases and multiply their exponents.

$\qquad = 3^4 c^{20} d^{12}$ Do the multiplications: $5 \cdot 4 = 20$ and $3 \cdot 4 = 12$.

$\qquad = 81 c^{20} d^{12}$ Find the power: $3^4 = 81$.

EXAMPLE 9 *Applying three properties.* Simplify: $(2a^2)^2(4a^3)^3$.

Solution

$$(2a^2)^2(4a^3)^3 = 2^2(a^2)^2 \cdot 4^3(a^3)^3 \qquad \text{To raise } 2a^2 \text{ and } 4a^3 \text{ to powers, raise the factors of each product to the appropriate power.}$$

$$= 2^2 a^{2\cdot2} \cdot 4^3 \cdot a^{3\cdot3} \qquad \text{To raise } a^2 \text{ and } a^3 \text{ to powers, keep the bases and multiply the exponents.}$$

$$= 2^2 a^4 \cdot 4^3 a^9 \qquad \text{Do the multiplications: } 2\cdot2 = 4 \text{ and } 3\cdot3 = 9.$$

$$= (2^2 \cdot 4^3)(a^4 \cdot a^9) \qquad \text{Change the order of the factors and group like bases.}$$

$$= (2^2 \cdot 4^3)(a^{4+9}) \qquad \text{To multiply } a^4 \cdot a^9, \text{ keep the base and add the exponents.}$$

$$= (2^2 \cdot 4^3)a^{13} \qquad \text{Do the addition: } 4 + 9 = 13.$$

$$= (4 \cdot 64)a^{13} \qquad \text{Find the powers: } 2^2 = 4 \text{ and } 4^3 = 64.$$

$$= 256a^{13} \qquad \text{Do the multiplication: } 4 \cdot 64 = 256.$$

Self Check

Simplify: $(4y^3)^2(3y^4)^3$.

Answer: $432y^{18}$

STUDY SET   Section 8.7

VOCABULARY   *In Exercises 1–6, fill in the blanks to make a true statement.*

1. In x^n, x is called the _____ and n is called the _____.

2. x^2 is the second _____ of x, or we can read it as "x _____."

3. $x^m \cdot x^n$ is the _____ of two exponential expressions with _____ bases.

4. $(x^m)^n$ is a power of a _____.

5. $(2x)^n$ is a _____ raised to a power.

6. In x^{m+n}, $m + n$ is the _____ of m and n.

CONCEPTS

7. Represent each repeated multiplication using exponents.
   a. $x \cdot x \cdot x \cdot x \cdot x \cdot x \cdot x$
   b. $x \cdot x \cdot y \cdot y \cdot y$
   c. $3 \cdot 3 \cdot 3 \cdot 3 \cdot a \cdot a \cdot b \cdot b \cdot b$

8. Write each exponential expression as repeated multiplication.
   a. $a^3 b^5$
   b. $(x^2)^3$
   c. $(2a)^3$

9. Write a product of two exponential expressions with like variable bases. Then simplify it using a property of exponents.

10. Write a power of a product and then simplify it using a property of exponents.

11. Write a power of a power and then simplify it using a property of exponents.

12. What algebraic property allows us to change the order of the factors of a multiplication?

13. Simplify each expression using a property of exponents.
   a. $x^m x^n$
   b. $(x^m)^n$
   c. $(ax)^n$

14. In each case, tell how the expression has been *improperly* simplified.
   a. $2^3 \cdot 2^4 = 2^{12}$
   b. $3^3 \cdot 3^4 = 9^7$
   c. $(2^3)^4 = 2^7$

15. Write each expression without an exponent.
   a. 2^1 **b.** $(-10)^1$ **c.** x^1

16. Find each power.
   a. 2^3 **b.** 4^3 **c.** 5^3

410   *Chapter 8   An Introduction to Algebra*

17. Simplify each expression, if possible.
 a. $x \cdot x$ and $x + x$
 b. $x \cdot x^2$ and $x + x^2$
 c. $x^2 \cdot x^2$ and $x^2 + x^2$

18. Simplify each expression, if possible.
 a. $a \cdot a$ and $a - a$
 b. $2ab \cdot ab$ and $2ab - ab$
 c. $2ab \cdot 3ab$ and $2ab - 3ab$

19. Simplify each expression, if possible.
 a. $4x \cdot x$ and $4x + x$
 b. $4x \cdot 3x$ and $4x + 3x$
 c. $4x^2 \cdot 3x$ and $4x^2 + 3x$

20. Simplify each expression, if possible.
 a. $ab(-2ab)$ and $ab - 2ab$
 b. $-2ab(3ab^2)$ and $-2ab + 3ab^2$
 c. $-2ab^2(-3ab^2)$ and $-2ab^2 - 3ab^2$

21. Evaluate the exponential expression x^{m+n} for $x = 3$, $m = 2$, and $n = 1$.

22. Evaluate the exponential expression $(x^m)^n$ for $x = 2$, $m = 3$, and $n = 2$.

NOTATION *In Exercises 23–26, complete each solution.*

23. $x^5 \cdot x^7 = x^{+}$ Since the bases are alike, apply the product rule for exponents.

$= x^{12}$ Do the addition.

24. $(x^5)^4 = x^{ \cdot}$ Apply the power rule for exponents.

$= x^{20}$ Do the multiplication.

25. $(2x^4)(8x^3) = (2 \cdot 8)(\underline{} \cdot \underline{})$ Apply the commutative and associative properties of multiplication.

$= 16x^{+}$ Apply the product rule for exponents.

$= 16x^7$ Do the addition.

26. $(2x^2)^3 = 2^3(x^2)^3$ Apply the product rule for powers.

$= 2^3 x^{ \cdot}$ Apply the power rule for exponents.

$= 2^3 x^{}$ Do the multiplication.

$= 8x^6$ Find the power.

PRACTICE *In Exercises 27–42, write each expression using one exponent.*

27. $x^2 \cdot x^3$

28. $t^4 \cdot t^3$

29. $x^3 x^7$

30. $y^2 y^5$

31. $f^5(f^8)$

32. $g^6(g^2)$

33. $n^{24} \cdot n^8$

34. $m^9 \cdot m^{61}$

35. $l^4 \cdot l^5 \cdot l$

36. $w^4 \cdot w \cdot w^3$

37. $x^6(x^3)x^2$

38. $y^5(y^2)(y^3)$

39. $2^4 \cdot 2^8$

40. $3^4 \cdot 3^2$

41. $5^6(5^2)$

42. $(8^3)8^4$

In Exercises 43–54, simplify each product.

43. $2x^2 \cdot 4x$

44. $5y \cdot 6y^3$

45. $5t \cdot t^9$

46. $f^4 \cdot 3f$

47. $-6x^3(4x^2)$

48. $-7y^5(5y^3)$

49. $-x \cdot x^3$

50. $8x^6(-x)$

51. $6y(2y^3)3y^4$

52. $2d(5d^4)(d^2)$

53. $-2t^3(-4t^2)(-5t^5)$

54. $-7k^5(-3k^3)(-2k^9)$

In Exercises 55–74, simplify each product.

55. $xy^2 \cdot x^2 y$

56. $s^2 t \cdot st$

57. $b^3 \cdot c^2 \cdot b^5 \cdot c^6$

58. $h^3 \cdot f^3 \cdot f^2 \cdot h^4$

59. $x^4 y(xy)$

60. $(ab)(ab^2)$

61. $a^2 b \cdot b^3 a^2$

62. $w^2 y \cdot yw^4$

63. $x^5 y \cdot y^6$

64. $a^7 \cdot b^2 a^4$

65. $3x^2 y^3 \cdot 6xy$

66. $25a^3 b \cdot 2ab^5$

67. $xy^2 \cdot 16x^3$

68. $mn^4 \cdot 8n^3$

69. $-6f^2 t(4f^4 t^3)$

70. $(-5a^2 b^2)(5a^3 b^6)$

71. $ab \cdot ba \cdot a^2 b$

72. $xy \cdot y^2 x \cdot x^2 y$

73. $-4x^2 y(-3x^2 y^2)$

74. $-2rt^4(-5r^2 t^2)$

In Exercises 75–86, simplify each expression.

75. $(x^2)^4$

76. $(y^6)^3$

77. $(m^{50})^{10}$

78. $(n^{25})^4$

79. $(2a)^3$

80. $(3x)^3$

81. $(xy)^4$

82. $(ab)^8$

83. $(3s^2)^3$

84. $(5f^6)^2$

85. $(2s^2 t^3)^2$

86. $(4h^5 y^6)^2$

In Exercises 87–98, simplify each expression.

87. $(x^2)^3(x^4)^2$ **88.** $(a^5)^2(a^3)^3$ **89.** $(c^5)^3 \cdot (c^3)^5$ **90.** $(y^2)^8 \cdot (y^8)^2$

91. $(2a^4)^2(3a^3)^2$ **92.** $(5x^3)^2(2x^4)^3$ **93.** $(3a^3)^3(2a^2)^3$ **94.** $(6t^5)^2(2t^2)^2$

95. $(x^2x^3)^{12}$ **96.** $(a^3a^3)^3$ **97.** $(2b^4b)^5$ **98.** $(3y^2y^5)^3$

WRITING *Write a paragraph using your own words.*

99. Explain the difference between x^2 and $2x$.

100. Explain why the properties of exponents do not apply to $x^3 + x^3$.

101. One of the properties of exponents is the power of a product is the product of the powers. Use a specific example to explain this property.

102. To find the result when *multiplying* two exponential expressions with like bases, we must *add* the exponents. Explain why this is so.

REVIEW

103. What is a variable?

104. When evaluated, what is the sign of $(-13)^5$?

105. Divide: $\dfrac{-25}{-5}$.

106. How much did the temperature change if it went from $-4°$ to $-17°$?

107. Evaluate: $2\left(\dfrac{12}{-3}\right) + 3(5)$

108. Solve: $-4 - 6 = x + 1$.

109. Solve: $-x = -12$.

110. Divide: $\dfrac{0}{10}$.

Variables

One of the major objectives of this course is for you to become comfortable working with **variables.** You will recall that a variable is a letter that stands for a number.

The application problems of Sections 8.1 and 8.2 were solved with the help of a variable. In these problems, we let the variable represent an unknown quantity such as the cost of a condominium or the number of people attending an open house. We then wrote an equation to describe the situation mathematically and solved the equation to find the value represented by the variable.

In Exercises 1–6, suppose that you are going to solve the following problems. What quantity should be represented by a variable? State your response in the form "Let $x = \ldots$."

1. The monthly cost to lease a van is $120 less than to buy it. To buy it, the monthly payments are $290. How much does it cost to lease the van each month?

2. One piece of pipe is 10 feet longer than another. Together, their lengths total 24 feet. How long is the shorter piece of pipe?

3. The length of a rectangular field is 50 feet. What is its width if it has a perimeter of 200 feet?

4. If one hose can fill a vat in 2 hours and another can fill it in 3 hours, how long will it take to fill the vat if both hoses are used?

5. Find the distance traveled by a motorist in three hours if her average speed was 55 miles per hour.

6. In what year was a couple married if their 50th anniversary was in 1988?

Variables can also be used to state properties of mathematics in a concise, "shorthand" notation. In Exercises 7–12, state each property using mathematical symbols and the given variable(s).

7. Use the variables a and b to state that two numbers can be added in either order to get the same sum.

8. Use the variable x to state that when 0 is subtracted from a number, the result is the same number.

9. Use the variable b to state that the result when dividing a number by 1 is the same number.

10. Use the variable x to show that the sum of a number and 1 is greater than the number.

11. Using the variable n, state the fact that when 1 is subtracted from any number, the difference is less than the number.

12. State the fact that the product of any number and 0 is 0, using the variable a.

CHAPTER REVIEW

SECTION 8.1	Solving Equations by Addition and Subtraction

CONCEPTS

An *equation* is a statement that two expressions are equal.

In the equation $x + 5 = 7$, x is the *variable* or the *unknown*. Two equations with exactly the same solutions are called *equivalent equations*.

To solve an equation, isolate the variable on one side of the equation by undoing the operation performed on it. This is accomplished using the opposite operation.

If the same number is added to both sides of an equation, an equivalent equation results:
If $a = b$, then $a + c = b + c$.

If the same number is subtracted from both sides of an equation, an equivalent equation results:
If $a = b$, then $a - c = b - c$.

To solve a problem, follow these steps:
1. Analyze the problem.
2. Form an equation.
3. Solve the equation.
4. State the conclusion.
5. Check the result.

REVIEW EXERCISES

1. Tell whether the given number is a solution of the equation.
  a. $x + 2 = 13$; $x = 5$ **b.** $x - 3 = 1$; $x = 4$

2. Identify the variable in each equation.
  a. $y - 12 = 50$ **b.** $114 = 4 - t$

3. Solve and check each equation.
  a. $x - 7 = 2$ **b.** $x - 11 = 20$
  c. $225 = y - 115$ **d.** $101 = p - 32$

4. Solve and check each equation.
  a. $x + 9 = 18$ **b.** $b + 12 = 26$
  c. $175 = p + 55$ **d.** $212 = m + 207$

5. FINANCING A newly married couple made a \$25,500 down payment on a \$122,750 house. How much did they need to borrow?

6. DOCTOR'S CLIENTELE After moving his office, a doctor lost 13 patients. If he had 172 patients left, how many did he have originally?

SECTION 8.2	Solving Equations by Division and Multiplication

If both sides of an equation are divided by the same nonzero number, an equivalent equation results:
If $a = b$, then $\frac{a}{c} = \frac{b}{c}$. ($c \neq 0$)

If both sides of an equation are multiplied by the same nonzero number, an equivalent equation results:
If $a = b$, then $a \cdot c = b \cdot c$. ($c \neq 0$)

7. Solve and check each equation.
  a. $3x = 12$ **b.** $15y = 45$
  c. $105 = 5r$ **d.** $224 = 16q$

8. Solve and check each equation.
  a. $\frac{x}{7} = 3$ **b.** $\frac{a}{3} = 12$
  c. $15 = \frac{s}{21}$ **d.** $25 = \frac{d}{17}$

9. CARPENTRY If you cut an 18-foot board into 3 equal pieces, how long will each piece be?

10. JEWELRY Four sisters split the cost of a gold chain evenly. How much did the chain cost if each sister's share was $32?

Algebraic Expressions and Formulas

When we replace the variable, or variables, in an algebraic expression with a specific number and then apply the rules for order of operations, we are *evaluating the algebraic expression*.

11. RETAINING WALL Illustration 1 shows the design for a retaining wall. The relationships between the lengths of its important parts are given in words. Use a variable and algebraic expressions to describe the height and the lengths of the upper and lower bases. Suppose engineers determine that a 10-foot-high wall is needed. Find the lengths of the upper and lower bases.

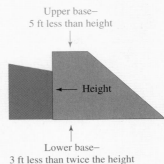

Upper base–
5 ft less than height

← Height

Lower base–
3 ft less than twice the height

ILLUSTRATION 1

12. Evaluate each algebraic expression.

a. $-2x + 6$ for $x = -3$

b. $\dfrac{6 - a}{1 + a}$ for $a = -2$

c. $b^2 - 4ac$ for $a = 4$, $b = 6$, and $c = -4$

d. $\dfrac{-2k^3}{1 - 2 - 3}$ for $k = -2$

A *formula* is a general rule that describes a known relationship between two or more variables.

Uniform motion:
  Distance = rate · time

13. DISTANCE TRAVELED Complete Illustration 2 by finding the distance traveled for a given time at a given rate.

	Rate (mph)	Time (hr)	Distance traveled
Monorail	65	2	
Subway	38	3	
Train	x	6	
Bus	55	t	

ILLUSTRATION 2

Formulas from business:
  Sale price = original
    price − discount

  Retail price = cost + markup

  Profit = revenue − costs

14. SALE PRICE Find the sale price of a trampoline that normally sells for $315 if a $37 discount is being offered.

15. RETAIL PRICE Find the retail price of a car if the dealer pays $14,505 and the markup is $725.

16. ANNUAL PROFIT The bar chart in Illustration 3 shows revenue and costs for a plastics company for the years 1994 to 1996, in millions of dollars.
  a. In which year was there the most revenue?
  b. Which year had the largest profit?
  c. What can you say about costs over this three-year span?

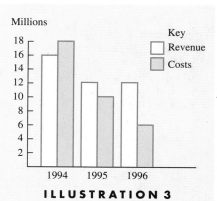

ILLUSTRATION 3

Formulas from science:
  Distance = rate · time

  $C = \dfrac{5(F - 32)}{9}$

  Distance fallen = 16 · (time)²

17. TEMPERATURE CONVERSION At a summer resort, visitors can relax by taking a dip in a swimming pool or a lake. The pool water is kept at a constant temperature of 77° F. The water in the lake is 23° C. Which water is warmer, and by how much?

18. DISTANCE FALLEN A steelworker accidentally dropped his hammer while working atop a new high-rise building. How far will the hammer fall in 3 seconds?

Formulas from mathematics:

  $\text{Mean} = \dfrac{\text{sum of values}}{\text{number of values}}$

19. AVERAGE YEARS OF EXPERIENCE Three generations of Smiths now operate a family-owned real estate office. The grandparents, who started the business, have been realtors for 40 years. Their son and daughter-in-law joined the company 18 years ago. Their grandson has worked as a realtor for 4 years. What is the average number of years experience for an employee of Smith Realty?

SECTION 8.4 — Simplifying Algebraic Expressions and the Distributive Property

To *simplify an algebraic expression,* we use properties of algebra to write the expression in simpler form.

The *distributive property:*
If a, b, and c are numbers, then
  $a(b + c) = ab + ac$
  $(b + c)a = ba + ca$
  $a(b - c) = ab - ac$
  $(b - c)a = ba - ca$
  $a(b + c + d) = ab + ac + ad$

20. Simplify each expression.
  a. $-2(5x)$ **b.** $-7x(-6y)$ **c.** $4d \cdot 3e \cdot 5$ **d.** $(4s)8$
  e. $-1(-e)(2)$ **f.** $7x \cdot 7y$ **g.** $4 \cdot 3k \cdot 7$ **h.** $(-10t)(-10)$

21. Multiply to remove parentheses.
  a. $4(y + 5)$ **b.** $-5(6t + 9)$
  c. $(-3 - 3x)7$ **d.** $-3(4e - 8x - 1)$

22. Simplify each expression.
  a. $-(6t - 4)$ **b.** $-(5 + x)$ **c.** $-(6t - 3s + 1)$ **d.** $-(-5a - 3)$

SECTION 8.5 — Combining Like Terms

A *term* is a number or a product of a number and one or more variables.

23. Identify the second term and the coefficient of the third term.
  a. $5x^2 - 4x + 8$ **b.** $7y - 3y + x - y$

In a term that is a product of a number and one or more variables, the factor that is a number is called the numerical *coefficient* of the term.

Like terms, or *similar terms*, are terms with exactly the same variables and exponents.

To *combine like terms*, add their numerical coefficients and keep the variables and exponents.

24. Tell whether x is used as a factor or a term.
   a. $5x - 6y^2$ **b.** $2b - x + 6$ **c.** $6xy$ **d.** $-36 + x + b$

25. Tell whether the following are like terms.
   a. $4x, -5x$ **b.** $4x, 4x^2$
   c. $3xy, xy$ **d.** $-5b^2c, -5bc^2$

26. Simplify by combining like terms.
   a. $3x + 4x$ **b.** $6r - 9r$
   c. $-3t - 6t$ **d.** $2z + (-5z)$
   e. $6x - x$ **f.** $-6y - 7y - (-y)$
   g. $5w - 8 - 4w + 3$ **h.** $-5x - 5y - x + 7y$

27. Simplify by combining like terms.
   a. $-45d - 2a + 4a - d$ **b.** $5y + 8h - 3 + 7h + 5y + 2$

28. Simplify each expression.
   a. $7(y + 6) + 3(2y + 2)$ **b.** $-4(t - 7) - (t + 6)$
   c. $5x - 2(x - 6)$ **d.** $6f + 7(12 - 8f)$

The perimeter of a rectangle is given by
   $P = 2l + 2w$
The perimeter of a square is given by
   $P = 4s$

29. HOLIDAY LIGHTS
To decorate a house, lights will be hung around the entire home, as shown in Illustration 4. They will also be placed around the two 5-foot-by-5-foot windows in the front. How many feet of lights will be needed?

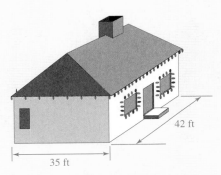

42 ft

35 ft

ILLUSTRATION 4

SECTION 8.6

Simplifying Expressions to Solve Equations

To *solve an equation* means to find all values of the variable that, when substituted into the original equation, make the equation a true statement.

30. Is $x = -3$ a solution of $-4x + 6 = 2(x + 12)$?

31. Solve each equation. Check all answers.
   a. $3x - 4x = -8$ **b.** $5a - 1 - 3a = -32 - 5$
   c. $7x + 6 = 5x - 6 - 2x$ **d.** $5(y - 15) = 0$
   e. $-3(2x + 4) - 4 = -40$ **f.** $4 + \dfrac{x}{5} - 6 = -1$
   g. $-6(2x + 3) = -(5x - 3)$ **h.** $2(3 - 2x) - 5(1 - 3x) = -21$

32. List the steps of the strategy for solving equations.

Exponents represent repeated multiplication.

The *product rule* for exponents:
$x^m x^n = x^{m+n}$

The *power rule* for exponents:
$(x^m)^n = x^{m \cdot n}$

The *power rule for products:*
$(xy)^m = x^m y^m$

33. a. Show the repeated multiplication that $(4h)^3$ represents.
b. Write this expression using exponents: $5 \cdot 5 \cdot d \cdot d \cdot d \cdot m \cdot m \cdot m \cdot m$

34. Simplify each expression.
a. $h^6 h^4$ **b.** $t^3(t^5)$ **c.** $w^2 \cdot w \cdot w^4$ **d.** $4^7 \cdot 4^5$

35. Simplify each product.
a. $2b^2 \cdot 4b^5$ **b.** $-6x^3(4x)$ **c.** $-2f^2(-4f)(3f^4)$ **d.** $-ab \cdot b \cdot a$

e. $xy^4 \cdot xy^2$ **f.** $(mn)(mn)$ **g.** $3z^3 \cdot 9m^3 z^4$ **h.** $-5cd(4c^2 d^5)$

36. Simplify each expression.
a. $(v^3)^4$ **b.** $(3y)^3$ **c.** $(5t^4)^2$ **d.** $(2a^4 b^5)^3$

37. Simplify each expression.
a. $(c^4)^5 (c^2)^3$ **b.** $(3s^2)^3 (2s^3)^2$
c. $(c^4 c^3)^2$ **d.** $(2xx^2)^3$

In Problems 1–6, solve each equation.

1. $10 = x + 6$

2. $y - 12 = 18$

3. $5t = 55$

4. $\dfrac{q}{3} = 27$

5. $500 + x = 700$

6. $100x = 2{,}400$

In Problems 7–8, use the five steps given in the strategy for problem-solving to solve each problem.

7. PARKING   After many student complaints, a college decided to commit funds to double the number of parking spaces on campus. This increase would bring the total number of spaces up to 6,200. How many parking spaces does the college have at this time?

8. LIBRARY   A library building is 6 years shy of its 200th birthday. How old is the building at this time?

9. Evaluate $\dfrac{x - 16}{x}$ for $x = 4$.

10. DISTANCE TRAVELED   Find the distance traveled by a motorist who departed from home at 9:00 A.M. and arrived at his destination at noon, traveling at a rate of 55 miles per hour.

In Problems 11–14, simplify by removing parentheses.

11. $5(5x + 1)$

12. $-6(7 - x)$

13. $-(6y + 4)$

14. $3(2a + 3b - 7)$

15. Tell whether x is used as a factor or as a term.
   a. $5xy$ **b.** $8y + x + 6$

16. Simplify by combining like terms.
   a. $-20y + 6x - 8y + 4x$
   b. $-t - t - t$

17. Identify each term in this algebraic expression:
   $8x^2 - 4x - 6$

18. Simplify.
   a. $7x + 4x$ **b.** $3c \cdot 4e \cdot 2$
   c. $6x - x$ **d.** $-5y(-6)$

19. Simplify the expression $4(y + 3) - 5(2y + 3)$.

20. Solve each equation.
   a. $3x - 4 - 3 = 2x + 3x + 11$
   b. $6r + 3 - 2r = -9$

21. Solve the equation $2(4x - 1) = 3(4 - 2x)$.

22. a. What is the value of k dimes?
b. What is the value of $p + 2$ twenty-dollar bills?

23. Simplify each expression.
a. $h^2 h^4$
b. $-7x^3(4x^2)$
c. $b^2 \cdot b \cdot b^5$
d. $-3g^2 k^3(-8g^3 k^{10})$

24. Simplify each expression.
a. $(f^3)^5$
b. $(2a^2 b)^2$
c. $(x^2)^3(x^3)^3$
d. $(x^2 x^3)^3$

25. Explain the difference between an equation and an expression.

26. Explain what it means to *solve an equation*.

27. Show how to check to see if $x = -7$ is a solution of $2x + 1 = -15$.

28. Explain what is *wrong* with the following solution: $5^4 \cdot 5^3 = 25^7$

In Exercises 1–2, consider the number 7,535,670.

1. Round to the nearest hundred.

2. Round to the nearest ten thousand.

In Exercises 3–6, do each operation.

3.
$$\begin{array}{r} 5,679 \\ +3,458 \end{array}$$

4.
$$\begin{array}{r} 7,697 \\ -4,375 \end{array}$$

5.
$$\begin{array}{r} 5,345 \\ \times \quad 46 \end{array}$$

6. $35\overline{)30,625}$

In Exercises 7–8, refer to the rectangular swimming pool shown in Illustration 1.

7. Find the perimeter of the pool.

8. Find the area of the pool.

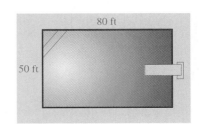

ILLUSTRATION 1

In Exercises 9–12, find the prime factorization of each number.

9. 168

10. 225

11. 180

12. 720

In Exercises 13–16, evaluate each expression.

13. $8 + (-2)(-5)$

14. $(-2)^4 - 3^3$

15. $\dfrac{2(-7) + 3(2)}{2(-4)}$

16. $\dfrac{2(3^2 + 4^2)}{-2(3) + 1}$

In Exercises 17–20, do the operations.

17. $\dfrac{10}{21} \cdot \dfrac{3}{10}$

18. $\dfrac{22}{25} \div \dfrac{11}{5}$

19. $\dfrac{11}{12} + \dfrac{2}{3}$

20. $\dfrac{11}{12} - \dfrac{2}{3}$

In Exercises 21–24, evaluate each expression when $x = 4$.

21. $2x - 1$

22. $\dfrac{9x}{2} - x$

23. $3x - x^3$

24. $x + 2(x - 7)$

In Exercises 25–28, simplify each expression.

25. $-3(5x)$

26. $-4x(-7x)$

27. $-2(3x - 4)$

28. $-5(3x - 2y + 4)$

In Exercises 29–32, combine like terms.

29. $-3x + 8x$ **30.** $4a^2 - (-3a^2)$ **31.** $4x - 3y - 5x + 2y$ **32.** $-2(3x - 4) + 2x$

In Exercises 33–38, solve and check each equation.

33. $3x + 2 = -13$ **34.** $-5z - 7 = 18$ **35.** $\dfrac{y}{4} - 1 = -5$ **36.** $\dfrac{n}{5} + 1 = -11$

37. $6x - 12 = 2x + 4$ **38.** $3(2y - 8) = -2(y - 4)$

39. OBSERVATION HOURS   To get a Masters degree in learning disabilities, a student must have 100 hours of observation time. If the student has already observed for 37 hours, how many more 3-hour shifts must he observe?

40. GEOMETRY   A rectangle is four times as long as it is wide. If its perimeter is 210 feet, find its dimensions.

In Exercises 41–48, simplify each product.

41. $p^3 p^5$ **42.** $-3t^2 \cdot 5t^7$ **43.** $(-x^2 y^3)(x^3 y^4)$ **44.** $(3a^2)^4$

45. $(-n^2)^3$ **46.** $(3x^3 y)^2$ **47.** $(2p^3)^2(3p^2)^3$ **48.** $(-x^2)^3(2x)^3$

Introduction to Geometry

9

423

GEOMETRY COMES FROM THE GREEK WORDS GEO
(MEANING EARTH) AND METRON *(MEANING MEASURE)*.

9.1 | *Some Basic Definitions*

In this section, you will learn about

- **Points, lines, and planes**
- **Angles**
- **Adjacent and vertical angles**
- **Complementary and supplementary angles**

INTRODUCTION. In this chapter, we will study two-dimensional geometric figures such as rectangles and circles. In daily life, it is often necessary to find the perimeter or area of one of these figures. For example, to find the amount of fencing that is needed to enclose a circular garden, we must find the perimeter of a circle (called its *circumference*). To find the amount of paint needed to paint a room, we must find the area of its four rectangular walls.

We will also study three-dimensional figures such as cylinders and spheres. To find the amount of space enclosed within these figures, we must find their volumes.

Points, lines, and planes

Geometry is based on three undefined words: **point, line,** and **plane.** Although we will make no attempt to define these words formally, we can think of a point as a geometric figure that has position but no length, width, or depth. Points are always labeled with capital letters. Point *A* is shown in Figure 9-1(a).

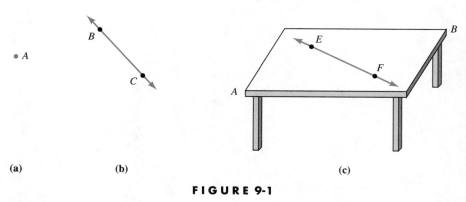

(a)   (b)   (c)

FIGURE 9-1

A line is infinitely long but has no width or depth. Figure 9-1(b) shows line BC, passing through points B and C. A plane is a flat surface, like a table top, that has length and width but no depth. In Figure 9-1(c), line EF lies in the plane AB.

As Figure 9-1(b) illustrates, points B and C determine exactly one line, the line BC. In Figure 9-1(c), the points E and F determine exactly one line, the line EF. In general, any two points will determine exactly one line.

Other geometric figures can be created by using parts or combinations of points, lines, and planes.

| Line segment | The **line segment** AB, denoted as $\overline{AB}$, is the part of a line that consists of points A and B and all points in between (see Figure 9-2). Points A and B are the **endpoints** of the segment. |

Line segment AB ($\overline{AB}$)

FIGURE 9-2

Every line segment has a **midpoint,** which divides the segment into two parts of equal length. In Figure 9-3, M is the midpoint of segment AB, because the measure of $\overline{AM}$ (denoted as m($\overline{AM}$)) is equal to the measure of $\overline{MB}$ (denoted as m($\overline{MB}$)).

$$m(\overline{AM}) = 4 - 1$$
$$= 3$$

and

$$m(\overline{MB}) = 7 - 4$$
$$= 3$$

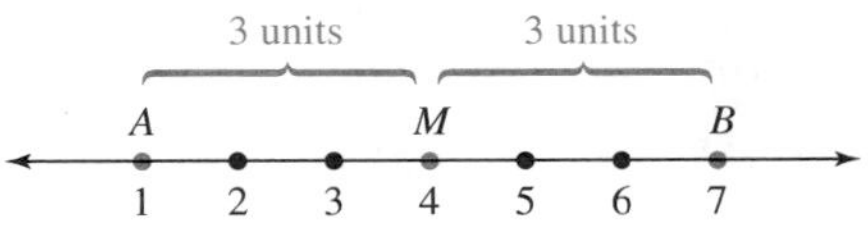

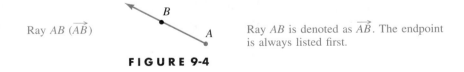

FIGURE 9-3

Since the measure of both segments is 3 units, m($\overline{AM}$) = m($\overline{MB}$).

When two line segments have the same measure, we say that they are **congruent.** Since m($\overline{AM}$) = m($\overline{MB}$), we can write

$$\overline{AM} \cong \overline{MB} \quad \text{Read} \cong \text{as "is congruent to."}$$

Another geometric figure is the *ray.*

| Ray | A **ray** is the part of a line that begins at some point (say, A) and continues forever in one direction. See Figure 9-4. Point A is the **endpoint** of the ray. |

Ray AB ($\overrightarrow{AB}$)

Ray AB is denoted as $\overrightarrow{AB}$. The endpoint is always listed first.

FIGURE 9-4

Angles

| Angle | An **angle** is a figure formed by two rays with a common endpoint. The common endpoint is called the **vertex,** and the rays are called **sides.** |

The angle in Figure 9-5 can be denoted as

$$\angle BAC, \quad \angle CAB, \quad \angle A, \quad \text{or} \quad \angle 1 \quad \text{The symbol } \angle \text{ means angle.}$$

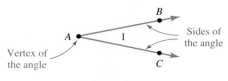

Vertex of the angle

Sides of the angle

FIGURE 9-5

 WARNING! When using three letters to name an angle, be sure the letter name of the vertex is the middle letter.

One unit of measurement of an angle is the **degree.** It is $\frac{1}{360}$ of a full revolution. We can use a **protractor** to measure angles in degrees. See Figure 9-6.

Angle	Measure in degrees
$\angle ABC$	30°
$\angle ABD$	60°
$\angle ABE$	110°
$\angle ABF$	150°
$\angle ABG$	180°

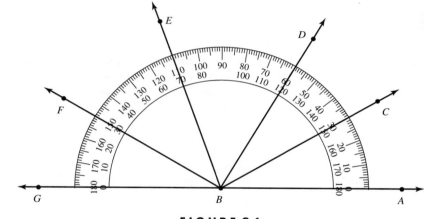

FIGURE 9-6

If we read the protractor from left to right, we can see that the measure of $\angle GBF$ (denoted as m($\angle GBF$)) is 30°.

When two angles have the same measure, we say that they are congruent. Since m($\angle ABC$) = 30° and m($\angle GBF$) = 30°, we can write

$$\angle ABC \cong \angle GBF$$

We classify angles according to their measure, as in Figure 9-7.

Classification of angles

Acute angles: Angles whose measures are greater than 0° but less than 90°.

Right angles: Angles whose measures are 90°.

Obtuse angles: Angles whose measures are greater than 90° but less than 180°.

Straight angles: Angles whose measures are 180°.

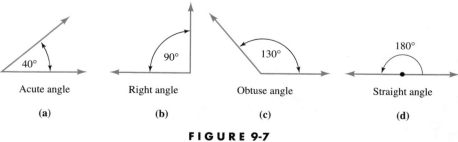

Acute angle	Right angle	Obtuse angle	Straight angle
(a)	**(b)**	**(c)**	**(d)**

FIGURE 9-7

EXAMPLE 1 **_Classifying angles._** Classify each angle in Figure 9-8 as an acute angle, a right angle, an obtuse angle, or a straight angle.

Solution Since $m(\angle 1) < 90°$, it is an acute angle.

Since $m(\angle 2) > 90°$ but less than $180°$, it is an obtuse angle.

Since $m(\angle BDE) = 90°$, it is a right angle.

Since $m(\angle ABC) = 180°$, it is a straight angle.

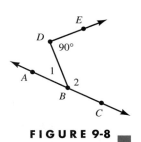

FIGURE 9-8 ■

Adjacent and vertical angles

Two angles that have a common vertex and are side-by-side are called **adjacent angles.**

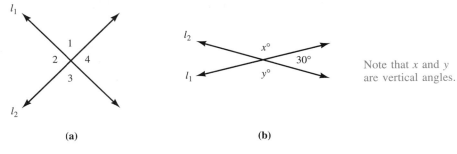

EXAMPLE 2 **_Evaluating angles._** Two angles with measures of $x°$ and $35°$ are adjacent angles. Use the information in Figure 9-9 to find x.

Solution
Since the sum of the measures of the angles is $80°$, we have

$$x + 35 = 80$$
$$x + 35 - 35 = 80 - 35 \quad \text{Subtract 35 from both sides.}$$
$$x = 45 \quad\quad 35 - 35 = 0 \text{ and } 80 - 35 = 45.$$

Thus, $x = 45$.

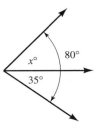

FIGURE 9-9

Self Check
In the figure below, find x.

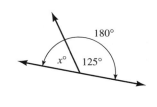

Answer: 55 ■

When two lines intersect, pairs of nonadjacent angles are called **vertical angles.** In Figure 9-10(a), $\angle 1$ and $\angle 3$ are vertical angles, as are $\angle 2$ and $\angle 4$.

To illustrate that vertical angles always have the same measure, we refer to Figure 9-10(b) with angles having measures of $x°$, $y°$, and $30°$. Since the measure of any straight angle is $180°$, we have

$$30 + x = 180 \quad\quad \text{and} \quad\quad 30 + y = 180$$
$$x = 150 \quad | \quad\quad\quad y = 150 \quad \text{Subtract 30 from both sides.}$$

Since x and y are both 150, $x = y$.

l_1

1
2   4
3

l_2

(a)

l_2

$x°$
$30°$

l_1 $y°$

Note that x and y are vertical angles.

(b)

FIGURE 9-10

Property of vertical angles	Vertical angles are congruent (have the same measure).

EXAMPLE 3	***Evaluating angles.*** In Figure 9-11, find **a.** m(∠1) and **b.** m(∠3).	

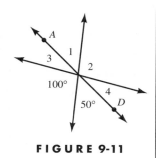

FIGURE 9-11

Solution

a. The 50° angle and ∠1 are vertical angles. Since vertical angles are congruent, m(∠1) = 50°.

b. Since AD is a line, the sum of the measures of ∠3, the 100° angle, and the 50° angle is 180°. If m(∠3) = x, we have

$$x + 100 + 50 = 180$$
$$x + 150 = 180 \quad \text{100 + 50 = 150.}$$
$$x = 30 \quad \text{Subtract 150 from both sides.}$$

Thus, m(∠3) = 30°.

EXAMPLE 4	***Evaluating angles.*** In Figure 9-12, find x.

FIGURE 9-12

Solution

Since the angles are vertical angles, they have equal measures.

$$4x - 20 = 3x + 15$$
$$x - 20 = 15 \quad \text{To eliminate } 3x \text{ from the right-hand side, subtract } 3x \text{ from both sides.}$$
$$x = 35 \quad \text{To undo the subtraction of 20, add 20 to both sides.}$$

Thus, x = 35.

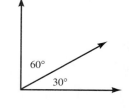
Complementary and supplementary angles

Complementary and supplementary angles	Two angles are **complementary angles** when the sum of their measures is 90°. Two angles are **supplementary angles** when the sum of their measures is 180°.

EXAMPLE 5	***Complementary and supplementary angles.***

a. Angles of 60° and 30° are complementary angles, because the sum of their measures is 90°. Each angle is the complement of the other.

b. Angles of 130° and 50° are supplementary, because the sum of their measures is 180°. Each angle is the supplement of the other.

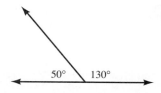

c. The definition of supplementary angles requires that the sum of *two* angles be 180°. Three angles of 40°, 60°, and 80° are not supplementary even though their sum is 180°.

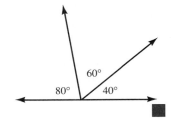

EXAMPLE 6   *Finding the complement and supplement of an angle.*

a. Find the complement of a 35° angle.

b. Find the supplement of a 105° angle.

Solution

a. See Figure 9-13. Let x represent the complement of the 35° angle. Since the angles are complementary, we have

$$x + 35 = 90$$
$$x = 55 \quad \text{Subtract 35 from both sides.}$$

The complement of 35° is 55°.

b. See Figure 9-14. Let y represent the supplement of the 105° angle. Since the angles are supplementary, we have

$$y + 105 = 180$$
$$y = 75 \quad \text{Subtract 105 from both sides.}$$

The supplement of 105° is 75°.

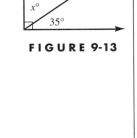

FIGURE 9-13

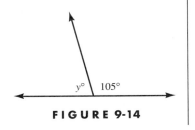

FIGURE 9-14

Self Check

a. Find the complement of a 50° angle.

b. Find the supplement of a 50° angle.

Answer: 40°, 130°

STUDY SET   Section 9.1

VOCABULARY   *In Exercises 1–12, fill in the blanks to make a true statement.*

1. A line _____ has two endpoints.

2. Two points _____ at most one line.

3. A _____ divides a line segment into two parts of equal length.

4. An angle is measured in _____.

5. A _____ is used to measure angles.

6. An _____ angle is less than 90°.

7. A _____ angle measures 90°.

8. An _____ angle is greater than 90° but less than 180°.

9. The measure of a straight angle is _____.

10. Adjacent angles have the same vertex and are _____.

11. The sum of two _____ angles is 180°.

12. The sum of two complementary angles is _____.

CONCEPTS In Exercises 13–20, refer to Illustration 1 and tell whether each statement is true. If a statement is false, explain why.

13. $\overrightarrow{GF}$ has point G as its endpoint.

14. $\overline{AG}$ has no endpoints.

15. Line CD has three endpoints.

16. Point D is the vertex of $\angle DGB$.

17. m($\angle AGC$) = m($\angle BGD$)

18. $\angle AGF \cong \angle BGE$

19. $\angle FGB \cong \angle EGA$

20. $\angle AGC$ and $\angle CGF$ are adjacent angles.

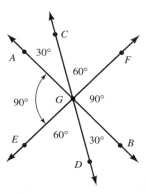

ILLUSTRATION 1

In Exercises 21–28, refer to Illustration 1 and tell whether each angle is an acute angle, a right angle, an obtuse angle, or a straight angle.

21. $\angle AGC$ **22.** $\angle EGA$ **23.** $\angle FGD$ **24.** $\angle BGA$

25. $\angle BGE$ **26.** $\angle AGD$ **27.** $\angle DGC$ **28.** $\angle DGB$

In Exercises 29–32, refer to Illustration 2 and tell whether each statement is true. If a statement is false, explain why.

29. $\angle AGF$ and $\angle DGC$ are vertical angles.

30. $\angle FGE$ and $\angle BGA$ are vertical angles.

31. m($\angle AGB$) = m($\angle BGC$).

32. $\angle AGC \cong \angle DGF$.

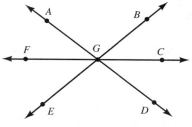

ILLUSTRATION 2

In Exercises 33–38, refer to Illustration 3 and tell whether each pair of angles are congruent.

33. $\angle 1$ and $\angle 2$

34. $\angle FGB$ and $\angle CGE$

35. $\angle AGB$ and $\angle DGE$

36. $\angle CGD$ and $\angle CGB$

37. $\angle AGF$ and $\angle FGE$

38. $\angle AGB$ and $\angle BGD$

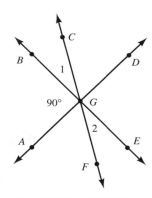

ILLUSTRATION 3

In Exercises 39–46, refer to Illustration 3 and tell whether each statement is true.

39. $\angle 1$ and $\angle CGD$ are adjacent angles.

40. $\angle 2$ and $\angle 1$ are adjacent angles.

41. $\angle FGA$ and $\angle AGC$ are supplementary.

42. $\angle AGB$ and $\angle BGC$ are complementary.

43. $\angle AGF$ and $\angle 2$ are complementary.

44. $\angle AGB$ and $\angle EGD$ are supplementary.

45. $\angle EGD$ and $\angle DGB$ are supplementary.

46. $\angle DGC$ and $\angle AGF$ are complementary.

NOTATION In Exercises 47–50, fill in the blanks to make a true statement.

47. The symbol $\angle$ means _____.

48. The symbol $\overline{AB}$ is read as "_____ AB."

49. The symbol $\overrightarrow{AB}$ is read as "_____ AB."

50. The symbol _____ is read as "is congruent to."

51. $\overline{AC}$ **52.** $\overline{BE}$

53. $\overline{CE}$ **54.** $\overline{BD}$

55. $\overline{CD}$ **56.** $\overline{DE}$

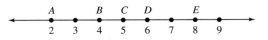

ILLUSTRATION 4

In Exercises 57–58, refer to Illustration 4 and find each midpoint.

57. Find the midpoint of $\overline{AD}$.   **58.** Find the midpoint of $\overline{BE}$.

In Exercises 59–62, use a protractor to measure each angle.

59. 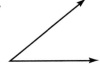   **60.** 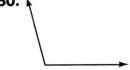   **61.** 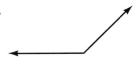   **62.**

In Exercises 63–70, find x.

63. 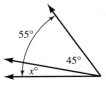   **64.** 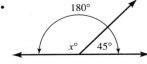   **65.**   **66.**

67. 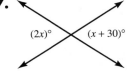   **68.** 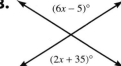   **69.** 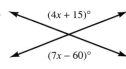   **70.**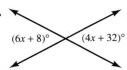

In Exercises 71–74, let x represent the unknown angle measure. Draw a diagram, write an appropriate equation, and solve it for x.

71. Find the complement of a 30° angle.   **72.** Find the supplement of a 30° angle.

73. Find the supplement of a 105° angle.   **74.** Find the complement of a 75° angle.

In Exercises 75–78, refer to Illustration 5, in which m($\angle 1$) = 50°. Find the measure of each angle or sum of angles.

75. $\angle 4$

76. $\angle 3$

77. m($\angle 1$) + m($\angle 2$) + m($\angle 3$)

78. m($\angle 2$) + m($\angle 4$)

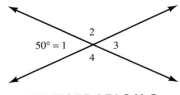

ILLUSTRATION 5

In Exercises 79–82, refer to Illustration 6, in which m($\angle$1) + m($\angle$3) + m($\angle$4) = 180°, $\angle$3 $\cong$ $\angle$4, and $\angle$4 $\cong$ $\angle$5. Find the measure of each angle.

79. $\angle$1   **80.** $\angle$2

81. $\angle$3   **82.** $\angle$6

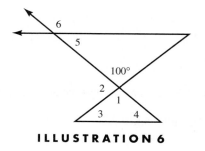

ILLUSTRATION 6

APPLICATIONS

83. Cite five examples in real life where you see lines.

84. Cite two examples in real life where you see right angles.

85. Cite two examples in real life where you see acute angles.

86. Cite two examples in real life where you see obtuse angles.

87. MUSICAL INSTRUMENTS Suppose that you are a beginning band teacher describing the correct posture needed to play various instruments. Use the diagrams in Illustration 7 to approximate the angle measure at which each instrument should be held in relation to the student's body: **a.** flute   **b.** clarinet   **c.** trumpet

88. PHRASES Explain what you think each of these phrases means. How is geometry involved?
   a. The president did a complete 180-degree flip on the subject of a tax cut.
   b. The rollerblader did a "360" as she jumped off the ramp.

ILLUSTRATION 7

WRITING   *Write a paragraph using your own words.*

89. Explain why an angle measuring 105° cannot have a complement.

90. Explain why an angle measuring 210° cannot have a supplement.

REVIEW   *In Exercises 91–98, do the calculations.*

91. Find 2^4.

92. Add: $\dfrac{1}{2} + \dfrac{2}{3} + \dfrac{3}{4}$

93. Subtract: $\dfrac{3}{4} - \dfrac{1}{8} - \dfrac{1}{3}$

94. Multiply: $\dfrac{5}{8} \cdot \dfrac{2}{15} \cdot \dfrac{6}{5}$

95. Divide: $\dfrac{12}{17} \div \dfrac{4}{34}$

96. $3 + 2 \cdot 4$

97. $5 \cdot 3 + 4 \cdot 2$

98. Find 30% of 60.

Parallel and Perpendicular Lines

In this section, you will learn about

- **Parallel and perpendicular lines**
- **Transversals and angles**
- **Properties of parallel lines**

INTRODUCTION. In this section, we will consider *parallel* and *perpendicular* lines. Since parallel lines are always the same distance apart, the railroad tracks shown in Figure 9-15(a) illustrate one application of parallel lines.

Since perpendicular lines meet and form right angles, the monument and the ground shown in Figure 9-15(b) illustrate one application of perpendicular lines.

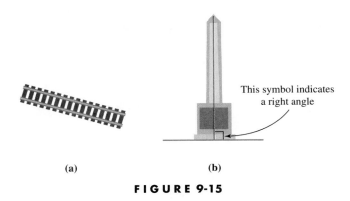

This symbol indicates a right angle

(a)   (b)

FIGURE 9-15

Parallel and perpendicular lines

If two lines lie in the same plane, they are called **coplanar.** Two coplanar lines that do not intersect are called **parallel lines.** See Figure 9-16(a).

Parallel lines	**Parallel lines** are coplanar lines that do not intersect.

If lines l_1 (l sub 1) and l_2 (l sub 2) are parallel, we can write $l_1 \parallel l_2$, where the symbol $\parallel$ is read as "is parallel to."

Perpendicular lines	**Perpendicular lines** are lines that intersect and form right angles.

In Figure 9-16(b), $l_1 \perp l_2$, where the symbol $\perp$ is read as "is perpendicular to."

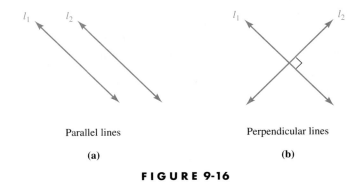

Parallel lines   Perpendicular lines

(a)   (b)

FIGURE 9-16

Transversals and angles

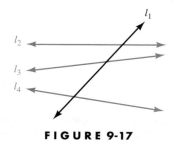

FIGURE 9-17

A line that intersects two or more coplanar lines is called a **transversal.** For example, line l_1 in Figure 9-17 is a transversal intersecting lines l_2, l_3, and l_4.

When two lines are cut by a transversal, the following types of angles are formed.

Alternate interior angles:

$\angle 4$ and $\angle 5$

$\angle 3$ and $\angle 6$

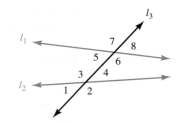

Corresponding angles:

$\angle 1$ and $\angle 5$

$\angle 3$ and $\angle 7$

$\angle 2$ and $\angle 6$

$\angle 4$ and $\angle 8$

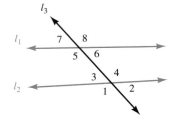

Interior angles:

$\angle 3$, $\angle 4$, $\angle 5$, and $\angle 6$

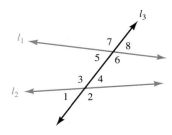

EXAMPLE 1	***Identifying angles.*** In Figure 9-18, identify **a.** all pairs of alternate interior angles, **b.** all pairs of corresponding angles, and **c.** all interior angles.

Solution　**a.** Pairs of alternate interior angles are

$\angle 3$ and $\angle 5$, $\angle 4$ and $\angle 6$

b. Pairs of corresponding angles are

$\angle 1$ and $\angle 5$, $\angle 4$ and $\angle 8$, $\angle 2$ and $\angle 6$, $\angle 3$ and $\angle 7$

c. Interior angles are

$\angle 3$, $\angle 4$, $\angle 5$, and $\angle 6$

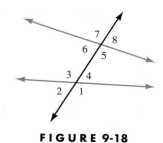

FIGURE 9-18

Properties of parallel lines

1. If two parallel lines are cut by a transversal, alternate interior angles are congruent. (See Figure 9-19.) If $l_1 \parallel l_2$, then $\angle 2 \cong \angle 4$ and $\angle 1 \cong \angle 3$.

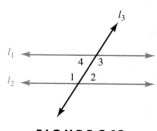

FIGURE 9-19

2. If two parallel lines are cut by a transversal, corresponding angles are congruent. (See Figure 9-20.) If $l_1 \parallel l_2$, then $\angle 1 \cong \angle 5$, $\angle 3 \cong \angle 7$, $\angle 2 \cong \angle 6$, and $\angle 4 \cong \angle 8$.

3. If two parallel lines are cut by a transversal, interior angles on the same side of the transversal are supplementary. (See Figure 9-21.) If $l_1 \parallel l_2$, then $\angle 1$ is supplementary to $\angle 2$ and $\angle 4$ is supplementary to $\angle 3$.

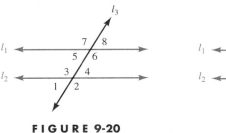

FIGURE 9-20   **FIGURE 9-21**

4. If a transversal is perpendicular to one of two parallel lines, it is also perpendicular to the other line. (See Figure 9-22.) If $l_1 \parallel l_2$ and $l_3 \perp l_1$, then $l_3 \perp l_2$.

5. If two lines are parallel to a third line, they are parallel to each other. (See Figure 9-23.) If $l_1 \parallel l_2$ and $l_1 \parallel l_3$, then $l_2 \parallel l_3$.

FIGURE 9-22   **FIGURE 9-23**

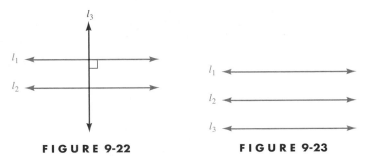

EXAMPLE 2   **Evaluating angles.**
See Figure 9-24. If $l_1 \parallel l_2$ and m($\angle 3$) = 120°, find the measures of the other angles.

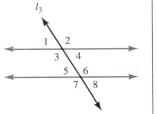

FIGURE 9-24

Solution

m($\angle 1$) = 60° $\angle 3$ and $\angle 1$ are supplementary.

m($\angle 2$) = 120°   Vertical angles are congruent: m($\angle 2$) = m($\angle 3$).

m($\angle 4$) = 60°   Vertical angles are congruent: m($\angle 4$) = m($\angle 1$).

m($\angle 5$) = 60°   If two parallel lines are cut by a transversal, alternate interior angles are congruent: m($\angle 5$) = m($\angle 4$).

m($\angle 6$) = 120°   If two parallel lines are cut by a transversal, alternate interior angles are congruent: m($\angle 6$) = m($\angle 3$).

m($\angle 7$) = 120°   Vertical angles are congruent: m($\angle 7$) = m($\angle 6$).

m($\angle 8$) = 60°   Vertical angles are congruent: m($\angle 8$) = m($\angle 5$).

Self Check

If $l_1 \parallel l_2$ and m($\angle 8$) = 50°, find the measures of the other angles. (See Figure 9-24.)

Answer: m($\angle 5$) = 50°, m($\angle 7$) = 130°, m($\angle 6$) = 130°, m($\angle 3$) = 130°, m($\angle 4$) = 50°, m($\angle 1$) = 50°, m($\angle 2$) = 130°

EXAMPLE 3   **Identifying congruent angles.**   See Figure 9-25. If $\overline{AB} \parallel \overline{DE}$, which pairs of angles are congruent?

Solution   Since $\overline{AB} \parallel \overline{DE}$, corresponding angles are congruent. So we have

$$\angle A \cong \angle 1 \qquad \text{and} \qquad \angle B \cong \angle 2$$

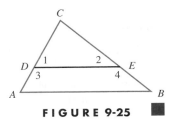

FIGURE 9-25

EXAMPLE 4 *Using algebra in ge-ometry.* In Figure 9-26, $l_1 \parallel l_2$. Find x.

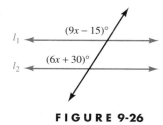

FIGURE 9-26

Solution

The angles involving x are corresponding angles. Since $l_1 \parallel l_2$, all pairs of corresponding angles are congruent.

$9x - 15 = 6x + 30$ The angle measures are equal.

$3x - 15 = 30$ Subtract $6x$ from both sides.

$3x = 45$ To undo the subtraction of 15, add 15 to both sides.

$x = 15$ To undo the multiplication by 3, divide both sides by 3.

Thus, $x = 15$.

Self Check

In the figure below, $l_1 \parallel l_2$. Find y.

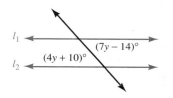

Answer: 8

EXAMPLE 5 *Using algebra in geometry.* In Figure 9-27, $l_1 \parallel l_2$. Find x.

Solution Since the angles are interior angles on the same side of the transversal, they are supplementary.

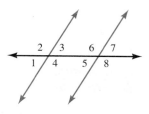

$3x - 80 + 3x + 20 = 180$ The sum of the measures of two supplementary angles is 180°.

$6x - 60 = 180$ Combine like terms.

$6x = 240$ To undo the subtraction of 60, add 60 to both sides.

$x = 40$ To undo the multiplication by 6, divide both sides by 6.

FIGURE 9-27

Thus, $x = 40$.

STUDY SET Section 9.2

VOCABULARY *In Exercises 1–6, fill in the blanks to make a true statement.*

1. Two lines in the same plane are _____.

2. _____ lines do not intersect.

3. If two lines intersect and form right angles, they are _____.

4. A _____ intersects two or more coplanar lines.

5. In Illustration 1, $\angle 4$ and $\angle 6$ are _____ angles.

6. In Illustration 1, $\angle 2$ and $\angle 6$ are _____ angles.

CONCEPTS *In Exercises 7–12, answer each question.*

7. Which pairs of angles shown in Illustration 1 are alternate interior angles?

8. Which pairs of angles shown in Illustration 1 are corresponding angles?

ILLUSTRATION 1

9. Which angles shown in Illustration 1 are interior angles?

10. In Illustration 2, $l_1 \parallel l_2$. What can you conclude about l_1 and l_3?

11. In Illustration 3, $l_1 \parallel l_2$ and $l_2 \parallel l_3$. What can you conclude about l_1 and l_3?

12. In Illustration 4, $\overline{AB} \parallel \overline{DE}$. What pairs of angles are congruent?

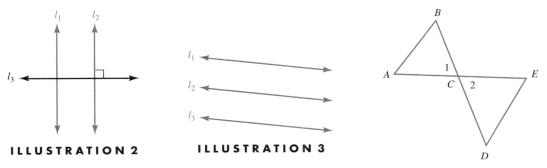

ILLUSTRATION 2   ILLUSTRATION 3

ILLUSTRATION 4

NOTATION   *In Exercises 13–16, fill in the blanks to make a true statement.*

13. The symbol ⌐ indicates _____.

14. The symbol $\parallel$ is read as _____.

15. The symbol $\perp$ is read as _____.

16. The symbol l_1 is read as _____.

PRACTICE   *In Exercises 17–20, find the measures of the missing angles.*

17. In Illustration 5, $l_1 \parallel l_2$ and m($\angle 4$) = 130°. Find the measures of the other angles.

18. In Illustration 6, $l_1 \parallel l_2$ and m($\angle 2$) = 40°. Find the measures of the other angles.

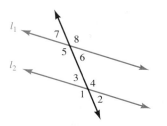

ILLUSTRATION 5   ILLUSTRATION 6

19. In Illustration 7, $l_1 \parallel \overline{AB}$. Find the measure of each angle.

20. In Illustration 8, $\overline{AB} \parallel \overline{DE}$. Find m($\angle B$), m($\angle E$), and m($\angle 1$).

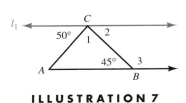

ILLUSTRATION 7

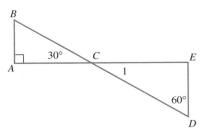

ILLUSTRATION 8

In Exercises 21–24, $l_1 \parallel l_2$. Find x.

21.

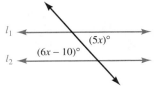

22.

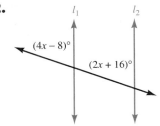

23.

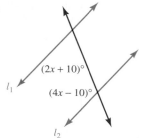

$(2x + 10)°$
$(4x - 10)°$
l_1
l_2

24.

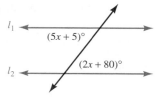

l_1
$(5x + 5)°$
$(2x + 80)°$
l_2

In Exercises 25–28, find x.

25. $l_1 \parallel \overline{CA}$

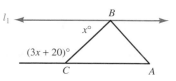

l_1
B
$x°$
$(3x + 20)°$
C
A

26. $\overline{AB} \parallel \overline{DE}$

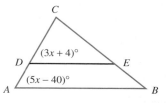

C
D
$(3x + 4)°$
E
A
$(5x - 40)°$
B

27. $\overline{AB} \parallel \overline{DE}$

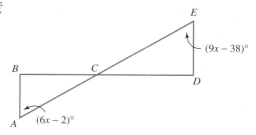

E
$(9x - 38)°$
B
C
D
A
$(6x - 2)°$

28. $\overline{AC} \parallel \overline{BD}$

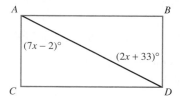

A
B
$(7x - 2)°$
$(2x + 33)°$
C
D

APPLICATIONS

29. BUILDING CONSTRUCTION   List five examples where you would see parallel lines in building construction.

30. PLUMB LINES   What is a plumb line? What geometric principle does it illustrate?

31. BUILDING CONSTRUCTION   List five examples where you would see perpendicular lines in building construction.

32. HANGING WALLPAPER   Explain why the concepts of perpendicular and parallel are both important when hanging wallpaper.

33. TOOLS   See Illustration 9. What geometric concepts do the tools show?
 a. Scissors
 b. Rake

34. PARKING DESIGN   Using terms from this chapter, write a paragraph describing the parking layout shown in Illustration 10.

ILLUSTRATION 9

North side of street

Planter

South side of street

ILLUSTRATION 10

35. Why do you think that ∠4 and ∠6 shown in Illustration 1 are called alternate interior angles?

36. Why do you think that ∠4 and ∠8 shown in Illustration 1 are called corresponding angles?

37. Are pairs of alternate interior angles always congruent? Explain.

38. Are pairs of interior angles always supplementary? Explain.

REVIEW

39. Find 60% of 120.

40. 80% of what number is 400?

41. What percent of 500 is 225?

42. Simplify: $3.45 + 7.37 \cdot 2.98$

43. Is every whole number an integer?

44. Multiply: $2\frac{1}{5} \cdot 4\frac{3}{7}$

9.3   *Polygons*

In this section, you will learn about

- **Polygons**
- **Triangles**
- **Properties of isosceles triangles**
- **The sum of the measures of the angles of a triangle**
- **Quadrilaterals**
- **Properties of rectangles**
- **The sum of the measures of the angles of a polygon**

INTRODUCTION.   In this section, we will discuss figures called *polygons*. We see these shapes every day. For example, the walls in most buildings are rectangular in shape. We also see rectangular shapes in doors, windows, and sheets of paper.

The gable ends of many houses are triangular in shape, as are the sides of the Great Pyramid in Egypt. Triangular shapes are especially important because triangles are rigid and contribute strength and stability to walls and towers.

The designs used in tile or linoleum floors often use the shapes of a pentagon or a hexagon. Stop signs are always in the shape of an octagon.

Polygons

| **Polygon** | A **polygon** is a closed geometric figure with at least three line segments for its sides. |

The figures in Figure 9-28 are **polygons.** They are classified according to the number of sides they have. The points where the sides intersect are called **vertices.**

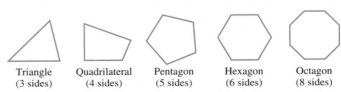

| Triangle | Quadrilateral | Pentagon | Hexagon | Octagon |
| (3 sides) | (4 sides) | (5 sides) | (6 sides) | (8 sides) |

FIGURE 9-28

| **EXAMPLE 1** | **Vertices of a polygon.** Give the number of vertices of
a. a triangle and **b.** a hexagon. | **Self Check**
Give the number of vertices of |

Solution

a. From Figure 9-28, we see that a triangle has three angles and therefore three vertices.

b. From Figure 9-28, we see that a hexagon has six angles and therefore six vertices.

Self Check
Give the number of vertices of

a. a quadrilateral

b. a pentagon

Answer: 4, 5

From the results of Example 1, we see that the number of vertices of a polygon is equal to the number of its sides.

Triangles

A **triangle** is a polygon with three sides. Figure 9-29 illustrates some common triangles. The slashes on the sides of a triangle indicate which sides are of equal length.

Equilateral triangle
(all sides equal length)

Isosceles triangle
(at least two sides of
equal length)

Scalene triangle
(no sides equal length)

Right triangle
(has a right angle)

FIGURE 9-29

 WARNING! Since equilateral triangles have at least two sides of equal length, they are also isosceles. However, isosceles triangles are not necessarily equilateral.

Since every angle of an equilateral triangle has the same measure, an equilateral triangle is also **equiangular.**

In an isosceles triangle, the angles opposite the sides of equal length are called **base angles,** the sides of equal length form the **vertex angle,** and the third side is called the **base.**

The longest side of a right triangle is called the **hypotenuse,** and the other two sides are called **legs.** The hypotenuse of a right triangle is always opposite the 90° angle

Properties of isosceles triangles

1. Base angles of an isosceles triangle are congruent.

2. If two angles in a triangle are congruent, the sides opposite the angles have the same length, and the triangle is isosceles.

| **EXAMPLE 2** | **Determining whether a triangle is isosceles.**
Is the triangle in Figure 9-30
an isosceles triangle? |

Solution

$\angle A$ and $\angle B$ are angles of the triangle. Since $m(\angle A) = m(\angle B)$, we know that $m(\overline{AC}) = m(\overline{BC})$ and that $\triangle ABC$ (read as "triangle ABC") is isosceles.

FIGURE 9-30

Self Check
In the figure below, $l_1 \parallel \overline{AB}$. Is the triangle an isosceles triangle?

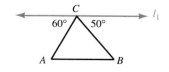

Answer: no

The sum of the measures of the angles of a triangle

If you draw several triangles and carefully measure each angle with a protractor, you will find that the sum of the angle measures in each triangle is 180°.

Angles of a triangle	The sum of the angle measures of any triangle is 180°.

EXAMPLE 3 *Sum of the angles of a triangle.* See Figure 9-31. Find x.

FIGURE 9-31

Solution

Since the sum of the angle measures of any triangle is 180°, we have

$x + 40 + 90 = 180$

$x + 130 = 180$ $40 + 90 = 130.$

$x = 50$ To undo the addition of 130, subtract 130 from both sides.

Thus, $x = 50$.

Self Check

In the figure below, find y.

Answer: 90

EXAMPLE 4 *Vertex angle of an isosceles triangle.* See Figure 9-32. If one base angle of an isosceles triangle measures 70°, how large is the vertex angle?

Solution Since one of the base angles measures 70°, so does the other. If we let x represent the measure of the vertex angle, we have

$x + 70 + 70 = 180$ The sum of the measures of the angles of a triangle is 180°.

$x + 140 = 180$ $70 + 70 = 140.$

$x = 40$ To undo the addition of 140, subtract 140 from both sides.

The vertex angle measures 40°.

FIGURE 9-32

Quadrilaterals

A **quadrilateral** is a polygon with four sides. Some common quadrilaterals are shown in Figure 9-33.

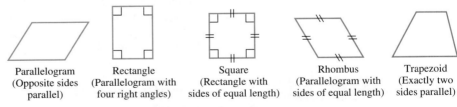

Parallelogram (Opposite sides parallel)   Rectangle (Parallelogram with four right angles)   Square (Rectangle with sides of equal length)   Rhombus (Parallelogram with sides of equal length)   Trapezoid (Exactly two sides parallel)

FIGURE 9-33

Properties of rectangles

1. All angles of a rectangle are right angles.

2. Opposite sides of a rectangle are parallel.

3. Opposite sides of a rectangle are of equal length.

4. The diagonals of a rectangle are of equal length.

5. If the diagonals of a parallelogram are of equal length, the parallelogram is a rectangle.

EXAMPLE 5 ***Squaring a foundation.*** A carpenter intends to build a shed with an 8-by-12-foot base. How can he make sure that the rectangular foundation is "square"?

Solution See Figure 9-34. The carpenter can use a tape measure to find the lengths of diagonals *AC* and *BD*. If these diagonals are of equal length, the figure will be a rectangle and have four right angles. Then the foundation will be "square."

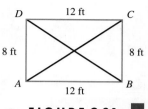

FIGURE 9-34 ∎

EXAMPLE 6 ***Properties of rectangles and triangles.*** In rectangle *ABCD* (Figure 9-35), the length of $\overline{AC}$ is 20 centimeters. Find each measure: **a.** m($\overline{BD}$), **b.** m(∠1), and **c.** m(∠2).

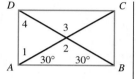

FIGURE 9-35

Solution

a. Since the diagonals of a rectangle are of equal length, m($\overline{BD}$) is also 20 centimeters.

b. We let m(∠1) = x. Then, since the angles of a rectangle are right angles, we have

$$x + 30 = 90$$
$$x = 60 \quad \text{To undo the addition of 30, subtract 30 from both sides.}$$

Thus, m(∠1) = 60°.

c. We let m(∠2) = y. Then, since the sum of the angle measures of a triangle is 180°, we have

$$30 + 30 + y = 180$$
$$60 + y = 180 \quad 30 + 30 = 60.$$
$$y = 120 \quad \text{To undo the addition of 60, subtract 60 from both sides.}$$

Thus, m(∠2) = 120°.

Self Check

In rectangle *ABCD*, the length of $\overline{DC}$ is 16 centimeters. Find each measure:

a. m($\overline{AB}$)

b. m(∠3)

c. m(∠4)

Answer: 16 cm, 120°, 60° ∎

The parallel sides of a trapezoid are called **bases,** the nonparallel sides are called **legs,** and the angles on either side of a base are called **base angles.** If the nonparallel sides are the same length, the trapezoid is an **isosceles trapezoid.** In an isosceles trapezoid, the base angles are congruent.

EXAMPLE 7 ***Cross section of a drainage ditch.*** A cross section of a drainage ditch (Figure 9-36) is an isosceles trapezoid with $\overline{AB} \parallel \overline{CD}$. Find *x* and *y*.

Solution Since the figure is an isosceles trapezoid, its nonparallel sides have the same length. So m($\overline{AD}$) and m($\overline{BC}$) are equal, and x = 8.

Since the base angles of an isosceles trapezoid are congruent, m(∠D) = m(∠C). So y = 120.

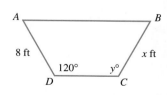

FIGURE 9-36

∎

The sum of the measures of the angles of a polygon

We have seen that the sum of the angle measures of any triangle is 180°. Since a polygon with n sides can be divided into $n - 2$ triangles, the sum of the angle measures of the polygon is $(n - 2)180°$.

Angles of a polygon	The sum S of the measures of the angles of a polygon with n sides is given by the formula
	$$S = (n - 2)180°$$

EXAMPLE 8   ***Sum of the angles of a pentagon.***   Find the sum of the angle measures of a pentagon.

Self Check

Find the sum of the angle measures of a quadrilateral.

Solution

Since a pentagon has 5 sides, we substitute 5 for n in the formula and simplify.

$$S = (n - 2)180°$$
$$S = (5 - 2)180°$$
$$= 3(180°)$$
$$= 540°$$

The sum of the angles of a pentagon is 540°.

Answer: 360°

STUDY SET   Section 9.3

VOCABULARY   *In Exercises 1–16, fill in the blanks to make a true statement.*

1. A polygon with four sides is called a _____ .

2. A _____ is a polygon with three sides.

3. A _____ is a polygon with six sides.

4. A polygon with five sides is called a _____ .

5. An eight-sided polygon is an _____ .

6. The points where the sides of a polygon intersect are called _____ .

7. A triangle with three sides of equal length is called an _____ triangle.

8. An _____ triangle has two sides of equal length.

9. The longest side of a right triangle is the _____ .

10. The _____ angles of an _____ triangle have the same measure.

11. A _____ with a right angle is a rectangle.

12. A rectangle with all sides of equal length is a _____ .

13. A _____ is a parallelogram with four sides of equal length.

14. A _____ has two sides that are parallel and two sides that are not parallel.

15. The legs of an _____ trapezoid have the same length.

16. The _____ of a polygon is the distance around it.

CONCEPTS In Exercises 17–24, give the number of sides of each polygon and classify it as a triangle, quadrilateral, pentagon, hexagon, or octagon. Then give the number of vertices.

17.

18.

19.

20.

21.

22.

23.

24.

In Exercises 25–32, classify each triangle as an equilateral triangle, an isosceles triangle, a scalene triangle, or a right triangle.

25.

26.

27.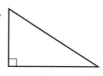

28.

29.

30.

31.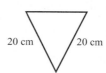

32.

In Exercises 33–40, classify each quadrilateral as a rectangle, a square, a rhombus, or a trapezoid.

33.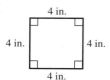

34.

35.

36.

37.

38.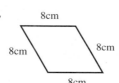

39.

40.

NOTATION In Exercises 41–42, fill in the blanks to make a true statement.

41. The symbol △ means _____ .

42. The symbol m(∠1) means the _____ of _____ .

PRACTICE In Exercises 43–48, the measures of two angles of △ABC (shown in Illustration 1) are given. Find the measure of the third angle.

43. m(∠A) = 30° and m(∠B) = 60°.
m(∠C) = _____ .

44. m(∠A) = 45° and m(∠C) = 105°.
m(∠B) = _____ .

45. m(∠B) = 100° and m(∠A) = 35°.
m(∠C) = _____ .

46. m(∠B) = 33° and m(∠C) = 77°.
m(∠A) = _____ .

47. m(∠A) = 25.5° and m(∠B) = 63.8°.
m(∠C) = _____.

48. m(∠B) = 67.25° and m(∠C) = 72.5°.
m(∠A) = _____.

ILLUSTRATION 1

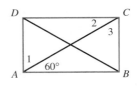

ILLUSTRATION 2

In Exercises 49–52, refer to rectangle ABCD, shown in Illustration 2.

49. m(∠1) = _____.

50. m(∠3) = _____.

51. m(∠2) = _____.

52. If m($\overline{AC}$) is 8 cm, then m($\overline{BD}$) = _____.

In Exercises 53–56, find the sum of the angle measures of each polygon.

53. A hexagon

54. An octagon

55. A decagon (10 sides)

56. A dodecagon (12 sides)

In Exercises 57–60, find the number of sides of the polygon with the given angle measure sum.

57. 900°

58. 1,260°

59. 2,160°

60. 3,600°

APPLICATIONS

61. Give three uses of triangles in everyday life.

62. Give three uses of rectangles in everyday life.

63. Give three uses of squares in everyday life.

64. Give a use of a trapezoid in everyday life.

WRITING *Write a paragraph using your own words.*

65. Explain why a square is a rectangle.

66. Explain why a trapezoid is not a parallelogram.

REVIEW

67. Find 20% of 110.

68. Find 15% of 50.

69. Find 20% of $\frac{6}{11}$.

70. Find 30% of $\frac{3}{5}$.

71. What percent of 200 is 80?

72. What percent of 500 is 100?

73. 20% of what number is 500?

74. 30% of what number is 21?

75. Simplify: 0.85 ÷ 2(0.25).

76. Simplify: 3.25 + 12 ÷ 0.4 · 2.

9.4 *Properties of Triangles*

In this section, you will learn about

- **Congruent triangles**
- **Similar triangles**
- **The Pythagorean theorem**

INTRODUCTION. We can often use proportions and triangles to measure distances indirectly. For example, by using a proportion, Eratosthenes (275–195 B.C.) was able to estimate the circumference of the earth to a remarkable degree of accuracy. On a sunny day, we can use properties of similar triangles to calculate the height of a tree while staying safely on the ground. By using a theorem proved by the Greek mathematician Pythagoras (about 500 B.C.), we can calculate the length of the third side of a right triangle whenever we know the lengths of two sides.

Congruent triangles

Triangles that have the same size and the same shape are called **congruent triangles.** In Figure 9-37, triangles *ABC* and *DEF* are congruent:

$$\triangle ABC \cong \triangle DEF \quad \text{Read as "Triangle } ABC \text{ is congruent to triangle } DEF \text{."}$$

Corresponding angles and corresponding sides of congruent triangles are called **corresponding parts.** The notation $\triangle ABC \cong \triangle DEF$ shows which vertices are corresponding parts.

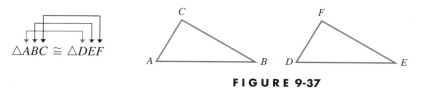

FIGURE 9-37

Corresponding parts of congruent triangles always have the same measure. In the congruent triangles shown in Figure 9-37,

$$m(\angle A) = m(\angle D), \qquad m(\angle B) = m(\angle E), \qquad m(\angle C) = m(\angle F),$$
$$m(\overline{BC}) = m(\overline{EF}), \qquad m(\overline{AC}) = m(\overline{DF}), \qquad m(\overline{AB}) = m(\overline{DE})$$

EXAMPLE 1 *Corresponding parts of congruent triangles.* Name the corresponding parts of the congruent triangles in Figure 9-38.

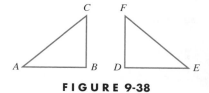

FIGURE 9-38

Solution The corresponding angles are

$$\angle A \text{ and } \angle E, \quad \angle B \text{ and } \angle D, \quad \angle C \text{ and } \angle F$$

Since corresponding sides are always opposite corresponding angles, the corresponding sides are

$$\overline{BC} \text{ and } \overline{DF}, \qquad \overline{AC} \text{ and } \overline{EF}, \qquad \overline{AB} \text{ and } \overline{ED}$$

We will discuss three ways of showing that two triangles are congruent.

SSS property If three sides of one triangle are congruent to three sides of a second triangle, the triangles are congruent.

The triangles in Figure 9-39 are congruent because of the SSS property.

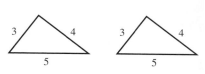

FIGURE 9-39

SAS property	If two sides and the angle between them in one triangle are congruent, respectively, to two sides and the angle between them in a second triangle, the triangles are congruent.

The triangles in Figure 9-40 are congruent because of the SAS property.

FIGURE 9-40

ASA property	If two angles and the side between them in one triangle are congruent, respectively, to two angles and the side between them in a second triangle, the triangles are congruent.

The triangles in Figure 9-41 are congruent because of the ASA property.

FIGURE 9-41

EXAMPLE 2 **Determining whether triangles are congruent.** Explain why the triangles in Figure 9-42 are congruent.

Solution Since vertical angles are congruent,

$$m(\angle 1) = m(\angle 2)$$

From the figure, we see that

$$m(\overline{AC}) = m(\overline{EC}) \quad \text{and} \quad m(\overline{BC}) = m(\overline{DC})$$

Since two sides and the angle between them in one triangle are congruent, respectively, to two sides and the angle between them in a second triangle, the triangles are congruent by the SAS property.

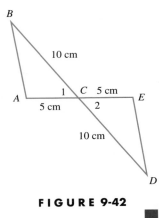

FIGURE 9-42

Similar triangles

If two angles of one triangle are congruent to two angles of a second triangle, the triangles will have the same shape. Triangles with the same shape are called **similar triangles.** In Figure 9-43, $\triangle ABC \sim \triangle DEF$ (read the symbol $\sim$ as "is similar to").

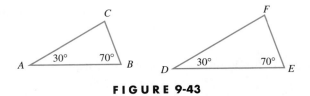

FIGURE 9-43

WARNING! Note that congruent triangles are always similar, but similar triangles are not always congruent.

Property of similar triangles	If two triangles are similar, all pairs of corresponding sides are in proportion.

In the similar triangles shown in Figure 9-43, the following proportions are true.

$$\frac{\overline{AB}}{\overline{DE}} = \frac{\overline{BC}}{\overline{EF}}, \qquad \frac{\overline{BC}}{\overline{EF}} = \frac{\overline{CA}}{\overline{FD}}, \qquad \text{and} \qquad \frac{\overline{CA}}{\overline{FD}} = \frac{\overline{AB}}{\overline{DE}}$$

EXAMPLE 3   ***Finding the height of a tree.***   A tree casts a shadow 18 feet long at the same time as a woman 5 feet tall casts a shadow that is 1.5 feet long. (See Figure 9-44.) Find the height of the tree.

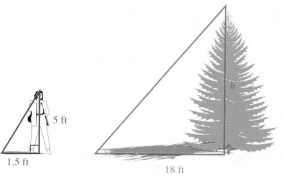

FIGURE 9-44

Solution The figure shows the triangles determined by the tree and its shadow and the woman and her shadow. Since the triangles have the same shape, they are similar, and the lengths of their corresponding sides are in proportion. If we let h represent the height of the tree, we can find h by solving the following proportion.

$\dfrac{h}{5} = \dfrac{18}{1.5}$ $\dfrac{\text{Height of the tree}}{\text{Height of the woman}} = \dfrac{\text{shadow of the tree}}{\text{shadow of the woman}}$.

$1.5h = 5(18)$ In a proportion, the product of the extremes is equal to the product of the means.

$1.5h = 90$ Do the multiplication: $5(18) = 90$.

$h = 60$ To undo the multiplication by 1.5, divide both sides by 1.5 and simplify.

The tree is 60 feet tall. ■

The Pythagorean theorem

In the movie *The Wizard of Oz,* the scarecrow was in search of a brain. To prove that he had found one, he recited the Pythagorean theorem.

In a right triangle, the square of the hypotenuse is equal to the sum of squares of the other two sides.

Pythagorean theorem	If the length of the hypotenuse of a right triangle is c and the lengths of its legs are a and b (as in Figure 9-45), then $$a^2 + b^2 = c^2$$	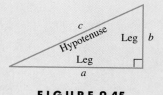

FIGURE 9-45

EXAMPLE 4 *Constructing a high-ropes adventure course.* A builder of a high-ropes adventure course wants to secure the pole shown in Figure 9-46 by attaching a cable from the ground anchor 20 feet from its base to a point 15 feet up the pole. How long should the cable be?

Solution

We can use the Pythagorean theorem with $a = 20$ and $b = 15$.

$c^2 = a^2 + b^2$

$c^2 = 20^2 + 15^2$ Substitute 20 for a and 15 for b.

$c^2 = 400 + 225$

$c^2 = 625$

$\sqrt{c^2} = \sqrt{625}$ Since equal positive numbers have equal square roots, take the positive square root of both sides.

$c = 25$ $\sqrt{c^2} = c$ and $\sqrt{625} = 25$.

The cable will be 25 feet long.

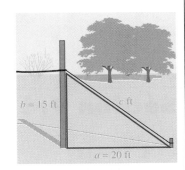

FIGURE 9-46

Self Check

A 26-foot ladder rests against the side of a building. If the base of the ladder is 10 feet from the wall, how far up the side of the building will the ladder reach?

Answer: 24 ft

Accent on Technology   *Finding the width of a television screen*

The size of a television screen is the diagonal measure of its rectangular screen. (See Figure 9-47.) To find the width of a 27-inch screen that is 17 inches high, we use the Pythagorean theorem with $c = 27$ and $b = 17$.

$$c^2 = a^2 + b^2$$
$$27^2 = a^2 + 17^2$$
$$27^2 - 17^2 = a^2$$
$$\sqrt{27^2 - 17^2} = \sqrt{a^2}$$ Take the positive square root of both sides.
$$\sqrt{27^2 - 17^2} = a$$ $\sqrt{a^2} = a$.

Evaluate: $\sqrt{27^2 - 17^2}$

Keystrokes: (2 7 x^2 − 1 7 x^2) $\sqrt{\ }$

To the nearest inch, the width is 21 inches.

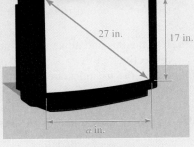

FIGURE 9-47

$$20.97617696$$

It is also true that

If the square of one side of a triangle is equal to the sum of the squares of the other two sides, the triangle is a right triangle.

EXAMPLE 5 *Determining whether a triangle is a right triangle.* Determine whether a triangle with sides of 5, 12, and 13 meters is a right triangle.

Self Check

Decide whether a triangle with sides of 9, 40, and 41 meters is a right triangle.

Solution

We check to see whether the square of one side is equal to the sum of the squares of the other two sides.

$13^2 \overset{?}{=} 5^2 + 12^2$ The longest side is the hypotenuse.

$169 \overset{?}{=} 25 + 144$

$169 = 169$

Since $13^2 = 5^2 + 12^2$, the triangle is a right triangle.

Answer: yes

STUDY SET   Section 9.4

VOCABULARY   *In Exercises 1–4, fill in the blanks to make a true statement.*

1. _____ triangles are the same size and the same shape.

2. All _____ parts of congruent triangles have the same measure.

3. If two triangles are _____, they have the same shape.

4. The _____ is the longest side of a right triangle.

CONCEPTS   *In Exercises 5–8, tell whether each statement is true. If a statement is false, tell why.*

5. If three sides of one triangle are the same length as three sides of a second triangle, the triangles are congruent.

6. If two sides of one triangle are the same length as two sides of a second triangle, the triangles are congruent.

7. If two sides and an angle of one triangle are congruent, respectively, to two sides and an angle of a second triangle, the triangles are congruent.

8. If two angles and the side between them in one triangle are congruent, respectively, to two angles and the side between them in a second triangle, the triangles are congruent.

9. Are the triangles shown in Illustration 1 congruent?

10. Are the triangles shown in Illustration 2 congruent?

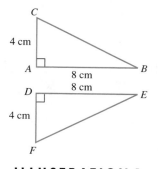

ILLUSTRATION 1

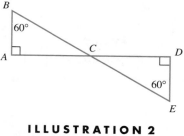

ILLUSTRATION 2

11. In a proportion, the _____ of the means is equal to the product of the _____.

12. If two angles of one triangle are congruent to two angles of a second triangle, the triangles are _____.

13. Are the triangles shown in Illustration 3 similar?

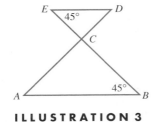

ILLUSTRATION 3

14. Are the triangles shown in Illustration 4 similar?

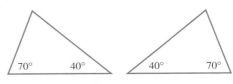

ILLUSTRATION 4

15. If x and y represent the lengths of two legs of a right triangle and z represents the length of the hypotenuse, the Pythagorean theorem states that $z^2 = x^2 + y^2$.

16. A triangle with sides of 3, 4, and 5 centimeters is a right triangle.

NOTATION *In Exercises 17–18, fill in the blanks to make a true statement.*

17. The symbol $\cong$ is read as "_____."

18. The symbol $\sim$ is read as "_____."

PRACTICE *In Exercises 19–20, name the corresponding parts of the congruent triangles.*

19. Refer to Illustration 5. (The slashes indicate pairs of congruent sides.)

$\overline{AC}$ corresponds to ___.

$\overline{DE}$ corresponds to ___.

$\overline{BC}$ corresponds to ___.

$\angle A$ corresponds to ___.

$\angle E$ corresponds to ___.

$\angle F$ corresponds to ___.

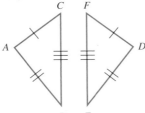

ILLUSTRATION 5

20. Refer to Illustration 6.

$\overline{AB}$ corresponds to ___.

$\overline{EC}$ corresponds to ___.

$\overline{AC}$ corresponds to ___.

$\angle D$ corresponds to ___.

$\angle B$ corresponds to ___.

$\angle 1$ corresponds to ___.

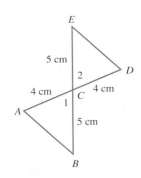

ILLUSTRATION 6

In Exercises 21–28, determine whether each pair of triangles is congruent. If they are, tell why.

21.

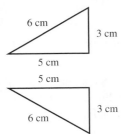

22.

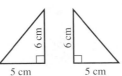

23.

24.

25.

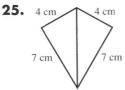

26.

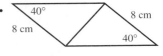

27.

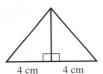

28.

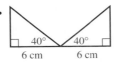

In Exercises 29–32, find x.

29.

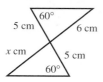

30.

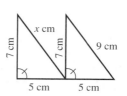

31.

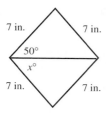

32.

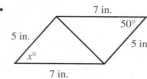

In Exercises 33–34, tell whether the triangles are similar.

33.

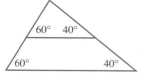

34.

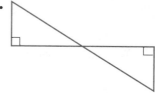

In Exercises 35–40, refer to Illustration 7 and find the length of the unknown side.

35. $a = 3$ and $b = 4$. Find c.

36. $a = 12$ and $b = 5$. Find c.

37. $a = 15$ and $c = 17$. Find b.

38. $b = 45$ and $c = 53$. Find a.

39. $a = 5$ and $c = 9$. Find b.

40. $a = 1$ and $b = 7$. Find c.

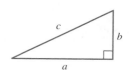

ILLUSTRATION 7

In Exercises 41–44, the length of the three sides of a triangle are given. Determine whether the triangle is a right triangle.

41. 8, 15, 17

42. 6, 8, 10

43. 7, 24, 26

44. 9, 39, 40

APPLICATIONS In Exercises 45–54, use a calculator to help solve each problem. If an answer is not exact, give the answer to the nearest tenth.

45. HEIGHT OF A TREE The tree in Illustration 8 casts a shadow of 24 feet when a 6-foot man casts a shadow of 4 feet. Find the height of the tree.

46. HEIGHT OF A BUILDING A man places a mirror on the ground and sees the reflection of the top of a building, as shown in Illustration 9. Find the height of the building.

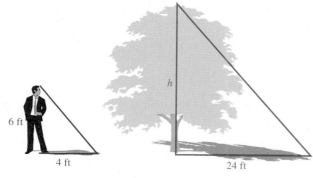

ILLUSTRATION 8

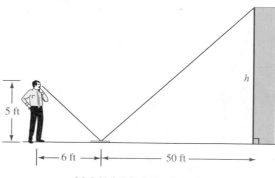

ILLUSTRATION 9

47. WIDTH OF A RIVER   Use the dimensions in Illustration 10 to find w, the width of the river.

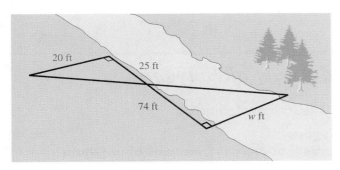

ILLUSTRATION 10

48. FLIGHT PATH   The airplane in Illustration 11 ascends 200 feet as it flies a horizontal distance of 1,000 feet. How much altitude is gained as it flies a horizontal distance of 1 mile? (*Hint:* 1 mile = 5,280 feet.)

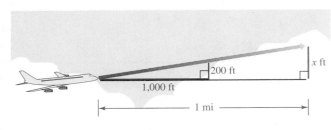

ILLUSTRATION 11

49. FLIGHT PATH   An airplane descends 1,200 feet as it files a horizontal distance of 1.5 miles. How much altitude is lost as it flies a horizontal distance of 5 miles?

50. GEOMETRY   If segment DE in Illustration 12 is parallel to segment AB, $\triangle ABC$ will be similar to $\triangle DEC$. Find x.

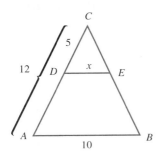

ILLUSTRATION 12

51. ADJUSTING A LADDER   A 20-foot ladder reaches a window 16 feet above the ground. How far from the wall is the base of the ladder?

52. LENGTH OF GUY WIRES   A 30-foot tower is to be fastened by three guy wires attached to the top of the tower and to the ground at positions 20 feet from its base. How much wire is needed?

53. BASEBALL   A baseball diamond is a square with each side 90 feet long. (See Illustration 13.) How far is it from home plate to second base?

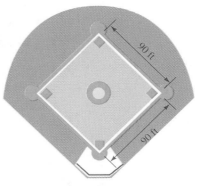

ILLUSTRATION 13

54. TELEVISION   What size is the television screen shown in Illustration 14?

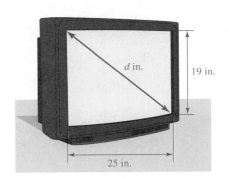

ILLUSTRATION 14

WRITING   *Write a paragraph using your own words.*

55. Explain the Pythagorean theorem.

56. Explain the procedure used to solve the equation $c^2 = 64$.

REVIEW   *In Exercises 57–60, estimate the answer to each problem.*

57. $\dfrac{0.95 \cdot 3.89}{2.997}$

58. 21% of 42

59. 32% of 60

60. $\dfrac{4.966 + 5.001}{2.994}$

In Exercises 61–62, simplify each expression.

61. $2 + 4 \cdot 3^2$

62. $3 - (5 - 2)^2 + 2^2$

9.5 Perimeters and Areas of Polygons

In this section, you will learn how to find

- **Perimeters of polygons**
- **Perimeters of figures that are combinations of polygons**
- **Areas of polygons**
- **Areas of figures that are combinations of polygons**

INTRODUCTION. In this section, we will discuss how to find perimeters and areas of polygons. Finding perimeters is important when estimating the cost of fencing or estimating the cost of woodwork in a house. Finding areas is important when calculating the cost of carpeting, the cost of painting a house, or the cost of fertilizing a yard.

Perimeters of polygons

Recall that the **perimeter** of a polygon is the distance around it. Since a square has four sides of equal length s, its perimeter P is $s + s + s + s$, or $4s$.

| Perimeter of a square | If a square has a side of length s, its perimeter P is given by the formula $$P = 4s$$ | |

EXAMPLE 1   *Perimeter of a square.*   Find the perimeter of a square whose sides are 7.5 meters long.

Solution

Since the perimeter of a square is given by the formula $P = 4s$, we substitute 7.5 for s and simplify.

$P = 4s$

$P = 4(7.5)$

$P = 30$

The perimeter is 30 meters.

Self Check

Find the perimeter of a square whose sides are 23.75 centimeters long.

Answer: 95 cm

Since a rectangle has two lengths l and two widths w, its perimeter P is $l + l + w + w$, or $2l + 2w$.

| Perimeter of a rectangle | If a rectangle has length l and width w, its perimeter P is given by the formula $$P = 2l + 2w$$ | 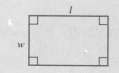 |

EXAMPLE 2 *Perimeter of a rectangle.*
Find the perimeter of the rectangle
in Figure 9-48.

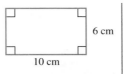

6 cm

10 cm

FIGURE 9-48

Self Check
Find the perimeter of the isosceles
trapezoid below.

10 cm

8 cm   8 cm

12 cm

Solution
Since the perimeter is given by the formula $P = 2l + 2w$,
we substitute 10 for l and 6 for w and simplify.

$$P = 2l + 2w$$
$$P = 2(10) + 2(6)$$
$$= 20 + 12$$
$$= 32$$

The perimeter is 32 centimeters.

Answer: 38 cm ■

EXAMPLE 3 *Converting units.* Find the perimeter
of the rectangle in Figure 9-49, in meters.

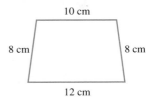

3 m

80 cm

FIGURE 9-49

Self Check
Find the perimeter of the triangle
below, in inches.

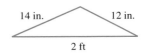

14 in.   12 in.

2 ft

Solution
Since 1 meter = 100 centimeters, we can convert 80 centimeters to meters by multiplying 80 centimeters by the unit conversion factor $\frac{1\,m}{100\,cm}$.

$$80 \text{ cm} = 80 \text{ cm} \cdot \frac{1 \text{ m}}{100 \text{ cm}} \qquad \tfrac{1\,m}{100\,cm} = 1.$$

$$= \frac{80}{100} \text{ m} \qquad \text{The units of centimeters divide out.}$$

$$= 0.8 \text{ m} \qquad \text{Do the division: } 80 \div 100 = 0.8.$$

We can now substitute 3 for l and 0.8 for w to get

$$P = 2l + 2w$$
$$P = 2(3) + 2(0.8)$$
$$= 6 + 1.6$$
$$= 7.6$$

The perimeter is 7.6 meters.

Answer: 50 in. ■

EXAMPLE 4 *Finding the base of an isosceles
triangle.* The perimeter of the isos-
celes triangle in Figure 9-50 is 50 meters. Find the length of
its base.

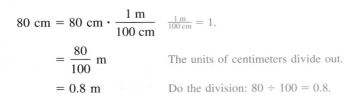

12 m   12 m

x m

FIGURE 9-50

Self Check
The perimeter of an isosceles triangle is 60 meters. If one of its
sides of equal length is 15 meters
long, how long is its base?

Solution
Two sides are 12 meters long, and the perimeter is 50 meters.
If x represents the length of the base, we have

$$12 + 12 + x = 50$$
$$24 + x = 50 \quad 12 + 12 = 24.$$
$$x = 26 \quad \text{To undo the addition of 24, subtract 24 from both sides.}$$

The length of the base is 26 meters.

Answer: 30 m ■

Perimeters of figures that are combinations of polygons

Accent on Technology   *Perimeter of a figure*

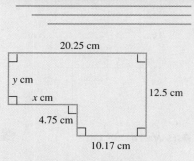

FIGURE 9-51

See Figure 9-51. To find the perimeter, we need to know the values of x and y. Since the figure is a combination of two rectangles, we can use a calculator to see that

$$x = 20.25 - 10.17 \qquad \text{and} \qquad y = 12.5 - 4.75$$
$$= 10.08 \qquad\qquad\qquad = 7.75$$

The perimeter P of the figure is

$$P = 20.25 + 12.5 + 10.17 + 4.75 + x + y$$
$$P = 20.25 + 12.5 + 10.17 + 4.75 + 10.08 + 7.75$$

Evaluate: $20.25 + 12.5 + 10.17 + 4.75 + 10.08 + 7.75$

Keystrokes: 2 0 . 2 5 + 1 2 . 5 + 1 0 . 1 7 +
4 . 7 5 + 1 0 . 0 8 + 7 . 7 5 =

| 65.5 |

The perimeter is 65.5 centimeters.

Areas of polygons

Recall that the **area** of a polygon is the measure of the amount of surface it encloses. Area is measured in square units, such as square inches or square centimeters. See Figure 9-52.

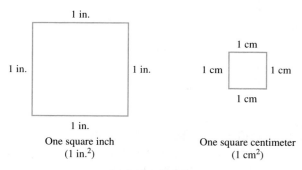

One square inch
(1 in.²)

One square centimeter
(1 cm²)

FIGURE 9-52

In everyday life, we commonly use areas. For example,

- To carpet a room, we buy square yards.
- A can of paint will cover a certain number of square feet.
- To measure real estate, we often use square miles.
- We buy house roofing by the "square." One square is 100 square feet.

The rectangle shown in Figure 9-53 has a length of 10 centimeters and a width of 3 centimeters. If we divide the rectangle into squares as shown in the figure, each square represents an area of 1 square centimeter—a surface enclosed by a square measuring 1 centimeter on each side. Because there are 3 rows with 10 squares in each row, there are 30 squares. Since the rectangle encloses a surface area of 30 squares, its area is 30 square centimeters, often written as 30 cm^2.

This example illustrates that to find the area of a rectangle, we multiply its length by its width.

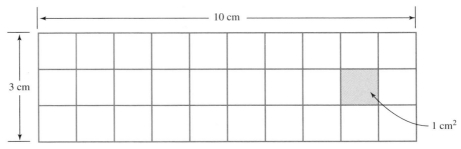

F I G U R E   9-53

 WARNING! Do not confuse the concepts of perimeter and area. Perimeter is the distance around a polygon. It is measured in linear units, such as centimeters, feet, or miles. Area is a measure of the surface enclosed within a polygon. It is measured in square units, such as square centimeters, square feet, or square miles.

In practice, we do not find areas by counting squares in a figure. Instead, we use formulas for finding areas of geometric figures.

Figure	Name	Formula for area
	Square	$A = s^2$, where s is the length of one side.
	Rectangle	$A = lw$, where l is the length and w is the width.
	Parallelogram	$A = bh$, where b is the length of the base and h is the height. (A height is always perpendicular to the base.)
	Triangle	$A = \frac{1}{2}bh$, where b is the length of the base and h is the height. The segment that represents the height is called an **altitude.**
	Trapezoid	$A = \frac{1}{2}h(b_1 + b_2)$, where h is the height of the trapezoid and b_1 and b_2 represent the length of each base.

EXAMPLE 5   *Area of a square.*   Find the area of the square in Figure 9-54.

Solution
We can see that the length of one side of the square is 15 centimeters. We can find its area by using the formula $A = s^2$ and substituting 15 for s.

$A = s^2$
$A = (15)^2$ Substitute 15 for s.
$A = 15 \cdot 15$ $15^2 = 15 \cdot 15$.
$A = 225$ $15 \cdot 15 = 225$.

The area of the square is 225 cm².

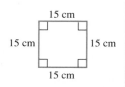

FIGURE 9-54

Self Check

Find the area of the square shown below.

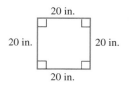

Answer: 400 in.²

EXAMPLE 6   *Number of square feet in 1 square yard.*   Find the number of square feet in 1 square yard. (See Figure 9-55.)

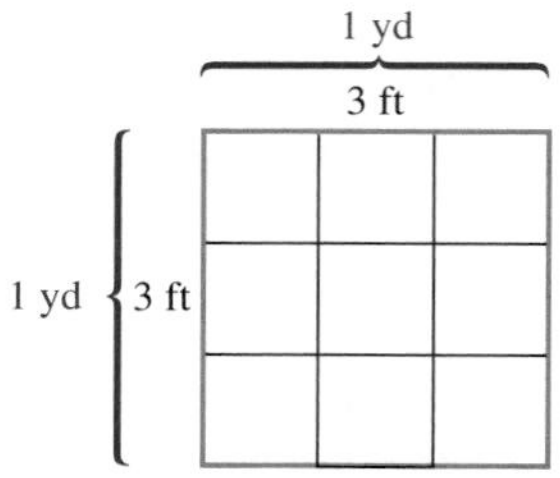

Solution
Since 3 feet = 1 yard, each side of 1 square yard is 3 feet long.

$1 \text{ yd}^2 = (1 \text{ yd})^2$
$= (3 \text{ ft})^2$ Substitute 3 feet for 1 yard.
$= 9 \text{ ft}^2$ $(3 \text{ ft})^2 = (3 \text{ ft})(3 \text{ ft}) = 9 \text{ ft}^2$.

There are 9 square feet in 1 square yard.

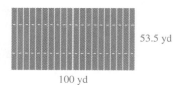

FIGURE 9-55

Self Check

Find the number of square centimeters in 1 square meter.

Answer: 10,000 cm²

EXAMPLE 7   *Area of a football field.*   Find the area of a rectangular football field in square feet. Disregard the end zones. (See Figure 9-56.)

Solution
To find the area in square yards, we substitute 100 for l and 53.5 for w in the formula $A = lw$.

$A = lw$
$A = (100)(53.5)$
$= 5,350$

The area is 5,350 square yards. Since there are 9 square feet per square yard, we can convert this number to square feet by multiplying 5,350 square yards by $\frac{9 \text{ ft}^2}{1 \text{ yd}^2}$.

$5,350 \text{ yd}^2 = 5,350 \text{ yd}^2 \cdot \dfrac{9 \text{ ft}^2}{1 \text{ yd}^2}$

$= 5,350 \cdot 9 \text{ ft}^2$ The units of square yards divide out.
$= 48,150 \text{ ft}^2$ $5,350 \cdot 9 = 48,150$.

The area of a football field is 48,150 ft².

FIGURE 9-56

Self Check

Find the area of a rectangle with dimensions of 6 inches by 2 feet, in square inches.

Answer: 144 in.²

| **EXAMPLE 8** | *Area of a parallelogram.* Find the area of the parallelogram in Figure 9-57. |

FIGURE 9-57

Solution

The length of the base of the parallelogram is

5 feet + 25 feet = 30 feet

The height is 12 feet. To find the area, we substitute 30 for *b* and 12 for *h* in the formula for the area of a parallelogram and simplify.

$$A = bh$$
$$A = 30 \cdot 12$$
$$= 360$$

The area of the parallelogram is 360 ft².

Self Check
Find the area of the parallelogram below.

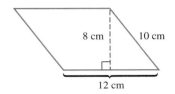

Answer: 96 cm²

| **EXAMPLE 9** | *Area of a triangle.* Find the area of the triangle in Figure 9-58. |

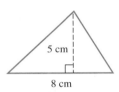

FIGURE 9-58

Solution

The area of a triangle is found by substituting 8 for *b* and 5 for *h* in the formula for the area of a triangle.

$$A = \frac{1}{2}bh$$

$$A = \frac{1}{2}(8)(5) \qquad \text{Substitute for } b \text{ and } h.$$

$$= 4(5) \qquad \text{Do the multiplication: } \frac{1}{2}(8) = 4.$$

$$= 20$$

The area of the triangle is 20 cm².

Self Check
Find the area of the triangle below.

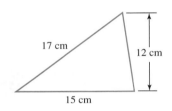

Answer: 90 cm²

| **EXAMPLE 10** | *Area of a triangle.* Find the area of the triangle in Figure 9-59. |

Solution In this case, the altitude falls outside the triangle.

$$A = \frac{1}{2}bh$$

$$A = \frac{1}{2}(9)(13) \qquad \begin{array}{l}\text{Substitute 9 for } b \\ \text{and 13 for } h.\end{array}$$

$$= \frac{1}{2}\left(\frac{9}{1}\right)\left(\frac{13}{1}\right) \qquad \begin{array}{l}\text{Write 9 as } \frac{9}{1} \text{ and} \\ 13 \text{ as } \frac{13}{1}.\end{array}$$

$$= \frac{117}{2} \qquad \text{Multiply the fractions.}$$

$$= 58.5 \qquad \text{Do the division.}$$

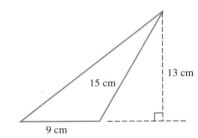

FIGURE 9-59

The area of the triangle is 58.5 cm².

EXAMPLE 11 ***Area of a trapezoid.*** Find the area of the trapezoid in Figure 9-60.

Solution

In this example, $b_1 = 10$ and $b_2 = 6$. It is incorrect to say that $h = 1$, because the height of 1 foot must be expressed as 12 inches to be consistent with the units of the bases.

Thus, we substitute 10 for b_1, 6 for b_2, and 12 for h in the formula for finding the area of a trapezoid and simplify.

$$A = \frac{1}{2}h(b_1 + b_2)$$

$$A = \frac{1}{2}(12)(10 + 6)$$

$$= \frac{1}{2}(12)(16)$$

$$= 6(16)$$

$$= 96$$

The area of the trapezoid is 96 in.^{2}

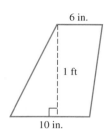

6 in.

1 ft

10 in.

FIGURE 9-60

Self Check

Find the area of the trapezoid.

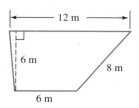

12 m

6 m

8 m

6 m

Answer: 54 m^2

Areas of figures that are combinations of polygons

EXAMPLE 12 ***Carpeting a room.*** A living room/dining room area has the floor plan shown in Figure 9-61. If carpet costs $29 per square yard, including pad and installation, how much will it cost to carpet the room? (Assume no waste.)

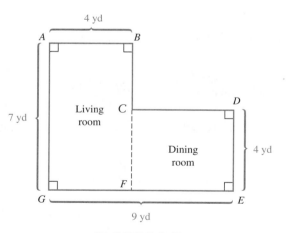

4 yd

A   B

7 yd

Living   C
room

D

Dining
room

4 yd

F

G   E

9 yd

FIGURE 9-61

Solution First we must find the total area of the living room and the dining room:

$$A_{\text{total}} = A_{\text{living room}} + A_{\text{dining room}}$$

Since CF divides the space into two rectangles, the areas of the living room and the dining room are found by multiplying their respective lengths and widths.

$$\text{Area of living room} = lw$$
$$= 7 \cdot 4$$
$$= 28$$

The area of the living room is 28 yd^2.

To find the area of the dining room, we find its length by subtracting 4 yards from 9 yards to obtain 5 yards, and note that its width is 4 yards.

$$\text{Area of dining room} = lw$$
$$= 5 \cdot 4$$
$$= 20$$

The area of the dining room is 20 yd^2.

The total area to be carpeted is the sum of these two areas.

$$A_{\text{total}} = A_{\text{living room}} + A_{\text{dining room}}$$
$$A_{\text{total}} = 28 \text{ yd}^2 + 20 \text{ yd}^2$$
$$= 48 \text{ yd}^2$$

At \$29 per square yard, the cost to carpet the room will be 48 · \$29, or \$1,392. ■

EXAMPLE 13   ***Area of one side of a tent.***   Find the area of one side of the tent in Figure 9-62.

Solution   Each side is a combination of a trapezoid and a triangle. Since the bases of each trapezoid are 30 feet and 20 feet and the height is 12 feet, we substitute 30 for b_1, 20 for b_2, and 12 for h into the formula for the area of a trapezoid.

$$A_{\text{trap.}} = \frac{1}{2}h(b_1 + b_2)$$

$$A_{\text{trap.}} = \frac{1}{2}(12)(30 + 20)$$

$$= 6(50)$$

$$= 300$$

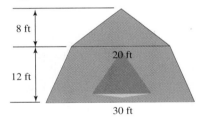

FIGURE 9-62

The area of the trapezoid is 300 ft^2.

Since the triangle has a base of 20 feet and a height of 8 feet, we substitute 20 for b and 8 for h in the formula for the area of a triangle.

$$A_{\text{triangle}} = \frac{1}{2}bh$$

$$A_{\text{triangle}} = \frac{1}{2}(20)(8)$$

$$= 80$$

The area of the triangle is 80 ft^2.

The total area of one side of the tent is

$$A_{\text{total}} = A_{\text{trap.}} + A_{\text{triangle}}$$
$$A_{\text{total}} = 300 \text{ ft}^2 + 80 \text{ ft}^2$$
$$= 380 \text{ ft}^2$$

The total area is 380 ft^2. ■

VOCABULARY *In Exercises 1–6, fill in the blanks to make a true statement.*

1. The distance around a polygon is called the _____ .

2. The perimeter of a polygon is measured in _____ units.

3. The measure of the surface enclosed by a polygon is called its _____ .

4. If each side of a square measures 1 foot, the area enclosed by the square is 1 _____ .

5. The area of a polygon is measured in _____ units.

6. The segment that represents the height of a triangle is called an _____ .

CONCEPTS *In Exercises 7–14, sketch and label each of the figures described.*

7. Two different rectangles, each having a perimeter of 40 in.

8. Two different rectangles, each having an area of 40 in.^{2}.

9. A square with an area of 25 m^2.

10. A square with a perimeter of 20 m.

11. A parallelogram with an area of 15 yd^2.

12. A triangle with an area of 20 ft^2.

13. A figure consisting of a combination of two rectangles whose total area is 80 ft^2.

14. A figure consisting of a combination of a rectangle and a square whose total area is 164 ft^2.

NOTATION *In Exercises 15–22, fill in the blanks to make a true statement.*

15. The formula for the perimeter of a square is _____ .

16. The formula for the perimeter of a rectangle is _____ .

17. The symbol 1 in.^{2} means _____ .

18. One square meter is expressed as _____ .

19. The formula for the area of a square is _____ .

20. The formula for the area of a rectangle is _____ .

21. The formula $A = \frac{1}{2}bh$ gives the area of a _____ .

22. The formula $A = \frac{1}{2}h(b_1 + b_2)$ gives the area of a _____ .

PRACTICE *In Exercises 23–28, find the perimeter of each figure.*

23.

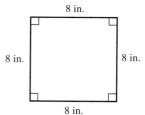

24.

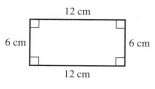

25.

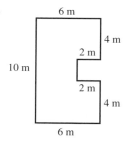

26.

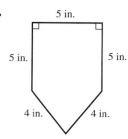

27.

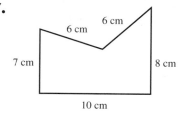

28.

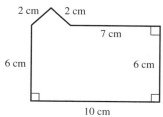

In Exercises 29–32, solve each problem.

29. Find the perimeter of an isosceles triangle with a base of length 21 centimeters and sides of length 32 centimeters.

30. The perimeter of an isosceles triangle is 80 meters. If the length of one side is 22 meters, how long is the base?

31. The perimeter of an equilateral triangle is 85 feet. Find the length of each side.

32. An isosceles triangle with sides of 49.3 inches has a perimeter of 121.7 inches. Find the length of the base.

In Exercises 33–46, find the area of the shaded part of each figure.

33.
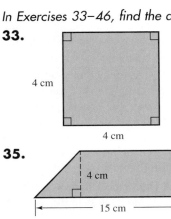
4 cm
4 cm

34.
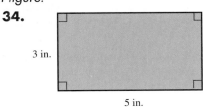
3 in.
5 in.

35.
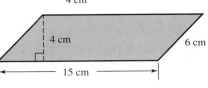
4 cm
6 cm
15 cm

36.
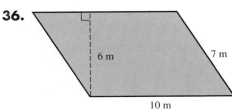
6 m
7 m
10 m

37.
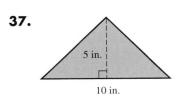
5 in.
10 in.

38.
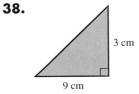
3 cm
9 cm

39.
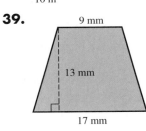
9 mm
13 mm
17 mm

40.
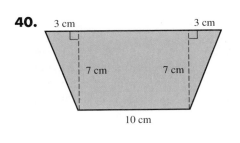
3 cm   3 cm
7 cm   7 cm
10 cm

41.
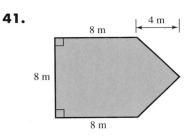
8 m
4 m
8 m
8 m

42.
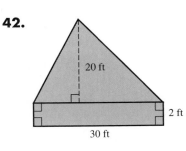
20 ft
2 ft
30 ft

43.
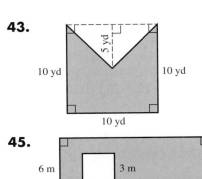
5 yd
10 yd   10 yd
10 yd

44.
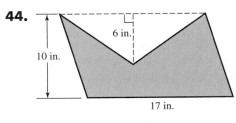
6 in.
10 in.
17 in.

45.
6 m
3 m
3 m
14 m

46.
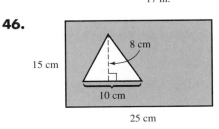
8 cm
15 cm
10 cm
25 cm

47. How many square inches are in 1 square foot?

48. How many square inches are in 1 square yard?

APPLICATIONS ▦ *Use a calculator to help solve each problem.*

49. FENCING A YARD   A man wants to enclose a rectangular yard with fencing that costs $2.44 a foot, including installation. Find the cost of enclosing the yard if its dimensions are 110 ft by 85 ft.

50. FRAMING A PICTURE   Find the cost of framing a rectangular picture with dimensions of 24 inches by 30 inches if framing material costs $8.46 per foot, including matting.

51. PLANTING A SCREEN   A woman wants to plant a pine-tree screen around three sides of her backyard. (See Illustration 1.) If she plants the trees 3 feet apart, how many trees will she need?

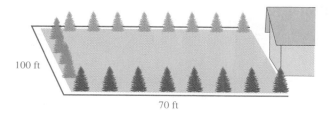

100 ft

70 ft

ILLUSTRATION 1

52. PLANTING MARIGOLDS   A gardener wants to plant a border of marigolds around the garden shown in Illustration 2 to keep out rabbits. How many plants will she need if she allows 6 inches between plants?

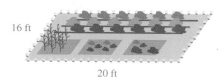

16 ft

20 ft

ILLUSTRATION 2

53. BUYING A FLOOR   Which is more expensive: A ceramic-tile floor costing $3.75 per square foot or linoleum costing $34.95 per square yard?

54. BUYING A FLOOR   Which is cheaper: A hardwood floor costing $5.95 per square foot or a carpeted floor costing $37.50 per square yard?

55. CARPETING A ROOM   A rectangular room is 24 feet long and 15 feet wide. At $30 per square yard, how much will it cost to carpet the room? (Assume no waste.)

56. CARPETING A ROOM   A rectangular living room measures 30 by 18 feet. At $32 per square yard, how much will it cost to carpet the room? (Assume no waste.)

57. TILING A FLOOR   A rectangular basement room measures 14 by 20 feet. Vinyl floor tiles that are 1 ft² cost $1.29 each. How much will the tile cost to cover the floor? (Disregard any waste.)

58. PAINTING A BARN   The north wall of a barn is a rectangle 23 feet high and 72 feet long. There are five windows in the wall, each 4 by 6 feet. If a gallon of paint will cover 300 ft², how many gallons of paint must the painter buy to paint the wall?

59. MAKING A SAIL   If nylon is $12 per square yard, how much would the fabric cost to make a triangular sail with a base of 12 feet and a height of 24 feet?

60. PAINTING A GABLE   The gable end of a warehouse is an isosceles triangle with a height of 4 yards and a base of 23 yards. It will require one coat of primer and one coat of finish to paint the triangle. Primer costs $17 per gallon, and the finish paint costs $23 per gallon. If one gallon covers 300 square feet, how much will it cost to paint the gable, excluding labor?

61. SODDING A LAWN   A landscaper charges $1.17 per square foot to sod a lawn. If the lawn is in the shape of a trapezoid, as shown in Illustration 3, what will it cost to sod the lawn?

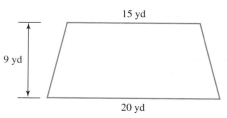

15 yd

9 yd

20 yd

ILLUSTRATION 3

62. COVERING A SWIMMING POOL   A swimming pool has the shape shown in Illustration 4. How many square meters of plastic sheeting will be needed to cover the pool? How much will the sheeting cost if it is $2.95 per square meter? (Assume no waste.)

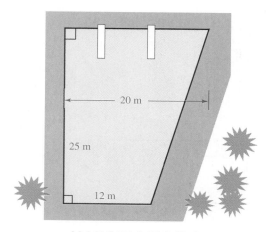

20 m

25 m

12 m

ILLUSTRATION 4

63. CARPENTRY How many sheets of 4-foot-by-8-foot sheetrock are needed to drywall the inside walls on the first floor of the barn shown in Illustration 5? (Assume that the carpenters will cover each wall entirely and then cut out areas for the doors and windows.)

ILLUSTRATION 5

64. CARPENTRY If it costs $90 per square foot to build a one-story home in northern Wisconsin, estimate the cost of building the house with the floor plan shown in Illustration 6.

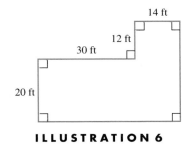

ILLUSTRATION 6

WRITING *Write a paragraph using your own words.*

65. Explain the difference between perimeter and area.

66. Why is it necessary that area be measured in square units?

REVIEW *In Review Exercises 67–72, do the calculations. Write all improper fractions as mixed numbers.*

67. $\dfrac{3}{4} + \dfrac{2}{3}$

68. $\dfrac{7}{8} - \dfrac{2}{3}$

69. $3\dfrac{3}{4} + 2\dfrac{1}{3}$

70. $7\dfrac{5}{8} - 2\dfrac{5}{6}$

71. $7\dfrac{1}{2} \div 5\dfrac{2}{5}$

72. $5\dfrac{3}{4} \cdot 2\dfrac{5}{6}$

9.6 Circles

In this section, you will learn about

- **Circles**
- **Circumference of a circle**
- **Area of a circle**

INTRODUCTION. In this section, we will discuss circles, one of the most useful geometric figures. In fact, the discovery of fire and the circular wheel were two of the most important events in the history of the human race.

Circles

Circle

A **circle** is the set of all points in a plane that lie a fixed distance from a point called its **center.**

A segment drawn from the center of a circle to a point on the circle is called a **radius.** (The plural of *radius* is *radii.*) From the definition, it follows that all radii of the same circle are the same length.

A **chord** of a circle is a line segment connecting two points on the circle. A **diameter** is a chord that passes through the center of the circle. Since a diameter D of a circle is twice as long as a radius r, we have

$$D = 2r$$

Each of the previous definitions is illustrated in Figure 9-63, in which O is the center of the circle.

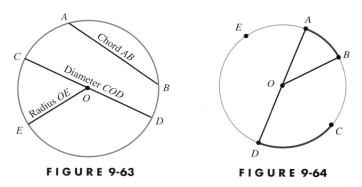

FIGURE 9-63   FIGURE 9-64

Any part of a circle is called an **arc.** In Figure 9-64, the part of the circle from point A to point B is $\overset{\frown}{AB}$, read as arc AB. $\overset{\frown}{CD}$ is the part of the circle from point C to point D. An arc that is half of a circle is a **semicircle.**

Semicircle	A **semicircle** is an arc of a circle whose endpoints are the endpoints of a diameter.

If point O is the center of the circle in Figure 9-64, $\overline{AD}$ is a diameter and $\overset{\frown}{AED}$ is a semicircle. The middle letter E is used to distinguish semicircle $\overset{\frown}{AED}$ from semicircle $\overset{\frown}{ABCD}$.

An arc that is shorter than a semicircle is a **minor arc.** An arc that is longer than a semicircle is a **major arc.** In Figure 9-64,

$\overset{\frown}{AB}$ is a minor arc   and $\overset{\frown}{ABCDE}$ is a major arc.

Circumference of a circle

Since early history, mathematicians have known that the ratio of the distance around a circle (the circumference) divided by the length of its diameter is approximately 3. First Kings, Chapter 7 of the Bible describes a round bronze tank that was 15 feet from brim to brim and 45 feet in circumference, and $\frac{45}{15} = 3$. Today, we have a better value for this ratio, known as π (pi). If C is the circumference of a circle and D is the length of its diameter, then

$$\pi = \frac{C}{D}, \qquad \text{where } \pi = 3.141592653589. \ldots \qquad \frac{22}{7} \text{ and } 3.14 \text{ are often used as estimates of } \pi.$$

If we multiply both sides of $\pi = \frac{C}{D}$ by D, we have the following formula.

Circumference of a circle	The circumference of a circle is given by the formula $C = \pi D$ where C is the circumference and D is the length of the diameter.

Since a diameter of a circle is twice as long as a radius r, we can substitute $2r$ for D in the formula $C = \pi D$ to obtain another formula for the circumference C:

$$C = 2\pi r$$

EXAMPLE 1   *Circumference of a circle.*   Find the circumference of a circle with diameter of 10 centimeters. (See Figure 9-65.)

Solution

We substitute 10 for D in the formula for the circumference of a circle.

$$C = \pi D$$

$$C = \pi(10)$$

$$C \approx 31.41592653589$$

To the nearest tenth, the circumference is 31.4 centimeters.

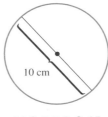

10 cm

FIGURE 9-65

Self Check

To the nearest hundredth, find the circumference of a circle with a radius of 12 meters.

Answer: 75.40 m

Accent on Technology   **Calculating revolutions of a tire**

To calculate how many times a 15-inch tire rotates when a car makes a 25-mile trip, we first find the circumference of the tire.

$$C = \pi D$$

$$C = \pi(15) \qquad \text{Substitute 15 for } D, \text{ the diameter of the tire.}$$

$$C \approx 47.1238898$$

The circumference of the tire is 47.1238898 inches.

We then change 25 miles to inches.

$$25 \text{ miles} \cdot \frac{5,280 \text{ feet}}{1 \text{ mile}} \cdot \frac{12 \text{ inches}}{1 \text{ foot}} = 25(5,280)(12) \text{ inches}$$

$$= 1,584,000 \text{ inches}$$

Finally, we divide 1,584,000 inches by 47.1238898 inches to get

$$\frac{\text{Total distance}}{\text{Circumference of tire}} = \frac{1,584,000}{47.1238898}$$

$$= 33,613.52398$$

To do this work on a scientific calculator, we press these keys.

Evaluate: $\dfrac{25 \cdot 5,280 \cdot 12}{15 \cdot \pi}$

Keystrokes: (2 5 × 5 2 8 0 × 1 2) ÷ π ÷ 1 5 =

```
33613.52398
```

The tire makes about 33,614 revolutions.

EXAMPLE 2   *Perimeter of a figure.*   Find the perimeter of the figure shown in Figure 9-66.

Solution   The figure is a combination of three sides of a rectangle and a semicircle. The perimeter of the rectangular part is

$$P_{\text{rectangular part}} = 8 + 6 + 8 = 22$$

The perimeter of the semicircle is one-half of the circumference of a circle with a 6-meter diameter.

$$P_{\text{semicircle}} = \frac{1}{2}\pi D$$

$$= \frac{1}{2}\pi(6) \qquad \text{Substitute 6 for } D.$$

$$\approx 9.424777961 \qquad \text{Use a calculator.}$$

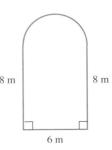

8 m   8 m

6 m

FIGURE 9-66

The total perimeter is the sum of the two parts.

$$P_{\text{total}} \approx 22 + 9.424777961$$
$$\approx 31.424777961$$

To the nearest hundredth, the perimeter of the figure is 31.42 meters.

Area of a circle

If we divide the circle shown in Figure 9-67(a) into an even number of pie-shaped pieces and then rearrange them as shown in Figure 9-67(b), we have a figure that looks like a parallelogram. The figure has a base that is one-half the circumference of the circle, and its height is about the same length as a radius of the circle.

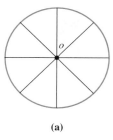 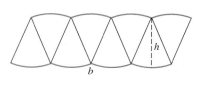

(a)   (b)

FIGURE 9-67

If we divide the circle into more and more pie-shaped pieces, the figure will look more and more like a parallelogram, and we can find its area by using the formula for the area of a parallelogram.

$$A = bh \qquad \text{The formula for the area of a parallelogram.}$$

$$= \frac{1}{2}Cr \qquad \text{Substitute } \tfrac{1}{2} \text{ of the circumference for } b, \text{ and } r \text{ for the height.}$$

$$= \frac{1}{2}(2\pi r)r \qquad C = 2\pi r.$$

$$= \pi r^2 \qquad \tfrac{1}{2} \cdot 2 = 1 \text{ and } r \cdot r = r^2.$$

Area of a circle — The **area of a circle** with radius r is given by the formula
$$A = \pi r^2$$

EXAMPLE 3 *Area of a circle.* To the nearest tenth, find the area of the circle in Figure 9-68.

Solution
Since the length of the diameter is 10 centimeters and the length of a diameter is twice the length of a radius, the length of the radius is 5 centimeters. To find the area of the circle, we substitute 5 for r in the formula for the area of a circle.

$$A = \pi r^2$$
$$A = \pi(5)^2$$
$$= 25\pi$$
$$\approx 78.53981634 \quad \text{Use a calculator.}$$

To the nearest tenth, the area is 78.5 cm².

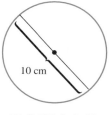

10 cm

FIGURE 9-68

Self Check
To the nearest tenth, find the area of a circle with a diameter of 12 feet.

Answer: 113.1 ft²

Orange paint is available in gallon containers at $19 each, and each gallon will cover 375 ft². To calculate how much the paint will cost to cover a helicopter pad 60 feet in diameter, we first calculate the area of the helicopter pad.

$$A = \pi r^2$$
$$A = \pi (30)^2 \qquad \text{Substitute one-half of 60 for } r.$$
$$= 900\pi \qquad 30 \cdot 30 = 900.$$
$$\approx 2{,}827.433388 \qquad \text{Use a calculator.}$$

The area of the pad is 2,827.433388 ft². Since each gallon of paint will cover 375 ft², we can find the number of gallons of paint needed by dividing 2,827.433388 by 375.

$$\text{Number of gallons needed} \approx \frac{2{,}827.433388}{375}$$
$$\approx 7.539822369$$

Because paint only comes in full gallons, the painter will need to purchase 8 gallons. The cost of the paint will be 8($19), or $152. To do this work on a calculator, we press these keys.

Evaluate: $\dfrac{(30)^2 \cdot \pi}{375}$

Keystrokes: $\boxed{3}\ \boxed{0}\ \boxed{x^2}\ \boxed{\times}\ \boxed{\pi}\ \boxed{=}\ \boxed{\div}\ \boxed{3}\ \boxed{7}\ \boxed{5}\ \boxed{=}$ $\boxed{7.539822369}$

Round to 8 gallons.

Keystrokes: $\boxed{8}\ \boxed{\times}\ \boxed{1}\ \boxed{9}\ \boxed{=}$ $\boxed{152}$

The cost of the paint will be $152.

EXAMPLE 4   *Finding areas.*   Find the shaded area in Figure 9-69.

Solution   The figure is a combination of a triangle and two semicircles. By the Pythagorean theorem, the hypotenuse h of the right triangle is

$$h = \sqrt{6^2 + 8^2} = \sqrt{36 + 64} = \sqrt{100} = 10$$

The area of the triangle is

$$A_{\text{right triangle}} = \frac{1}{2}bh = \frac{1}{2}(6)(8) = \frac{1}{2}(48) = 24$$

The area enclosed by the smaller semicircle is

$$A_{\text{smaller semicircle}} = \frac{1}{2}\pi r^2 = \frac{1}{2}\pi(4)^2 = 8\pi$$

The area enclosed by the larger semicircle is

$$A_{\text{larger semicircle}} = \frac{1}{2}\pi r^2 = \frac{1}{2}\pi(5)^2 = 12.5\pi$$

The total area is

$$A_{\text{total}} = 24 + 8\pi + 12.5\pi \approx 88.4026494 \qquad \text{Use a calculator.}$$

To the nearest hundredth, the area is 88.40 in.²

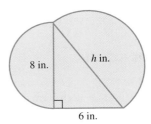

FIGURE 9-69

VOCABULARY *In Exercises 1–8, fill in the blanks to make a true statement.*

1. A segment drawn from the center of a circle to a point on the circle is called a _____.

2. A segment joining two points on a circle is called a _____.

3. A _____ is a chord that passes through the center of a circle.

4. An arc that is one-half of a complete circle is a _____.

5. An arc that is shorter than a _____ is called a minor arc.

6. An arc that is longer than a semicircle is called a _____ arc.

7. The distance around a circle is called its _____.

8. The surface enclosed by a circle is called its _____.

CONCEPTS *In Exercises 9–14, refer to Illustration 1.*

9. Name each radius.

10. Name each diameter.

11. Name each chord.

12. Name each minor arc.

13. Name each semicircle.

14. Name each major arc.

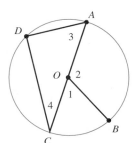

ILLUSTRATION 1

15. If you know the radius of a circle, how can you find its diameter?

16. If you know the diameter of a circle, how can you find its radius?

NOTATION *In Exercises 17–20, fill in the blanks to make a true statement.*

17. The symbol $\overset{\frown}{AB}$ is read as _____.

18. To the nearest hundredth, the value of π is _____.

19. The formula for the circumference of a circle is _____ or _____.

20. The formula $A = \pi r^2$ gives the area of a _____.

PRACTICE

21. To the nearest hundredth, find the circumference of a circle with a diameter of 12 inches.

22. To the nearest hundredth, find the circumference of a circle with a radius of 20 feet.

23. Find the diameter of a circle with a circumference of 36π meters.

24. Find the radius of a circle with a circumference of 50π meters.

In Exercises 25–28, find the perimeter of each figure to the nearest hundredth.

25.

8 ft
3 ft

26.
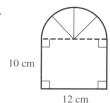
10 cm
12 cm

27.
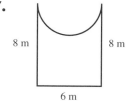
8 m   8 m
6 m

28.
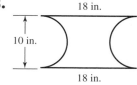
18 in.
10 in.
18 in.

In Exercises 29–30, find the area of each circle to the nearest tenth.

29.

3 in.

30.

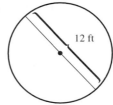

12 ft

In Exercises 31–34, find the total area of each figure to the nearest tenth.

31.

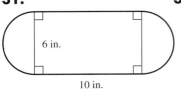

6 in.

10 in.

32.

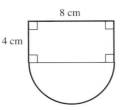

8 cm

4 cm

33.

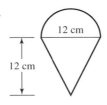

12 cm

12 cm

34.

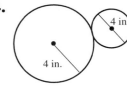

4 in.

4 in.

4 in.

In Exercises 35–38, find the area of each shaded region to the nearest tenth.

35.

4 in.

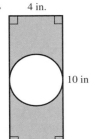

10 in

36.

8 in.

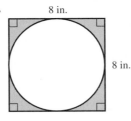

8 in.

37.

$r = 4$ in.

$h = 9$ in.

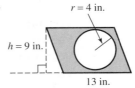

13 in.

38.

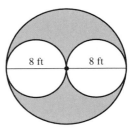

8 ft   8 ft

APPLICATIONS *In Exercises 39–46, give each answer to the nearest hundredth.*

39. AREA OF ROUND LAKE   Round Lake has a circular shoreline 2 miles in diameter. Find the area of the lake.

40. TAKING A WALK   Sam is planning to hike around Round Lake. in Exercise 39. How far will he walk?

41. JOGGING   Joan wants to jog 10 miles on a circular track $\frac{1}{4}$ mile in diameter. How many times must she circle the track?

42. FIXING THE ROTUNDA   The rotunda at a state capitol is a circular area 100 feet in diameter. The legislature wishes to appropriate money to have the floor of rotunda tiled. The lowest bid is $83 per square yard, including installation. How much must the legislature spend?

43. BANDING THE EARTH   A steel band is drawn tightly about the earth's equator. The band is then loosened by increasing its length by 10 feet, and the resulting slack is distributed evenly along the band's entire length. How far above the earth's surface is the band? (*Hint:* You don't need to know the earth's circumference.)

44. CONCENTRIC CIRCLES   Two circles are called **concentric circles** if they have the same center. Find the area of the band between two concentric circles if their diameters are 10 centimeters and 6 centimeters.

45. ARCHERY   See Illustration 2. What percentage of the area of the target is the bullseye?

46. LANDSCAPE DESIGN   See Illustration 3. How much of the lawn does not get watered by the sprinklers at the center of each circle?

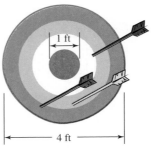

1 ft

4 ft

ILLUSTRATION 2

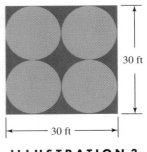

30 ft

30 ft

ILLUSTRATION 3

47. Explain what is meant by the circumference of a circle.
48. Explain what is meant by the area of a circle.
49. Explain the meaning of π.
50. Distinguish between a major arc and a minor arc.

REVIEW *In Exercises 51–56, solve each problem.*

51. Change $\dfrac{9}{10}$ to a percent.

52. Change $\dfrac{7}{8}$ to a percent.

53. Find 30% of 1,600.

54. Find $\dfrac{1}{2}$% of 520.

55. COST OF A DRESS   Maria bought a dress for 25% off the regular price of $98. How much did she pay?

56. COST OF A SHIRT   Bill bought a shirt on sale for $17.50. Find its original cost if it was on sale at 30% off.

9.7   *Surface Area and Volume*

In this section, you will learn about

- **Volumes of solids**
- **Surface areas of rectangular solids**
- **Volumes and surface areas of spheres**
- **Volumes of cylinders**
- **Volumes of cones**
- **Volumes of pyramids**

INTRODUCTION.   In this section, we will discuss a measure of capacity called **volume.** Volumes are measured in cubic units, such as cubic inches, cubic yards, or cubic centimeters. For example,

- We buy gravel or topsoil by the cubic yard.
- We measure the capacity of a refrigerator in cubic feet.
- We often measure amounts of medicine in cubic centimeters.

We will also discuss surface area. The ability to compute surface area is necessary to solve problems such as calculating the amount of material necessary to make a cardboard box or a plastic beach ball.

Volumes of solids

A **rectangular solid** and a **cube** are two common geometric solids. (See Figure 9-70.)

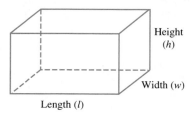
A rectangular solid

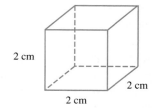
A cube

FIGURE 9-70

The **volume** of a rectangular solid is a measure of the space it encloses. Two common units of volume are cubic inches (in.³) and cubic centimeters (cm³). (See Figure 9-71.)

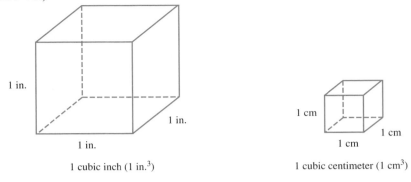

FIGURE 9-71

If we divide the rectangular solid shown in Figure 9-72 into cubes, each cube represents a volume of 1 cm³. Because there are 2 levels with 12 cubes on each level, the volume of the rectangular solid is 24 cm³.

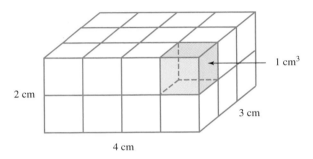

FIGURE 9-72

In practice, we do not find volumes by counting cubes. Instead, we use the following formulas.

Figure	Name	Volume	Figure	Name	Volume
	Cube	$V = s^3$		Cylinder	$V = \pi r^2 h$
	Rectangular Solid	$V = lwh$		Cone	$V = \dfrac{1}{3}\pi r^2 h$
	Prism	$V = Bh^*$		Pyramid	$V = \dfrac{1}{3}Bh^*$

(*continued*)

*B represents the area of the base.

Figure	Name	Volume	Figure	Name	Volume
	Sphere	$V = \dfrac{4}{3}\pi r^3$			

 WARNING! The height of a geometric solid is always measured along a line perpendicular to its base. In each of the solids in Figure 9-73, h is the height.

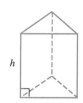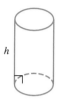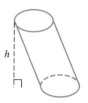

FIGURE 9-73

EXAMPLE 1 ***Number of cubic inches in one cubic foot.*** How many cubic inches are there in 1 cubic foot? (See Figure 9-74.)

Solution
Since a cubic foot is a cube with each side measuring 1 foot, each side also measures 12 inches. Thus, the volume in cubic inches is

$V = s^3$ The formula for the volume of a cube.

$V = (12)^3$ Substitute 12 for s.

$\quad = 1{,}728$

There are 1,728 cubic inches in 1 cubic foot.

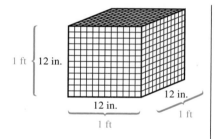

FIGURE 9-74

Self Check
How many cubic centimeters are in 1 cubic meter?

Answer: $1{,}000{,}000 \text{ cm}^3$ ■

EXAMPLE 2 ***Volume of an oil storage tank.*** An oil storage tank is in the form of a rectangular solid with dimensions of 17 by 10 by 8 feet. (See Figure 9-75.) Find its volume.

Solution
To find the volume, we substitute 17 for l, 10 for w, and 8 for h in the formula $V = lwh$ and simplify.

$V = lwh$

$V = (17)(10)(8)$

$\quad = 1{,}360$

The volume is $1{,}360 \text{ ft}^3$.

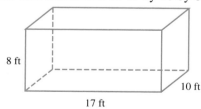

FIGURE 9-75

Self Check
Find the volume of a rectangular solid with dimensions of 8 by 12 by 20 meters.

Answer: $1{,}920 \text{ m}^3$ ■

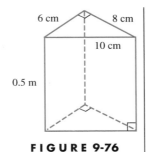

FIGURE 9-76

EXAMPLE 3 *Volume of a triangular prism.* Find the volume of the triangular prism in Figure 9-76.

Solution

The volume of the prism is the area of its base multiplied by its height. Since there are 100 centimeters in 1 meter, the height in centimeters is

$$0.5 \text{ m} = 0.5(\mathbf{1 \text{ m}})$$
$$= 0.5(\mathbf{100 \text{ cm}}) \quad \text{Substitute 100 centimeters for 1 meter.}$$
$$= 50 \text{ cm}$$

Since the area of the triangular base is 24 square centimeters and the height of the prism is 50 centimeters, we have

$$V = Bh$$
$$V = 24(\mathbf{50})$$
$$= 1,200$$

The volume of the prism is 1,200 cm³.

Self Check

Find the volume of the triangular prism below.

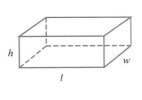

Answer: 120 cm³

Surface areas of rectangular solids

The **surface area** of a rectangular solid is the sum of the areas of its six faces. (See Figure 9-77.)

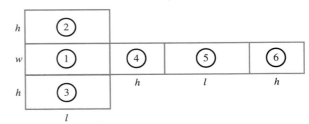

FIGURE 9-77

$$SA = A_{\text{rectangle 1}} + A_{\text{rectangle 2}} + A_{\text{rectangle 3}} + A_{\text{rectangle 4}} + A_{\text{rectangle 5}} + A_{\text{rectangle 6}}$$
$$= lw + lh + lh + hw + lw + hw$$
$$= 2lw + 2lh + 2hw \quad \text{Combine like terms.}$$

Surface area of a rectangular solid	The surface area of a rectangular solid is given by the formula $$SA = 2lw + 2lh + 2hw$$ where *l* is the length, *w* is the width, and *h* is the height.

EXAMPLE 4 *Surface area of an oil tank.* An oil storage tank is in the form of a rectangular solid with dimensions of 17 by 10 by 8 feet. (See Figure 9-78.) Find the surface area of the tank.

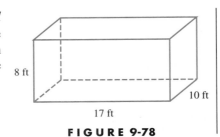

FIGURE 9-78

Self Check

Find the surface area of a rectangular solid with dimensions of 8 by 12 by 20 meters.

Solution

To find the surface area, we substitute 17 for l, 10 for w, and 8 for h in the formula for surface area and simplify.

$$SA = 2lw + 2lh + 2hw$$
$$SA = 2(17)(10) + 2(17)(8) + 2(8)(10)$$
$$= 340 + 272 + 160$$
$$= 772$$

The surface area is 772 ft^2.

Answer: 992 m^2 ■

FIGURE 9-79

Volumes and surface areas of spheres

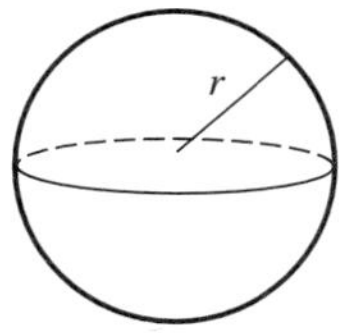

A **sphere** is a hollow, round ball. (See Figure 9-79.) The points on a sphere all lie at a fixed distance r from a point called its *center*. A segment drawn from the center of a sphere to a point on the sphere is called a *radius*.

Accent on Technology **Filling a water tank**

See Figure 9-80. To calculate how many cubic feet of water are needed to fill a spherical water tank with a radius of 15 feet, we substitute 15 for r in the formula for the volume of a sphere and simplify.

$$V = \frac{4}{3}\pi r^3$$

$$V = \frac{4}{3}\pi(15)^3$$

$$= \frac{4}{3}\pi(3{,}375)$$

$$= 4{,}500\pi$$

$$\approx 14{,}137.16694 \quad \text{Use a calculator.}$$

To do the arithmetic with a calculator, press these keys.

Evaluate: $\frac{4}{3}\pi(15)^3$

Keystrokes: $\boxed{1}$ $\boxed{5}$ $\boxed{y^x}$ $\boxed{3}$ $\boxed{=}$ $\boxed{\times}$ $\boxed{4}$ $\boxed{\div}$ $\boxed{3}$ $\boxed{=}$ $\boxed{\times}$ $\boxed{\pi}$ $\boxed{=}$

$$\boxed{14137.16694}$$

FIGURE 9-80

To the nearest tenth, 14,137.2 ft^3 of water will be needed to fill the tank.

There is a formula to find the surface area of a sphere.

Surface area of a sphere

The surface area of a sphere with radius r is given by the formula

$$SA = 4\pi r^2$$

EXAMPLE 5   *Manufacturing beach balls.*   A beach ball is to have a diameter of 16 inches. (See Figure 9-81.) How many square inches of material will be needed to make the ball? (Ignore any waste.)

Solution   Since a radius r of the ball is one-half the diameter, $r = 8$ inches. We can now substitute 8 for r in the formula for the surface area of a sphere.

$$SA = 4\pi r^2$$
$$SA = 4\pi(8)^2$$
$$SA = 256\pi$$
$$\approx 804.2477193$$

A little more than 804 in.^{2} of material will be needed to make the ball.

FIGURE 9-81

Volumes of cylinders

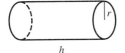

FIGURE 9-82

A **cylinder** is a hollow figure like a piece of pipe. (See Figure 9-82.)

EXAMPLE 6 Find the volume of the cylinder in Figure 9-83.

Solution Since a radius is one-half of the diameter of the circular base, $r = 3$ cm. From the figure, we see that the height of the cylinder is 10 cm. So we can substitute 3 for r and 10 for h in the formula for the volume of a cylinder.

$$V = \pi r^2 h$$
$$V = \pi(3)^2(10)$$
$$= 90\pi$$
$$\approx 282.7433388$$

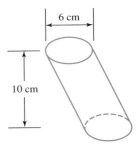

FIGURE 9-83

To the nearest hundredth, the volume of the cylinder is 282.74 cm^3.

Accent on Technology **Volume of a silo**

The silo in Figure 9-84 is a cylinder 50 feet tall topped with a **hemisphere** (a half-sphere). To find the volume of the silo, we add the volume of the cylinder to the volume of the dome.

$$\text{Volume}_{\text{cylinder}} + \textbf{volume}_{\textbf{dome}} = (\text{Area}_{\text{cylinder's base}})(\textbf{height}_{\textbf{cylinder}}) + \frac{1}{2}(\textbf{volume}_{\textbf{sphere}})$$

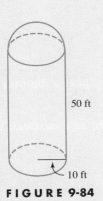

50 ft

10 ft

FIGURE 9-84

$$= \pi r^2 h + \frac{1}{2}\left(\frac{4}{3}\pi r^3\right)$$

$$= \pi r^2 h + \frac{2\pi r^3}{3} \qquad \frac{1}{2}\left(\frac{4}{3}\pi r^3\right) = \frac{1}{2}\cdot\frac{4}{3}\pi r^3 = \frac{4}{6}\pi r^3 = \frac{2\pi r^3}{3}.$$

$$= \pi(10)^2(50) + \frac{2\pi(10)^3}{3} \qquad \text{Substitute 10 for } r \text{ and 50 for } h.$$

$$= 5,000\pi + \frac{2,000}{3}\pi$$

$$= \frac{17,000}{3}\pi \qquad 5,000\pi + \frac{2,000}{3}\pi = \frac{15,000}{3}\pi + \frac{2,000}{3}\pi.$$

$$\approx 17,802.35837$$

To do the arithmetic with a scientific calculator, press these keys.

Evaluate: $\pi(10)^2(50) + \dfrac{2\pi(10)^3}{3}$

Keystrokes: $\boxed{\pi}\ \boxed{\times}\ \boxed{1}\ \boxed{0}\ \boxed{x^2}\ \boxed{\times}\ \boxed{5}\ \boxed{0}\ \boxed{=}\ \boxed{+}\ \boxed{(}\ \boxed{2}\ \boxed{\times}\ \boxed{\pi}\ \boxed{\times}$
$\boxed{1}\ \boxed{0}\ \boxed{y^x}\ \boxed{3}\ \boxed{\div}\ \boxed{3}\ \boxed{)}\ \boxed{=}$ $\qquad\boxed{17802.35837}$

The volume of the silo is approximately 17,802 ft^3.

EXAMPLE 7

Machining a block of metal.
See Figure 9-85. Find the volume that
is left when the hole is drilled through
the metal block.

Solution

We must find the volume of the rectangular solid and then subtract the
volume of the cylinder.

$$V_{\text{rect. solid}} = lwh$$
$$V_{\text{rect. solid}} = 12(12)(18)$$
$$= 2{,}592$$

$$V_{\text{cylinder}} = \pi r^2 h$$
$$V_{\text{cylinder}} = \pi(4)^2(18)$$
$$= 288\pi$$
$$\approx 904.7786842$$

$$V_{\text{drilled block}} = V_{\text{rect. solid}} - V_{\text{cylinder}}$$
$$\approx 2{,}592 - 904.7786842$$
$$\approx 1{,}687.221316$$

To the nearest hundredth, the volume is 1,687.22 cm^3.

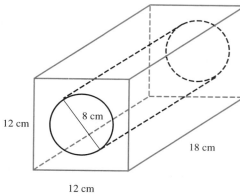

12 cm · 8 cm · 18 cm · 12 cm

FIGURE 9-85

Volumes of cones

Two **cones** are shown in Figure 9-86. Each cone has a height h and a radius r, which
is the radius of the circular base.

FIGURE 9-86

EXAMPLE 8

Volume of a cone. To the nearest tenth, find the volume of the cone in Figure
9-87.

Solution

Since the radius is one-half of the diameter, $r = 4$ cm. We then substitute 4 for r and
6 for h in the formula for the volume of a cone.

$$V = \frac{1}{3}Bh$$

$$V = \frac{1}{3}\pi r^2 h$$

$$V = \frac{1}{3}\pi(4)^2(6)$$

$$V = 32\pi$$
$$\approx 100.5309649$$

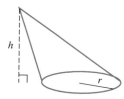

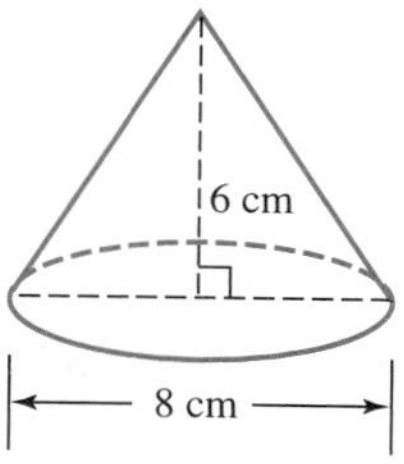

6 cm · 8 cm

FIGURE 9-87

To the nearest tenth, the volume is 100.5 cubic centimeters.

Volumes of pyramids

Two **pyramids** with a height h are shown in Figure 9-88.

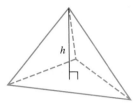

The base is a triangle.

(a)

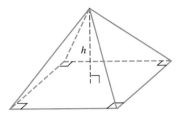

The base is a square.

(b)

FIGURE 9-88

EXAMPLE 9   ***Volume of a pyramid.***   Find the volume of a pyramid with a square base with each side 6 meters long and a height of 9 meters.

Solution
Since the base is a square with each side 6 meters long, the area of the base is 6^2 m², or 36 m². We can then substitute 36 for the area of the base and 9 for the height in the formula for the volume of a pyramid.

$$V = \frac{1}{3}Bh$$

$$V = \frac{1}{3}(36)(9)$$

$$= 108$$

The volume of the pyramid is 108 m³.

Self Check
Find the volume of the pyramid shown below.

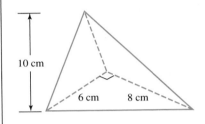

Answer: 80 cm³

STUDY SET   Section 9.7

VOCABULARY   *In Exercises 1–10, fill in the blanks to make a true statement.*

1. The space contained within a geometric solid is called its _____ .

2. A _____ is like a hollow shoe box.

3. A _____ is a rectangular solid with all sides of equal length.

4. The volume of a _____ with each side 1 inch long is 1 cubic inch.

5. The _____ of a rectangular solid is the sum of the areas of its faces.

6. The point that is equidistant from every point on a sphere is its _____ .

7. A _____ is a hollow figure like a drinking straw.

8. A _____ is one-half of a sphere.

9. A _____ looks like a witch's pointed hat.

10. A figure with a polygon for its base that rises to a point is called a _____ .

11. A rectangular solid

12. A prism

13. A sphere

14. A cylinder

15. A cone

16. A pyramid

17. Write the formula for finding the surface area of a rectangular solid.

18. Write the formula for finding the surface area of a sphere.

19. How many cubic feet are in 1 cubic yard?

20. How many cubic inches are in 1 cubic yard?

21. How many cubic decimeters are in 1 cubic meter?

22. How many cubic millimeters are in 1 cubic centimeter?

In Exercises 23–24, tell what geometric concept (perimeter, circumference, area, volume, or surface area) should be applied to find each of the following.

23. a. size of a room to be air conditioned
   b. amount of land in a national park
   c. amount of space in a refrigerator freezer
   d. amount of cardboard in a shoe box
   e. distance around a checkerboard
   f. amount of material used to make a basketball

24. a. amount of cloth in a car cover
   b. size of a trunk of a car
   c. amount of paper used for a postage stamp
   d. amount of storage in a cedar chest
   e. amount of beach available for sunbathing
   f. distance the tip of a propeller travels

NOTATION Fill in the blanks to make a true statement.

25. The symbol in.³ is read as _____.

26. One cubic centimeter is represented as

_____.

PRACTICE In Exercises 27–38, find the volume of each solid. If an answer is not exact, round to the nearest hundredth.

27. A rectangular solid with dimensions of 3 by 4 by 5 centimeters.

28. A rectangular solid with dimensions of 5 by 8 by 10 meters.

29. A prism whose base is a right triangle with legs 3 and 4 meters long and whose height is 8 meters.

30. A prism whose base is a right triangle with legs 5 and 12 feet long and whose height is 10 feet.

31. A sphere with a radius of 9 inches.

32. A sphere with a diameter of 10 feet.

33. A cylinder with a height of 12 meters and a circular base with a radius of 6 meters.

34. A cylinder with a height of 4 meters and a circular base with diameter of 18 meters.

35. A cone with a height of 12 centimeters and a circular base with diameter of 10 centimeters.

36. A cone with a height of 3 inches and a circular base with radius of 4 inches.

37. A pyramid with a square base 10 meters on each side and a height of 12 meters.

38. A pyramid with a square base 6 inches on each side and a height of 4 inches.

In Exercises 39–42, find the surface area of each solid. If an answer is not exact, round to the nearest hundredth.

39. A rectangular solid with dimensions of 3 by 4 by 5 centimeters.

40. A cube with a side 5 centimeters long.

41. A sphere with a radius of 10 inches.

42. A sphere with a diameter of 12 meters.

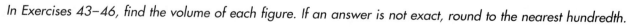 In Exercises 43–46, find the volume of each figure. If an answer is not exact, round to the nearest hundredth.

43.

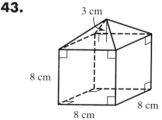

44.

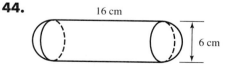

45.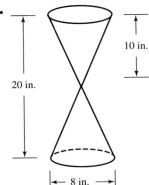

46.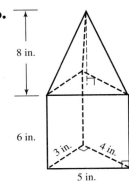

APPLICATIONS   *In Exercises 47–52, if an answer is not exact, round to the nearest hundredth.*

47. VOLUME OF A SUGAR CUBE   A sugar cube is $\frac{1}{2}$ inch on each edge. How much volume does it occupy?

48. VOLUME OF A CLASSROOM   A classroom is 40 feet long, 30 feet wide, and 9 feet high. Find the number of cubic feet of air in the room.

49. VOLUME OF AN OIL TANK   A cylindrical oil tank has a diameter of 6 feet and a length of 7 feet. Find the volume of the tank.

50. VOLUME OF A DESSERT   A restaurant serves pudding in a conical dish that has a diameter of 3 inches.

If the dish is 4 inches deep, how many cubic inches of pudding are in each dish?

51. HOT-AIR BALLOONS   The lifting power of a spherical balloon depends on its volume. How many cubic feet of gas will a balloon hold if it is 40 feet in diameter?

52. VOLUME OF A CEREAL BOX   A box of cereal measures 3 by 8 by 10 inches. The manufacturer plans to market a smaller box that measures $2\frac{1}{2}$ by 7 by 8 inches. By how much will the volume be reduced?

WRITING   *Write a paragraph using your own words.*

53. What is meant by the *volume* of a cube?

54. What is meant by the *surface area* of a cube?

REVIEW   *Do the operations.*

55. $4(6 + 4) - 2^2$

56. $-5(5 - 2)^2 + 3$

57. $5 + 2(6 + 2^3)$

58. $3(6 + 3^4) - 2^4$

59. BUYING PENCILS   Carlos bought 6 pencils at $0.60 each and a notebook for $1.25. He gave the clerk a $5 bill. How much change did he receive?

60. BUYING CLOTHES   Mary bought 3 pairs of socks at $3.29 each and a pair of shoes for $39.95. Can she buy these clothes with three $20 bills?

61. BUYING GOLF EQUIPMENT   George bought 3 packages of golf balls for $1.99 each, a package of tees for $0.49, and a golf glove for $6.95. How much did he spend?

62. BUYING MUSIC   Lisa bought 4 compact discs at $9.99 each, 3 tapes for $6.95 each, and a carrying case for $10.25. How much did she spend?

Formulas

A **formula** is a mathematical expression that is used to express a relationship between quantities. We have studied formulas used in mathematics, business, and science.

Write a formula describing the mathematical relationship between the given quantities.

1. Distance traveled (*d*), rate traveled (*r*), time traveling at that rate (*t*)

2. Sale price (*s*), original price (*p*), discount (*d*)

3. Perimeter of a rectangle (*P*), length of the rectangle (*l*), width of the rectangle (*w*)

4. Amount of interest earned (*I*), principal (*P*), interest rate (*r*), time the money is invested (*t*)

Use a formula to solve each problem.

5. Find the area (*A*) of the triangular lot in Illustration 1.

6. Find the volume (*V*) of the ice chest in Illustration 2.

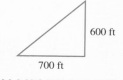

600 ft

700 ft

ILLUSTRATION 1

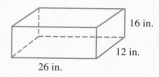

16 in.

12 in.

26 in.

ILLUSTRATION 2

7. Find the retail price (*p*) of a cookware set that costs the store owner $45.50 and is marked up $35.

8. Find the profit (*p*) made by a school candy sale if revenue was $14,500 and costs were $10,200.

9. Find the distance (*d*) that a rock falls in 3 seconds after being dropped from the edge of a cliff.

10. Find the temperature in degrees Celsius (*C*) if the temperature in degrees Fahrenheit is 59.

Sometimes we use the same formula to answer several related questions.

11. Find the interest earned by each account. Enter the answers in the table.

Type of account	Principal	Annual rate earned	Time invested	Interest earned
Savings	$ 5,000	5%	3 yr	
Passbook	$ 2,250	2%	1 yr	
Trust fund	$10,000	6.25%	10 yr	

| SECTION 9.1 | *Some Basic Definitions* |

CONCEPTS

In geometry, we study *points, lines,* and *planes.*

A *line segment* is a part of a line with two endpoints. A *ray* is a part of a line with one endpoint.

An *angle* is a figure formed by two rays with a common endpoint. The common endpoint is called the *vertex* of the angle.

A *protractor* is used to find the measure of an angle.

An *acute angle* is greater than 0° but less than 90°. A *right angle* measures 90°. An *obtuse angle* is greater than 90° but less than 180°. A *straight angle* measures 180°.

Two angles that have the same vertex and are side-by-side are called *adjacent angles.*

REVIEW EXERCISES

1. In Illustration 1, identify a point, a line, and a plane.

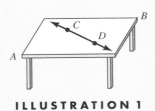

ILLUSTRATION 1

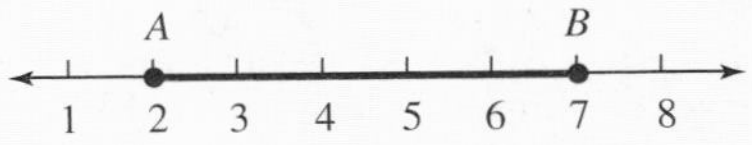

ILLUSTRATION 2

2. In Illustration 2, find m($\overline{AB}$).

3. In Illustration 3, give four ways to name the angle.

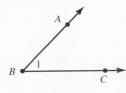

ILLUSTRATION 3

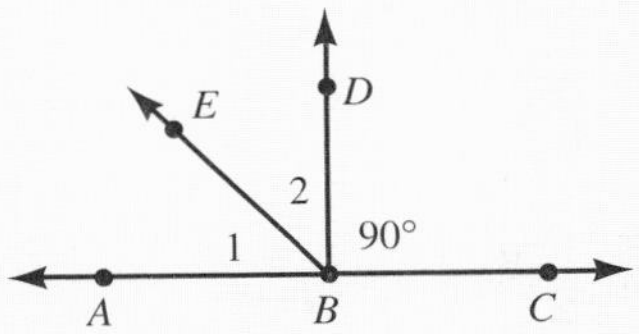

ILLUSTRATION 4

4. In Illustration 3, use a protractor to find the measure of the angle.

5. In Illustration 4, identify each acute angle, right angle, obtuse angle, and straight angle.

6. The measures of several angles are given. Identify each angle as an acute angle, a right angle, an obtuse angle, or a straight angle.
  a. m($\angle A$) = 150°   **b.** m($\angle B$) = 90°
  c. m($\angle C$) = 180°   **d.** m($\angle D$) = 25°

7. The two angles shown in Illustration 5 are adjacent angles. Find x.

8. Line AB is shown in Illustration 6. Find y.

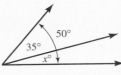

ILLUSTRATION 5

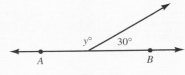

ILLUSTRATION 6

When two lines intersect, pairs of nonadjacent angles are called *vertical angles.*

Vertical angles have the same measure.

9. In Illustration 7, find **a.** m(∠1) and **b.** m(∠2).

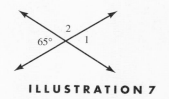

ILLUSTRATION 7

If the sum of two angles is 90°, the angles are *complementary.* If the sum of two angles is 180°, the angles are *supplementary.*

10. Find the complement of an angle that measures 50°.

11. Fund the supplement of an angle that measures 140°.

12. Are angles measuring 30°, 60°, and 90° supplementary?

SECTION 9.2 — *Parallel and Perpendicular Lines*

Parallel lines do not intersect. *Perpendicular* lines intersect and make right angles.

A line that intersects two or more *coplanar* lines is called a *transversal.*

13. Which part of Illustration 8 represents parallel lines?

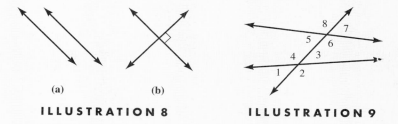

(a)   **(b)**

ILLUSTRATION 8   **ILLUSTRATION 9**

When a transversal intersects two coplanar lines, *alternate interior angles* and *corresponding* angles are formed.

14. Identify all pairs of alternate interior angles shown in Illustration 9.

15. Identify all pairs of corresponding angles shown in Illustration 9.

16. Identify all pairs of vertical angles shown in Illustration 9.

17. In Illustration 10, $l_1 \parallel l_2$. Find the measure of each angle.

18. In Illustration 11, $\overline{DC} \parallel \overline{AB}$. Find the measure of each angle.

If two parallel lines are cut by a transversal,
1. alternate interior angles are congruent (have equal measures).
2. corresponding angles are congruent.
3. interior angles on the same side of the transversal are supplementary.

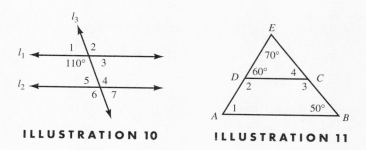

ILLUSTRATION 10   **ILLUSTRATION 11**

19. In Illustration 12, $l_1 \parallel l_2$. Find x.

20. In Illustration 13, $l_1 \parallel l_2$. Find x.

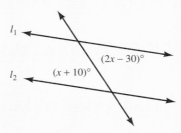

ILLUSTRATION 12

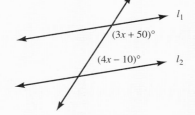

ILLUSTRATION 13

Polygons

A *polygon* is a closed geometric figure. The points at which the sides intersect are called *vertices*.

Polygons are classified as follows:

Number of sides	Name
3	triangle
4	quadrilateral
5	pentagon
6	hexagon
8	octagon

An *equilateral triangle* has three sides of equal length.
An *isosceles triangle* has at least two sides of equal length.
A *scalene triangle* has no sides of equal length.
A *right triangle* has one right angle.

In an isosceles triangle, the angles opposite the sides of equal length are called *base angles*. The third angle is called the *vertex angle*. The third side is called the *base*.

Properties of isosceles triangles:
1. The base angles are congruent.
2. If two angles in a triangle are congruent, the sides opposite the angles are congruent, and the triangle is isosceles.

21. Identify each polygon as a triangle, quadrilateral, pentagon, hexagon, or octagon.

a.

b.

c.

d.

e.

22. Give the number of vertices of each polygon.
   a. Triangle
   b. Quadrilateral
   c. Octagon
   d. Hexagon

23. Classify each of the triangles as an equilateral triangle, an isosceles triangle, a scalene triangle, or a right triangle.

a.

8 in.   8 in.

b.

6 cm   7 cm

9 cm

c.

5 cm   5 cm

5 cm

d.

90°

24. Determine whether each triangle is isosceles.

a.

50°   50°

b.

60°

50°

70°

The sum of the angle measures of any triangle is 180°.

25. In each triangle, find *x*.

a.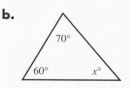

b.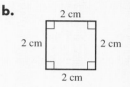

26. If one base angle of an isosceles triangle measures 65°, how large is is the vertex angle?

27. If one base angle of an isosceles triangle measures 60°, what can you conclude about the triangle?

Quadrilaterals are classified as follows:

Property	Name
Opposite sides parallel	parallelogram
Parallelogram with four right angles	rectangle
Rectangle with all sides equal	square
Parallelogram with sides of equal length	rhombus
Exactly two sides parallel	trapezoid

Properties of rectangles:
1. All angles are right angles.
2. Opposite sides are parallel.
3. Opposite sides are of equal length.
4. Diagonals are of equal length.
5. If the diagonals of a parallelogram are of equal length, the parallelogram is a rectangle.

28. Classify each quadrilateral as a parallelogram, rectangle, square, rhombus, or trapezoid.

a.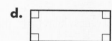

b.

c.

d.

e.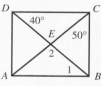

f.

29. In Illustration 14, the length of diagonal $\overline{AC}$ of rectangle *ABCD* is 15 centimeters. Find each measure.
  a. m($\overline{BD}$)   **b.** m(∠1)   **c.** m(∠2)

30. In Illustration 14, *ABCD* is a rectangle. Classify each statement as true or false.
  a. m($\overline{AB}$) = m($\overline{DC}$)   **b.** m($\overline{AD}$) = m($\overline{DC}$)
  c. Triangle *ABE* is isosceles.   **d.** m($\overline{AC}$) = m($\overline{BD}$)

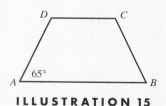

ILLUSTRATION 14   **ILLUSTRATION 15**

The parallel sides of a trapezoid are called *bases*. The nonparallel sides are called *legs*. If the legs of a trapezoid are of equal length, it is *isosceles*. In an isosceles trapezoid, the angles opposite the sides of equal length are *base angles*, and they are congruent.

31. In Illustration 15, *ABCD* is an isosceles trapezoid. Find each measure.
  a. m(∠B)   **b.** m(∠C)

The sum of the measures of the angles of a polygon is given by the formula

$$S = (n - 2)180°$$

32. Find the sum of the angle measures of each polygon.
   a. quadrilateral
   b. hexagon

Properties of Triangles

If two triangles have the same size and the same shape, they are *congruent triangles.*

Corresponding parts of congruent triangles have the same measure.

33. In Illustration 16, complete the list of corresponding parts.
$\angle A$ corresponds to _____
$\angle B$ corresponds to _____
$\angle C$ corresponds to _____
$\overline{AC}$ corresponds to _____
$\overline{AB}$ corresponds to _____
$\overline{BC}$ corresponds to _____

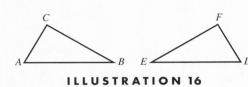

ILLUSTRATION 16

Three ways to show that two triangles are congruent are
1. The SSS property
2. The SAS property
3. The ASA property

34. Tell whether the triangles in each pair are congruent. If they are, tell why.

a.

b.

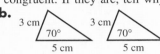

c.

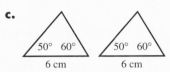

d.

If two triangles have the same shape, they are said to be *similar.* If two angles of one triangle have the same measure as two angles of a second triangle, the triangles are similar.

35. Tell whether the triangles in each pair are similar.

a.

b.

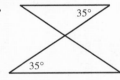

36. If a tree casts a 7-foot shadow at the same time a 6-foot man casts a 2-foot shadow, how tall is the tree?

The Pythagorean theorem:
If the length of the *hypotenuse* of a right triangle is c, and the lengths of its legs are a and b, then

$$a^2 + b^2 = c^2$$

37. Refer to Illustration 17 and find the length of the unknown side.
   a. If $a = 5$ and $b = 12$, find c.
   b. If $a = 8$ and $c = 17$, find b.

38. ▦ To the nearest tenth, find the height of the television screen shown in Illustration 18.

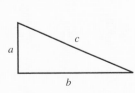

ILLUSTRATION 17

ILLUSTRATION 18

Perimeters and Areas of Polygons

The *perimeter* of a polygon is the distance around it.

39. Find the perimeter of a square with sides 18 inches long.

40. Find the perimeter of a rectangle that is 3 meters long and 1.5 meters wide.

41. Find the perimeter of each polygon.

a.

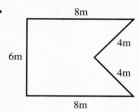

b.

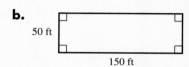

The *area* of a polygon is the measure of the surface it encloses.

Formulas for area:

Figure	Area
Square	$A = s^2$
Rectangle	$A = lw$
Parallelogram	$A = bh$
Triangle	$A = \frac{1}{2}bh$
Trapezoid	$A = \frac{1}{2}h(b_1 + b_2)$

42. Find the area of each polygon.

a.

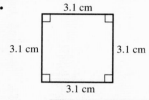

b.

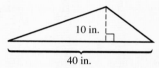

c.

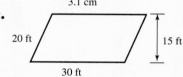

d.

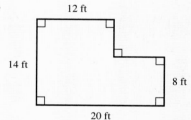

e.

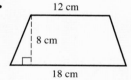

f.

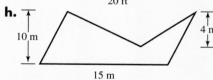

g.

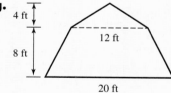

h.

43. How many square feet are there in 1 square yard?

44. How many square inches are in 1 square foot?

Circles

A *circle* is the set of all points in a plane that lie a fixed distance from a point called its *center*.
The fixed distance is the circle's *radius*.

A *chord* of a circle is a line segment connecting two points on the circle.
A *diameter* is a chord that passes through the circle's center.

The *circumference* (perimeter) of a circle is given by the formulas
$$C = \pi D \quad \text{or} \quad C = 2\pi r$$
$$\pi = 3.14159\ldots$$

The *area* of a circle is given by the formula
$$A = \pi r^2$$

45. Refer to Illustration 19.
 a. Name each chord.
 b. Name each diameter.
 c. Name each radius.
 d. Name the center.

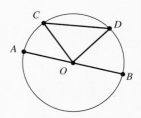

ILLUSTRATION 19

 In Problems 46–49, use a calculator to find each answer to the nearest tenth.

46. Find the circumference of a circle with a diameter of 21 centimeters.

47. Find the perimeter of the figure shown in Illustration 20.

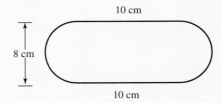

48. Find the area of a circle with a diameter of 18 inches.

ILLUSTRATION 20

49. Find the area of the figure shown in Illustration 20.

Surface Area and Volume

The *volume* of a solid is a measure of the space it occupies.

Figure	Volume
Cube	$V = s^3$
Rectangular solid	$V = lwh$
Prism	$V = Bh*$
Sphere	$V = \frac{4}{3}\pi r^3$
Cylinder	$V = \pi r^2 h$
Cone	$V = \frac{1}{3}\pi r^2 h$
Pyramid	$V = \frac{1}{3}Bh*$

*B represents the area of the base.

50. Find the volume of each solid to the nearest unit.
 a.

 b.

 c.

 d.

 e.

 f.

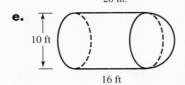
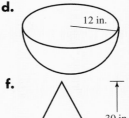
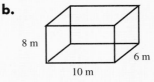

g.

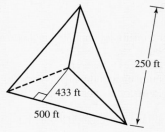

250 ft

433 ft

500 ft

h.

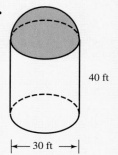

40 ft

|← 30 ft →|

51. How many cubic inches are there in 1 cubic foot?

52. How many cubic feet are there in 2 cubic yards?

The *surface area* of a rectangular solid is the sum of the areas of its six faces.

The surface area of a sphere is given by the formula
$$SA = 4\pi r^2$$

53. To the nearest tenth, find the surface area of each solid.

a.

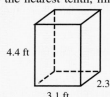

4.4 ft

2.3 ft

3.1 ft

b.

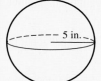

5 in.

1. Find m($\overline{AB}$).

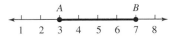

2. Which point is the vertex of $\angle ABC$?

Identify each statement as true or false.

3. An angle of 47° is an acute angle.

4. An angle of 90° is a straight angle.

5. An angle of 180° is a right angle.

6. An angle of 132° is an obtuse angle.

7. Find *x*.

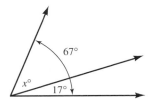

8. Find *y*.

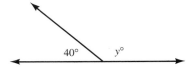

9. Find *y*.

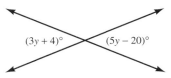

10. Find *x*.

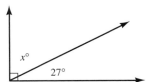

11. Find the complement of an angle measuring 67°.

12. Find the supplement of an angle measuring 117°.

In Problems 13–16, refer to Illustration 1, in which $l_1 \parallel l_2$.

13. m($\angle 1$) = _____.

14. m($\angle 2$) = _____.

15. m($\angle 3$) = _____.

16. Find *x*.

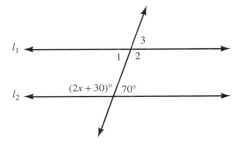

ILLUSTRATION 1

17. Complete the chart.

Polygon	Number of sides
Triangle	
Quadrilateral	
Hexagon	
Pentagon	
Octagon	

18. Complete the chart about triangles.

Property	Kind of triangle
All sides of equal length	
No sides of equal length	
Two sides of equal length	

In Problems 19–20, refer to Illustration 2.

19. Find m(∠A).

20. Find m(∠C).

21. If the measures of two angles in a triangle are 65° and 85°, find the measure of the third angle.

22. Find the sum of the measures of the angles in a decagon (a ten-sided polygon).

23. In Illustration 3, *ABCD* is a rectangle. State three pairs of line segments with equal length.

24. In Illustration 4, *ABCD* is an isosceles trapezoid. Find *x*.

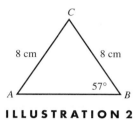

ILLUSTRATION 2

ILLUSTRATION 3

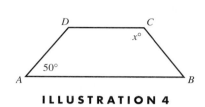

ILLUSTRATION 4

In Problems 25–26, refer to Illustration 5, in which △ABC ≅ △DEF.

25. Find m($\overline{DE}$).

26. Find m(∠E).

In Problems 27–28, refer to Illustration 6, in which m(∠A) = m(∠D) and m(∠C) = m(∠F).

27. Find *x*.

28. Find *y*.

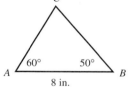

ILLUSTRATION 5

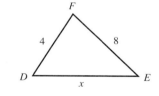

ILLUSTRATION 6

In Problems 29–36, use a calculator to help solve each problem. Give each answer to the nearest tenth.

29. A baseball diamond is a square with each side 90 feet long. How far is it from third base to first base?

30. Find the area of a triangle with a base 44.5 centimeters long and a height of 17.6 centimeters.

31. Find the area of a trapezoid with bases that are 12.2 feet and 15.7 feet long, and a height of 6 feet.

32. Find the circumference of a circle with a diameter that is 6 feet long.

33. Find the area of a circle with a diameter that is 6 feet long.

34. Find the volume of a rectangular solid with dimensions 4.3 by 5.7 by 6.5 meters.

35. Find the volume of a sphere that is 8 meters in diameter.

36. Find the volume of a 10-foot-tall pyramid with a rectangular base 5 feet long and 4 feet wide.

Chapters 1–9 Cumulative Review Exercises

In Exercises 1–2, evaluate each expression when $x = 2$, $y = -3$, and $z = -4$.

1. $x + y - z$

2. $\dfrac{x + y}{y - z}$

In Exercises 3–6, solve each equation. Check each solution.

3. $3(x + 2) = 13$

4. $\dfrac{3t + 11}{5} = t + 3$

5. $3(p + 15) + 4(11 - p) = 0$

6. $\dfrac{2}{3}q - 1 = -6$

In Exercises 7–12, simplify each expression.

7. $x^4 x^5$

8. $(x^3 x^6)^3$

9. $(p^2)^3 (p^3)^2$

10. $(a^2 b^3)^4$

11. $3a + 4b - 5a + 2b$

12. $3(3p - 2) - 5p$

13. Multiply: $\dfrac{2}{3} \cdot \dfrac{1}{5}$.

14. Divide: $\dfrac{15}{8} \div \dfrac{10}{1}$.

15. Subtract: $\dfrac{3}{4} - \dfrac{3}{5}$.

16. Simplify: $\dfrac{7 - \dfrac{2}{3}}{4\dfrac{5}{6}}$.

17. Change $\dfrac{6}{15}$ to a decimal.

18. Evaluate: $\sqrt{121} - \sqrt{64}$.

19. Add: $2\dfrac{7}{8} + 3\dfrac{1}{3}$.

20. Evaluate $-3x^2 - 2x$ when $x = -2$.

21. Subtract: $52.8 - 16.97$.

22. Multiply: $2.3(0.07)$.

23. Add: $23.1 + 76 + 15.701$.

24. Divide: $5.724 \div 1.06$.

25. Find 37% of 460.

26. What number is 15% of 450?

27. 150 is what percent of 600?

28. 48 is 20% of what number?

In Exercises 29–36, use a calculator to help solve each problem.

29. SALES TAX   If the sales tax rate is $6\frac{1}{4}$%, how much sales tax will be added to the price of a new car selling for $18,550?

30. SALES TAX   Find the retail price of a notebook computer if the 5% sales tax on its sale amounted to $190.

31. INCOME TAX   On a taxable income of $47,000, a man paid $10,124.22 in income tax. Find his tax rate.

32. ELECTION RESULTS   In a two-candidate race, 72,667 votes were cast, and the winner got 52% of the vote. How many votes did she win by?

33. FINDING THE SALE PRICE   A pair of $69.90 shoes are on sale at 30% off. Find the discount and the sale price.

34. FINDING THE DISCOUNT RATE   A coat that regularly sells for $115 is marked down to $74.75. Find the discount rate.

35. PAYING OFF A LOAN   To pay for tuition, a college student borrows $1,500 for two months. If the interest rate is 9%, how much will the student have to repay when the loan comes due?

36. SAVING FOR RETIREMENT   When he got married, a man invested $5,000 in an account that guaranteed to pay 8% interest, compounded monthly, for 50 years. At the end of 50 years, how much will his account be worth?

In Exercises 37–38, write each phrase as a ratio.

37. 3 centimeters to 7 centimeters

38. 13 weeks to 1 year

In Exercises 39–40, solve each proportion.

39. $\dfrac{t}{4} = \dfrac{15}{12}$

40. $\dfrac{-x}{14} = \dfrac{13}{28}$

41. QUALITY CONTROL   In a manufacturing process, 98% of the parts are to be within specifications. How many defective parts would be expected in a run of 1,500 pieces?

42. GAS CONSUMPTION   If a bus gets 9 miles per gallon of gas, how far can it go on 50 gallons?

In Exercises 43–64, make each conversion.

43. 168 inches = _____ feet

44. 15 yards = _____ inches

45. 3 miles = _____ yards

46. 212 ounces = _____ pounds

47. 25 cups = _____ fluid ounces

48. 30 gallons = _____ quarts

49. 738 minutes = _____ hours

50. 10 meters = _____ centimeters

51. 20 decimeters = _____ millimeters

52. 5 kilometers = _____ dekameters

53. 5 kilograms = _____ centigrams

54. 7,500 kilograms = _____ metric tons

55. 20 grams = _____ centigrams

56. 600 milligrams = _____ centigrams

57. 5 kiloliters = _____ liters

58. 600 milliliters = _____ liter

59. 66.04 centimeters = _____ inches

60. 500 yards = _____ meters

61. 600 ounces = _____ grams

62. 99.88 kilograms = _____ pounds

63. 75° C = _____ F

64. 113° F = _____ C

65. How many degrees are in a right angle?

66. How many degrees are in an acute angle?

67. Find the supplement of an angle of 105°.

68. Find the complement of an angle of 75°.

In Exercises 69–72, refer to Illustration 1, in which $l_1 \parallel l_2$. Find the measure of each angle.

69. m(∠1)

70. m(∠2)

71. m(∠3)

72. m(∠4)

In Exercises 73–76, refer to Illustration 2, in which AB ∥ DE and m(AC) = m(BC). Find the measure of each angle.

73. m(∠1)

74. m(∠C)

75. m(∠2)

76. m(∠3)

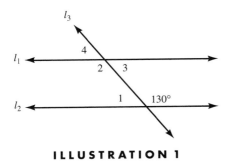

ILLUSTRATION 1

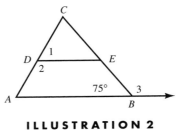

ILLUSTRATION 2

77. Find the sum of the angles of a pentagon (a five-sided polygon).

78. If two sides of a right triangle measure 5 meters and 12 meters, how long is the hypotenuse?

79. DEER POPULATION To estimate the deer population in a forest that covers 275 acres, the forestry department surveys 5 acres and finds a population of 17 animals. Give an estimate of the population of deer in the forest.

80. DOSAGE The recommended dosage of a medicine is 150 units for a 120-pound person. How much should be administered to a 75-pound child?

In Exercises 81–88, if an answer is not exact, round to the nearest hundredth.

81. Find the perimeter and area of a rectangle with dimensions of 9 meters by 12 meters.

82. Find the area of a triangle with a base that is 14 feet long and an altitude of 18 feet.

83. Find the area of a trapezoid that has bases that are 12 inches and 14 inches long and a height of 7 inches.

84. Find the circumference and area of a circle with a diameter of 14 centimeters.

85. Find the volume of a rectangular solid with dimensions of 5 meters by 6 meters by 7 meters.

86. Find the volume of a sphere with a diameter of 10 inches.

87. Find the volume of a cone that has a circular base 8 meters in diameter, and a height of 9 meters.

88. Find the volume of a cylindrical pipe that is 20 feet long and 6 inches in diameter.

APPENDIX I ROOTS AND POWERS

n	n^2	$\sqrt{n}$	n^3	$\sqrt[3]{n}$	n	n^2	$\sqrt{n}$	n^3	$\sqrt[3]{n}$
1	1	1.000	1	1.000	51	2,601	7.141	132,651	3.708
2	4	1.414	8	1.260	52	2,704	7.211	140,608	3.733
3	9	1.732	27	1.442	53	2,809	7.280	148,877	3.756
4	16	2.000	64	1.587	54	2,916	7.348	157,464	3.780
5	25	2.236	125	1.710	55	3,025	7.416	166,375	3.803
6	36	2.449	216	1.817	56	3,136	7.483	175,616	3.826
7	49	2.646	343	1.913	57	3,249	7.550	185,193	3.849
8	64	2.828	512	2.000	58	3,364	7.616	195,112	3.871
9	81	3.000	729	2.080	59	3,481	7.681	205,379	3.893
10	100	3.162	1,000	2.154	60	3,600	7.746	216,000	3.915
11	121	3.317	1,331	2.224	61	3,721	7.810	226,981	3.936
12	144	3.464	1,728	2.289	62	3,844	7.874	238,328	3.958
13	169	3.606	2,197	2.351	63	3,969	7.937	250,047	3.979
14	196	3.742	2,744	2.410	64	4,096	8.000	262,144	4.000
15	225	3.873	3,375	2.466	65	4,225	8.062	274,625	4.021
16	256	4.000	4,096	2.520	66	4,356	8.124	287,496	4.041
17	289	4.123	4,913	2.571	67	4,489	8.185	300,763	4.062
18	324	4.243	5,832	2.621	68	4,624	8.246	314,432	4.082
19	361	4.359	6,859	2.668	69	4,761	8.307	328,509	4.102
20	400	4.472	8,000	2.714	70	4,900	8.367	343,000	4.121
21	441	4.583	9,261	2.759	71	5,041	8.426	357,911	4.141
22	484	4.690	10,648	2.802	72	5,184	8.485	373,248	4.160
23	529	4.796	12,167	2.844	73	5,329	8.544	389,017	4.179
24	576	4.899	13,824	2.884	74	5,476	8.602	405,224	4.198
25	625	5.000	15,625	2.924	75	5,625	8.660	421,875	4.217
26	676	5.099	17,576	2.962	76	5,776	8.718	438,976	4.236
27	729	5.196	19,683	3.000	77	5,929	8.775	456,533	4.254
28	784	5.292	21,952	3.037	78	6,084	8.832	474,552	4.273
29	841	5.385	24,389	3.072	79	6,241	8.888	493,039	4.291
30	900	5.477	27,000	3.107	80	6,400	8.944	512,000	4.309
31	961	5.568	29,791	3.141	81	6,561	9.000	531,441	4.327
32	1,024	5.657	32,768	3.175	82	6,724	9.055	551,368	4.344
33	1,089	5.745	35,937	3.208	83	6,889	9.110	571,787	4.362
34	1,156	5.831	39,304	3.240	84	7,056	9.165	592,704	4.380
35	1,225	5.916	42,875	3.271	85	7,225	9.220	614,125	4.397
36	1,296	6.000	46,656	3.302	86	7,396	9.274	636,056	4.414
37	1,369	6.083	50,653	3.332	87	7,569	9.327	658,503	4.431
38	1,444	6.164	54,872	3.362	88	7,744	9.381	681,472	4.448
39	1,521	6.245	59,319	3.391	89	7,921	9.434	704,969	4.465
40	1,600	6.325	64,000	3.420	90	8,100	9.487	729,000	4.481
41	1,681	6.403	68,921	3.448	91	8,281	9.539	753,571	4.498
42	1,764	6.481	74,088	3.476	92	8,464	9.592	778,688	4.514
43	1,849	6.557	79,507	3.503	93	8,649	9.644	804,357	4.531
44	1,936	6.633	85,184	3.530	94	8,836	9.695	830,584	4.547
45	2,025	6.708	91,125	3.557	95	9,025	9.747	857,375	4.563
46	2,116	6.782	97,336	3.583	96	9,216	9.798	884,736	4.579
47	2,209	6.856	103,823	3.609	97	9,409	9.849	912,673	4.595
48	2,304	6.928	110,592	3.634	98	9,604	9.899	941,192	4.610
49	2,401	7.000	117,649	3.659	99	9,801	9.950	970,299	4.626
50	2,500	7.071	125,000	3.684	100	10,000	10.000	1,000,000	4.642

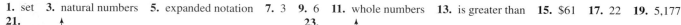

Study Set   Section 1.1   (page 7)

1. set   **3.** natural numbers   **5.** expanded notation   **7.** 3   **9.** 6   **11.** whole numbers   **13.** is greater than   **15.** $61   **17.** 22   **19.** 5,177
21.   **23.**

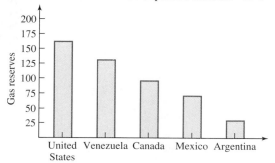

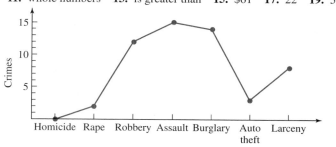

25. 2 hundreds + 4 tens + 5 ones; two hundred forty-five   **27.** 3 thousands + 6 hundreds + 9 ones; three thousand six hundred nine
29. 3 ten-thousands + 2 thousands + 5 hundreds; thirty-two thousand five hundred   **31.** 1 hundred thousand + 4 thousands +
4 hundreds + 1 one; one hundred four thousand four hundred one   **33.** 425   **35.** 2,736   **37.** 456   **39.** 27,598   **41.** 660   **43.** 138
45. 863   **47.** 79,590   **49.** 80,000   **51.** 5,926,000   **53.** 5,900,000   **55.** $419,160   **57.** $419,000
59.

No. 201		March 9 , 19 97

Payable to _____ Davis Chevrolet _____ $ 15,601 $\frac{00}{100}$

Fifteen thousand six hundred one and $\frac{NO}{100}$ _____ DOLLARS

Don Smith

45-365-02

61. a. This diploma awarded this twenty-seventh day of June,
one thousand nine hundred ninety-six.   **b.** The suggested
contribution is eight hundred fifty dollars a plate, or an entire
table may be purchased for five thousand two hundred fifty
dollars.   **63.** 299,800,000 m/sec

Study Set   Section 1.2   (page 18)

1. addends   **3.** rectangle   **5.** square   **7.** commutative   **9.** perimeter   **11.** commutative property of addition   **13.** associative
property of addition   **15.** commutative property of addition   **17.** 0   **19.** parentheses   **21.** 33 plus 12 equals 45   **23.** 25   **25.** 38
27. 461   **29.** 150   **31.** 363   **33.** 979   **35.** 1,985   **37.** 10,000   **39.** 15,907   **41.** 1,861   **43.** 5,312   **45.** 88 ft   **47.** 68 in.   **49.** 3
51. 25   **53.** 103   **55.** 5   **57.** 24   **59.** 118   **61.** 958   **63.** 1,689   **65.** 10,457   **67.** 303   **69.** 6   **71.** $18   **73.** $213   **75.** 10,057 mi
77. 62 ft   **79.** 196 in.   **81.** 43,731   **83.** 35 hr   **85.** 33 points   **91.** 3 thousands + 1 hundred + 2 tens + 5 ones   **93.** 6,354,780
95. 6,350,000

Study Set   Section 1.3   (page 31)

1. addition   **3.** multiplication   **5.** associative; multiplication   **7.** quotient   **9.** 4 · 8   **11.** 5 · 8 = 8 · 5   **13.** 0   **15.** ×, ·, ( )
17. 28   **19.** 84   **21.** 324   **23.** 180   **25.** 7,623   **27.** 1,060   **29.** 2,576   **31.** 20,079   **33.** 2,919,952   **35.** 50,712,116   **37.** $132
39. 406 mi   **41.** 125,800   **43.** 312   **45.** yes   **47.** 84 in.^{2} **49.** 144 in.^{2} **51.** 8   **53.** 3   **55.** 12   **57.** 13   **59.** 73   **61.** 41
63. 205   **65.** 210   **67.** quotient = 8; remainder = 25   **69.** quotient = 20; remainder = 3   **71.** quotient = 30; remainder = 13
73. quotient = 31; remainder = 28   **75.** 4   **77.** 59,375 gal   **79.** 1,260 mi, 97,200 mi^2 **81.** 64   **83.** 388 ft^2 **85.** 440 ft
87. 1,053 ft^2; 702 ft^2; 351 ft^2 **89.** $41   **95.** 8   **97.** 46,000   **99.** 872   **101.** $22

Estimation   (page 36)

1. no   **3.** no   **5.** no   **7.** approx. 8,900 mi   **9.** approx. 30 bags

Study Set   Section 1.4   (page 42)

1. factors   **3.** even or exactly   **5.** prime; itself   **7.** even   **9.** factor; prime   **11.** base; exponent   **13.** 1 · 27 or 3 · 9   **15.** 44   **17.** 1
and 11   **19.** 1 and 37   **21.** Each is prime.   **23.** yes   **25.** 90   **27.** 605   **29.** no   **31.** 2   **33.** 2 and 5   **35.** 3 · 5 · 2 · 5; 5 · 3 · 5 · 2;

they are the same **37. a.** 6, 4 **b.** 12, 2 **39.** 2 **41.** $7 \cdot 7 \cdot 7$ **43.** $3 \cdot 3 \cdot 3 \cdot 3 \cdot 3$ **45.** 2^5 **47.** 5^4 **49.** 1, 2, 5, 10 **51.** 1, 2, 4, 5, 8, 10, 20, 40 **53.** 1, 2, 3, 6, 9, 18 **55.** 1, 2, 4, 11, 22, 44 **57.** 1, 7, 11, 77 **59.** $3 \cdot 13$ **61.** $2 \cdot 3 \cdot 5$ **63.** $2 \cdot 3^4$ **65.** $2^2 \cdot 5 \cdot 11$ **67.** 81 **69.** 32 **71.** 144 **73.** 4,096 **75.** 72 **77.** 3,456 **79.** 12,812,904 **81.** 1,162,213 **85.** $5^1, 5^2, 5^3, 5^4$ **91.** 868 **93.** 136 **95.** 1,596 **97.** 41 **99.** $7,600

Study Set   Section 1.5   (page 49)

1. grouping **3.** denominator **5.** 3; square, multiply, subtract **7.** multiply, subtract **9.** $2 \cdot 3^2 = 2 \cdot 9$; $(2 \cdot 3)^2 = 6^2$ **11.** parentheses **13.** braces **15.** 4; 20 **17.** 9; 36 **19.** 27 **21.** 23 **23.** 15 **25.** 10 **27.** 5 **29.** 25 **31.** 18 **33.** 813 **35.** 24 **37.** 13 **39.** 10 **41.** 214 **43.** 17 **45.** 75 **47.** 17 **49.** 191 **51.** 3 **53.** 29 **55.** 14 **57.** 64 **59.** 3 **61.** 21 **63.** 11 **65.** 1 **67.** 158 m **69.** 122 ft **71.** 496 **73.** 2,845 **75.** $75 **77.** 4 **79.** 79° **81.** 71 **87.** 674 **89.** 1,119 **91.** 9,591 **93.** 725

Key Concept   (page 52)

1. addition, subtraction, power **3.** multiplication, subtraction, division **5.** 5 **7.** 0 **9.** 206

Chapter 1 Review   (page 53)

1. a. 2, 5, 9 **b.** 0, 2, 5, 9 **2. a.**

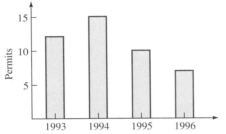
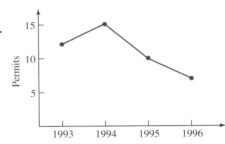

3. a. 6 **b.** 7 **4. a.** 5 hundred thousands + 7 ten thousands + 3 hundreds + 2 ones **b.** 3 ten millions + 7 millions + 3 hundred thousands + 9 thousands + 5 tens + 4 ones **5. a.** 3,207 **b.** 23,253,412 **6. a.** > **b.** < **7. a.** 2,507,300 **b.** 2,510,000 **c.** 2,507,350 **d.** 2,500,000 **8. a.** 13 **b.** 13 **c.** 14 **d.** 14 **e.** 20 **f.** 18 **9. a.** 348 **b.** 11,925 **c.** 518 **d.** 6,000 **10.** 2,746 ft **11. a.** 3 **b.** 4 **c.** 17 **d.** 54 **e.** 74 **f.** 2,075 **12.** $45 **13.** $785 **14.** 147 **15. a.** 56 **b.** 56 **c.** 0 **d.** 7 **e.** 210 **f.** 210 **16. a.** 3,297 **b.** 178,704 **c.** 31,684 **d.** 455,544 **17.** $342 **18. a.** 32 cm² **b.** 2,496 in.² **19.** Santiago **20.** 720 **21. a.** 21 **b.** 37 **c.** 19 R6 **d.** 23 R27 **22.** 16 **23.** 25 **24. a.** 1, 2, 3, 6, 9, 18 **b.** 1, 5, 25 **25. a.** prime **b.** composite **c.** neither **d.** neither **e.** composite **f.** prime **26. a.** odd **b.** even **c.** even **d.** odd **27. a.** $2 \cdot 3 \cdot 7$ **b.** $3 \cdot 5^2$ **28. a.** 6^4 **b.** $5^3 \cdot 13^2$ **29. a.** 125 **b.** 121 **c.** 200 **d.** 2,700 **30. a.** 49 **b.** 32 **c.** 75 **d.** 4 **e.** 38 **f.** 24 **g.** 8 **h.** 24 **i.** 1 **j.** 3 **k.** 19 **l.** 7 **31. a.** 77 **b.** 78 **32.** 288 ft

Chapter 1 Test   (page 57)

1. 0, 1, 2, 3, 4 **2.** 5 thousands + 2 hundreds + 6 tens + 6 ones **3.** 7,507 **4.** 35,000,000
5.

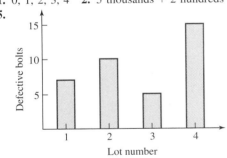

6.

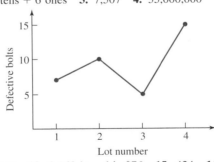

7. > **8.** < **9.** 762 **10.** 248 **11.** 8,100 **12.** 942 **13.** 2,168 in. **14.** $76 **15.** 424 **16.** 26,791 **17.** 72 **18.** 114 R57 **19.** 80 cm, 391 cm² **20.** 47 **21.** $2^2 \cdot 3^2 \cdot 5 \cdot 7$ **22.** 44 **23.** 29 **24.** 26 **25.** 39

Study Set   Section 2.1   (page 66)

1. line **3.** graph; dot **5.** inequality **7.** absolute **9.** integers **11. a.** < **b.** > **c.** <; > **13.** yes **15.** $15 - 8$ **17.** $15 > 12$ **19. a.** −225, $225 deposit **b.** −10, 10 seconds after liftoff **c.** −3, 3 degrees above normal **d.** −12,000, a surplus of $12,000 **e.** −2, 2 lengths ahead of the second-place finisher **21.** It is negative. **23.** −4 **25.** −8 and 2 **27.** −7 **29.** $6 - 4, -6, -(-6)$ (answers may vary) **31.** 9 **33.** 8 **35.** 14 **37.** −20 **39.** −6 **41.** 203 **43.** 4 **45.** 12
47.  **49.** ◄─┼─┼─┼─┼─┼─┼─┼─┼─┼─┼─► −5 −4 −3 −2 −1 0 1 2 3 4 5 **51.** < **53.** < **55.** > **57.** >
59. ≤ **61.** ≤ **63.** $540 > 391$ **65.** $-500 < 0$ **67.** $49 < 50$ **69.** $475 > 0$ **71.** 2, 3, 2, 0, −3, −7 **73.** peaks: 2, 4, 0; valleys: −3, −5, −2 **75. a.** −1 (1 below par) **b.** −2 (2 below par) **c.** Most of the scores are below par. **77. a.** −10° to −20° **b.** 10°

c. 10° **79. a.** 200 yr **b.** the birth of Christ **c.** A.D. **d.** B.C. **81.** **87.** 23,500

89. true **91.** associative property of multiplication

Study Set   Section 2.2   (page 76)

1. identity **3.** 3 **5.** −2 **7.** yes **9.** 0 **11.** commutative property of addition **13.** subtract; smaller; larger **15.** −18 **17.** 5
19. 11 **21.** 23 **23.** 0 **25.** −14 **27.** 9 **29.** −10 **31.** 1 **33.** −7 **35.** −20 **37.** 15 **39.** 8 **41.** 2 **43.** −10 **45.** 9 **47.** 8
49. −21 **51.** 3 **53.** −10 **55.** −4 **57.** 7 **59.** −21 **61.** −7 **63.** 9 **65.** 0 **67.** 0 **69.** 5 **71.** 0 **73.** −3 **75.** −10
77. −1 **79.** −17 **81.** 3G, −3G **83.** no; $70 shortfall each month **85. a.** 85 ft below sea level **b.** 20 ft **c.** −45 and −65 ft
87. −1, 0 **89.** 7 ft over flood stage **91.** −8,346 **93.** −1,032 **95.** profit: $10 million **101.** 15 ft^2 **103.** 16 ft **105.** 5^3

Study Set   Section 2.3   (page 83)

1. difference **3.** opposites **5.** subtraction **7.** 6 **9.** addition **11.** brackets **13.** −8 − (−4) **15.** 7 **17.** −3; 2 **19.** −2; −10; 6
21. 9 **23.** −13 **25.** −10 **27.** −1 **29.** 0 **31.** 8 **33.** 5 **35.** −4 **37.** −4 **39.** −20 **41.** 0 **43.** 0 **45.** −15 **47.** −9 **49.** 3
51. 9 **53.** −2 **55.** −10 **57.** −14 **59.** 3 **61.** −8 **63.** −18 **65.** −6 **67.** 10 **69.** −4 **71.** −120 ft **73.** 16 points **75.** −8
77. 1,007 ft **79.** −4 yd **81. a.** **b.** 37 ft **83.** −2,447 **85.** 20,503 **87.** −1,676 **89.** No; he

will be $244 overdrawn (−244). **95.** 5,990 **97.** 1, 2, 4, 5, 10, 20 **99.** 156 oranges **101.** 4 thousands + 5 hundreds + 2 ones

Study Set   Section 2.4   (page 92)

1. factors; product **3.** integers **5.** 3; exponent **7.** unlike **9.** commutative **11.** −9, the opposite of that number **13.** pos · pos,
pos · neg, neg · pos, neg · neg **15. a.** negative **b.** positive **17. a.** 3 **b.** 12 **c.** 5 **d.** 9 **e.** 10 **f.** 25 **21.** 6 **23.** 54
25. −15 **27.** −36 **29.** 56 **31.** −20 **33.** −120 **35.** 0 **37.** 6 **39.** 7 **41.** −23 **43.** −48 **45.** 40 **47.** −30 **49.** −60
51. −1 **53.** −18 **55.** 0 **57.** 0 **59.** 60 **61.** 16 **63.** −125 **65.** −8 **67.** 81 **69.** −1 **71.** 1 **73.** 49, −49 **75.** −144, 144
77. a. plan #1 −30 lb, plan #2 −28 lb **b.** plan #1; the workout time is double that of plan #2 **79. a.** high 2, low −3 **b.** high 4,
low −6 **81.** −20° **83.** −20 ft **85.** −59,812 **87.** 43,046,721 **89.** −25,728 **91.** 390,625 **93.** −$43,515 **99.** 45 **101.** 2,100
103. 24 yd **105.** 2, 3, 5, 7, 11, 13, 17, 19, 23, 29

Study Set   Section 2.5   (page 99)

1. quotient; divisor **3.** absolute **5.** 5(−5) = −25 **7.** 0(?) = −6 **9.** $\frac{-20}{5}$ = −4 or $\frac{-20}{-4}$ = 5 **11.** positive **13. a.** always true
b. sometimes true **c.** always true **15.** −7 **17.** 2 **19.** 5 **21.** 3 **23.** −20 **25.** −2 **27.** 0 **29.** undefined **31.** −5 **33.** 1
35. −1 **37.** 10 **39.** −4 **41.** −3 **43.** 5 **45.** −4 **47.** −5 **49.** −4 **51.** −4° **53.** 1,000 ft **55.** 6 **57.** $15 **59.** −542
61. −16 **63.** $1,740 **69.** 14 **71.** 2 · 3 · 5 · 7 **73.** true **75.** 81

Study Set   Section 2.6   (page 105)

1. order; operations **3.** grouping **5.** 3; power, multiplication, subtraction **7.** multiplication; subtraction **9.** The base of the first
exponential expression is 3; the base of the second is −3. **11.** 4; 20; −20 **13.** 9; −36 **15.** −7 **17.** 1 **19.** −21 **21.** −14
23. −7 **25.** −5 **27.** 12 **29.** −14 **31.** 30 **33.** 2 **35.** −5 **37.** −3 **39.** 4 **41.** 0 **43.** −14 **45.** 19 **47.** 4 **49.** −3
51. 25 **53.** −48 **55.** 44 **57.** 91 **59.** −9 **61.** −5 **63.** 17 **65.** 11 **67.** −200 **69.** −320 **71.** −9,000 **73.** −1,200
75. 19 **77.** 4 yd **79.** 60 cents gain **81.** −1,707 **83.** −15 **89.** 4

Key Concept   (page 108)

1. −5 **3.** −30 **5.** +10 **7.** −205 **9.** **11.** −8 < −4

13. Same sign: Add their absolute values and attach their common sign to the sum. Different signs: Subtract their absolute values, the
smaller from the larger, and attach the sign of the number with the larger absolute value to that result. **15.** Same sign: The product is
positive. Different signs: The product is negative.

Chapter 2 Review   (page 109)

1. a. **b.** **2. a.** < **b.** < **c.** >
d. < **3. a.** true **b.** true **c.** true **d.** false **4. a.** −$1,200 **b.** −10 sec **5. a.** 4 **b.** 0 **c.** 43 **d.** −12 **6. a.** negative
b. the opposite **c.** negative **d.** minus **7. a.** 12 **b.** −8 **c.** 8 **d.** 0
8. a. **b.**

9. a. -10 **b.** -5 **c.** -83 **d.** -8 **10. a.** -1 **b.** 2 **c.** -9 **d.** 12 **e.** 11 **f.** -11 **g.** -3 **h.** 13 **11. a.** -4 **b.** -20 **c.** 0 **d.** 0 **12. a.** 11 **b.** -4 **c.** -7 **d.** -10 **13.** 65 ft **14. a.** -3 **b.** -21 **c.** 4 **d.** -16 **e.** -6 **f.** 6 **g.** -3 **h.** 30 **15.** adding; opposite **16. a.** -4 **b.** 15 **c.** 6 **d.** -8 **17.** -225 ft **18.** -1 **19.** Alaska: $180°$; Virginia: $140°$ **20. a.** -45 **b.** 18 **c.** -14 **d.** 32 **e.** -100 **f.** 1 **g.** -25 **h.** -150 **21. a.** -36 **b.** -36 **c.** 0 **d.** 1 **22.** $-3, -6, -9$ **23. a.** 25 **b.** -32 **c.** 64 **d.** -64 **24.** negative **25.** first expression: base of 2; second: base of -2; $-4, 4$ **26.** $5(-3)$ **27. a.** -2 **b.** -5 **c.** -8 **d.** 10 **28. a.** 0 **b.** undefined **c.** 1 **d.** 10 **29.** 2 min **30. a.** -22 **b.** 4 **c.** -43 **d.** 8 **e.** 41 **f.** 0 **g.** -11 **h.** 32 **31. a.** 12 **b.** -16 **c.** 48 **d.** 1 **32. a.** -1 **b.** -4 **33. a.** -70 **b.** 20 **c.** $-7,000$ **d.** $1,100$

Chapter 2 Test (page 113)

1. a. $>$ **b.** $<$ **c.** $<$ **2.** $\ldots, -3, -2, -1, 0, 1, 2, 3, \ldots$ **3.** $-3°$ **4.** $-5,$

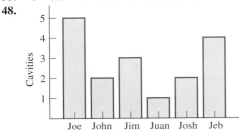

5. a. -13 **b.** -1 **6. a.** -70 **b.** -48 **c.** 16 **7.** $(-4)(5) = -20$ **8. a.** -8 **b.** undefined **c.** -5 **9.** \$3 million **10. a.** 6 **b.** 7 **c.** 6 **11. a.** 16 **b.** -16 **c.** 49 **12.** -27 **13.** 1 **14.** -34 **15.** 2 **16.** 4 **17.** $1,050$ ft **18.** 18

Cumulative Review Exercises (page 115)

1. $1, 2, 5, 9$ **2.** $0, 1, 2, 5, 9$ **3.** $-2, -1$ **4.** $-2, -1, 0, 1, 2, 5, 9$ **5.** 6 **6.** 3 **7.** $7,326,500$ **8.** $7,330,000$ **9.** 786 **10.** $3,806$ **11.** $4,684$ **12.** $13,136$ **13.** 104 ft **14.** 65 **15.** $11,745$ **16.** 13 **17.** $307,329$ **18.** 467 **19.** 595 ft^2 **20.** $1, 2, 3, 6, 9, 18$ **21.** prime, odd **22.** composite, even **23.** even **24.** odd **25.** $2^3 \cdot 3^2 \cdot 7$ **26.** 11^4 **27.** 175 **28.** $1,960$ **29.** 50 **30.** 2 **31.** no **32.** yes **33.** ![number line from -3 to 3 with dots] **34.** ![number line from -4 to 2] **35.** -5 **36.** -5 **37.** -2 **38.** 12 **39.** 24 **40.** -35 **41.** 2 **42.** -5 **43.** 26 **44.** -35 **45.** -3 **46.** 4 **47.** \$126,037 **48.**

![Bar graph titled "Cavities" with bars for Joe (5), John (2), Jim (3), Juan (1), Josh (2), Jeb (4)]

Study Set Section 3.1 (page 122)

1. numerator; denominator **3.** dividing; same **5.** equivalent **7.** terms **9.** They indicate that the 5's were divided by 5 and that each result is 1. **11.** equivalent fractions: $\frac{2}{6} = \frac{1}{3}$ **13. a.** In the second case, the numerator and denominator were prime factored. **b.** yes **15.** The 2's in the numerator and denominator aren't common factors. **17.** $\frac{3}{8} = \frac{3 \cdot 4}{8 \cdot 4} = \frac{12}{32}$ **19.** $3; 2; 3; 2; 3; 2$ **21.** $\frac{1}{3}$ **23.** $-\frac{1}{3}$ **25.** $\frac{2}{3}$ **27.** $\frac{5}{2}$ **29.** $-\frac{1}{2}$ **31.** $\frac{6}{7}$ **33.** $\frac{5}{9}$ **35.** $\frac{6}{7}$ **37.** in lowest terms **39.** $\frac{3}{8}$ **41.** $\frac{5}{7}$ **43.** $\frac{4}{5}$ **45.** $\frac{6}{7}$ **47.** $\frac{7}{8}$ **49.** $-\frac{1}{3}$ **51.** $\frac{3}{5}$ **53.** $\frac{5}{4}$ **55.** 2 **57.** $\frac{35}{40}$ **59.** $\frac{28}{35}$ **61.** $-\frac{45}{54}$ **63.** $-\frac{15}{30}$ **65.** $\frac{4}{14}$ **67.** $\frac{54}{60}$ **69.** $-\frac{25}{20}$ **71.** $-\frac{6}{45}$ **73.** $\frac{15}{5}$ **75.** $-\frac{48}{8}$ **77.** $\frac{36}{9}$ **79.** $-\frac{4}{2}$ **81.** $\frac{3}{5}$ **83.** $\frac{3}{4}$ **85.** $\frac{7}{10}; \frac{1}{8}$ **87.** It gives the dates of the full, half, and new moons. **89.** one quarter turn to left; three-quarters of a turn to the right **91.**

Snacks
Potato chips
Peanuts
Pretzels
Tortilla chips

97. $1, 2, 3, 4, 5$ **99.** $564,000$ **101.** 11 **103.** 100

Study Set Section 3.2 (page 131)

1. multiply **3.** product **5.** base; height **7.** numerators **9. a.** $\frac{1}{4}$ **b.** $12, 1, \frac{1}{12}$ **11.** positive **13. a.** true **b.** false **15.** $7; 15; 3; 3$ **17.** $\frac{1}{8}$ **19.** $\frac{21}{128}$ **21.** $\frac{4}{7}$ **23.** $\frac{77}{60}$ **25.** $-\frac{1}{5}$ **27.** $\frac{2}{9}$ **29.** $\frac{2}{3}$ **31.** 1 **33.** $\frac{1}{20}$ **35.** $\frac{1}{30}$ **37.** 15 **39.** -12 **41.** $\frac{1}{2}$ **43.** $\frac{8}{3}$ **45.** $-\frac{16}{3}$ **47.** $-\frac{3}{2}$ **49.** $\frac{5}{6}$ **51.** $\frac{4}{9}$ **53.** $\frac{25}{81}$ **55.** $\frac{16}{9}$ **57.** $-\frac{1}{64}$ **61.** 15 ft^2 **63.** $\frac{15}{2}$ yd^2 **65.** 290 **67.** the mallard stamp **71.** 42 ft^2 **73.** $147,600,000$ mi^2 **79.** $6,800$ **81.** 26 **83.** 5^3

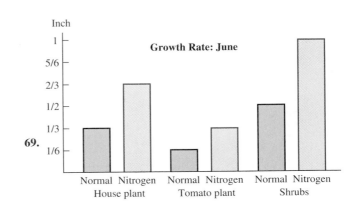

69. Bar graph titled "Growth Rate: June" with y-axis in Inch ($1/6, 1/3, 1/2, 2/3, 5/6, 1$). Categories: Normal/Nitrogen House plant, Normal/Nitrogen Tomato plant, Normal/Nitrogen Shrubs.

Study Set Section 3.3 (page 139)

1. reciprocals **3.** $\frac{2}{3}, \frac{5}{3}$ **5.** $4 \div \frac{1}{3}, 12$ **7.** 1 **9. a.** 5 **b.** $15 \cdot \frac{1}{3} = 5$ **c.** $\frac{1}{3}$ **11.** 9; 10; 9; 10; 5; 5; 9; 9; 5 **13.** $\frac{5}{6}$ **15.** $\frac{27}{16}$ **17.** 1 **19.** $\frac{2}{3}$ **21.** 36 **23.** $-\frac{27}{8}$ **25.** $\frac{2}{15}$ **27.** $-\frac{15}{2}$ **29.** $\frac{1}{14}$ **31.** $\frac{2}{5}$ **33.** $\frac{1}{14}$ **35.** 50 **37.** -6 **39.** $-\frac{5}{2}$ **41.** 104 **43.** 56 **45.** route 1 **47.** 7,855 **53.** is less than **55.** positive **57.** false **59.** -29

Study Set Section 3.4 (page 147)

1. least **3.** higher; same **5.** numerators; common **7.** $\frac{5}{8}$ **9.** The denominators are unlike. **11.** 4 **13.** 60 **15.** $\frac{5}{1}$ **17.** 3; 3; 6; 5; 6; 5 **19.** $\frac{9}{12}$ **21.** $\frac{10}{2}$ **23.** $\frac{49}{28}$ **25.** $\frac{25}{45}$ **27.** $\frac{4}{7}$ **29.** $\frac{20}{103}$ **31.** $\frac{5}{8}$ **33.** $\frac{3}{10}$ **35.** $\frac{22}{15}$ **37.** $-\frac{23}{24}$ **39.** $\frac{47}{50}$ **41.** $-\frac{11}{16}$ **43.** $\frac{5}{21}$ **45.** $-\frac{1}{18}$ **47.** $\frac{9}{56}$ **49.** $\frac{26}{45}$ **51.** $-\frac{13}{5}$ **53.** $-\frac{23}{4}$ **55.** $\frac{47}{60}$ **57.** $\frac{3}{4}$ **59.** $\frac{19}{48}$ **61.** $-\frac{43}{45}$ **63.** $\frac{26}{75}$ **65.** $\frac{17}{54}$ **67.** $\frac{5}{36}$ **69. a.** $\frac{7}{32}$ in. **b.** $\frac{3}{32}$ in. **71.** $\frac{17}{24}$, no **73.** $\frac{1}{16}$ lb, undercharge **75.** $\frac{4}{5}, \frac{3}{4}, \frac{5}{8}$ **77.** $\frac{7}{10}$ **79.** $\frac{1}{6}$ hp **85.** 4 **87.** 576 **89.** 11 **91.** $P = 2l + 2w$

Study Set Section 3.5 (page 155)

1. mixed **3.** graph **5. a.** $-5\frac{10}{1}$ **b.** $-1\frac{7}{8}$ **7. a.** $-2\frac{2}{3}$ **b.** $-3\frac{1}{3}$ **9.** $-\frac{4}{5}, -\frac{2}{5}, \frac{1}{5}$ **13.**  **15.** 8; 8; 4; 4; 6

17. $3\frac{3}{4}$ **19.** $5\frac{4}{5}$ **21.** $-3\frac{1}{3}$ **23.** $10\frac{7}{12}$ **25.** $\frac{13}{5}$ **27.** $\frac{104}{5}$ **29.** $-\frac{56}{9}$ **31.** $\frac{602}{9}$ **37.** $3\frac{3}{4}$ **39.** $10\frac{1}{2}$ **41.** 14 **43.** $-13\frac{3}{4}$ **45.** $-8\frac{1}{3}$ **47.** $\frac{35}{72}$ **49.** $-1\frac{1}{4}$ **51.** $\frac{25}{9} = 2\frac{7}{9}$ **53.** $-\frac{64}{27} = -2\frac{10}{27}$ **55.** $1\frac{9}{11}$ **57.** $-\frac{9}{10}$ **59.** 12 **61.** $\frac{5}{16}$ **63.** $-\frac{2}{3}$ **65.** $2\frac{1}{2}$ **67.** -2 **69.** 64 calories **71.** \$2.72 **73.** 675 **75.** $2\frac{3}{4}$ in., $1\frac{1}{4}$ in. **77.** $42\frac{5}{8}$ in.^{2} **79.** 602 **85.** 36 **87.** 4(8) **89.** $\frac{23}{5}$ **91.** 7

Study Set Section 3.6 (page 163)

1. commutative **3.** $4 + \frac{3}{5}$ **5.** $14 + 53 + \frac{5}{6} + \frac{1}{6}$ **7.** 30 **9.** $17\frac{1}{2}$ **11.** 70; 39; 70; 39; 7; 5; 7; 5; 35; 35; 31 **13.** $4\frac{2}{5}$ **15.** $5\frac{1}{7}$ **17.** $7\frac{1}{2}$ **19.** $5\frac{11}{30}$ **21.** $1\frac{1}{4}$ **23.** $1\frac{11}{24}$ **25.** $9\frac{3}{10}$ **27.** $3\frac{5}{14}$ **29.** $129\frac{11}{15}$ **31.** $398\frac{7}{12}$ **33.** $73\frac{5}{9}$ **35.** $23\frac{8}{21}$ **37.** $11\frac{1}{30}$ **39.** $1\frac{1}{16}$ **41.** $2\frac{1}{2}$ **43.** $26\frac{7}{24}$ **45.** $10\frac{7}{16}$ **47.** $20\frac{5}{18}$ **49.** $6\frac{1}{3}$ **51.** $\frac{1}{4}$ **53.** $3\frac{12}{35}$ **55.** $3\frac{5}{8}$ **57.** $4\frac{1}{3}$ **59.** $3\frac{7}{8}$ **61.** $53\frac{5}{12}$ **63.** $160\frac{1}{8}$ **65.** $-5\frac{1}{4}$ **67.** $-5\frac{7}{8}$ **69.** $2\frac{3}{4}$ mi **71.** $6\frac{1}{12}$ cups **73.** $48\frac{1}{2}$ ft **75. a.** $16\frac{1}{2}, 16\frac{1}{2}; 5\frac{1}{5}, 5\frac{1}{5}$ **b.** $21\frac{7}{10}$ mi **77. a.** 20¢ **b.** 30¢ **79.** $181\frac{2}{3}$ ft **85.** 52 **87.** $\frac{13}{20}$ **89.** 6 **91.** the amount of surface a figure encloses

Study Set Section 3.7 (page 171)

1. complex; fractions **3.** $\frac{2}{3} \div \frac{1}{5}$ **5.** 15 **7.** negative **9.** 60 **11.** $\frac{3}{4}, \frac{4}{3}, 4, 4$ **13.** $\frac{1}{3}$ **15.** $\frac{31}{45}$ **17.** $\frac{37}{40}$ **19.** $\frac{3}{10}$ **21.** $-1\frac{27}{40}$ **23.** $\frac{3}{4}$ **25.** $-\frac{3}{64}$ **27.** $-1\frac{1}{6}$ **29.** $8\frac{1}{2}$ **31.** $\frac{49}{4}$ **33.** $\frac{121}{16}$ **35.** 102 **37.** 36 **39.** $8\frac{1}{4}$ in. **41.** $\frac{5}{6}$ **43.** $-1\frac{1}{3}$ **45.** $10\frac{1}{2}$ **47.** $\frac{4}{9}$ **49.** 3 **51.** 5 **53.** -20 **55.** 11 **57.** $\frac{3}{7}$ **59.** $-\frac{3}{8}$ **61.** $8\frac{1}{2}$ **63.** $14\frac{11}{20}$ mi **65.** yes **67.** $10\frac{1}{2}$ mi **69.** 6 sec **75.** 144 **77.** 37 **79.** 8 **81.** $2^5 \cdot 3^2$

Key Concept (page 176)

1. 5, 5, 5 **2.** 5, 5, 5, 5, 5 **3.** 6, 6, 6, 6

Chapter 3 Review (page 177)

1. $\frac{7}{24}$ **2.** The figure is not divided into equal parts. **3.** $-\frac{2}{3}, \frac{-2}{3}$ **4.** equivalent fractions: $\frac{6}{8} = \frac{3}{4}$ **5.** The numerator and denominator of the fraction are being divided by 2. **6.** The numerator and denominator of the fraction are being divided by 2. The answer to each division is 1. **7. a.** $\frac{1}{3}$ **b.** $\frac{5}{12}$ **c.** $\frac{3}{4}$ **d.** $\frac{11}{18}$ **8. a.** $\frac{5}{6}$ **b.** $\frac{5}{7}$ **c.** $\frac{2}{3}$ **d.** $\frac{3}{7}$ **9.** The numerator and denominator of the original fraction are being multiplied by 2 to obtain an equivalent fraction in higher terms. **10. a.** $\frac{12}{18}$ **b.** $-\frac{6}{16}$ **c.** $\frac{21}{45}$ **d.** $\frac{36}{9}$ **11. a.** $\frac{1}{6}$ **b.** $-\frac{14}{45}$ **c.** $\frac{5}{12}$ **d.** $\frac{1}{5}$ **e.** $\frac{21}{9}$ **f.** $\frac{9}{4}$ **g.** 1 **h.** 1 **12. a.** true **b.** false **13. a.** $\frac{2}{9}$ **b.** $-\frac{8}{21}$ **c.** $\frac{1}{5}$ **d.** $-\frac{5}{9}$ **14. a.** $\frac{9}{16}$ **b.** $-\frac{125}{8}$ **c.** $\frac{4}{9}$ **d.** $-\frac{8}{125}$ **15.** 30 lb **16.** 60 in.^{2} **17. a.** 8 **b.** $-\frac{12}{11}$ **c.** $\frac{1}{5}$ **d.** $\frac{7}{8}$ **18. a.** $\frac{25}{66}$ **b.** $\frac{7}{2}$ **c.** $\frac{3}{32}$ **d.** $\frac{5}{2}$ **19. a.** $-\frac{3}{2}$ **b.** $-\frac{8}{5}$ **c.** $\frac{4}{9}$ **d.** -6 **20.** 12 **21. a.** $\frac{5}{7}$ **b.** $-\frac{6}{5}$ **c.** $\frac{1}{2}$ **d.** $\frac{5}{4}$ **22.** The denominators are not the same. **23.** 90 **24. a.** $\frac{5}{6}$ **b.** $\frac{1}{40}$ **c.** $-\frac{29}{24}$ **d.** $\frac{20}{7}$ **e.** $-\frac{11}{50}$ **f.** $\frac{25}{6}$ **g.** $-\frac{23}{6}$ **h.** $\frac{47}{60}$ **25.** $\frac{7}{32}$ in. **26.** the second hour **27.** $2\frac{1}{6} = \frac{13}{6}$ **28. a.** $3\frac{1}{5}$ **b.** $-3\frac{11}{12}$ **c.** 1 **d.** $2\frac{1}{3}$ **29. a.** $\frac{75}{8}$ **b.** $-\frac{11}{5}$ **c.** $\frac{201}{2}$ **d.** $\frac{199}{100}$ **31. a.** $-\frac{3}{10}$ **b.** $\frac{21}{22}$ **c.** 40 **d.** $-2\frac{1}{2}$ **32.** $48\frac{1}{8}$ in. **33. a.** $3\frac{23}{40}$ **b.** $6\frac{1}{6}$ **c.** $1\frac{1}{12}$ **d.** $1\frac{1}{16}$ **34.** $39\frac{11}{12}$ gal **35. a.** $82\frac{5}{18}$ **b.** $113\frac{3}{20}$ **c.** $31\frac{11}{24}$ **d.** $16\frac{3}{4}$ **36. a.** $20\frac{1}{2}$ **b.** $34\frac{5}{8}$ **37. a.** $\frac{8}{9}$ **b.** $\frac{19}{72}$ **38. a.** $-\frac{12}{17}$ **b.** $-\frac{2}{5}$ **39. a.** $\frac{23}{10} = 2\frac{3}{10}$ **b.** $11\frac{1}{6}$

Chapter 3 Test (page 183)

1. a. $\frac{4}{5}$ **b.** $\frac{1}{5}$ **2. a.** $\frac{3}{4}$ **b.** $\frac{2}{5}$ **3.** $-\frac{3}{20}$ **4.** 40 **5.** 6 **6.** $\frac{1}{30}$ **7.** $\frac{21}{24}$ **8.**

$$-1\frac{1}{7} \qquad \frac{7}{6} \qquad 2\frac{4}{5}$$

number line from -2 to 3

9. \$1$\frac{1}{2}$ million **10.** $\frac{35}{8} = 4\frac{3}{8}$ **11.** $161\frac{11}{36}$ **12.** $37\frac{5}{12}$ **13.** $\frac{11}{7}$ **14.** $11\frac{3}{4}$ in. **15.** perimeter: $53\frac{1}{3}$ in.; area: $106\frac{2}{3}$ in.^{2} **16.** $\frac{13}{24}$ **17.** $-\frac{20}{21}$ **18.** $-\frac{5}{3}$ **19. a.** dividing the numerator and denominator of a fraction by the same number **b.** equivalent fractions: $\frac{1}{2} = \frac{2}{4}$ **c.** multiplying the numerator and denominator of a fraction by the same number **20.** 144

Study Set Section 4.1 (page 191)

1. tens, ones, tenths, hundredths, thousandths, ten thousandths **3.** rounding **5. a.** thirty-two and four hundred fifteen thousandths **b.** 32 **c.** $\frac{415}{1,000}$ **d.** $30 + 2 + \frac{4}{10} + \frac{1}{100} + \frac{5}{1,000}$ **9. a.** true **b.** false **c.** true **d.** true **11.** $\frac{47}{100}$, 0.47 **13.** _____|_____ 0.3

15. 9,816.0245 **17.** fifty and one tenth; $50\frac{1}{10}$ **19.** negative one hundred thirty-seven ten thousandths; $-\frac{137}{10,000}$ **21.** three hundred four and three ten-thousandths; $304\frac{3}{10,000}$ **23.** negative seventy-two and four hundred ninety-three thousandths; $-72\frac{493}{1,000}$ **25.** -0.39 **27.** 6.187 **29.** 506.1 **31.** 77.2 **33.** -0.14 **35.** 33.00 **37.** 3.233 **39.** 55.039 **41.** 39 **43.** 2,988 **45. a.** \$3,090 **b.** \$3,090.30 **47.** < **49.** > **51.** 132.64, 132.6401, 132.6499 **53.** A: granule; B: clay; C: silt; D: sand **55.** The horse begins strong, slows down for the middle two splits, and finishes strong. **57. a.** 0.91 **b.** 0.30 **c.** 1,609.34 **d.** 453.59 **e.** 28.35 **f.** 3.79 **61.** 50, 5, 55, 500, 1 **67.** $164\frac{11}{20}$ **69.** $\frac{32}{243}$ **71.** 72 in.^{2} **73.** -1

Study Set   Section 4.2   (page 198)

1. sum **3.** unwritten **5. a.** 0.47 **b.** $\frac{3}{10}, \frac{17}{100}$ **c.** $\frac{47}{100}$ **d.** 0.47 **e.** They are the same. **7.** 39.9 **9.** 54.72 **11.** 15.9 **13.** 0.23064 **15.** 288.46 **17.** 58.04 **19.** 9.53 **21.** 70.29 **23.** 4.977 **25.** 0.19 **27.** -10.9 **29.** 38.29 **31.** -14.3 **33.** -0.0355 **35.** -16.6 **37.** 47.91 **39.** 2.598 **41.** 11.01 **43.** 4.1 **45.** 35.85 **47.** -57.47 **49.** 6.2 **51.** 8.03 **53. a.** 53.044 sec **b.** 102.38 **55.** 103.4 in. **57.** 1.8; Texas **61.** 43.99 sec **63.** 765.69, \$740.69 **65.** 1994: profit of \$1.9 million; 1995: loss of \$0.4 million; 1996: profit of \$1.7 million **67.** 8,156.9343 **69.** 1,932.645 **71.** 2,529.0582 **75.** -125 **77.** $-\frac{5}{6}$

Study Set   Section 4.3   (page 206)

1. factors; product **3.** whole; sum **5.** larger **7. a.** $\frac{21}{1,000}$ **b.** $\frac{21}{1,000} = 0.021$. They are the same. **9.** 0.08 **11.** -0.15 **13.** 0.98 **15.** 0.072 **17.** 12.32 **19.** -0.0049 **21.** -0.084 **23.** -8.6265 **25.** 9.6 **27.** -56.7 **29.** 12.24 **31.** -18.183 **33.** 0.024 **35.** -16.5 **37.** 42 **39.** 6,716.4 **41.** -0.56 **43.** 8,050 **45.** 980 **47.** -200 **49.** 0.01, 0.04, 0.09, 0.16, 0.25, 0.36, 0.49, 0.64, 0.81 **51.** 1.44 **53.** 1.69 **55.** -17.48 **57.** 14.24 **59.** 0.84 **61.** -3.872 **63.** 38.16 **65.** 25.5 **67. b.** \$14,075 **69.** -0.75 in. **71.** 136.4 lb **73.** \$52.00; \$52.50; \$6.66 **75.** 160.6 m **77.** 15.29694 **79.** 631.2722 **81.** \$102.65 **87.** 7,300 **89.** -6 **91.** the absolute value of negative three **93.** -1

Estimation   (page 215)

1. approx. \$240 **3.** approx. 2 cubic feet less **5.** approx. 30 **7.** approx. \$330 **9.** approx. \$520 **11.** not reasonable

Study Set   Section 4.4   (page 215)

1. dividend; divisor; quotient **3.** whole; dividend; right; answer **5.** true **7.** 10 **9.** Use multiplication to see whether $0.9 \cdot 2.13 = 1.917$. **11.** moving the decimal points in the divisor and dividend two places to the right **13.** 4.5 **15.** -9.75 **17.** 6.2 **19.** 32.1 **21.** 2.46 **23.** -7.86 **25.** 2.66 **27.** 7.17 **29.** 130 **31.** 1,050 **33.** 0.6 **35.** 0.6 **37.** 5.3 **39.** -2.4 **41.** 13.60 **43.** 0.79 **45.** 0.07895 **47.** -0.00064 **49.** 0.0348 **51.** 4.504 **53.** -0.96 **55.** 1,027.19 **57.** 3.5 **59.** 5.2 **61.** 0.37 mi **63.** 280 **65.** 11 hr later: 6 P.M. **67.** 567 **69.** cook: \$11.28; server: \$9.57; manager: \$13.74 **71.** 7.24 **73.** -3.96 **79.** $\frac{7}{6}$ **81.** . . . , $-3, -2, -1, 0, 1, 2, 3, . . .$ **83.** $38\frac{1}{2}$ **85.** 42.05

Study Set   Section 4.5   (page 223)

1. repeating **3.** real **5.** as a fraction and as division, $7 \div 8$ **7.** smaller **13.** It is a repeating decimal. **15.** 0.5 **17.** -0.625 **19.** 0.5625 **21.** -0.53125 **23.** 0.55 **25.** 0.775 **27.** -0.015 **29.** 0.002 **31.** $0.\overline{6}$ **33.** $0.\overline{45}$ **35.** $-0.58\overline{3}$ **37.** $0.0\overline{3}$ **39.** 0.23 **41.** 0.38 **43.** 0.152 **45.** 0.370 **47.** 1.33 **49.** -3.09 **51.** 3.75 **53.** -8.67 **55.** 12.6875 **57.** 203.73 **59.** < **61.** < **63.** $-\frac{153}{40}$ **65.** $\frac{37}{90}$ **67.** $\frac{3}{22}$ **69.** $-\frac{1}{90}$ **71.** -2.55 **73.** 0.068 **75.** 7.11 **77.** -1.7 **79.** 18.1 **81.** $0.\overline{2277}$ **83.** 34.72 **85.** 0.0625, 0.375, 0.5625, 0.9375 **87.** $\frac{3}{40}$ in. **89.** 23.4 sec, 23.8 sec, 24.2 sec, 32.6 sec **91.** 93.6 in.^{2} **95.** -1 **97.** 0, 1, 2, 3, 4, 5, 6, 7 **99.** 15

Study Set   Section 4.6   (page 229)

1. square root **3.** radical sign; positive **5.** radicand **7.** 25; 25 **9.** $7^2, (-7)^2$ **11.** $\frac{3}{4}$ **13.** $\sqrt{6}, \sqrt{11}, \sqrt{23}, \sqrt{27}$ **15. a.** 1 **b.** 0 **17. a.** 2.4 **b.** 5.76 **c.** 0.24 **21. a.** 4, 5 **b.** 9, 10 **23.** $-7; 8$ **25.** 4 **27.** -11 **29.** -0.7 **31.** 0.5 **33.** 0.3 **35.** $-\frac{1}{9}$ **37.** $-\frac{4}{3}$ **39.** $\frac{2}{5}$ **41.** 31 **43.** -20 **45.** $-\frac{7}{20}$ **47.** -70 **49.** 2.56 **51.** -3.6 **53.** 1, 1.414, 1.732, 2, 2.236, 2.449, 2.646, 2.828, 3, 3.162 **55.** 37 **57.** 61 **59.** 3.87 **61.** 8.12 **63.** 4.904 **65.** -3.332 **67. a.** 5 ft **b.** 10 ft **69.** 127.3 ft **71.** 41-inch **73.** 4,899 **75.** -0.0333 **81.** 82.35 **83.** 16 **85.** 1, 2, 3, 4, 5, 6, . . . **87.** $-\frac{17}{144}$

Key Concept   (page 233)

1. 1, 2, 3, 4, 5, . . . **2.** 0, 1, 2, 3, 4, 5, . . . **3.** . . ., $-3, -2, -1, 0, 1, 2, 3, . . .$ **4.** the set of all fractions; any number that can be expressed as either a terminating or a repeating decimal **5.** nonterminating, nonrepeating decimals; a number that can't be written as a fraction **6.** true **7.** false **8.** false **9.** false **10.** true **11.** true **12.** true **13.** false **14.** false **15.** true **16. a.** 10 **b.** 0, 10 **c.** $-2, 0, 10$ **d.** $-2, -1.2, -\frac{7}{8}, 0, 1\frac{2}{3}, 2.75, 10$ **e.** $\sqrt{23}, 1.161661666 . . .$ **f.** All of them are real numbers.

Chapter 4 Review (page 234)

1. $0.67, \frac{67}{100}$ **2.** 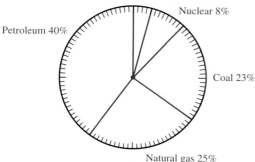 0.8 **3.** $10 + 6 + \frac{4}{10} + \frac{5}{100} + \frac{2}{1,000} + \frac{3}{10,000}$ **4. a.** two and three tenths, $2\frac{3}{10}$ **b.** negative fifteen and fifty-nine hundredths, $-15\frac{59}{100}$ **c.** six hundred one ten thousandths, $\frac{601}{10,000}$ **d.** one one hundred thousandth, $\frac{1}{100,000}$ **7.** Washington, Diaz, Chou, Singh, Gerbac **8.** true **9. a.** $<$ **b.** $>$ **c.** $=$ **d.** $<$ **10. a.** 4.58 **b.** 3,706.090 **c.** -0.1 **d.** 88.1 **11. a.** 66.7 **b.** 45.188 **12. a.** 15.17 **b.** 27.71 **13. a.** -7.7 **b.** 3.1 **c.** -4.8 **d.** -29.09 **14. a.** -25.6 **b.** 4.939 **15.** $48.21 **16.** 8.15 in. **17. a.** -0.24 **b.** 2.07 **c.** -17.05 **d.** 197.945 **e.** 0.00006 **f.** 4.2 **18. a.** 90,145.2 **b.** 2,897 **19. a.** 0.04 **b.** 0.0225 **c.** 10.89 **d.** 0.001 **20. a.** -10.61 **b.** 25.82 **21.** 692.25 **22.** 68.62 in.^{2} **23.** 0.07 in. **24. a.** 1.25 **b.** -10.45 **c.** 1.29 **d.** 4.103 **25. a.** -2.9 **b.** 0.053 **c.** 63 **d.** 0.81 **26. a.** 12.9 **b.** -667.3 **27.** 20.22 **28.** $8.34 **29. a.** 0.8976 **b.** -0.00112 **30.** 13.95 **31.** 14 **32.** 9.5 **33. a.** 0.875 **b.** -0.4 **c.** 0.5625 **d.** 0.06 **34. a.** $0.\overline{54}$ **b.** $-0.\overline{6}$ **35. a.** 0.58 **b.** 1.03 **36. a.** $>$ **b.** $>$ **38. a.** $\frac{11}{15}$ **b.** -6.24 **c.** 93 **d.** 39.564 **39.** 33.49 **40.** 34.88 in.^{2} **41.** $8^2; (-8)^2$ **42. a.** 7 **b.** -4 **c.** 10 **d.** 0.3 **e.** $\frac{8}{5}$ **f.** 0.9 **g.** $-\frac{1}{6}$ **h.** 0 **43.** 9 and 10 **44.** It differs by 0.11. **46. a.** -30 **b.** 2.5 **c.** -27 **d.** 1.5 **47. a.** 4.36 **b.** 7.68

Chapter 4 Test (page 239)

1. $\frac{79}{100}, 0.79$ **2.** Selway, Monroe, Paston, Covington, Cadia **3.** $\frac{271}{1,000}$ **4.** 33.050 **5.** $208.75 **6. a.** 0.567909 **b.** 0.458 **7.** 1.02 in. **8. a.** 10.75 **b.** 6.121 **c.** 0.1024 **d.** 14.07 **9.** 6.48 mi^2 **10.** 0.004 in. **11.** 3.588 **12. a.** 0.34 **b.** $0.41\overline{6}$ **13.** -2.29 **14.** $1.\overline{18}$ **16.** $\frac{41}{30}$ **17.** 0.42 g **18.** $10^2; (-10)^2$ **20. a.** 11 **b.** $-\frac{1}{30}$ **21. a.** $>$ **b.** $>$ **c.** $>$ **d.** $>$ **22. a.** -0.2 **b.** 1.3

Cumulative Review Exercises (page 241)

1. 5,434,700 **2.** 5,430,000 **3.** 8,136 **4.** 3,519 **5.** 299,320 **6.** 991 **7.** 450 ft **8.** 11,250 ft^2 **9.** $2^2 \cdot 3 \cdot 7$ **10.** $2 \cdot 3^2 \cdot 5^2$ **11.** $2^3 \cdot 3^2 \cdot 5$ **12.** $2^4 \cdot 3^2 \cdot 5^2$ **13.** 16 **14.** -35 **15.** 2 **16.** 2 **17.** $\frac{3}{4}$ **18.** $\frac{5}{2}$ **19.** $\frac{3}{4}$ **20.** $-\frac{3}{5}$ **21.** $\frac{4}{5}$ **22.** $\frac{1}{2}$ **23.** $1\frac{5}{12}$ **24.** $\frac{11}{15}$ **25.** $\frac{23}{6}$ **26.** $-\frac{53}{8}$ **27.** $-\frac{65}{6}$ **28.** $\frac{101}{20}$ **29.** $1\frac{1}{4}$ **30.** $5\frac{1}{4}$ **31.** $-6\frac{3}{7}$ **32.** $-4\frac{11}{25}$ **33.** $9\frac{11}{12}$ **34.** $5\frac{11}{15}$ **35.** $\frac{2}{7}$ **36.** $-1\frac{9}{29}$ **37.** 37.4 **38.** 95.0 **39.** 22.465 **40.** 12.229 **41.** 11.73 **42.** 62.3 **43.** 0.8 **44.** $0.7\overline{3}$ **45.** 7 **46.** $\frac{25}{4}$ **47.** 4.16 **48.** 0.03

Study Set   Section 5.1 (page 249)

1. percent **3.** 100 **5.** right **7. a.** $0.84, 84\%, \frac{21}{25}$ **b.** 16% **9.** $\frac{17}{100}$ **11.** $\frac{1}{20}$ **13.** $\frac{3}{5}$ **15.** $\frac{5}{4}$ **17.** $\frac{1}{150}$ **19.** $\frac{21}{400}$ **21.** $\frac{3}{500}$ **23.** $\frac{19}{1,000}$ **25.** 0.19 **27.** 0.06 **29.** 0.408 **31.** 2.5 **33.** 0.0079 **35.** 0.0025 **37.** 93% **39.** 61.2% **41.** 3.14% **43.** 843% **45.** 5,000% **47.** 910% **49.** 17% **51.** 16% **53.** 40% **55.** 105% **57.** 62.5% **59.** 18.75% **61.** $66\frac{2}{3}\%$ **63.** $83\frac{1}{3}\%$ **65.** 11.11% **67.** 55.56% **69. a.** $\frac{15}{184}$ **b.** 8% **71. a.** $\frac{9}{22}$ **b.** 41% **73. a.** $\frac{5}{29}$ **b.** 17% **c.** 24% **75.** 5 ft **77.** 0.9944 **79.** as a decimal; 89.6% **81.** 0.27% **87.** $-\frac{1}{12}$ **89.** -1 **91.** 68.25 in.^{2}

Study Set   Section 5.2 (page 258)

1. $A = 0.10 \cdot 50$ **3.** $48 = p \cdot 47$ **5.** circle **7. a.** 0.12 **b.** 0.056 **c.** 1.25 **d.** 0.0025 **9.** more **11.** 19% **13.** 25 **15.** 50 **17.** 90 **19.** 80% **21.** 65 **23.** 0.096 **25.** 1.25% **27.** 44 **29.** 43.5 **31.** 107.1 **33.** 99 **35.** 60 **37.** 31.25% **39.** **41.** 120 **43.** $525 billion **45.** 24 oz **47.** yes **49.** 30; 12 **51.** 2.7 in. **53.** 5% **55.** 12% **61.** 18.17 **63.** 5.001 **65.** 34,546.4

Petroleum 40% Hydroelectric 4% Nuclear 8% Coal 23% Natural gas 25%

Study Set   Section 5.3 (page 266)

1. commission **3.** discount **5.** The number of members has doubled. **7.** $0.75 **9.** 8% **11.** $47.34, $2.84, $50.18 **13.** $150 **15.** 8%, 1.2% **17.** 7.5% **19.** 96 calories **21.** $768, $32,768 **23.** 10% **25.** 31% **27.** $2,955 **29.** 1.5% **31.** $12,000 **33.** $39.95, 25% **35.** $187.49 **37.** $349.97, 13% **39.** $3.60, 23%, $11.88 **45.** -50 **47.** $\frac{13}{36}$ **49.** 173.4 **51.** 13

Estimation (page 270)

1. 164 **3.** $60 **5.** $54,000 **7.** 21 **9.** 3,100

Study Set   Section 5.4   (page 276)

1. principal   **3.** interest   **5.** simple   **7. a.** 2   **b.** 4   **c.** 365   **d.** 12   **9. a.** 0.07   **b.** 0.098   **c.** 0.0625   **11. a.** compound interest   **b.** $1,000   **c.** 4   **d.** $50   **e.** 1 year   **13.** multiplication   **15.** $5,300   **17.** $1,472   **19.** $4,262.14   **21.** $10,000, 0.0725, 2 yr, $1,450   **23.**   **25.** $18.828 million   **27.** $755.83   **29.** $1,271.22   **31.** $570.65   **33.** $30,915.66   **39.** $\frac{1}{2}$ **41.** $\frac{29}{35}$ **43.** $8\frac{1}{3}$ **45.** 12

Loan Application Worksheet

1. Amount of loan (principal)   *$1,200.00*

2. Length of loan (time)   *2 YEARS*

3. Annual percentage rate   *8%*

4. Interest charged $192

5. Total amount to be repaid $1,392

6. Check method of repayment:
   ☐ 1 lump sum   ☑ monthly payments

   Borrower agrees to pay   *24*   equal payments of   *$58*   to repay loan.

Key Concept   (page 279)

1. 198.4   **3.** 62.5   **5.** 17%   **7.** 3,000

Chapter 5 Review   (page 280)

1. a. 39%, 0.39, $\frac{39}{100}$ **b.** 111%, 1.11, $\frac{111}{100} = 1\frac{11}{100}$ **2.** 61%   **3. a.** $\frac{3}{20}$ **b.** $\frac{6}{5}$ **c.** $\frac{37}{400}$ **d.** $\frac{1}{1,000}$ **4. a.** 0.27   **b.** 0.08   **c.** 1.55   **d.** 0.018   **5. a.** 83%   **b.** 62.5%   **c.** 5.1%   **d.** 600%   **6. a.** 50%   **b.** 80%   **c.** 87.5%   **d.** 6.25%   **7. a.** $33\frac{1}{3}$%   **b.** $83\frac{1}{3}$%   **8. a.** 55.56%   **b.** 266.67%   **9.** 63%   **10.** amount: 15, base: 45, percent: $33\frac{1}{3}$%   **11.** $A = 32\% \cdot 96$ **12. a.** 200   **b.** 125   **c.** 1.75%   **d.** 2,100   **e.** 121   **f.** 30   **13.** 14.4 gal nitro, 0.6 gal methane   **14.** 68   **15.** 87%   **16.** $5.43   **17.**

Miscellaneous 14%

Solid waste disposal 3%

Industrial processes 8%

Fuel combustion in homes, offices, electrical plants 12%

Transportation vehicles 63%

18. 139,531,200 mi^2 **19.** $3.30; $63.29   **20.** 4%   **21.** $40.20   **22.** 25%   **23.** 9.6%   **24.** $50, $189.99, 26%   **25.** $6,000, 8%, 2 years, $960   **26.** $10,308.22   **27.** $134.69   **28.** $2,142.45   **29.** $6,076.45   **30.** $43,265.78

Chapter 5 Test   (page 285)

1. 199%, $\frac{199}{100}$, 1.99   **2.** 61%, $\frac{61}{100}$, 0.61   **3. a.** 0.67   **b.** 0.123   **c.** 0.0975   **4. a.** 25%   **b.** 62.5%   **c.** 12%   **5. a.** 19%   **b.** 347%   **c.** 0.5%   **6. a.** $\frac{11}{20}$ **b.** $\frac{1}{10,000}$ **c.** $\frac{5}{4}$ **7.** 23.33%   **8.** 60%   **9.** $66\frac{2}{3}$%   **10.** 25%   **11. a.** 1.02 in.   **b.** 32.98 in.   **12.** 6.5%   **13.** $3.81   **14.** 30%   **15.** 90   **16.** 21   **17.** 144   **18.** 79%   **19.** $35.92   **20.** $41,440   **21.** $11.95, $3, 20%   **22.** 22%   **23.** $150   **24.** $5,079.60

Study Set   Section 6.1   (page 292)

1. comparison; quotient   **3.** equal   **7.** $\frac{2}{5}$ **9.** $\frac{5}{7}$ **11.** $\frac{1}{2}$ **13.** $\frac{2}{3}$ **15.** $\frac{2}{7}$ **17.** $\frac{1}{3}$ **19.** $\frac{1}{5}$ **21.** $\frac{3}{7}$ **23.** $\frac{3}{4}$ **25.** $1,800   **27.** $\frac{1}{3}$ **29.** $8,725   **31.** $\frac{336}{1,745}$ **33.** $\frac{1}{16}$ **35.** $\frac{\$21.59}{17\ gal}$; $1.27 per gal   **37.** 7¢ per oz   **39.** the 6-oz can   **41.** the first student   **43.** $\frac{11,880\ gal}{27\ min}$; 440 gal per min   **45.** 5%   **47.** 325 mi, 65 mph   **49.** the second car   **53.** 45.537   **55.** 192.7012   **57.** $1\frac{3}{4}$

Study Set   Section 6.2   (page 299)

1. proportion; ratios   **3.** means   **5.** 3(21) = 7(9)   **7.** 18x; 288; 18; 18   **9.** no   **11.** yes   **13.** no   **15.** yes   **17.** 4   **19.** 6   **21.** 3   **23.** 9   **25.** 1   **27.** −2   **29.** 18   **31.** −3.1   **33.** $17   **35.** $6.50   **37.** They are the same.   **39.** 24   **41.** about $4\frac{1}{4}$ **43.** 47   **45.** $7\frac{1}{2}$ gal   **47.** $309   **49.** 49 ft $3\frac{1}{2}$ in.   **51.** 162, 114, 51   **53.** not exactly, but close   **57.** 90%   **59.** $\frac{1}{3}$ **61.** 480   **63.** $73.50   **65.** $88.70

Study Set   Section 6.3   (page 308)

1. American   **3.** 1   **5.** cups, pints, quarts   **7.** 1   **9.** 5,280   **11.** 16   **13.** 8   **15.** 1   **17.** 24   **19.** 36; 36   **21.** 2; 4; 12   **23.** $2\frac{5}{8}$ in.
25. 1 in.   **27.** 11 in.   **29.** $4\frac{3}{4}$ in.   **31.** 48 in.   **33.** 42 in.   **35.** 2 ft   **37.** 288 in.   **39.** 2.5 yd   **41.** $4\frac{2}{3}$ ft   **43.** 15 ft   **45.** $2\frac{1}{3}$ yd
47. 3 mi   **49.** 2,640 ft   **51.** 5 lb   **53.** 3.5 tons   **55.** 24,800 lb   **57.** 6 pt   **59.** 2 gal   **61.** 2 pt   **63.** 4 hr   **65.** 5 days   **67.** 150 yd
69. 0.1375 mi   **71.** 128 oz   **73.** 2.5 tons   **75.** 68   **77.** 71.875 gal   **79.** 2.6 hr   **83.** 3,700   **85.** 3,673.26   **87.** 0.101   **89.** 0.1

Study Set   Section 6.4   (page 316)

1. tens   **3.** thousands   **5.** hundredths   **7.** mass   **9.** 10   **11.** $\frac{1}{100}$ **13.** $\frac{1}{1,000}$ **15.** 1,000   **17.** 1,000   **19.** 1,000   **21.** $\frac{1}{100}$ **23.** 1
25. 1; 100   **27.** 1,000; 1; 1,000   **29.** 156 mm   **31.** 21 mm   **33.** 28 cm   **35.** 16 cm   **37.** 300   **39.** 570   **41.** 3.1   **43.** 7,680,000
45. 0.472   **47.** 4.532   **49.** 0.0325   **51.** 37.5   **53.** 125   **55.** 675,000   **57.** 6.383   **59.** 0.63   **61.** 69.5   **63.** 5.689   **65.** 5.762
67. 0.000645   **69.** 0.65823   **71.** 3,000   **73.** 2,000   **75.** 1,000,000   **77.** 0.5   **79.** 3,000   **81.** 5,000   **83.** 20,000   **85.** 500 dm
87. 400,000 cg   **89.** 20 dL   **91.** 4   **93.** 3 g   **97.** $23.99   **99.** $402   **101.** $1\frac{9}{35}$ **103.** 27

Study Set   Section 6.5   (page 322)

1. Fahrenheit   **3.** $F = \frac{9}{5}C + 32$ **5.** 0.3048; 1,371.6   **7.** 0.264   **9.** 91.4   **11.** 147.6   **13.** 39,372   **15.** 127   **17.** 1   **19.** 11,350
21. 17.6   **23.** 0.6   **25.** 0.1   **27.** 243.4   **29.** 710   **31.** 0.5   **33.** 10°   **35.** 122°   **37.** 14°   **39.** −20.6°   **41.** 5 mi   **43.** 144.8 km
45. 70 mph   **47.** 16 lb   **49.** the 30-lb box   **51.** the 3-quart package   **53.** 28° C   **55.** −5° C and 0° C   **59.** $\frac{29}{15} = 1\frac{14}{15}$ **61.** $\frac{4}{5}$
63. 8.05   **65.** 15.6

Key Concept   (page 325)

2. 375   **3.** 10,800 ft

Chapter 6 Review   (page 326)

1. a. $\frac{1}{3}$ **b.** $\frac{1}{4}$ **2.** the 8-oz can   **3.** $7.75   **4.** $493\frac{1}{3}$ mph   **5. a.** 75   **b.** 15   **6. a.** no   **b.** yes   **7. a.** yes   **b.** no   **8. a.** 4.5   **b.** 16
c. 2.3   **d.** 7.2   **e.** −8   **f.** −0.12   **9.** 192.5 mi   **10.** 300   **11.** $1\frac{1}{2}$ in.   **12. a.** 15 ft   **b.** 216 in.   **c.** 5.5 ft   **d.** 306 in.   **e.** 1.75 mi
f. 1,760 yd   **13. a.** 2 lb   **b.** 275.2 oz   **c.** 96,000 oz   **d.** 2.25 tons   **14. a.** 80 fl oz   **b.** 0.5 gal   **c.** 68 c   **d.** 5.5 qt   **e.** 40 pt
f. 56 c   **15. a.** 1,200 sec   **b.** 15 min   **c.** $8\frac{1}{3}$ days   **d.** 360 min   **e.** 108 hr   **f.** 86,400 sec   **16.** $484\frac{2}{3}$ yd   **17.** 100   **18.** 4 cm
19. a. 4.75 m   **b.** 8,000 mm   **c.** 0.03 km   **d.** 2,000 dm   **e.** 50 hm   **f.** 25 hm   **20. a.** 70 mg   **b.** 8 g   **c.** 5.425 kg   **d.** 0.005425 t
e. 7.5 g   **f.** 0.05 kg   **21.** 50   **22. a.** 1.5 L   **b.** 3.25 kL   **c.** 1,000 dL   **d.** 40 cL   **e.** 20 hL   **f.** 400 mL   **23.** 200 cL   **24.** 82.02 ft
25. the World Trade Center   **26.** 482.79 km   **27.** 198.12 cm   **28. a.** 850.5 g   **b.** 33 lb   **c.** 11,000 g   **d.** 910 kg   **29.** 184.8 lb
30. LaCroix   **31.** the 5-liter bottle   **32.** the 2-qt bottle   **33.** 25° C   **34.** 104° F   **35.** 30° C   **36.** 10° C

Chapter 6 Test   (page 331)

1. $\frac{3}{4}$ **2.** $\frac{1}{6}$ **3.** the 2-pound can   **4.** 22.5 kwh per day   **5.** no   **6.** yes   **7.** 6   **8.** yes   **9.** 15   **10.** 63.24   **11.** −0.21   **12.** 0.2
13. $3.43   **14.** $1\frac{2}{3}$ c   **15.** 15 ft   **16.** 6 yd   **17.** 160 oz   **18.** 3,200 lb   **19.** 128 fl oz   **20.** 10,080 min   **21.** 500 cm   **22.** 500 mm
23. 10,000 m   **24.** 0.08 kg   **25.** 0.7 hL   **26.** 7.5 g   **27.** the 100-yd race   **28.** Jim   **29.** the one-liter bottle   **30.** 104° F

Cumulative Review Exercises   (page 333)

1. 6,246,000   **2.** 6,000,000   **3.** 22 m   **4.** 52 in.   **5.** 24 m²   **6.** 169 in.²   **7.** $2^3 \cdot 3 \cdot 5$ **8.** $2^3 \cdot 3 \cdot 5^2$ **9.** $\frac{1}{15}$ **10.** $\frac{14}{45}$ **11.** $\frac{11}{10} = 1\frac{1}{10}$
12. $\frac{9}{40}$ **13.** 57.5   **14.** 57.54   **15.** 351.053   **16.** 107.26   **17.** 2.1303   **18.** 4.67   **19.** 0.6   **20.** $0.\overline{35}$ **21.** 5.6   **22.** −4.8   **23.** 11
24. $\frac{9}{2}$ **25.** 0.5   **26.** 29   **27.** 783   **28.** 12%   **29.** 72   **30.** 96   **31.** $14,000   **32.** $18,555.34   **33.** 2   **34.** 54   **35.** $5.25 per hour
36. $2.25 per pound   **37.** 60 in.   **38.** $6\frac{2}{3}$ ft   **39.** 6.5 gal   **40.** 6.25 lb   **41.** 400 cm   **42.** 6.5 m   **43.** 4,000 g   **44.** 0.85 cm

Study Set   Section 7.1   (page 342)

1. (a)   **3.** (c)   **5.** (d)   **7.** histogram   **9.** $6.95   **11.** $4.86   **13.** $10,918   **15.** $2,594.50   **17.** nuclear energy   **19.** 49%   **21.** about
60%   **23.** 1980   **25.** 1970   **27.** 320 thousand metric tons   **29.** reckless driving and failure to yield   **31.** reckless driving
33. seniors   **35.** $50   **37.** French and German   **39.** English   **41.** 51.4%   **43.** about 8%   **45.** about 11%   **47.** 4.6%   **49.** $190
51. miners'   **53.** miners   **55.** 1   **57.** 1   **59.** runner 1 was running; runner 2 was stopped   **61.** 27   **63.** 90   **71.** −7   **73.** $\frac{25}{36}$
75. 11, 13, 17, 19, 23, 29   **77.** 4

Study Set   Section 7.2   (page 351)

1. mean   **3.** median   **5.** the number of values   **7.** 8   **9.** 35   **11.** 19   **13.** 9   **15.** 6   **17.** 17.5   **19.** 3   **21.** none   **23.** 22.7
25. about 63¢   **27.** 60¢   **29.** 50¢   **31.** about 61°   **33.** 64°   **35.** 2,670 mi   **37.** 89 mi   **39.** Median and mode are 85.
41. same average (56); sister's scores more consistent   **45.** 3^4 **47.** $\frac{1}{6}$ **49.** 6   **51.** $\frac{19}{10} = 1\frac{9}{10}$

Key Concept (page 353)

1. 13.7 **2.** 15 **3.** 15 **4.** no **5.** 6 **6.** 6 **7.** 6 **8.** yes

Chapter 7 Review (page 354)

1. $-18°$ **2.** 30 mph **3. a.** about 4.9 billion **b.** 1989–1991 **c.** 1991–1993 **d.** about 60% **4. a.** about 830 million
b. about 865 million **c.** 1987 **d.** about 1,770 million **5.** 180 **6.** 160 **7.** yes **8.** median **9.** 1.2 oz **10.** 1.138 oz

Chapter 7 Test (page 357)

1. about $1,659 **2.** about $11 **3.** about 4.1% **4.** about 1.2% **5.** about 19% **6.** about 6% **7.** about 270,000 **8.** about 240,000
9. 1985 **10.** 2.4% **11.** A **12.** C **13.** E **14.** bicyclist 1 **15.** 7.5 **16.** 7.5 **17.** 5 **18.** mean

Study Set   Section 8.1   (page 365)

1. equal **3.** solution; root **5.** equivalent **7.** $y + c$ **9.** addition of 6; subtract 6 from both sides **11.** 8; 8 **13.** yes **15.** no
17. yes **19.** yes **21.** yes **23.** yes **25.** no **27.** no **29.** no **31.** yes **33.** 10 **35.** 7 **37.** 3 **39.** 4 **41.** 13 **43.** 75 **45.** 740
47. 339 **49.** 3 **51.** 5 **53.** 9 **55.** 10 **57.** 1 **59.** 56 **61.** 84 **63.** 105 **65.** 4 **67.** 12 **69.** 8 **71.** 47 **75.** 62 **77.** $80
79. $39 **81.** $190 **87.** 325,780 **89.** 90 **91.** 3

Study Set   Section 8.2   (page 372)

1. division **3.** divide; 6 **5.** It is being multiplied by 4. Divide by 4. **7. a.** Subtract 5 from both sides. **b.** Add 5 to both sides.
c. Divide both sides by 5. **d.** Multiply both sides by 5. **9.** $\frac{x}{z} = \frac{y}{z}$ **11.** 3; 3 **13.** 1 **15.** 6 **17.** 3 **19.** 6 **21.** 14 **23.** 42
25. 225 **27.** 169 **29.** 1 **31.** 3 **33.** 2 **35.** 1 **37.** 40 **39.** 1,200 **43.** 390 wpm **45.** $40,000 **47.** 480 **51.** 48 cm
53. $2^3 \cdot 3 \cdot 5$ **55.** 72 **57.** 77

Study Set   Section 8.3   (page 381)

1. formula **3.** substitute **5.** $2 - 8 + 10$; it looks like subtraction **7. a.** x = length part 1; $x - 40$ = length part 2; $x + 16$ = length
part 3 **b.** 20 in. and 76 in. **9.** $22; $27; $(p + 2)$; $(10p + 2)$ **11.** 50 mi; 48 mi; $3t$ mi; 5 mi; $3x$ mi **13.** speedometer: rate;
odometer: distance; clock: time; $d = rt$ **15.** The rate is expressed in miles per hour and the time in minutes. **17. a.** $d = rt$
b. $C = \frac{5(F - 32)}{9}$ **c.** $d = 16t^2$ **19.** 17 **21.** 4 **23.** 40 **25.** -6 **27.** -6 **29.** 23 **31.** -8 **33.** 100 **35.** -28 **37.** 3 **39.** -3
41. -7 **43.** -18 **45.** 25 **47.** 21 **49.** -5 **51.** -29 **53.** -45 **55.** 70 cents **57.** $8,200 **59.** $23 **61.** 300 mi **63.** $-10°$ C
65. 240 m^2 **67.** 64 ft **69. a.** 1993: broke even; 1994: lost 5 million; 1995: $5 million profit **b.** 1996: $10 million profit
71. $D = B - C$; $20; $12; $42 **73.** 4 yards per carry **75.** 40 **77.** 4 **83.** 17, 37, 41 **85.** 7 **87.** division by 3 **89.** 3

Study Set   Section 8.4   (page 390)

1. distributive property; removed **3.** equivalent **5.** $x(y + z) = xy + xz$ **7.** $(w + 7)5$ **9.** 5; 6, 6, 2, 3 **11.** $-y - 9$
13. distributing the -9 **15.** -5 **17.** $-9, -9; -45y$ **19.** $12x$ **21.** $-30y$ **23.** $100t$ **25.** $12s$ **27.** $14c$ **29.** $-40h$ **31.** $-42xy$
33. $16rs$ **35.** $30xy$ **37.** $-30br$ **39.** $80c$ **41.** $-8e$ **43.** $4x + 4$ **45.** $16 - 4x$ **47.** $-6e - 6$ **49.** $-16q + 48$ **51.** $12 + 20s$
53. $42 + 24d$ **55.** $-25r + 30$ **57.** $-24 - 18d$ **59.** $9x - 21y + 6$ **61.** $9z + 9x + 15y$ **63.** $-x - 3$ **65.** $-4t - 5$ **67.** $3w + 4$
69. $-5x + 4y - 1$ **71.** $2(5 + 4x)$ **73.** $(-4 - 3x)5$ **75.** $-3(4y - 2)$ **77.** $3(4 - 7t - 5s)$ **83.** analyze, form, solve, state, check
85. multiplication, division, subtraction, addition **87.** $>$ **89.** carpeting, painting

Study Set   Section 8.5   (page 398)

1. term **3.** perimeter **5.** distributive **7.** equivalent; combining **9. a.** term **b.** factor **c.** factor **d.** factor
11. $a(b + c) = ab + ac$ **13. a.** 11 **b.** 8 **c.** -4 **d.** 1 **e.** -1 **f.** 102 **15.** It helps identify the like terms. **17.** $(2d + 15)$ mi
19. To combine like terms, add the coefficients of the like terms. **21.** 7 **23.** $3b, 3b$ **25.** $2y$ **27.** $15t$ **29.** $4s$ **31.** x **33.** $4d$
35. $-4e$ **37.** $-7x$ **39.** $-6z$ **41.** cannot be simplified **43.** $3x^2, -5x, 4$ **45.** 5, $5t$, $-8t$, 4 **47.** 2 **49.** 2, 5 **51.** $11t + 12$
53. $2w - 5$ **55.** $-7r + 11R$ **57.** $-50d$ **59.** $8x - 4y - 9$ **61.** $9x + 34$ **63.** $-22s + 23$ **65.** $19e - 21$ **67.** $2t + 8$ **69.** $3x + 8$
71. $10y - 32$ **73.** 360 ft **79.** 2 **81.** $2^2 \cdot 5^2$ **83.** absolute value

Study Set   Section 8.6   (page 403)

1. solve **3.** distributive **5.** addition; subtraction **7.** addition property of equality **9.** subtraction property of equality **11. a.** -6
b. $2x + 2$ **c.** $x = -3$ **13. a.** $2x + 4$ **b.** $x = 2$ **15.** $2x$; 5; 2, 2 **17.** 9; 45; 45, 45; $5x$; 5, 5 **19.** 3 **21.** -4 **23.** 6 **25.** 2
27. 8 **29.** 4 **31.** -4 **33.** 1 **35.** 0 **37.** -12 **39.** 14 **41.** -28 **43.** -3 **45.** -11 **47.** -3 **49.** -5 **51.** -1 **53.** -4
55. 3 **57.** no **59.** yes **65.** -16 **67.** -3 **69.** 5 **71.** positive

Study Set   Section 8.7   (page 410)

1. base; exponent   **3.** product; like   **5.** product   **7. a.** x^7 **b.** x^2y^3 **c.** $3^4a^2b^3$ **9.** $x^2 \cdot x^6 = x^8$ (answers may vary)   **11.** $(c^5)^2 = c^{10}$ (answers may vary)   **13. a.** x^{m+n} **b.** x^{mn} **c.** a^nx^n **15. a.** 2   **b.** -10 **c.** x **17. a.** x^2; $2x$ **b.** x^3; $x + x^2$ **c.** x^4; $2x^2$
19. a. $4x^2$; $5x$ **b.** $12x^2$; $7x$ **c.** $12x^3$; $4x^2 + 3x$ **21.** 27   **23.** 5, 7   **25.** x^4, x^3; 4, 3   **27.** x^5 **29.** x^{10} **31.** f^{13} **33.** n^{32} **35.** l^{10}
37. x^{11} **39.** 2^{12} **41.** 5^8 **43.** $8x^3$ **45.** $5t^{10}$ **47.** $-24x^5$ **49.** $-x^4$ **51.** $36y^8$ **53.** $-40t^{10}$ **55.** x^3y^3 **57.** b^8c^8 **59.** x^5y^2
61. a^4b^4 **63.** x^5y^7 **65.** $18x^3y^4$ **67.** $16x^4y^2$ **69.** $-24f^6t^4$ **71.** a^4b^3 **73.** $12x^4y^3$ **75.** x^8 **77.** m^{500} **79.** $8a^3$ **81.** x^4y^4
83. $27s^6$ **85.** $4s^4t^6$ **87.** x^{14} **89.** c^{30} **91.** $36a^{14}$ **93.** $216a^{15}$ **95.** x^{60} **97.** $32b^{25}$ **103.** a letter used to represent a number
105. 5   **107.** 7   **109.** 12

Key Concept   (page 413)

1. Let x = the monthly cost to lease the van.   **3.** Let x = the width of the field.   **5.** Let x = the distance traveled by the motorist.
7. $a + b = b + a$ **9.** $\frac{b}{1} = b$ **11.** $n - 1 < n$

Chapter 8 Review   (page 414)

1. a. no   **b.** yes   **2. a.** y **b.** t **3. a.** 9   **b.** 31   **c.** 340   **d.** 133   **4. a.** 9   **b.** 14   **c.** 120   **d.** 5   **5.** \$97,250   **6.** 185
7. a. 4   **b.** 3   **c.** 21   **d.** 14   **8. a.** 21   **b.** 36   **c.** 315   **d.** 425   **9.** 6 ft   **10.** \$128   **11.** upper base = $h - 5$, lower
base = $2h - 3$; upper base = 5 ft, lower base = 17 ft   **12. a.** 12   **b.** -8 **c.** 100   **d.** -4 **13.** 130 mi, 114 mi, $6x$ mi, $55t$ mi
14. \$278   **15.** \$15,230   **16. a.** 1994   **b.** 1996   **c.** They decreased.   **17.** The pool is 2° C warmer.   **18.** 144 ft   **19.** 24 yr
20. a. $-10x$ **b.** $42xy$ **c.** $60de$ **d.** $32s$ **e.** $2e$ **f.** $49xy$ **g.** $84k$ **h.** $100t$ **21. a.** $4y + 20$ **b.** $-30t - 45$ **c.** $-21 - 21x$
d. $-12e + 24x + 3$ **22. a.** $-6t + 4$ **b.** $-5 - x$ **c.** $-6t + 3s - 1$ **d.** $5a + 3$ **23. a.** $-4x$, 8   **b.** $-3y$, 1   **24. a.** factor
b. term   **c.** factor   **d.** term   **25. a.** yes   **b.** no   **c.** yes   **d.** no   **26. a.** $7x$ **b.** $-3r$ **c.** $-9t$ **d.** $-3z$ **e.** $5x$ **f.** $-12y$
g. $w - 5$ **h.** $-6x + 2y$ **27. a.** $-46d + 2a$ **b.** $10y + 15h - 1$ **28. a.** $13y + 48$ **b.** $-5t + 22$ **c.** $3x + 12$ **d.** $-50f + 84$
29. 194   **30.** yes   **31. a.** 8   **b.** -18 **c.** -3 **d.** 15   **e.** 4   **f.** 5   **g.** -3 **h.** -2 **33. a.** $4h \cdot 4h \cdot 4h$ **b.** $5^2 \cdot d^3 \cdot m^4$
34. a. h^{10} **b.** t^8 **c.** w^7 **d.** 4^{12} **35. a.** $8b^7$ **b.** $-24x^4$ **c.** $24f^7$ **d.** $-a^2b^2$ **e.** x^2y^6 **f.** m^2n^2 **g.** $27m^3z^7$ **h.** $-20c^3d^6$
36. a. v^{12} **b.** $27y^3$ **c.** $25t^8$ **d.** $8a^{12}b^{15}$ **37. a.** c^{26} **b.** $108s^{12}$ **c.** c^{14} **d.** $8x^9$

Chapter 8 Test   (page 419)

1. 4   **2.** 30   **3.** 11   **4.** 81   **5.** 200   **6.** 24   **7.** 3,100   **8.** 194 yr   **9.** -3 **10.** 165 mi   **11.** $25x + 5$ **12.** $-42 + 6x$
13. $-6y - 4$ **14.** $6a + 9b - 21$ **15. a.** factor   **b.** term   **16. a.** $-28y + 10x$ **b.** $-3t$ **17.** $8x^2$, $-4x$, -6 **18. a.** $11x$ **b.** $24ce$
c. $5x$ **d.** $30y$ **19.** $-6y - 3$ **20. a.** -9 **b.** -3 **21.** 1   **22. a.** $10k\text{¢}$ **b.** $\$20(p + 2)$ **23. a.** h^6 **b.** $-28x^5$ **c.** b^8 **d.** $24g^5k^{13}$
24. a. f^{15} **b.** $4a^2b^2$ **c.** x^{15} **d.** x^{15}

Cumulative Review Exercises   (page 421)

1. 7,535,700   **2.** 7,540,000   **3.** 9,137   **4.** 3,322   **5.** 245,870   **6.** 875   **7.** 260 ft   **8.** 4,000 ft^2 **9.** $2^3 \cdot 3 \cdot 7$ **10.** $3^2 \cdot 5^2$
11. $2^2 \cdot 3^2 \cdot 5$ **12.** $2^4 \cdot 3^2 \cdot 5$ **13.** 18   **14.** -11 **15.** 1   **16.** -10 **17.** $\frac{1}{7}$ **18.** $\frac{2}{5}$ **19.** $\frac{19}{12} = 1\frac{7}{12}$ **20.** $\frac{1}{4}$ **21.** 7   **22.** 14   **23.** -52
24. -2 **25.** $-15x$ **26.** $28x^2$ **27.** $-6x + 8$ **28.** $-15x + 10y - 20$ **29.** $5x$ **30.** $7a^2$ **31.** $-x - y$ **32.** $-4x + 8$ **33.** -5
34. -5 **35.** -16 **36.** -60 **37.** 4   **38.** 4   **39.** 21   **40.** 21 ft by 84 ft   **41.** p^8 **42.** $-15t^9$ **43.** $-x^5y^7$ **44.** $81a^8$ **45.** $-n^6$
46. $9x^6y^2$ **47.** $108p^{12}$ **48.** $-8x^9$

Study Set   Section 9.1   (page 429)

1. segment   **3.** midpoint   **5.** protractor   **7.** right   **9.** 180°   **11.** supplementary   **13.** true   **15.** false   **17.** true   **19.** true   **21.** acute
23. obtuse   **25.** right   **27.** straight   **29.** true   **31.** false   **33.** yes   **35.** yes   **37.** no   **39.** true   **41.** true   **43.** true   **45.** true
47. angle   **49.** ray   **51.** 3   **53.** 3   **55.** 1   **57.** B **59.** 40°   **61.** 135°   **63.** 10   **65.** 27.5   **67.** 30   **69.** 25   **71.** 60°   **73.** 75°
75. 130°   **77.** 230°   **79.** 100°   **81.** 40°   **87. a.** 80°   **b.** 30°   **c.** 65°   **91.** 16   **93.** $\frac{7}{24}$ **95.** 6   **97.** 23

Study Set   Section 9.2   (page 436)

1. coplanar   **3.** perpendicular   **5.** alternate interior   **7.** $\angle 4$ and $\angle 6$, $\angle 3$ and $\angle 5$ **9.** $\angle 3$, $\angle 4$, $\angle 5$, $\angle 6$ **11.** They are parallel.   **13. a.**
right angle   **15.** is perpendicular to   **17.** m($\angle 1$) = 130°, m($\angle 2$) = 50°, m($\angle 3$) = 50°, m($\angle 5$) = 130°, m($\angle 6$) = 50°, m($\angle 7$) = 50°,
m($\angle 8$) = 130°   **19.** m($\angle A$) = 50°, m($\angle 1$) = 85°, m($\angle 2$) = 45°, m($\angle 3$) = 135°   **21.** 10   **23.** 30   **25.** 40   **27.** 12
33. a. intersecting lines, vertical angles   **b.** parallel and perpendicular lines   **39.** 72   **41.** 45%   **43.** yes

Study Set   Section 9.3   (page 443)

1. quadrilateral   **3.** hexagon   **5.** octagon   **7.** equilateral   **9.** hypotenuse   **11.** parallelogram   **13.** rhombus   **15.** isosceles
17. 4, quadrilateral, 4   **19.** 3, triangle, 3   **21.** 5, pentagon, 5   **23.** 6, hexagon, 6   **25.** scalene triangle   **27.** right triangle
29. equilateral triangle   **31.** isosceles triangle   **33.** square   **35.** rhombus   **37.** rectangle   **39.** trapezoid   **41.** triangle   **43.** 90°
45. 45°   **47.** 90.7°   **49.** 30°   **51.** 60°   **53.** 720°   **55.** 1,440°   **57.** 7-sided polygon   **59.** 14-sided polygon   **67.** 22   **69.** $\frac{6}{55}$
71. 40%   **73.** 2,500   **75.** 0.10625

Study Set   Section 9.4   (page 450)

1. congruent   **3.** similar; same   **5.** true   **7.** false   **9.** yes   **11.** product; extremes   **13.** yes   **15.** true   **17.** is congruent to
19. $\overline{DF}$, $\overline{AB}$, $\overline{EF}$, $\angle D$, $\angle B$, $\angle C$ **21.** yes, SSS   **23.** not necessarily   **25.** yes, SSS   **27.** yes, SAS   **29.** 6 cm   **31.** 50°   **33.** yes
35. 5   **37.** 8   **39.** $\sqrt{56}$ **41.** yes   **43.** no   **45.** 36 ft   **47.** 59.2 ft   **49.** 4,000 ft   **51.** 12 ft   **53.** 127.3 ft   **57.** $1\frac{1}{3}$ **59.** 20   **61.** 38

Study Set   Section 9.5   (page 462)

1. perimeter   **3.** area   **5.** square   **7.** length 15 in. and width 5 in.; length 16 in. and width 4 in. (answers may vary)   **9.** sides of
length 5 m   **11.** base 5 yd and height 3 yd (answers may vary)   **13.** length 5 ft and width 4 ft; length 20 ft and width 3 ft (answers
may vary)   **15.** $P = 4s$ **17.** one square inch   **19.** $A = s^2$ **21.** triangle   **23.** 32 in.   **25.** 36 m   **27.** 37 cm   **29.** 85 cm   **31.** $28\frac{1}{3}$ ft
33. 16 cm²   **35.** 60 cm²   **37.** 25 in.²   **39.** 169 mm²   **41.** 80 m²   **43.** 75 yd²   **45.** 75 m²   **47.** 144   **49.** $951.60   **51.** 81
53. linoleum   **55.** $1,200   **57.** $361.20   **59.** $192   **61.** $1,658.48   **63.** 51   **67.** $1\frac{5}{12}$ **69.** $6\frac{1}{12}$ **71.** $1\frac{7}{18}$

Study Set   Section 9.6   (page 470)

1. radius   **3.** diameter   **5.** semicircle   **7.** circumference   **9.** OA, OC, and OB **11.** DA, DC, and AC **13.** $\overset{\frown}{ABC}$ and $\overset{\frown}{ADC}$
15. Double the radius.   **17.** arc AB **19.** $C = \pi D$ or $C = 2\pi r$ **21.** 37.70 in.   **23.** 36 m   **25.** 25.42 ft   **27.** 31.42 m
29. $A = 28.3$ in.²   **31.** 88.3 in.²   **33.** 128.5 cm²   **35.** 27.4 in.²   **37.** 66.7 in.²   **39.** 3.14 mi²   **41.** 12.73 times   **43.** 1.59 ft
45. 6.25%   **51.** 90%   **53.** 480   **55.** $73.50

Study Set   Section 9.7   (page 479)

1. volume   **3.** cube   **5.** surface area   **7.** cylinder   **9.** cone   **11.** $V = lwh$ **13.** $V = \frac{4}{3}\pi r^3$ **15.** $V = \frac{1}{3}Bh$
17. $SA = 2lw + 2lh + 2hw$ **19.** 27 ft³   **21.** 1,000 dm³   **23. a.** volume   **b.** area   **c.** volume   **d.** surface area   **e.** perimeter
f. surface area   **25.** 1 cubic inch   **27.** 60 cm³   **29.** 48 m³   **31.** 3,053.63 in.³   **33.** 1,357.17 m³   **35.** 314.16 cm³   **37.** 400 m³
39. 94 cm²   **41.** 1,256.64 in.²   **43.** 576 cm³   **45.** 335.10 in.³   **47.** 0.13 in.³   **49.** 197.92 ft³   **51.** 33,510.32 ft³   **55.** 36   **57.** 33
59. $0.15   **61.** $13.41

Key Concept   (page 482)

1. $d = rt$ **2.** $s = p - d$ **3.** $P = 2l + 2w$ **4.** $I = Prt$ **5.** 210,000 ft²   **6.** 4,992 in.³   **7.** $80.50   **8.** $4,300   **9.** 144 ft   **10.** 15° C
11. $750; $45; $6,250

Chapter 9 Review   (page 483)

1. points C and D, line CD, plane AB **2.** 5 units   **3.** $\angle ABC$, $\angle CBA$, $\angle B$, $\angle 1$ **4.** 48°   **5.** $\angle 1$ and $\angle 2$ are acute, $\angle ABD$ and $\angle CBD$
are right angles, $\angle CBE$ is obtuse, and $\angle ABC$ is a straight angle.   **6. a.** obtuse angle   **b.** right angle   **c.** straight angle   **d.** acute
angle   **7.** 15   **8.** 150   **9.** m($\angle 1$) = 65°; m($\angle 2$) = 115°   **10.** 40°   **11.** 40°   **12.** no   **13.** part a   **14.** $\angle 4$ and $\angle 6$, $\angle 3$ and $\angle 5$
15. $\angle 1$ and $\angle 5$, $\angle 4$ and $\angle 8$, $\angle 2$ and $\angle 6$, $\angle 3$ and $\angle 7$ **16.** $\angle 1$ and $\angle 3$, $\angle 2$ and $\angle 4$, $\angle 5$ and $\angle 7$, $\angle 6$ and $\angle 8$ **17.** m($\angle 1$) = 70°,
m($\angle 2$) = 110°, m($\angle 3$) = 70°, m($\angle 4$) = 110°, m($\angle 5$) = 70°, m($\angle 6$) = 110°, m($\angle 7$) = 70°   **18.** m($\angle 1$) = 60°, m($\angle 2$) = 120°,
m($\angle 3$) = 130°, m($\angle 4$) = 50°   **19.** 40   **20.** 20   **21. a.** octagon   **b.** pentagon   **c.** triangle   **d.** hexagon   **e.** quadrilateral   **22. a.** 3
b. 4   **c.** 8   **d.** 6   **23. a.** isosceles   **b.** scalene   **c.** equilateral   **d.** right triangle   **24. a.** yes   **b.** no   **25. a.** 90   **b.** 50   **26.** 50°
27. It is equilateral.   **28. a.** trapezoid   **b.** square   **c.** parallelogram   **d.** rectangle   **e.** rhombus   **f.** rectangle   **29. a.** 15 cm
b. 40°   **c.** 100°   **30. a.** true   **b.** false   **c.** true   **d.** true   **31. a.** 65°   **b.** 115°   **32. a.** 360°   **b.** 720°   **33.** $\angle A \leftrightarrow \angle D$; $\angle B \leftrightarrow \angle E$;
$\angle C \leftrightarrow \angle F$; $\overline{AC} \leftrightarrow \overline{DF}$; $\overline{AB} \leftrightarrow \overline{DE}$; $\overline{BC} \leftrightarrow \overline{EF}$ **34. a.** congruent, SSS   **b.** congruent, SAS   **c.** congruent, ASA   **d.** not necessarily
congruent   **35. a.** yes   **b.** yes   **36.** 21 ft   **37. a.** 13   **b.** 15   **38.** 31.3 in.   **39.** 72 in.   **40.** 9 m   **41. a.** 30 m   **b.** 36 m
42. a. 9.61 cm²   **b.** 7,500 ft²   **c.** 450 ft²   **d.** 200 in.²   **e.** 120 cm²   **f.** 232 ft²   **g.** 152 ft²   **h.** 120 m²   **43.** 9 ft²   **44.** 144 in.²
45. a. $\overline{CD}$, $\overline{AB}$ **b.** $\overline{AB}$ **c.** $\overline{OA}$, $\overline{OC}$, $\overline{OD}$, $\overline{OB}$ **d.** O **46.** 66.0 cm   **47.** 45.1 cm   **48.** 254.5 in.²   **49.** 130.3 cm²   **50. a.** 125 cm³
b. 480 m³   **c.** 600 in.³   **d.** 3,619 in.³   **e.** 1,518 ft³   **f.** 785 in.³   **g.** 9,020,833 ft³   **h.** 35,343 ft³   **51.** 1,728 in.³   **52.** 54 ft³
53. a. 61.8 ft²   **b.** 314.2 in.²

Chapter 9 Test   (page 491)

1. 4 units   **2.** B **3.** true   **4.** false   **5.** false   **6.** true   **7.** 50   **8.** 140   **9.** 12   **10.** 63   **11.** 23°   **12.** 63°   **13.** 70°   **14.** 110°
15. 70°   **16.** 40   **17.**

Polygon	Number of sides
Triangle	3
Quadrilateral	4
Hexagon	6
Pentagon	5
Octagon	8

18.

Property	Kind of triangle
All sides of equal length	equilateral triangle
No sides of equal length	scalene triangle
Two sides of equal length	isosceles triangle

19. 57° **20.** 66° **21.** 30° **22.** 1,440° **23.** m($\overline{AB}$) = m($\overline{DC}$), m($\overline{AD}$) = m($\overline{BC}$), and m($\overline{AC}$) = m($\overline{BD}$) **24.** 130° **25.** 8 in. **26.** 50° **27.** 6 **28.** 12 **29.** 127.3 ft **30.** 391.6 cm² **31.** 83.7 ft² **32.** 18.8 ft **33.** 28.3 ft² **34.** 159.3 m³ **35.** 268.1 m³ **36.** 66.7 ft³

Cumulative Review Exercises   (page 493)

1. 3 **2.** −1 **3.** $\frac{7}{3}$ **4.** −2 **5.** 89 **6.** $-\frac{15}{2}$ **7.** x^9 **8.** x^{27} **9.** p^{12} **10.** $a^8 b^{12}$ **11.** $-2a + 6b$ **12.** $4p - 6$ **13.** $\frac{2}{15}$ **14.** $\frac{3}{16}$ **15.** $\frac{3}{20}$ **16.** $1\frac{9}{29}$ **17.** 0.4 **18.** 3 **19.** $6\frac{5}{24}$ **20.** −8 **21.** 35.83 **22.** 0.161 **23.** 114.801 **24.** 5.4 **25.** 170.2 **26.** 67.5 **27.** 25% **28.** 240 **29.** $1,159.38 **30.** $3,800 **31.** about 21.5% **32.** 2,907 **33.** $20.97, $48.93 **34.** 35% **35.** $1,522.50 **36.** $269,390.92 **37.** $\frac{3}{7}$ **38.** $\frac{1}{4}$ **39.** 5 **40.** −6.5 **41.** 30 **42.** 450 mi **43.** 14 **44.** 540 **45.** 5,280 **46.** 13.25 **47.** 200 **48.** 120 **49.** 12.3 **50.** 1,000 **51.** 2,000 **52.** 500 **53.** 500,000 **54.** 7.5 **55.** 2,000 **56.** 60 **57.** 5,000 **58.** 0.6 **59.** 26 **60.** 457.2 **61.** 17,010 **62.** 219.7 **63.** 167° **64.** 45° **65.** 90 **66.** more than 0 but less than 90 **67.** 75° **68.** 15° **69.** 50° **70.** 130° **71.** 50° **72.** 50° **73.** 75° **74.** 30° **75.** 105° **76.** 105° **77.** 540° **78.** 13 m **79.** 935 **80.** 93.75 units **81.** 42 m, 108 m² **82.** 126 ft² **83.** 91 in.² **84.** 43.98 cm, 153.94 cm² **85.** 210 m³ **86.** 523.60 in.³ **87.** 150.80 m³ **88.** 3.93 ft³

INDEX